Handbook of Glass Properties

Handbook of Glass Properties

NAROTTAM P. BANSAL
and
R. H. DOREMUS

Materials Engineering Department
Rensselaer Polytechnic Institute
Troy, New York

1986

ACADEMIC PRESS, INC.

Harcourt Brace Jovanovich, Publishers

Orlando   San Diego   New York   Austin
London   Montreal   Sydney   Tokyo   Toronto

ACADEMIC PRESS, INC.
Orlando, Florida 32887

United Kingdom Edition published by
ACADEMIC PRESS INC. (LONDON) LTD.
24–28 Oval Road, London NW1 7DX

Library of Congress Cataloging in Publication Data

Bansal, N. P.
    Handbook of glass properties.

    Includes index.
    1. Glass–Handbooks, manuals, etc.   I. Doremus, R. H.
II. Title.   III. Series.
TA450.B27   1986   620.1'44   85-13326
ISBN   0–12–078140–9   (alk. paper)

PRINTED IN THE UNITED STATES OF AMERICA

86 87 88 89   9 8 7 6 5 4 3 2 1

Contents

Preface

For many years, members of the Glass Division of the American Ceramic Society have urged the collection of property data for glasses into a handbook. In response to this need, and with the support of Academic Press, we have assembled the text and property values in this volume. It is incomplete — commercial glasses from European and Japanese manufacturers, nonsilicate glasses, and doubtless much other information have been slighted. We present this volume as a beginning toward the collection of critically selected and correlated data on silicate glasses and welcome suggestions for corrections and the most desirable additions, which we hope to incorporate in a second volume.

The typing of the text, references, and table titles was ably carried out by Kathleen Curtice and Nancy Fowler. Mrs. Shashi Bansal helped a great deal with assembling references and tables. Marc Doremus transliterated Russian names.

We gratefully acknowledge the use of facilities of the Materials Engineering Department of Rensselaer Polytechnic Institute.

Narottam P. Bansal is presently at NASA–Lewis Research Center, Cleveland, Ohio.

Chapter 1

Introduction

Silicate glass is used throughout modern technology. Traditional uses for containers, windows, lamps, and optical components are being supplemented by a multitude of specialized applications in communications, electronics, and composites. Fiber optic waveguides, laser optics for initiating fusion reactions, and containers for radioactive waste are recent developments that demonstrate the importance, versatility, and promise of glass for additional uses.

Data on properties of silicate glasses are widely scattered in books, journal articles, and company data sheets. The purpose of this handbook is to gather data on glass of interest to technologists and scientists.

In Chapter 2 vitreous silica is discussed and its properties tabulated, because it is compositionally the simplest silicate glass and has commercial importance. Many properties of silica can guide understanding of properties of multicomponent silicate glasses. The silicon–oxygen network is the fundamental structural component of all silicate glasses and determines many properties.

Commercial silicate glasses are described and their properties are listed in Chapter 3. These multicomponent glasses often have many components; compositions are given for most glasses. These compositions are representative only and not exact, because in commercial practice, especially for large-tonnage glasses such as soda-lime, wide variations in compositions are tolerated. These variations are frustrating for controlled studies of properties and for some applications but must be recognized. Most of the property values for commercial glasses were taken from manufacturers' data sheets.

In the following chapters property values for silicate glasses are tabulated, with values of the same property for different compositions listed in a particular chapter. Emphasis is on simple compositions, especially binary and ternary oxides. A list of systems is given in the Appendix, together with the properties. These glasses have been melted on a small scale, in contrast to the large, continuous tanks used to melt high tonnage commercial glasses.

In a final chapter, factors for estimating properties of silicate glasses are discussed and tabulations of them are referred to.

Each set of tabulated data has been examined carefully, and only results judged to be reliable are recorded. If several investigators reported results for the same property and compositions, we tried to judge the most reliable and report it. For some properties and compositions with many measurements of roughly the same reliability, we plotted or averaged the results to obtain the most trustworthy values. We chose this method to provide the reader with the best data possible, and to avoid the need for the reader to make this kind of evaluation.

Values are reported to the accuracy judged justified by the data, and no error limits are included. If the error was judged to be greater than about plus or minus four digits in a particular integer, that integer was not given.

A list of reference books and articles is given at the end of this chapter, within some brief descriptions of their contents. Two books need special mention. The "Handbook of Glass Data," organized by Professor Mazurin of the Leningrad Institute of Silicate Chemistry, has much data on properties of vitreous silica and binary silicates, mainly in graphical form. Future volumes on ternary glasses are planned. "Properties of Glass," by George Morey, although published in 1954, is still a valuable resource for tables of glass properties and their discussions of their measurement. These books have been extensively referred to in gathering the property values for the present volume.

BIBLIOGRAPHY

History of Glass

Charleston, R. J. (1980). "Masterpieces of Glass. A World History From the Museum of Glass." Abrams, New York.
Douglas, R. W., and Frank, S. (1972). "A History of Glass Making." Foulis, Henley-on-Thames, England.
Polak, A. (1975). "Glass: Its Tradition and Its Makers." Putnam, New York.
Zerwick, C. (1980). "A Short History of Glass." Corning Museum of Glass, Corning, New York.

Monographs on Glass

Babcock, C. L. (1977). "Silicate Glass Technology Methods." Wiley, New York. Special emphasis on elastic, optical, and thermal properties.
Doremus, R. H. (1973). "Glass Science." Wiley, New York.
Dunken, H. H. (1981). "Physikalische Chemie der Glasoberflache." VEB Dtsch. Verlag Grundstoffind., Leipzig. A very thorough review and discussion of glass surfaces, including information on structure, composition, and chemical properties of glass.
Holloway, D. G. (1973). "The Physical Properties of Glass." Wykeham Publ., London. An introduction to the structure and properties of glass.
McMillan, P. W. (1979). "Glass Ceramics," 2nd Ed. Academic Press, New York.
Morey, G. W. (1954). "The Properties of Glass." Reinhold, New York. See text.

Scholze, H. (1965). "Glas: Natur, Struktur, und Eigenschaften." Friedr. Vieweg and Sohn, Braunschweig. Discussions of many properties of glass.
Volf, W. B. (1984). "Chemical Approach to Glass." Elsevier, Amsterdam.
Wong, J., and Angell, C. A. (1976). "Glass Structure by Spectroscopy." Dekker, New York.
Zarzycki, J. (1982). "Les Verres et L'État Vitreux." Masson, Paris.

Article

Boyd, D. C., and Thompson, D. A. (1980). Glass, in "Encyclopedia of Chemical Technology," 3rd Ed., Vol. 11, pp. 807–880. Wiley, New York.

Collections of Review Articles

Mackenzie, J. D., ed. (1960, 1962, 1964). "Modern Aspects of the Vitreous State," Vols. 1–3. Butterworth, London.
Pye, L. D., Stevens, H. J., and LaCourse, W. C., eds. (1972). "Introduction to Glass Science." Plenum, New York.
Tomozawa, M., and Doremus, R. H., eds. (1977, 1979, 1982, 1985). "Glass I, II, III, IV," Treatise on Materials Science and Technology, Vols. 12, 17, 22, and 26. Academic Press, New York.
Uhlmann, D. R., and Kreidl, N. J., eds. (1980–1985). "Glass: Science and Technology," Vols. 1, 2, 3, and 5. Academic Press, New York.

Reference Works

Eitel, W. (1965, 1976). "Silicate Science," Vols. II, III, and VIII. Academic Press, New York. Summaries of articles on glass.
Mazurin, O. V. (1971, 1975, 1977, 1979, 1980). "Properties of Glass and Glassy Melts," Vols. 1, 2, 3(1), 3(2), and 4. Publisher "Science," Leningrad. (In Russ.) A comprehensive collection of data, mostly in tables, on properties of simple, binary and ternary oxide glasses and melts.
Mazurin, O. V., Stretsina, M. V., and Shvaiko-Shvaikovskaya, T. P. (1983, 1985). "Handbook of Glass Data," Parts A and B. Elsevier, Amsterdam.

Journals on Glass

Glass Technology and *Physics and Chemistry of Glasses.* Society of Glass Technology, Sheffield, England.
Glastechnische Berichte. Deutsche Glastechnische Gesellschaft, Frankfurt.
Journal of Non-Crystalline Solids. North-Holland, Amsterdam.
Journal of the American Ceramic Society. Columbus, Ohio.
Soviet Journal of Glass Physics and Chemistry. In Russian. (Also translation into English, Consultants Bureau, New York.)
Yogyo Kyokaishi, Journal of the Ceramic Society of Japan. Mostly in Japanese, with English abstracts. Tokyo, Japan.

Part I

Commercial Glasses

Chapter **2**

Vitreous Silica

INTRODUCTION **I**

Vitreous silica (SiO_2) is the simplest silicate glass and for some properties such as strength and optical properties can be used as a starting point for the understanding of other silicate glasses. Other properties, such as viscosity, thermal expansion, and electrical conductivity, are strongly dependent on glass composition; even for these properties there are similarities in the behavior of vitreous silica and multicomponent silicate glasses.

Vitreous silica is also called silica glass, fused silica, fused quartz, or simply quartz; the terms vitreous silica and silica glass are preferred because different methods of manufacture and starting materials give the same final structure; quartz is undesirable because of the confusion with crystalline quartz.

In this chapter impurity compositions, uses, methods of manufacture, and the structure of vitreous silica are discussed briefly, and then properties are described under the categories thermodynamic, mechanical, thermal, electrical, optical, chemical, and diffusion.

TYPES AND IMPURITY COMPOSITIONS **II**

The main types, impurity concentrations, and designations of vitreous silica are given in Table 2.1. A list of manufacturers and suppliers in the United States is given in Table 2.2.

Types I and II are based on natural quartz crystals, either chunks or sand, as raw material, so the impurity content is variable but can be reduced by pretreatment, especially of sand. Types III and IV are purer but much more expensive.

7

TABLE 2.1

Different Types of Vitreous Silica

Type	Method of manufacture	Maximum impurity concentration (ppm)									Manufacturing designations				
		Al	Fe	Ca	Mg	K	Na	Li	Cl	OH	General Electric	Thermal Syndicate	Heraeus	Corning	Quartz et Silice
I	Electrical fusion of quartz crystal	150	7	12	7	4	12	12	50	4	101 204 125	IR Vitreosil	Infrasil		Pursil
II	Flame fusion of quartz crystal	Similar to I									400		OG Vitreosil	Herosil Homosil Vitrasil		
III	Flame hydrolysis of $SiCl_4$	10	6	4	3	2	2	1	60	1200		Spectrosil	Suprasil I	7940	Tetrasil
IV	Vapor phase oxidation of $SiCl_4$	Similar to III							500	low		Spectrosil WF	Suprasil W	7943	

TABLE 2.2

Manufacturers and Suppliers of Vitreous Silica in the United States

Amersil, Inc., Hillside, New Jersey (Heraeus, West Germany)
Corning Glass Works, Corning, New York
General Electric Company, Willoughby, Ohio
Thermal American, Montville, New Jersey (Thermal Syndicate, England)

USES　III

Vitreous silica is one of the purest materials available commercially. It is used for crucibles and furnaces for high-temperature processing, especially of semiconductors; for lamp envelopes; for electrical transducers and insulators; and for optical components such as fiber optics, telescope mirrors, lenses, and prisms. These uses derive from the properties of high purity, high-temperature stability, low thermal shock, low electrical conductivity and dielectric loss, high chemical durability, and a wide range of high optical transparency. Even wider use of vitreous silica is limited by its relatively high cost, resulting from the high temperature needed to melt silica (nearly 2000°C) and the expense of the containers and power used.

Two newly developing applications of vitreous silica are of special importance and interest. Optical waveguide fibers of vitreous silica of very high transparency are replacing metallic cables in many communications systems (Kao, 1982). The high optical transparency, resistance to radiation damage, and easy formability of vitreous silica made it ideal for this application. The oxidation of silicon to form a thin layer of vitreous silica on its surface in electronic devices provides a stable, high-resistance insulator. The silica formed in this way has the same properties of bulk vitreous silica of high purity.

METHODS OF MANUFACTURE　IV

Vitreous silica crucibles are often made by arc melting of quartz sand that acts as a self-container; this method gives many bubbles. Ingots of vitreous silica are made by heating quartz sand or chunks in crucibles of graphite, molybdenum, or tungsten in vacuum or in an inert or reducing atmosphere. The temperature is usually 1800 to 2000°C; it must be greater than the melting temperature of cristobalite of 1740°C. The ingot is then cut up or drawn to other shapes as tubing or rod.

Very pure vitreous silica is made from vapor-phase oxidation or hydrolysis of silicon tetrachloride ($SiCl_4$). In one method the $SiCl_4$ is mixed with oxygen and natural gas, fed through a burner, and deposited on a substrate of preheated sand or on a rotating air-cooled mandrel of aluminum. A plasma torch can also be used.

Fibers of silica can be pulled from a preformed vitreous silica rod or other convenient shape.

Corning 7900 (Vycor) glass is made by etching a phase-separated borosilicate glass to give finely porous "thirsty glass" and then heating the porous material at $\sim 1000°C$ to form a dense, clear glass. It contains about 96% SiO_2, 4% B_2O_3, and 0.03% Na_2O, K_2O, and Al_2O_3.

V   STRUCTURE

The basic structural unit of vitreous silica is the silicon–oxygen tetrahedron, as in all silicates. In vitreous silica these tetrahedra are linked to other tetrahedra at their corners into a three-dimensional network. The structure has order on a short range (one or two coordination spheres of tetrahedra around a central one) but no long-range order beyond a few tetrahedra. This structure is best described as a random network (Mozzi and Warren, 1969); some workers have also proposed microcrystallite models.

VI   THERMODYNAMIC PROPERTIES

Some thermodynamic properties are given in Table 2.3. Schick (1960) has given details of calculations of thermodynamic properties of silica at high temperatures and experimental results for many different reactions involving silicon and oxygen. He used heat capacity values (in kilojoule per mole per degree kelvin) of silica calculated from the equation

$$C_p = 55.98 + (15.40 \times 10^{-3})T - (14.4 \times 10^5/T^2), \tag{1}$$

where T is the absolute temperature in kelvin. This equation follows closely the data chosen by Sosman (1927) (Table 2.4); these data are questionable

TABLE 2.3

Thermodynamic Properties of Vitreous Silica at 25°C

Heat of formation $\Delta H°$ (kJ mole^{-1})	Free energy of formation $\Delta G°$ (kJ mole^{-1})	Entropy of formation $\Delta S°$ (J mole^{-1} K^{-1})	Heat capacity C_p (J mole^{-1} K^{-1})	Reference
−903.49	−805.70	46.9	44.4	Wagman et al. (1982)

TABLE 2.4

Heat Capacity of Vitreous Silica (Type I)

T (°C)	C_p (J mole^{-1} K^{-1})	T (K)	C_p (J mole^{-1} K^{-1})
1700	76.4	50	6.71
1600	76.4	40	4.98
1500	76.2	30	3.19
1400	75.7	23	1.94
1300	75.2	19.0	1.286
1200	74.4	16.06	0.877
1100	73.9	14.01	0.628
1000	73.4	11.99	0.413
900	72.7	10.04	0.245
800	71.9	7.98	0.116
700	70.6	5.99	0.0418
600	69.1	4.03	9.07×10^{-3}
500	67.9	2.34	1.40×10^{-3}
400	64.4	1.0	1.69×10^{-4}
300	61.6	0.2	1.5×10^{-5}
200	57.1		
150	54.1		
100	50.5		
50	46.5		
25	44.5		
0	41.7		
−50	35.7		
−100	28.4		
−150	20.0		
−200	10.9		

above about 1100°C (see below), so more reliable values should be used above 1100°C for the most accurate work. Below the melting temperature of silicon (1412°C) the formation equation is

$$Si(solid) + O_2(gas) = SiO_2(vitreous). \tag{2}$$

The heat of formation ΔH_f at temperatures different from 25°C can be calculated from the relation

$$\Delta H_f = -903.49 + \int_{298}^{T} \Delta C_p \, dT, \tag{3}$$

where ΔC_p is the difference in heat capacities at constant pressure for reaction (2):

$$\Delta C_p = C_p(SiO_2) - C_p(O_2) - C_p(Si). \tag{4}$$

The entropy of formation ΔS_f can be calculated from the relation

$$\Delta S_f = \int_{0}^{T} \frac{\Delta C_p}{T} \, dT, \tag{5}$$

and finally the free energy of formation as a function of temperature is

$$\Delta G_f = \Delta H_f - T \Delta S_f. \tag{6}$$

A Melting Temperature

The best measurement of the melting temperature of cristobalite is probably that of Wagstaff (1734°C), although lower values near 1725°C are often quoted. Quartz melts at 1410°C and is the stable phase below 870°C.

In very pure silica, cristobalite is the stable phase above 870°C; small amounts of impurities stabilize tridymite from 870 to 1470°C, with a melting temperature of 1680°C. Each of these three phases have α and β modifications, which represent small structural variants. The α–β transformations occur at the following temperatures: quartz, 573°C; tridymite, 163°C; cristobalite ~270°C.

B Heat Capacity

The heat capacity at constant pressure C_p for type I (water-free) vitreous silica is given in Table 2.4. The values from −200 to 1100°C are taken from a compilation of Sosman (1927) from earlier data; they agree well with those of Lucks et al. (1960) from −160 to 500°C. The agreement is also good with the data of Casey et al. (1976) from 30 to 200°C, but above 200°C the data of Casey et al. are progressively lower at 428°C; for example, at 400°C, Casey et al. measured 62.8 J mole^{-1} K^{-1}.

Up to 1100°C the heat capacities of vitreous silica and crystalline cristobalite are nearly the same, except near the α–β transformation of cristobalite. Above about 1100°C the values chosen by Sosman for vitreous silica show an anomalous increase above those of cristobalite. He apparently derived these anomalous values from the results of Wietzel (1921), whereas at least one other set of data by Bornemann and Hengstenberg (1920) showed no anomalous increase up to 1400°C. Above 1100°C vitreous silica crystallizes (see crystallization rates below), and it is possible that the high values of Wietzel were influenced by crystallization. In any event it seems best to use the heat capacities of cristobalite for vitreous silica above 1100°C, and these are given in Table 2.4.

Casey et al. (1976) found that 0.12 wt. % OH in vitreous silica reduced the heat capacity about 0.5% at 300 K and 1.6% at 700 K, so that types II and III should have slightly lower heat capacities than those given in Table 2.4.

There have been a number of recent measurements of the heat capacity of vitreous silica below 10 K because these heat capacities are higher than expected from the Debye model. The values in Table 2.4 are taken from several measurements that agree well (Stephens, 1973, 1976; Fagaly and Bohn, 1978; Lasjaunias *et al.*, 1980; and other references in these papers).

Vapor Pressure C

Above about 1350°C vitreous silica begins to vaporize by the reaction

$$SiO_2 = SiO + \tfrac{1}{2}O_2. \tag{7}$$

The SiO exists only in the vapor; it forms SiO_2 on depositing on a solid surface. In flame-working vitreous silica one can observe a band of haze just outside the intensely heated region, which results from redeposited SiO_2 granules. The haze can be removed by gentle reheating in the flame. The vaporization reaction is much enhanced in vacuum or in a reducing atmosphere. An equation for the pressure of SiO in atm over vitreous silica under neutral conditions above 2000 K is (Schick, 1960)

$$\ln p(SiO) = 18.41[1 - (3160/T)], \tag{8}$$

with T in kelvin. The total pressure is just $1.5p(SiO)$. Below 2000 K the equation is

$$\ln p(SiO) = 19.72[1 - (3075/T)]. \tag{9}$$

Values from these equations are given in Table 2.5. In air of pressure p_t the SiO pressure over vitreous silica is (Schick, 1960)

$$p_{SiO}(p_t + p_{SiO})^{1/2} = e^{28.0(1 - 3110/T)}. \tag{10}$$

Rate of Crystallization D

The rate of crystallization of vitreous silica rises from zero at the melting point of cristobalite (about 1740°C) to a maximum of about 2×10^{-7} cm s^{-1} at about 1675°C and decreases to lower rates at lower temperatures, as given in Table 2.6. Nevertheless, surface crystallization or "devitrification" can be significant in practical uses of vitreous silica to temperatures as low as 1000°C, especially if the surface is in contact with some foreign substance that can act as a nucleation agent. The crystalline phase that forms is β-cristobalite, which has almost the same density and refractive index as vitreous silica, so it may not be noticed at higher temperatures. At about 270°C or below, the β-cristobalite transforms to the more dense α-cristobalite with a large volume change that causes cracking around the crystals, reducing the strength and optical clarity of the silica.

<table>
<tr><td colspan="4" align="center">TABLE 2.5</td><td colspan="4" align="center">TABLE 2.6</td></tr>
<tr><td colspan="4" align="center">Vapor Pressure of SiO Over Vitreous
Silica in Inert Atmosphere</td><td colspan="4" align="center">Rate of Crystallization of Cristobalite
in Vitreous Silica[a]</td></tr>
</table>

T (K)	$p(SiO)$ (atm)	T (K)	$p(SiO)$ (atm)	t (°C)	Rate (cm s^{-1})	t (°C)	Rate (cm s^{-1})
3180	1	2400	2.9×10^{-3}	1347	1.1×10^{-9}	1709	1.59×10^{-7}
3100	0.72	2200	3.2×10^{-4}	1408	6.2×10^{-9}	1723	9.0×10^{-8}
3000	0.40	2000	2.3×10^{-5}	1483	3.0×10^{-8}	1728	5.2×10^{-8}
2900	0.20	1900	5.1×10^{-6}	1540	6.8×10^{-8}	1733	3.5×10^{-8}
2800	0.10	1800	8.6×10^{-7}	1618	1.50×10^{-7}	1741	-6.5×10^{-8}
2600	0.019	1700	1.2×10^{-7}	1680	2.03×10^{-7}	1746	-1.05×10^{-7}
				1694	1.84×10^{-7}	1754	-2.18×10^{-7}

[a] Data from Wagstaff (1969).

E Surface Energy

The theoretically calculated surface energy of vitreous silica at 25°C is about 5.2 J m^{-2}; the calculated value of 3.5 J m^{-2} for soda-lime glass agrees with the lowest fracture energies of this glass found experimentally, as expected (Doremus, 1976). These values are much higher than the surface tension of a molten silica surface of about 0.3 J m^{-2} at 1800°C (Kingery, 1959) because the latter is rapidly covered with silanol groups (SiOH) after its formation.

VII MECHANICAL PROPERTIES

A Viscosity

The viscosity of vitreous silica with low impurities is given in Table 2.7. These data were carefully selected from many measurements as the most reliable. The measurements of Hofmaier and Urbain (1968) agree reasonably well with measurements of other authors over more limited temperature ranges. Hetherington et al. (1964) showed that at lower temperatures silica must be held for long times at the measuring temperatures before a reproducible value of viscosity can be measured.

The viscosity is decreased by the presence of small amounts of impurities, which is one of the main reasons for errors and divergent measurements. The results for the purest waterfree material can be represented by

$$\text{from 1600 to 2500°C} \quad \log_{10}\eta = -6.24 + (2.69 \times 10^4)/T, \quad (11)$$

$$\text{from 1000 to 1400°C} \quad \log_{10}\eta = -12.5 + (3.73 \times 10^4)/T, \quad (12)$$

TABLE 2.7

**Viscosity of Pure Vitreous Silica Containing
Less than 10 ppm Water**

t (°C)	Log viscosity (P)	t (°C)	Log viscosity (P)
1000	16.8	1600	8.12
1100	14.67	1800	6.74
1200	12.82	2000	5.60
1300	11.21	2200	4.64
1400	9.80	2400	3.82

where η is the viscosity in poise. As the amount of water in the silica increases, the viscosity below about 1500°C decreases. For 0.04-wt. % water,

$$\log_{10}\eta = -7.7 + (2.86 \times 10^4)/T, \tag{13}$$

and for 0.12-wt. % water,

$$\log_{10}\eta = -6.7 + (2.67 \times 10^4)/T. \tag{14}$$

Thus with 0.12-wt. % water the viscosity equation at low temperatures has the same slope as at higher temperatures.

Temperature points of vitreous silica for various water concentrations are given in Table 2.8.

Elastic Properties   B

The elastic moduli of vitreous silica increase with increasing temperature, as shown in Table 2.9, contrary to the usual decrease in modulus with increasing temperature. The modulus shows a maximum at about 1150°C, but above about 1000°C there can be viscous deformation during the measurement of modulus, so that values above this temperature probably are not true

TABLE 2.8

**Softening, Annealing, and Strain Temperatures for
Pure Vitreous Silicas with Different Water Content[a]**

	Hydroxyl wt. %		
	0.0003	0.04	0.12
Softening point ($10^{7.6}$ P)	1670°C	1596°C	1594°C
Annealing point (10^{13} P)	1190°C	1108°C	1082°C
Strain point ($10^{14.5}$ P)	1108°C	1015°C	987°C

[a] From Hetherington et al. (1964).

TABLE 2.9

**Elastic Properties of Vitreous Silica Containing
Less Than 100 ppm Water as a Function
of Temperature above Ambient**[a]

T (°C)	Young's modulus (GPa)	Shear modulus (GPa)	Poisson's ratio
25	72.9	31.3	0.165
100	74.0	31.6	0.171
200	75.1	32.0	0.173
400	77.2	32.8	0.177
600	78.7	33.3	0.182
800	80.0	33.7	0.187
1000	81.1	34.0	0.193
1100	(81.4)	(34.1)	(0.194)
1200	(81.5)	(34.1)	(0.195)
1250	(81.4)	(34.0)	(0.197)

[a] Parentheses indicate possible flow contributions
to deformations; see text. From Spinner (1962).

instantaneous elastic moduli. At low temperature the modulus shows a
minimum at about 80 K (Krause, 1971), (see Table 2.12).

The three moduli and Poisson's ratio are interrelated; any two values are
enough to calculate all (see Chapter 10 on elastic properties).

The shear modulus decreases as the pressure increases, which is also
anomalous (see Table 2.10). Here dG/dP and dE/dP are constant up to about 1
GPa, and above about 0.5 GPa dK/dP decreases, as shown in Table 2.11
(Peselnick et al., 1967).

TABLE 2.10

**Pressure Dependence of Elastic Moduli for
Vitreous Silica at 25°C**

Modulus derivative	Slope	Approximate limit of constant slope (GPa)
dG/dP (Shear)	-3.25	1
dK/dP (Bulk)	-6.15	0.5[a]
dE/dP (Young's)	-3.55	1

[a] See Table 2.11.

TABLE 2.11

**Effect of Pressure on
the Bulk Modulus of
Vitreous Silica**[a]

Pressure (GPa)	Bulk modulus (GPa)
0	36.5
0.2	35.3
0.5	33.4
0.7	32.4
0.8	31.9
0.9	31.4
1.0	31.0

[a] From Peselnick et al.
(1967).

At high strain the Young's modulus E of silica increases with strain ε (Hillig, 1962; Mallinder and Proctor, 1964) according to the equation

$$E = E_0(1 + 5.75\varepsilon), \tag{15}$$

where E_0 is the modulus at zero strain. One might expect most materials to decrease in modulus at high strains, although few measurements have been made at high enough strains to test this expectation. Soda-lime glass shows a decreasing modulus at high strains, so the increase of Eq. (12) is probably related to the other anomalous behavior of elastic properties of silica. Poisson's ratio also increases at high strains to a value of about 0.40 at 12% strain (Mallinder and Proctor, 1964).

The compressibility β is the reciprocal of the bulk modulus,

$$(1/V_0)(dV/dP) = -\beta. \tag{16}$$

Direct measurement of the compressibility of vitreous silica has given scattered and nonreproducible results (Ernsberger, 1980), probably because compressive deformation includes a time-dependent anelastic component.

The sonic velocity in glass v is directly related to the elastic modulus; for shear waves,

$$v = \sqrt{G/\rho}, \tag{17}$$

where ρ is the density. Modulus is often measured from the acoustic velocity. The modulus values in Tables 2.10 and 2.12 were calculated from measurements of acoustic velocity, using a value of 2.20 g cm^{-3} for ρ.

Strength and Fatigue   C

The fracture strength of a brittle solid is controlled by flaws on its surface (see Chapter 12 on strength). Thus the strength of vitreous silica is not a material property, but depends upon its thermal and handling history. A

TABLE 2.12

Shear Modulus of Vitreous Silica as a Function of Temperature below Ambient; Sample with Less Than 100 ppm Water[a]

T (K)	Shear modulus (GPa)	T (K)	Shear modulus (GPa)
300	31.3	120	30.4
280	31.0	80	30.3
240	30.8	40	30.5
200	30.6	4	30.8
160	30.5		

[a] From Krause (1971).

typical value of strength for as-received samples is 100 MPa, but can vary widely. Silica samples become weaker with time; again the amount of fatigue depends strongly on sample history, as well as ambient humidity and temperature.

D   Hardness

The resistance of glass to deformation by an indenter, or the hardness, is easy to measure but hard to reproduce or interpret. On Mho's scale vitreous silica has a hardness about the same as quartz, or seven. The wide disagreement on measurements of hardness of vitreous silica (Ernsberger, 1980) means that a useful value cannot be given.

VIII   THERMAL PROPERTIES

A   Density

The density of pure vitreous silica with low water content is 2.20 g cm^{-3}.

B   Coefficient of Thermal Expansion

Values from -50 to $1000°C$ are given in Table 2.13, as calculated by Sosman (1927) from measurements of several different investigators. Some more recent work agrees with these values; other workers have found somewhat different values (Otto and Thomas, 1963; Oishi and Kimur, 1969; Pavlova and Amatuni, 1975).

Below about $0°C$ the coefficient of thermal expansion depends strongly on the thermal history of the silica. Data for samples heated at $1000°C$ are given in Table 2.13; if the glass is heated at higher temperatures, the minimum in α shifts to higher temperatures and more positive values.

C   Thermal Conductivity

Values are given in Table 2.14. At temperatures above about $300°C$ radiation effects become important and results are uncertain. Wray and Connolly (1959) have reported thermal conductivities of vitreous silica to high temperatures, but these results have apparently not been confirmed.

TABLE 2.13

Linear Coefficient of Expansion of Vitreous Silica with Low Water Content (<0.01%)

Before Heating[a]		After Heating at 1000°C[b]	
Temperature (°C)	$\alpha = \dfrac{1}{L_0}\dfrac{dL}{dt} \times 10^6\ (\mathrm{K}^{-1})$	Temperature (K)	$\alpha = \dfrac{1}{L_0}\dfrac{dL}{dt} \times 10^6\ (\mathrm{K}^{-1})$
−50	0.25	250	0.36
0	0.38	200	0.26
25	0.41	150	−0.08
50	0.43	120	−0.30
100	0.48	100	−0.44
200	0.53	80	−0.57
300	0.56	60	−0.67
400	0.57	40	−0.70
500	0.57	30	−0.72
600	0.56	20	−0.60
800	0.54	10	−0.24
1000	0.54	5	−0.04

[a] From Sosman (1927).
[b] First eight values from White (1973); last four from Kurkjian *et al.* (1972).

TABLE 2.14

Thermal Conductivity of Vitreous Silica[a]

Temperature (°C)	Thermal conductivity (W m^{-1} K^{-1})	Temperature (K)	Thermal conductivity (W m^{-1} K^{-1})
−173	0.67	20	0.17
−150	0.84	10	0.13
−100	1.05	5	0.10
0	1.32	2	0.047
25	1.37	1	0.015
50	1.41	0.5	4.4×10^{-3}
100	1.48	0.2	1.0×10^{-3}
200	1.59	0.1	2.8×10^{-4}
300	1.71		

[a] Values in left column from Ratcliffe (1963) and Danielson (1982). Values in right column from Zeller and Pohl (1971).

ELECTRICAL PROPERTIES   IX

Electric Conductivity   A

Electrical current in vitreous silica is carried by monovalent cations, because these ions are the most mobile. The two most mobile monovalent cations are sodium and lithium, and in all silicas tested so far, either sodium

TABLE 2.15

Diffusion Coefficients of Sodium in Vitreous Silica[a]

t (°C)	D (cm^2 s^{-1})	t (°C)	D (cm^2 s^{-1})
1000	7.9×10^{-6}	400	1.6×10^{-9}
900	3.8×10^{-6}	300	5.4×10^{-11}
800	1.6×10^{-6}	250	3.5×10^{-12}
700	5.7×10^{-7}	200	2.0×10^{-13}
600	1.3×10^{-7}	170	2.6×10^{-14}
500	1.9×10^{-8}		

[a] From Frischat (1967, 1975).

ions alone or sodium and lithium ions together are the conducting species. Thus the electrical conductivity of vitreous silica depends directly on the concentrations of these ions in it.

The relation between electrical conductivity σ and diffusion coefficient D of a material containing one species of ionic conductor was derived by Einstein:

$$\sigma = Z^2 F^2 (Dc/RT) \tag{18}$$

where Z is the valence of the ion, F the Faraday, c the concentration of the ion, R the gas constant and T the temperature. If σ is in (Ω centimeter)$^{-1}$, D in square centimeters per second, C in moles per cubic centimeter, and T in kelvin,

$$\sigma = 1.120 \times 10^9 \, (Dc/T). \tag{19}$$

The diffusion coefficient of sodium in type I vitreous silica as a function of temperature is given in Table 2.15 (Frischat, 1967, 1975). The temperature dependence does not fit the simple Arrhenius equation $D = D_0 \exp(-Q/RT)$; the apparent activation energy varies from about 146 kJ mole^{-1} below 280°C to about 100 kJ mole^{-1} above about 600°C (cf. Table 2.16). The diffusion coefficient of lithium in vitreous silica is a factor of 6.7 less than that of sodium at 380°C (Doremus, 1969) and probably has about the same temperature dependence as sodium, although there have apparently been no measurements of lithium diffusion in silica as a function of temperature. If there are two

TABLE 2.16

**Arrhenius Parameters for the Diffusion of
Sodium (Table 2.15) in Vitreous Silica**[a]

t (°C)	D_0 (cm^2 s^{-1})	Q/R (K $\times 10^{-3}$)
600–1000	0.041	10.87
250–600	0.37	12.98
170–250	2.1	14.19

[a] From Frischat (1967, 1975).

conducting ions A and B, Eq. (19) must be modified to

$$\sigma = (1.120 \times 10^9/T)(D_A c_A + D_B c_B) \tag{20}$$

If the concentrations of sodium and lithium in the silica are known, the conductivity can be calculated from Eq. (20), the diffusion coefficients in Table 2.15, and the ratio of 6.7 between sodium and lithium diffusion coefficients.

The electrical conductivity of type I vitreous silica with about 2.3 ppm sodium and unknown lithium concentration as a function of temperature as measured by Owen and Douglas (1959) is given in Table 2.17. The temperature dependence of this conductivity is very close to that predicted by the data for sodium diffusion in Table 2.15 and Eq. (19). The temperature dependence of conductivity in silica in which all lithium was replaced by sodium (Doremus, 1969) was also close to that for the data in Table 2.15. These results increase the confidence of a calculation of conductivity from Eq. (20).

Electrical conductivity measurements on a silica of unknown alkali concentration at temperatures above 1000°C are also given in Table 2.17 (Veltri, 1963). Veltri also reported conductivity measurements at higher temperatures, but these appear to be unreliable and have not been confirmed.

Sometimes an additional factor f is introduced into Eq. (18). Careful comparison of the conductivity of silica containing a known sodium ion concentration and the diffusion coefficients of Table 2.15 (Frischat, 1967, 1975), showed that this f factor was unity within experimental error (Doremus, 1969).

Silica containing a large amount of OH ($>0.1\%$) appears to have a lower sodium-ion mobility (Frischat, 1975; Schaeffer et al., 1979); however, the sodium ions in this glass are nonuniformly distributed through the cross section of the glass (Doremus, 1969), and interpretation of diffusion profiles under these conditions requires information on this distribution, which these authors did not take into account. More work is needed to confirm the influence of OH on ionic mobilities.

TABLE 2.17

Electrical Conductivity of Type I Vitreous Silica[a]

t (°C)	$\log \sigma$ (Ω^{-1} cm^{-1})	t (°C)	$\log \sigma$ (Ω^{-1} cm^{-1})
800	−6.4	1000	−5.4
600	−7.4	1200	−4.8
500	−8.1	1400	−4.4
400	−9.0	1600	−4.1
300	−10.6		
200	−13.3		
150	−15.0		

[a] First seven values from Owen and Douglas (1959); next four values from Veltri (1963).

B Dielectric Properties

The real part ε' of the dielectric constant (see Chapter 14) of different silicas at room temperature is about 4 over a wide frequency range. At higher temperatures the dielectric constant increases to a maximum and then decreases; the maximum is at lower temperatures at lower frequencies. At a constant temperature the dielectric constant increases at lower frequencies, the increase starting at a lower frequency at lower temperatures. At microwave frequencies (e.g., 10^6 Hz) the dielectric constant is about 4 over a wide temperature range. The height and positions of these peaks and increases are influenced by impurity concentrations in the silica, especially OH and alkalis (Owen and Douglas, 1959).

The imaginary part of the dielectric constant ε'' is also influenced by frequency, temperature, and impurities in the silica. It is usually reported in terms of tan δ, where tan $\delta = \varepsilon''/\varepsilon'$. Near room temperature tan δ is usually about 10^{-4} for vitreous silica at 1 to 100 kHz. As the temperature increases tan δ increases logarithmically with $1/T$ (see Table 2.18), and also strongly at lower frequencies (Owen and Douglas, 1959). Heat treatment and impurities, especially water, also strongly influence tan δ. There is no good theoretical explanation of these effects, so it is difficult to predict ε'' and ε' for a particular vitreous silica, even if method of manufacture, heat treatment, and impurity content are known.

X OPTICAL PROPERTIES

The refractive index of vitreous silica is given in Table 2.19 as a function of wavelength and at three different temperatures (Wray and Neu, 1969). The authors reported two more significant figures and some additional wavelengths. At the accuracy given in the table, the refractive index does not seem to be greatly influenced by impurities or heat treatment, and agreement

TABLE 2.18

tan $\delta = \varepsilon''/\varepsilon'$ for Type I Vitreous Silica at
1 kHz as a Function of Temperature[a]

t (°C)	log tan δ	t (°C)	log tan δ
25	−4.0	500	0.4
250	−2.8	600	1.2
300	−2.0	700	1.7
400	−0.5	800	2.7

[a] From Owen and Douglas (1959).

TABLE 2.19

**Refractive Index of Vitreous Silica at
Three Temperatures as a Function
of Wavelength, Type III[a]**

λ (μm)	n		
	26°C	471°C	828°C
0.2302	1.520	1.529	1.536
0.2753	1.496	1.503	1.509
0.3022	1.487	1.494	1.499
0.3342	1.480	1.486	1.491
0.4047	1.470	1.476	1.480
0.4358	1.467	1.472	1.477
0.5461	1.460	1.466	1.470
0.5780	1.459	1.464	1.469
1.0140	1.450	1.456	1.460
1.470	1.445	1.450	1.454
1.660	1.443	1.448	1.452
1.981	1.439	1.444	1.447
2.553	1.429	1.435	1.439
3.00	1.420	1.425	1.429
3.37	1.410	1.415	1.419

λ (μm)	n		
	0°C	−100°C	−200°C
0.4047	1.469	1.469	1.468
0.4358	1.467	1.466	1.465
0.5016	1.462	1.461	1.461
0.5876	1.458	1.458	1.457
0.6678	1.456	1.455	1.455

[a] Top portion of table from Wray and Neu (1969); second from Waxler and Cleek (1973).

between measurements by different investigators is good. The next significant figure is not as reproducible, and is probably justified only for measurement on the same sample. The hydrostatic pressure dependence of the refractive index of silica dn/dP is linear up to about 4×10^8 Pa at $+9.4 \times 10^{-12}$ Pa^{-1}; above this pressure the increase in n becomes slightly greater than linear (Vedam et al., 1966).

The optical transmission of types I and III silica from 0.16 to 5 μm is given in Fig. 2.1. Electrical fusion gives rise to absorption in the ultraviolet from reduced centers, and there are water bands for flame fused silica in the 2–3.5-μm region. The absorption further into the infrared and reflectance in the ultraviolet has been discussed elsewhere (Sigel, 1977).

The influence of radiation on the optical absorption of silica and other glasses was reviewed by Friebele and Griscom (1979).

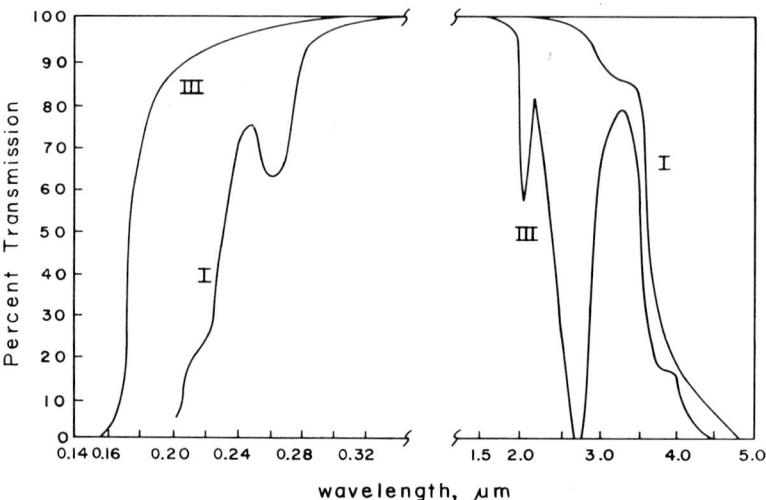

Fig. 2.1. Optimal transmission of different types (see Table 2.1) of vitreous silica. From Doremus (1973). © John Wiley & Sons, Inc.

XI CHEMICAL DURABILITY

The chemical durability is usually considered to be related to the rate of attack of water and aqueous solutions on glass. In vitreous silica this rate is directly related to the solubility of the silica in the contacting solution. The solubility of vitreous (or amorphous) silica in dilute aqueous solution as a function of pH is shown in Tables 2.20 and 2.21. Addition of other salts to the solution can reduce the solubility. This solubility behavior can be described by chemical equilibrium equations:

$$SiO_2 + 2H_2O = H_4SiO_4, \tag{21}$$

$$H_4SiO_4 = H_3SiO_4^- + H^+, \tag{22}$$

$$H_3SiO_4^- = H_2SiO_4^= + H^+. \tag{23}$$

There is always an equilibrium concentration of un-ionized silicic acid (H_4SiO_4) dissolved in solutions of different pH, of concentration about 1.2×10^{-4} g/g solution at 25°C. The equilibrium constants for Eqs. (22) and (23) are about $10^{-9.8}$ and $10^{-12.2}$ (Roller and Ervin, 1940) or

$$1.6 \times 10^{-10} = \frac{[H_3SiO_4^-][H^+]}{[H_4SiO_4]}, \tag{24}$$

$$6.9 \times 10^{-13} = \frac{[H_2SiO_4^=][H^+]}{[H_3SiO_4^-]}, \tag{25}$$

where the brackets denote thermodynamic activities, or concentrations in dilute solution. The concentration units are moles per liter. From these

TABLE 2.20

**Solubility of Vitreous (Amorphous)
Silica in Aqueous Solutions of
Different pH at 25°C[a]**

pH	Solubility (g SiO_2/g solution, $\times 10^4$)
2	1.5
3	1.5
4	1.3
6	1.2
8	1.2
9	1.4
9.5	1.8
10	3.1
10.6	8.8

[a] From Alexander et al. (1954).

TABLE 2.21

**Solubility of Vitreous Silica in
Aqueous Solutions of pH 6–8 at
Different Temperatures[a]**

t (°C)	Solubility (g SiO_2/g solution, $\times 10^4$)
0	0.7
25	1.2
50	2.2
100	4.0
200	10

[a] From Iler (1979).

equations and the solubility of H_4SiO_4, the concentrations of different ionic species at different pH can be calculated. Thermodynamic properties are given in Table 2.3 and Chapter 18.

The rate of dissolution of vitreous silica at pH 8.5 and 25°C is about 10^{-12} cm s^{-1} (Stöber, 1967). For other temperatures and higher pH values, an estimate of the dissolution can be found by considering this rate proportional to the solubilities in Tables 2.20 and 2.21. At lower pH values experimental measurements of dissolution rates are uncertain and may not be directly related to solubilities (Iler, 1979). At 80°C the rate of dissolution in 0.1 M NaOH is about 6×10^{-9} cm/s (Oka and Tomozawa, 1980).

DIFFUSION XII

Gases A

Gases dissolve molecularly in vitreous silica because of its open structure. Small molecules diffuse rapidly through the silica lattice. The permeability of a membrane to a gas is proportional to the product of the solubility (concentration dissolved) of the gas with its diffusion coefficient. See Chapter 15 for definitions of these terms and methods of measurement.

The solubilities of some gases in vitreous silica are given in Table 2.22 in terms of the ratio c_i/c_g of the concentration of gas dissolved in the glass to its concentration in the gas phase. As the size of the molecule increases, its solubility decreases. The temperature dependence of the solubilities is small in the range 200 to 1000°C, and different investigators do not agree on the

TABLE 2.22

Solubility of Gases in Vitreous Silica[a,b]

Gas	Molecular diameter (Å)	c_0/c_g (200–1000°C)	S (molecules cm^{-3} atm^{-1} at 200°C, $\times 10^{-17}$)
Helium	2.0	0.025	3.9
Hydrogen ⎱ Deuterium ⎰	2.5	0.03	4.7
Neon	2.4	0.019	3.1
Argon	3.2	0.01	1.5
Oxygen	3.2	0.01	1.5

[a] From Doremus (1962, 1973).
[b] c_i = concentration in the glass; c_g = concentration in the gas phase.

influence of temperature. The pressure dependence of the solubility ratio c_i/f_g, where f_g is the fugacity of the gas, is also small up to about 100 atm. Above this pressure, competition for solubility sites in the glass reduces the solubility (Shelby, 1979).

The diffusion coefficient of helium in vitreous silica at different temperatures is given in Table 2.23. The dependence of diffusion coefficient on temperature is

$$D = D'T \exp(-Q'/RT); \qquad (26)$$

the data are accurate enough to show that the pre-exponential factor of temperature is needed (Doremus, 1973; Shelby, 1979).

Molecular diffusion coefficients for different gases are compared in Table 2.24. As the molecules become larger, the diffusion coefficient decreases and its activation energy increases. The activation energies Q' in kilojoules per mole and molecular diameters d in Angstrom units are related by the

TABLE 2.23

Diffusion Coefficients of Helium in Vitreous Silica as a Function of Temperature[a]

t (°C)	D (cm^2 s^{-1}, $\times 10^6$)	t (°C)	D (cm^2 s^{-1}, $\times 10^6$)
24	0.024	490	9.0
78	0.11	605	16
112	0.22	700	24
148	0.37	814	36
191	0.73	860	40
284	2.0	1034	61
380	4.8		

[a] From Swets et al. (1961).

TABLE 2.24

Molecular Diffusion in Fused Silica

Molecule	Diameter (Å)	Diffusion coefficient (cm^2 s^{-1})		Activation energy Q' (kJ mole^{-1})
		25°C	1000°C	
Helium	2.0	2.4×10^{-8}	5.5×10^{-5}	20
Neon	2.4	5×10^{-12}	2.5×10^{-6}	37
Hydrogen (deuterium)	2.5	2.2×10^{-11}	7.3×10^{-6}	36
Argon	3.2		1.4×10^{-9}	111
Oxygen	3.2		6.6×10^{-9}	105
Water	3.3		$\sim 3 \times 10^{-7}$	71
Nitrogen	3.4			110
Krypton	4.2			~ 190
Xenon	4.9			~ 300

equation

$$Q' = 25(d - 1.1)^2. \qquad (27)$$

The effect of pressure on molecular diffusion coefficients in vitreous silica is small up to at least 100 atm. At higher pressures the diffusion coefficient appear to increase with increasing pressure (Shelby, 1979). The result is that the permeability coefficient K, which is proportional to D times the solubility (see Chapter 15 on diffusion), is not influenced by pressure up to at least 1000 atm, the decrease in solubility c_i/f_g at high pressures being just compensated by an increase in D.

Water reacts with the silica lattice as

$$H_2O + Si-O-Si = SiOH\ HOSi, \qquad (28)$$

and the reacted OH groups have a higher concentration in the glass than dissolved molecular water, so this reaction has a strong influence on the apparent diffusion of water in silica. See Doremus (1973) for a detailed discussion. Oxygen can react and exchange with oxygen in the silicate lattice, complicating the interpretation of tracer diffusion measurements (Doremus, 1973).

Ions   **B**

The diffusion coefficient of sodium ions in vitreous silica was given in Table 2.15 because of its relation to the electrical conductivity of the silica. As mentioned in that discussion, the activation for sodium diffusion was a function of temperature, and presumably the activation energies for diffu-

TABLE 2.25
Ratios of Ionic Mobilities in Type I Vitreous Silica[a]

Ions	Temperature (°C)	Ratio D_{Na}/D_x
Na–Li	380	6.7
Na–Ag	380	12
Na–K	380	~ 1000
Na–H	1000	$\sim 10{,}000$

[a] From Doremus (1973).

TABLE 2.26
Diffusion Coefficients of Various Ions in Vitreous Silica at 1000°C[a]

Ion	D (cm^2 s^{-1})
Sodium	7.9×10^{-6}
Lithium	1×10^{-6}
Silver	7×10^{-7}
Potassium	1×10^{-8}
Calcium	2×10^{-8}
Aluminum	1×10^{-13}
Phosphorus	8×10^{-14}
Nickel	1×10^{-15}
Arsenic	1×10^{-16}
Boron	1×10^{-17}

[a] Data from Doremus (1973) and Frischat (1975).

sion of other monovalent cations in vitreous silica are also functions of temperature.

The diffusion coefficients of a number of different substances in vitreous silica are compared in Tables 2.25 and 2.26.

BIBLIOGRAPHY

General

"Amersil Fused Silica and Quartz." Amersil, Inc., Hillsdale, New Jersey.
Danielson, P. (1980). Vitreous silica. In "Encyclopedia of Chemical Technology," 3rd Ed. Wiley, New York.
"Fused Quartz Catalog." General Electric Co., Willoughby, Ohio.
Iler, R. K. (1979). "The Chemistry of Silica." Wiley, New York.
Kao, C. K. (1982). "Optical Fiber Systems: Technology, Design and Applications." McGraw-Hill, New York.
Mozzi, R. L., and Warren, B. E. (1969). The structure of vitreous silica. *J. Appl. Cryst.* **2,** 164.
Sosman, R. B. (1927). "The Properties of Silica." Reinhold, New York.

Thermodynamic Properties

Schick, H. L. (1960). A thermodynamic analysis of the high-temperature vaporization properties of silica. *Chem. Rev.* **59,** 331.
Wagman, D. D., *et al.* (1982). "The NBS Tables of Thermodynamic Properties," Table 2-111. Am. Inst. Phys., New York.

Heat Capacity

Bornemann, K., and Hengstenberg, O. (1920). *Met. Erz.* **17,** 313–319.
Casey, D. N., Hetherington, G., Winterburn, J. A., and Yates, B. (1976). *Phys. Chem. Glasses* **17,** 77–82.
Fagaly, R. L., and Bohn, R. H. (1978), *J. Non-Cryst. Solids* **28,** 67–76.

Lasjaunias, J. C., Penn, G., Ravex, A., and Vandorpe, M. (1980). *J. Phys. Lett.* (*Orsay, Fr.*) **41**, L131–L133.
Lucks, C. F., Deems, H. W., and Wood, W. D. (1960). *Am. Ceram. Soc. Bull.* **39**, 313–319.
Sosman, R. B. (1927). pp. 310 ff.
Stephens, R. B. (1973). *Phys. Rev. B* **8**, 2896; **13**, 852 (1976).
Stephens, R. B. (1976). *Phys. Rev. B* **13**, 852.
Wietzel, R. (1921). *Z. Anorg. Allg. Chem.* **116**, 71–95.

Crystallization Rate

Wagstaff, F. E. (1969). *J. Am. Ceram. Soc.* **52**, 650.

Surface Energy

Doremus, R. H. (1976). *J. Appl. Phys.* **47**, 1833.
Kingery, W. D. (1959). *J. Am. Ceram. Soc.* **42**, 6.

Viscosity

Hetherington, G., Jack, K. H., and Kennedy, J. C. (1964). *Phys. Chem. Glasses* **5**, 130.
Hofmaier, G., and Urbain, G. (1968). "Science of Ceramics," Vol. 4, p. 25. Br. Ceram. Soc., Stoke-on-Trent, England.

Elastic Modulus

Ernsberger, F. G. (1980). *In* "Glass: Science and Technology" (D. R. Uhlmann and N. J. Kreidl, eds.), pp. 1–19. Academic Press, New York.
Hillig, W. B. (1962). *In* "Modern Aspects of the Vitreous State" (J. D. Mackenzie, ed.), p. 152. Butterworths, London.
Krause, J. T. (1971). *J. Appl. Phys.* **42**, 3035.
Mallinder, F. P., and Proctor, B. A. (1964). *Phys. Chem. Glasses* **5**, 91–103.
Peselnick, L., Meister, R., and Wilson, W. H. (1967). *J. Phys. Chem. Solids* **28**, 635.
Spinner, S. (1962). *J. Am. Ceram. Soc.* **45**, 394–397.

Coefficient of Thermal Expansion

Kurkjian, C. R., Krause, J. T., McSkimm, H. J., Andreatch, P., and Bateman, T. B. (1972). *In* "Amorphous Materials" (R. W. Douglas and B. Ellis, eds.), pp. 463–473. Wiley, New York.
Oishi, J., and Kumura, T. (1969). *Metrologia* **5**, 50–55.
Otto, J., and Thomas, W. (1963). *Z. Phys.* **175**, 337–344.
Pavlova, G. A., and Amatuni, A. N. (1975). *Izv. Akad. Nauk SSSR, Neog. Mat.* **11**, 1686–1689.
Sosman, R. B. (1927). "The Properties of Silica," pp. 362 ff. Reinhold, New York.
White, G. K. (1973). *J. Phys. D* **6**, 2070–2078.

Thermal Conductivity

Danielson, P. (1982). "Encyclopedia of Chemical Technology," p. 795. Wiley, New York.
Ratcliffe, E. H. (1963). *Glass Technol.* **4**, 113–128.
Wray, K. L., and Connolly, T. J. (1959). *J. Appl. Phys.* **30**, 1702–1705.
Zeller, R. C., and Pohl, R. O. (1971). *Phys. Rev. B* **4**, 2029–2041.

Electrical Properties

Doremus, R. H. (1969). *Phys. Chem. Glasses* **10**, 28.
Frischat, G. H. (1967). *Naturwissenschaften* **54**, 561–562.
Frischat, G. H. (1975). "Ionic Diffusion in Oxide Glasses," pp. 37 ff. Trans. Tech. Publ., Bay Village, Ohio.
Owen, A. E., and Douglas, R. W. (1959). *J. Soc. Glass Technol.* **43**, 159–178.
Schaeffer, H. A., Mecha, J., and Steinman, J. (1979). *J. Am. Ceram. Soc.* **62**, 343–346.
Veltri, R. D. (1963). *Phys. Chem. Glasses* **4**, 221–228.

Optical Properties

Friebele, E. J., and Griscom, D. L. (1979). Radiation effects in glass. *In* "Glass II" (M. Tomozawa and R. H. Doremus, eds.), Treatise on Materials Science and Technology, Vol. 17, pp. 257–351. Academic Press, New York.

Sigel, G. H. (1977). Optical absorption of glasses. *In* "Glass I" (M. Tomozawa and R. H. Doremus, eds.), Treatise on Materials Science and Technology, Vol. 12, pp. 5–89. Academic Press, New York.

Vedam, K., Schmidt, E. D. D., and Roy, R. (1966). *J. Am. Ceram. Soc.* **49,** 531–535.

Waxler, R. M., and Cleek, G. W. (1973). *J. Res. Natl. Bur. Stand., Sect. A* **77A,** 755–763.

Wong, J., and Angell, C. A. (1976). "Glass Structure by Spectroscopy." Dekker, New York.

Wray, J. H., and Neu, J. T. (1969). *J. Opt. Soc. Am.* **59,** 774–776.

Chemical Durability

Alexander, G. B., Heston, M. M., and Iler, R. K. (1954). *J. Phys. Chem.* **58,** 153.

Iler, R. K. (1979). "The Chemistry of Silica," pp. 30–49. Wiley, New York.

Oka, Y., and Tomozawa, M. (1980). *J. Non-Cryst. Solids* **42,** 535–544.

Roller, P. S., and Ervin, G. (1940). *J. Am. Chem. Soc.* **62,** 468.

Stöber, W. (1967). *Adv. Chem. Ser.* No. 67, p. 161.

Diffusion

See also references under electrical properties.

Barrer, R. M. (1941). "Diffusion in and Through Solids," pp. 117 ff. Cambridge Univ. Press, Cambridge, England.

Doremus, R. H. (1962). Diffusion in noncrystalline silicates. *In* "Modern Aspects of the Vitreous State" (J. D. Mackenzie, ed.), Vol. II, pp. 1–71. Butterworth, London.

Doremus, R. H. (1973). "Glass Science, " pp. 121–176. Wiley, New York.

Frischat, G. H. (1975). "Ionic Diffusion in Oxide Glasses." Trans. Tech. Publ., Bay Village, Ohio.

Shelby, J. E. (1979). Molecular solubility and diffusion. *In* "Glass II" (M. Tomozawa and R. H. Doremus, eds.), Treatise on Materials Science and Technology, Vol. 17, pp. 1–40. Academic Press, New York.

Swets, D. E., Lee, R. W., and Frank, R. C. (1961). *J. Chem. Phys.* **34,** 17.

Chapter **3**

Multicomponent Commercial Glasses

Synthetic glass has a long history, dating from at least 10,000 years ago. Mixtures of sand, ash, and limestone heated to about 1400°C form soda-lime glass. A large number of other oxides can be mixed into the melt to form the wide variety of commercial glasses available today. The compositions of some of the most common types of commercial glasses are given in Table 3.1, their uses in Table 3.2, and their properties in Tables 3.3–3.6. The compositions of commercial glasses for special purposes are given in Table 3.7, and some of their properties in Tables 3.8 and 3.9. Compositions of Schott optical glasses providing a range of refractive index from 1.465 to 1.668 are given in Table 3.10, and some properties of a few of these glasses are given in Table 3.11.

The only glasses that can be easily bought in small amounts in different shapes and sizes are soda-lime, Pyrex borosilicate, and vitreous silica. Other glasses, even those in Table 3.1, are usually available only on special order for large quantities. It would be valuable for users requiring small quantities of glass, for testing, and for research to have retail stocks of a variety of compositions.

By far the most common glass is based on the soda-lime silicate (sodium calcium silicate) system. All ancient glasses contained the oxides of sodium, calcium, and silicon. These glasses are cheap, chemically durable, and relatively easy to melt and form. Many minor additions to the basic composition (about 70% silica, 15% soda, Na_2O, 10% CaO + MgO, and 5% other oxides) are made to improve properties of melting and forming: alumina for improved chemical durability and reduced devitrification (crystallization), borates for easier working and lower thermal expansion, zinc oxide for lower melting temperatures, and arsenic and antimony oxides for fining (removal of

TABLE 3.1

Approximate Compositions of Important Commercial Silicate Glasses (in wt. %)

Glass type	Corning number	Kimble (Owens-Illinois) number	SiO_2	B_2O_3	Na_2O	K_2O	MgO	CaO	PbO	Al_2O_3
Soda-lime	0080	R-6	73		15		4	7		1
Pyrex borosilicate	7740	KG-33	81	13	4					2
Potash–soda–lead	0010	KG-1	62		7	7			22	2
Lime–magnesia aluminosilicate	1720	EZ-1	61	5	1		7	8		17
Sodium borosilicate	7050	K-705	68	24	6					2
E fiber glass			54	8	1		1	21		15

TABLE 3.2

Uses of Commercial Glasses

Soda-lime
   Containers—bottles, jars, tumblers
   Windows
   Lamps
   Lenses
   Table ware
Pyrex borosilicate
   Automobile headlights
   Cooking ware
   Laboratory ware
Potash–soda–lead
   Lamp tubing
   Sealing
Lime–magnesia aluminosilicate
   High temperatures—ignition tubes
   Cooking ware
   Molybdenum seals
Sodium borosilicate
   Sealing to tungsten and molybdenum
E glass
   Fibers—insulation, composites with polymers, fabrics
Vitreous silica
   Semiconductor crucibles
   Lamps
   Optics

TABLE 3.3

Viscosity of Commercial Glasses in Table 3.1 as a Function of Temperature[a]

	$\log \eta$ (P)				
t (°C)	R-6	KG-33	KG-1	EZ-1	K-705
1400	2.00	3.40	2.15	2.70	2.33
1300	2.28	3.70	2.40	3.30	2.60
1200	2.70	4.17	2.80	4.00	2.98
1100	3.23	4.74	3.25	4.80	3.45
1000	3.84	5.48	3.79	6.02	4.04
900	4.70	6.97	4.59	7.80	4.85
800	5.86	7.88	5.40	10.82	6.00
700	7.57	9.72	6.57	13.80	7.61
600	10.28	12.15	8.23		10.04
500	14.17	15.0	11.19		13.2
400			14.56		

[a] Source was data sheets, Owens-Illinois Glass Company, Toledo, Ohio.

TABLE 3.4

Density, Viscosity, and Mechanical Properties of the Commercial Silicate Glasses in Table 3.1

Glass type	Density (g cm^{-3})	Temperature points for				Young's modulus (GPa)	Poisson's ratio
		Working (log η = 4)	Softening (log η = 7.6)	Annealing (log η = 13.4)	Strain (log η = 14.6)		
Soda-lime	2.5	990	700	520	480	72	0.25
Pyrex borosilicate	2.23	1250	820	565	515	64	0.20
Potash–soda–lead	2.86	980	630	435	395	61	0.21
Lime–magnesia aluminosilicate	2.52	1200	915	710	670	88	0.25
Sodium borosilicate	2.24	1020	710	500	460	60	0.22
E fiber glass	2.61	1070	840	670	630	72	

TABLE 3.5
Thermal Properties of the Commercial Glasses in Table 3.1

Glass type	Coefficient thermal expansion 0–300°C ($\times 10^7$ °C^{-1})	Contraction coefficient, annealing point to 25°C	Thermal conductivity (W cm^{-1} °C^{-1}) 0°C	100°C	Thermal shock resistance Δt (°C) 0.32 cm thick	0.64 cm thick	1.26 cm thick	Mean heat capacity (J g^{-1} °C^{-1}), 25–175°
Soda-lime	92	114	10.0	11.2	60	50	35	0.87
Pyrex borosilicate	33	37	11.2	12.9	180	150	100	0.85
Potash–soda–lead	89	103	8.3	9.6	65	50	35	0.66
Lime–magnesia aluminosilicate	42	56	13.7	12.9	135	115	75	
Sodium borosilicate	46	57	10.4	12.1	120	100	70	
E fiber glass	60							

TABLE 3.6

Electrical, Optical, and Chemical Properties of Commercial Glasses in Table 3.1

Glass type	Log resistivity (Ω cm)		Activation energy for conduction (kJ mole^{-1})	Dielectric properties at 25°C and 10 Hz			Refractive index at 0.589 μm	Stress optical coefficient, Brewster's	Chemical durability in ($N/50$) H_2SO_4 (relative units)
	250°C	350°C		Dielectric constant	Loss factor (%)	Power factor (%)			
Soda-lime	6.6	5.2	87	7.3	5.1	0.70	1.52	2.6	7.8
Pyrex borosilicate	7.9	6.4	93	4.6	2.2	0.48	1.47	3.7	0.2
Potash–soda–lead	8.9	6.8	131	6.8	1.	0.16	1.54	3.0	17
Lime–magnesia aluminosilicate	10.5	8.7	112	6.3	2.3	0.37	1.53	2.7	0.6
Sodium borosilicate	8.7	6.9	106	4.9	1.4	0.30	1.48	3.9	15
E fiber glass				6.4			1.55		

TABLE 3.7

Compositions and Uses of Special Commercial Glasses

Corning number	Owens Illinois number	SiO$_2$	B$_2$O$_3$	Li$_2$O	Na$_2$O	K$_2$O	MgO	CaO	SrO	BaO	PbO	Al$_2$O$_3$	F	Other	Uses
0120	KG-12	56	5		4	9					28	2			Electronic ware, Dumet sealing
0137		51			1	13			5		29	1			Color television tubes, neck
0138		50			6	8	3	4			22	5			Color television tubes, funnel
1723		57	4				7	9		7		16			Electron tubes, resistors
1990		41		2	5	12					40				Iron sealing
3320	EN-3	76	14		4	2						3		1U$_3$O$_8$	Tungsten sealing
7040	K-704	70	22		5	3						3			Seals to Kovar alloy
7052	EN-1	65	18	1	2	3				3		7	0.3		Seals to Kovar alloy
7059	EM-1	50	14							25		11			Electronic substrates
7070	ES-1	71	26	0.5	1	0.5						1			Electronic components, low loss
7570		4	11								74	11			High lead, solder sealing
7720		74	14		4						6	2			Tungsten sealing
7800	N-51A	74	9		6	1		0.5		2		6			Pharmaceutical containers
8871	EG-4	42		1	2	6					49				Capacitors
9008		66		0.5	6	7				12	4	4	0.9		Television tubes, black and white
9068		63			7	9	1	2	10	2	2	2	0.2	0.5TiO$_2$ 0.5CeO$_2$	Color television, panel
9062		50			4	9					27	2		8FeO	Reed switches

TABLE 3.8

Density, Viscosity, and Mechanical Properties of the Commercial Glasses in Table 3.7

Corning number	Density (g cm^{-3})	Temperature points (°C)				Young's modulus (GPa)	Poisson's ratio
		Working	Softening	Annealing	Strain		
0120	3.05	980	630	435	395	59	0.22
0137	3.18	977	661	478	436		
0138	3.02	970	670	494	451	71	
1723	2.64	1175	910	710	670	86	0.25
1990	3.47	755	500	360	330	58	0.25
3320	2.27	1155	780	540	500	65	0.19
7040	2.24	1080	700	490	450	59	0.23
7052	2.28	1115	710	480	435	57	0.22
7059	2.76	1160	844	639	593	68	
7070	2.13	1070		495	455	51	0.22
7570	5.42		440	363	342	55	
7720	2.35	1140	755	525	485	63	0.20
7800	2.37	1189	795	576	533	70	
8871	3.84	785	525	385	350	58	0.26
9008	2.66	1002	646	446	408	70	
9068	2.68	1005	688	503	462	70	
9362	3.12	958	627	445	405		

bubbles). Soda-lime glass is often termed "soft" glass because of its relatively low softening temperatures.

Pyrex borosilicate glass was developed by Corning Glass Works to be more resistant to thermal shock and more chemically durable than soda-lime glass, yet still to melt at a similar temperature. The borate in this glass reduces its viscosity and the coefficient of thermal expansion, and it allows low sodium content, which increases chemical durability. Pyrex borosilicate glass is often called "hard" glass because of its higher softening temperature compared to soda-lime glass, and it is somewhat more expensive than soda-lime glass because of its somewhat higher melting temperature and more expensive borate raw material. The mirror for the Mount Palomar telescope was made of Pyrex borosilicate glass because of its low thermal expansion; nevertheless it is necessary to correct minute distortions of the mirror surface caused by temperature differences.

A variety of lead glasses are important as low-melting and solder glasses with a wide working temperature range. Lead glass for fine "crystal" contains much more lead than these glasses.

II MECHANICAL PROPERTIES

The viscosities of glasses in Table 3.1 are given in Table 3.3 as a function of temperature. Four temperature "points" for most of the glasses in the tables give a measure of their viscosities in important temperature ranges.

TABLE 3.9
Electrical, Optical, and Chemical Properties of Commercial Glasses in Table 3.7

Corning number	Log resistivity (Ω cm)		Activation energy for resistivity (kJ mole⁻¹)	Dielectric properties at 25°C and 10⁶ Hz			Refractive index at 0.565 μm	Stress optical coefficient, Brewster's	Chemical durability in (N/50) H_2SO_4
	250°C	350°C		Dielectric constant	Loss factor (%)	Power factor (%)			
0120	9.9	7.8	131	6.8	0.7	0.10	1.56	3.0	4.9
0137	10.1	8.4	106	8.6	0.1		1.57		
1723	13.5	11.3	137	6.3	1.0	0.16	1.55		
1990	10.1	7.7	143	8.3	0.33	0.04			
3320	8.6	7.1	93	4.9	1.5	0.30	1.48		
7040	9.2	7.4	112	5.0	1.5	0.3	1.48		
7052	9.0	7.2	112	5.1	1.3	0.26	1.48	4.1	2.9
7059	13.1	11.0	118	5.9	0.1		1.53		
7070	11.2	9.4	112	4.0	0.4	0.11	1.48	4.8	2.2
7570	10.6	8.7	118	15	3.3	0.22	1.86		
7720	8.8	7.2	100	4.7	1.3	0.27	1.49		
7800	7.2	5.8	88	5.9	3.7	0.63	1.50	3.5	0.36
8871	11.1	8.8	143	8.4	0.42	0.05			
9008	9.1	7.2	118	6.6	0.18		1.51		
9362	9.4	7.7	106						

TABLE 3.10

Optical Properties and Compositions of Schott Optical Glasses[a]

Number	n_D	ν	SiO$_2$	B$_2$O$_3$	Al$_2$O$_3$	Na$_2$O	K$_2$O	CaO	BaO	ZnO	PbO	TiO$_2$	Sb$_2$O$_3$	KHF$_2$
BK1	1.5101	63.5	71.4	6.5		5.2	13.9	2.0						
				Fluoride glasses										
FK3	1.4645	65.8	47.7	17.4	14.0	2.2	2.4							16.0
KK5	1.4875	70.4	56.9	15.7			5.6							21.6
						Lead glasses								
KF3	1.5145	54.7	70.0		1.5	16.6				5.0	3.5	1.2	0.5	
LLF2	1.5407	47.2	63.2			5.0	8.0				23.5			
LF7	1.5750	41.5	33.9			2.5	7.9				34.9			
F8	1.5955	39.2	50.2	0.4		3.8	5.6				39.7			
F2	1.6200	36.4	45.7			3.6	5.0				45.1			
SF2	1.6477	33.8	40.9			0.5	6.8				50.8			
SF5	1.6727	32.2	38.7			1.5	3.9				55.6			
SF10	1.7282	28.4	35.3			2.0	2.5				55.7	4.0		
SF14	1.7618	26.5	32.1		1.2	1.0	1.0				60.5	3.7		
SF11	1.7847	25.8	29.2		2.5	0.5					63.3	4.0		
SF6	1.8052	25.4	26.9		0.5	1.0					71.3			

						Barium glasses					
BaK1	1.5725	57.6	47.7	4.5	1.0	1.0	7.5	29.0	8.6		
SK5	1.5891	61.2	38.7	15.0	5.0			40.1			0.3
SK3	1.6088	58.9	35.0	11.9	4.5	0.5		45.9		0.6	1.6
SK1	1.6102	56.7	40.1	5.7	2.5			42.2	8.5		0.3
SK4	1.6272	58.6	33.2	1.9	5.5			47.8		0.3	0.3
SK10	1.6228	56.9	30.6	11.7	5.0	0.1		48.2	2.0	0.7	0.8
						Barium–lead glasses					
BaF4	1.6056	43.9	45.5		2.2	0.5	7.3	15.8	8.0	22.5	
BaF9	1.6433	48.0	35.3	5.3				36.1	8.2	12.3	0.2
BaSF2	1.6645	35.8	38.7			2.5	4.4	13.0	5.0	39.5	2.0
BaSF6	1.6676	41.9	34.9	4.5	0.5	1.1	1.9	5.6	25.8	4.6	18.0	2.5

[a] All glasses contain from 0.5 to 1% As_2O_3 for fining.

TABLE 3.11

Some Additional Properties of Certain Optical Glasses in Table 3.10

Schott number	Density (g cm^{-3})	Coefficient of thermal expansion ($\times 10^7$ °C^{-1}) for 25–300°C	Young's modulus (GPa)	Poisson's ratio	Heat capacity (J g^{-1} °C^{-1})
BK1	2.47	92	73	0.21	
SF2	3.88	98	54	0.22	
SF6		85			
BaK1	3.21	80	73	0.25	
SK5		72			
SK3		76			
SK1	3.53	76	78	0.27	0.58
BaF4	3.50				

The Young's modulus E and Poisson's ratio σ are given in the tables. The bulk modulus K and shear modulus G can be calculated from them as

$$K = E/3(1 - 2\sigma),\qquad(1)$$

$$G = E/2(1 + \sigma).\qquad(2)$$

The temperature dependence of the elastic properties of silicate glasses up to the glass transition temperature and at low temperatures is small (see data for vitreous silica in Chapter 2).

Strength of commercial glasses is discussed in Chapter 12.

III TEMPERATURE DEPENDENCE OF PROPERTIES

The temperature dependence of the heat capacity of silicate glasses below ambient is similar to that of vitreous silica (see Chapter 2). Above about 200°C to the glass transition temperature (or annealing point), the heat capacities of a soda-lime glass and Pyrex borosilicate glass are nearly constant at 1.01 J g^{-1} °C^{-1} and 1.11 J g^{-1} °C^{-1}, respectively. Above the annealing temperature to about 900°C the heat capacities of these glasses are again constant: soda-lime, 1.49 J g^{-1} °C^{-1}, and Pyrex borosilicate, 1.45 J g^{-1} °C^{-1} (Haggerty et al., 1968). The composition of the soda-lime glass was 70% SiO_2, 20% Na_2O, and 10% CaO. These results and others show that the heat capacity is not much dependent on composition in silicate glasses of commercial interest.

The temperature dependence of thermal conductivity of multicomponent silicate glasses of commercial interest is also similar to that of vitreous silica (see Chapter 2). Ratcliffe (1963) has given factors for the calculation of thermal conductivity of silicate glasses of different compositions and at different temperature.

The refractive index depends little on temperature, as shown in Chapter 2 for vitreous silica.

The electrical resistivity and dielectric properties are strongly temperature dependent. The resistivity ρ is given by the equation

$$\rho = \rho_0 \exp(E/RT), \tag{3}$$

where ρ_0 and E, the activation energy, are constants, R is the gas constant $(8.314 \text{ J mole}^{-1} \text{ K}^{-1})$, and T is the absolute temperature. Values of the activation energy E are given in Tables 3.6 and 3.9 together with log resistivities at 250 and 350°C. The resistivity at temperature T in K from about 100°C to the annealing temperature can be calculated from ρ_s, the (standard) resistivity at 250°C, and the activation energy E:

$$\log \rho = \log \rho_s + \frac{E}{2.303R}\left(\frac{1}{T} - \frac{1}{523}\right). \tag{4}$$

Extrapolation to lower temperatures is increasingly less reliable as the temperatures become lower.

The dielectric constant at high frequency increases slowly with increasing temperature; at lower frequencies the increase is sharper. The dielectric loss (or the power factor) increase strongly with increasing temperature and decreasing frequency. Numerical values depend upon the electrodes used and the surface condition of the glass; there are considerable differences between different investigators. [See Tomozawa (1977) for a review.]

OPTICAL PROPERTIES IV

The refractive index n_D in the tables is given at 25°C and a wavelength of 0.589 μm (sodium D line). The change of refractive index with wavelength in the visible spectral region is often estimated from the quantity v, which is defined as

$$v = (n_D - 1)/(n_F - n_C), \tag{5}$$

where n_F is the index at 0.486 μm and n_C at 0.656 μm. The larger v, the smaller is the change of refractive index with wavelength. As shown in Table 3.10, the larger the refractive index, the smaller is v, so the change of refractive index with wavelength usually increases with higher refractive index.

The refractive index changes nonlinearly with wavelength (λ); at wavelengths below about 0.45 μm it increases rather sharply with decreasing wavelength. A simple empirical formula that fits refractive index data on many commercial silicate glasses between about 0.3 and 1.0 μm is

$$n = A + B/\lambda, \tag{6}$$

where A and B are constants. For one commercial soda-lime glass similar to the 0080 and R-6 glasses in Table 3.1, $A = 1.493$ and $B = 0.0178 \ \mu m$. This is a Bausch and Lomb glass described by Morey (1954, p. 432) with composition 73 wt. % SiO_2, 1% B_2O_3, 15% $Na_2O + K_2O$, 10% $CaO + MgO$, and 1% ZnO. Small changes in composition can change A and B in the last decimal place.

It would be more useful to have values of A and B in Eq. (6) than to use the factor v of Eq. (5) as a measure of dispersion (wavelength dependence of refractive index). The factor B can be calculated from n_D and v by the following relation:

$$B = n_D / 0.5335v \quad \mu m. \tag{7}$$

Then

$$A = n_D - B/0.5893. \tag{8}$$

The optical transmission of a commercial soda-lime silicate glass is shown in Fig. 3.1. The glass appears transparent in the visible spectral region in thickness less than about 1 cm; much thicker specimens look green in transmission because iron (Fe^{3+}) impurity has an intense absorption band in the ultraviolet.

The absorption edges of the soda-lime glass in the ultraviolet and infrared are not intrinsic properties of the glass but result from impurities in the glass. In the ultraviolet, trivalent iron and other transition metals are responsible for absorption, whereas in the infrared, water (SiOH groups) causes the absorption edges. Optical transmission spectra of all commercial silicate glasses except vitreous silica resemble that in Fig. 3.1, with the edges depending on impurity contents. Considerable care is required to prepare glasses that have intrinsic adsorption edges (Sigel, 1977). These impurity-dependent edges mean that silicate glasses cannot be analyzed or identified from the adsorption edges or regions.

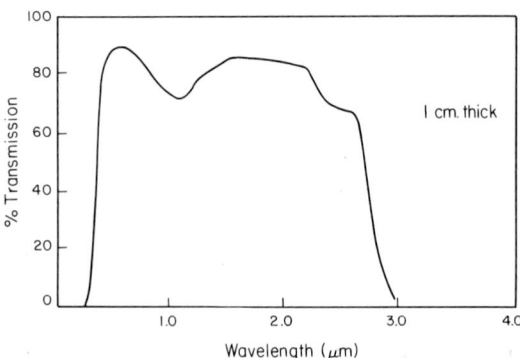

Fig. 3.1. Optical transmission of a commericial soda-lime silicate glass. From Holloway (1973).

Additional information on properties that also apply to commercial glasses is given in the chapters on these properties, especially strength and chemical durability.

BIBLIOGRAPHY

Haggerty, J. S., Cooper, A. R., and Heasley, J. H. (1968). Heat capacity of three inorganic glasses and supercooled liquids. *Phys. Chem. Glasses* **9,** 47–51.

Morey, G. W. (1954). "The Properties of Glass." Van Nostrand-Reinhold, Princeton, New Jersey.

Ratcliffe, E. H. (1963). Thermal conductivities of glasses...". *Glass Technol.* **4,** 113–128.

Sigel, G. H. (1977). Optical absorption of glasses. *In* "Glass I" (M. Tomozawa and R. H. Doremus, eds.), Treatise on Materials Science and Technology, Vol. 12, pp. 5–89. Academic Press, New York.

Tomozawa, M. (1977). Dielectric characteristics of glass. *In* "Glass I" (M. Tomozawa and R. H. Doremus, eds.), Treatise on Materials Science and Technology, Vol. 12, pp. 283–345. Academic Press, New York.

Part II
Thermodynamic and Thermal Properties

Chapter **4**
Density

The density is frequently measured and used to understand properties and composition of glass. Density values in grams per cubic centimeter at room temperature are tabulated in this section for compositions in mole percent of modifier oxide. Since the density is not a strong function of temperature, the temperature can be in the range 18–25°C without changing the density in the third decimal place.

The displacement method of Archimedes is most frequently used to measure density. It does not require a sample of known dimensions and is capable of good precision. A convenient sample is weighed in air g_a and in a liquid of known density ρ_L, giving weight g_L. The unknown density ρ is calculated from the relation

$$\rho = g_a \rho_L / (g_a - g_L). \tag{1}$$

In the pyknometric method a vessel (pyknometer) is weighed first (g_1) filled with a liquid of known density ρ_L and then with a sample of known weight g_s. The density of the sample is then

$$\rho = g_s \rho_L / (g_s + g_1 - g_2), \tag{2}$$

where g_2 is the weight with the sample and liquid in the pyknometer. It is difficult to obtain as precise measurements with this technique as with the displacement method.

II   TABLES AND FIGURES

Table 4.1 lists density values for single-component glasses. Tables 4.2–4.10 provide values for some binary silicate glasses, and Tables 4.11–4.97 and Figs. 4.1 and 4.2 values for ternary silicate glasses.

BIBLIOGRAPHY

Abou-El-Azm, A., and Hussein, A. L. (1962). *J. Chem. U.A.R.* **5,** 1.
Akimov, V. V. (1973). *Zh. Prik. Khim.* **46,** 2178.
Aleksandrov, V. I., Borik, M. A., Dechev, G. K., Markov, N. I., Myzina, V. A., Osiko, V. V., Tatarintsev, V. M., and Khodakovskaya, R. Y. (1980). *Sov. J. Glass Phys. Chem.* (*Engl. Transl.*) **6,** 117.
Alekseeva, Z. D. (1964). "Sbornik Khimiya Redkikh Elementov," p. 122. IZD Leningrad Univ., Leningrad.
Alekseeva, Z. D., and Polozok, N. V. (1972). *Inorg. Mater.* (*Engl. Transl.*) **8,** 135.
Aleinikov, F. K., Vaytkus, Y. P., and Zhitkyavichyute, I. I. (1967). *Tr. Mokslu Akad. Darb, Ser. B* **2**(49), 75.
Antonova, N. I. (1975). *Fiz. Khim. Stekla* **1,** 176.
Appen, A. A. (1953). *Zh. Prikl. Khim.* **26,** 9.
Appen, A. A., and Gan'Fu-Si (1959a). *J. Appl. Chem. USSR* (*Engl. Transl.*) **32,** 1006.
Appen, A. A., and Gan'Fu-Si (1959b). *J. Appl. Chem. USSR* (*Engl. Transl.*) **32,** 1013.
Aslanova, M. S., Sapozhkova, L. A., and Gordon, S. S. (1980). *Sov. J. Glass Phys. Chem.* (*Engl. Transl.*) **6,** 444.
Belyustin, A. A., and Volkov, S. E. (1967). *Sov. Electrochem.* (*Engl. Transl.*) **3,** 781.
Bihuniak, P. P., and Condrate, R. A. (1981). *J. Non-Cryst. Solids* **44,** 331.
Bogatyreva, V. V., Bogatyrev, Y. Z., and Solov'yeva, T. I. (1973). *Sov. J. Opt. Technol.* (*Engl. Transl.*) **40,** 495.
Bollin, P. L. (1972). *J. Am. Ceram. Soc.* **55,** 483.
Botvinkin, O. K., and Demichev, S. A. (1964). *Steklo* No. 2, p. 1.
Botvinkin, O. K., and Yaroker, K. G. (1966). *Steklo* No. 3, p. 1.
Botvinkin, O. K., and Yaroker, K. G. (1967). *Inorg. Mater.* (*Engl. Transl.*) **3,** 1427.
Brekhovskikh, S. M., and Sesorova, V. N. (1960). "Conditions of Glass Formation," p. 444. Moscow–Leningrad.
Brosset, C. (1962). *Int. Ceram. Congr., 8th, Copenhagen* p. 15.
Buchanan, R. C., and Zuegel, M. A. (1968). *J. Am. Ceram. Soc.* **51,** 28.
Chakraborty, M. R. (1969). *Am. Ceram. Soc. Bull.* **48,** 1076.
Chang, Y.-h., and Ying, C.-w. (1965). *J. Chin. Silic. Soc.* **4,** 1.
Charles, R. J. (1966). *J. Am. Ceram. Soc.* **49,** 55.
Cleek, G. W., and Babcock, C. L. (1973). "Properties of Glasses in Some Ternary Systems Containing BaO and SiO_2, No. 135. Natl. Bur. Stand., Washington, D.C.
Cousen, A., and Turner, W. E. S. (1928a). *Glastech. Ber.* **6,** 393.
Cousen, A., and Turner, W. E. S. (1928b). *J. Soc. Glass Technol.* **12**(47), 169.
Cousen, A., and Turner, W. E. S. (1928c). *J. Chem. Soc.* p. 2654.
Danilova, N. P. (1967). *Steklo* No. 1, p. 89.
Danilova, N. P., and Dubrovo, S. K. (1967). *J. Appl. Chem. USSR* (*Engl. Transl.*) **40,** 959.
Danilova, N. P., and Dubrovo, S. K. (1971). "Glass Formation of Silicates of Barium," 3509–71 Dep. VINITI.
Demkina, L. I. (1958). "Issledovaniye Zavisimosti Svoystv Stekol ot ikh Sostara, M.," p. 7. Prilozheniye.
Din, A., and Hennicke, H. W. (1974). *Glastech. Ber.* **47,** 14.

Dubrovo, S. K. and Kasimova, S. S. (1964). *Uzb. Khim. Zhur.* No. 8, p. 14.
Dubrovo, S. K., and Shmidt, Y. A. (1959). *J. Appl. Chem. USSR (Engl. Transl.)* **32**, 767.
Dubrovo, S. K., and Shnypikov, A. D. (1966). *Inorg. Mater. (Engl. Transl.)* **2**, 1417.
Dubrovo, S. K., Danilova, N. P., and Tsekhomskaya, T. S. (1965). *Issled. Obl. Khim. Silik. Okislov* p. 11.
Dubrovo, S. K., Zasolotskaya, M. V., and Tsekhomskaya, T. S. (1969). *J. Appl. Chem. USSR (Engl. Transl.)* **42**, 2295.
Ellern, G. A., and Pavlushkin, N. M. (1969). *Tr. Mosk. Khim.-Tekhnol. Inst.* No. 59, p. 30.
Ershov, O. S., Dimakov, I. V., Markova, T. P., and Shul'ts, M. M. (1972). *Inorg. Mater. (Engl. Transl.)* **8**, 1606.
Escola, P. (1922). *Am. J. Sci.* **4**, 331.
Evstropev, K. K., and Pavlovskii, V. K. (1967). *Inorg. Mater. (Engl. Transl.)* **3**, 592.
Faick, C. A., Young, J. C., Hubbard, D., and Finn, A. N. (1935). *J. Res. Natl. Bur. Stand. (U.S.)* **14**, 133.
Galant, E. I. (1961). *Dokl. Akad. Nauk SSSR* **141**, 417.
Galant, E. I., Makarova, T. M., Malchanov, V. S., and Tsekhomskii, V. A. (1966). *Opt.-Mekh. Prom.* No. 4, p. 32.
Goral'nik, A. S., Kul'bidkaya, M. N., Mikhaylov, I. G., Fershtat, L. N., and Shutilov, D. A. (1972). *Akystich. Zh.* **18**, 391.
Graham, P. W. L., and Rindone, G. E. (1967). *Phys. Chem. Glasses* **8**, 160.
Guckelsberger, K., and Lasjaunias, J. C. (1970). *C. R. Hebd. Seances Acad. Sci., Ser. B* **270**, 1427.
Hakim, R. M., and Uhlmann, D. R. (1967). *Phys. Chem. Glasses* **8**, 174.
Hamilton, E. H., and Cleek, G. W. (1958). *J. Res. Natl. Bur. Stand. (U.S.)* **61**, 89.
Hamilton, E. H., Cleek, G. W., and Grauer, O. H. (1958). *J. Am. Ceram. Soc.* **41**, 209.
Hanada, T., Goto, S., Ota, R., and Soga, N. (1981). *Nippon Kagaku Kaishi* **13**, 1577.
Haydenov, A. P. (1963). *Steklo* No. 3, p. 75.
Hirayama, C., and Berg, D. W. (1963). *Phys. Chem. Glasses* **46**, 85.
Huang, Y. Y., Sarkar, A., and Schultz, P. C. (1978). *J. Non-Cryst. Solids* **27**, 29.
Hurt, I. C., and Phillips, C. J. (1970). *J. Am. Ceram. Soc.* **53**, 269.
Imaoka, M., Hasegawa, H., Hamaguchi, Y., and Kurotaki, Y. (1971). *Yogyo Kyokaishi* **5**, 164.
Ishikawa, T., and Akagi, S. (1978). *Phys. Chem. Glasses* **19**, 108.
Iskhakov, K. S. (1971). *Uzb. Khim. Zh.* No. 2, p. 79.
Ivanov, A. O., and Galant, E. I. (1963). *Opt.-Mekh. Prom.* No. 3, p. 43.
Jabra, R., Pelous, J., and Phalippou, J. (1980). *J. Non-Cryst. Solids* **37**, 349.
Karapetyan, G. O., Livshits, V. Y., and Tennison, D. G. (1979). *Sov. J. Glass Phys. Chem. (Engl. Transl.)* **5**, 278.
Karapetyan, G. O., Livshits, V. Y., Tennison, D. G., and Zhukovskaya, O. V. (1980). *Sov. J. Glass Phys. Chem. (Engl. Transl.)* **6**, 310.
Karlsson, K. (1970). *Suom. Kemistil.* **43**, 479.
Kasimova, S. S. (1964a). *Steklo* No. 3, p. 87.
Kasimova, S. S. (1964b). Physico-chemical properties of sodium–strontium–silicate glasses. Candidate's Dissertation, Tashkent.
Kas'imova, S. S., and Makhkamova, D. (1970). *Uzb. Khim. Zh.* No. 4, p. 9.
Kee, H. K. (1968). *J. Korean Chem. Soc.* **12**, 65.
Komiyama, T. (1974). *Osaka Kogyo Gijutsu Shikensho Kiho* **25**, 84.
Kondrat'ev, Y. N., and Smirnova, L. A. (1970). *Inorg. Mater. (Engl. Transl.)* **6, 294.**
Konijnendijk, W. L. (1975). The structure of borosilicate glasses. Thesis, Eindhoven, Netherlands.
Kordes, E. (1939a). *Z. Phys. Chem.* **43**, 173.
Kordes, E. (1939b). *Z. Anorg. Allgem. Chem.* **241**, 1.
Kordes, E. (1939c). *Z. Phys. Chem.* **43**, 119.
Kordes, E., and Becker, H. (1949). *Z. Anorg. Allgem. Chem.* **260**, 185.
Kuznetsova, M. G. (1972). Investigations of the physico-chemical properties of the glass systems $RO-MnO-SiO_2$ and $RO-MnO-B_2O_3$. Candidate's Dissertation, Leningrad.
Lai, C. F., and Silverman, A. (1928). *J. Am. Ceram. Soc.* **11**, 535.

Lai, C. F., and Silverman, A. (1930). *J. Am. Ceram. Soc.* **13**, 393.

Larsen, E. S. (1909). *Am. J. Sci.* **28**, 263.

Layton, M. M., and Herczog, A. (1967). *J. Am. Ceram. Soc.* **50**, 369.

Lisenenkov, A. A., and Vasilev, A. I. (1979). *Sov. J. Glass Phys. Chem.* (*Engl. Transl.*) **5**, 485.

MacDowell, J. F. (1965). *Proc. Br. Ceram. Soc.* No. 3, p. 229.

Mackenzie, J. D. (1958). *J. Chem. Phys.* **29**, 605.

McVay, G. L., and Day, D. E. (1970). *J. Am. Ceram. Soc.* **53**, 508.

Matusita, K., and Mackenzie, J. D. (1979). *J. Non-Cryst. Solids* **30**, 285.

Mochida, N., Takahashi, K., and Shibusawa, S. (1980). *Yogyo Kyokaishi* **88**, 583.

Moore, H., and McMillan, P. W. (1956). *J. Soc. Glass Technol.* **40**, 139T.

Morey, G. W. (1954). "The Properties of Glass." Reinhold, New York.

Morey, G. W., and Merwin, H. E. (1932). *J. Opt. Soc. Am.* **22**, 632.

Moriya, T., Akao, Y., and Hatano, N. (1960). *Yogyo Kyokaishi* **68**, 145.

Nassau, K., Shiever, J. W., and Krause, J. T. (1975). *J. Am. Ceram. Soc.* **58**, 461.

Nemilov, S. V. (1973). *Russ. J. Phys. Chem.* (*Engl. Transl.*) **47**, 831.

Otto, K., and Milberg, M. E. (1967). *J. Am. Ceram. Soc.* **50**, 513.

Parfenov, A. I., Kilmov, A. F., and Mazurin, O. V. (1959). *Vestn. Leningr. Univ.* (*Fiz. Khim.*) No. 10, p. 129.

Parfenov, A. I., Shul'ts, M. M., Netzrasova, T. N., and Palozava, I. P. (1963). *Vestn. Leningr. Univ.* No. 4, p. 126.

Peddle, C. T. (1920). *J. Soc. Glass Technol.* **4**, 3.

Polukhin, V. N. (1960). *Opt.-Mekh. Prom.* No. 11, p. 18.

Rao, B. V. J. (1962a). *J. Am. Ceram. Soc.* **45**, 555.

Rao, B. V. J. (1962b). *J. Sci. Ind. Res., Sect. B* **21**, 108.

Rao, B. V. J. (1963a). *Phys. Chem. Glasses* **4**, 22.

Rao, B. V. J. (1963b). *J. Am. Ceram. Soc.* **46**, 107.

Rao, B. V. J. (1964). *Glass Technol.* **5**, 67.

Rao, B. V. J. (1965). *J. Am. Ceram. Soc.* **48**, 311.

Riebling, E. F. (1968). *J. Am. Ceram. Soc.* **51**, 406.

Riegel, E. R., and Sharp, D. (1934). *J. Am. Ceram. Soc.* **17**, 88.

Roedder, E. W. (1951). *Am. J. Sci.* **249**, 81.

Rothermel, D. L. (1967). *J. Am. Ceram. Soc.* **50**, 574.

Sasek, L., and Lisy, A. (1972a). *Sb. Vys. Sk. Chem.-Technol. Praze, Chem. Technol. Silik.* **L2**, 165.

Sasek, L., and Lisy, A. (1972b). *Sb. Vys. Sk. Chem.-Technol. Praze, Chem. Technol. Silik.* **L2**, 218.

Schroeder, J. (1980). *J. Non-Cryst. Solids* **40**, 549.

Schultz, P. C., and Dumbaugh, W. H. (1980). *J. Non-Cryst. Solids* **38/39**, 33.

Schultz, P. C., and Smyth, H. T. (1970). Corning Glass Works, New York.

Schwartz, M., and Mackenzie, J. D. (1966). *J. Am. Ceram. Soc.* **49**, 582.

Shabanova, E. B. (1967). *Tr. Gor'k. Politekh. Inst.* **23**, 38.

Shaw, R. R., and Uhlmann, D. R. (1969). *J. Non-Cryst. Solids* **1**, 474.

Shaw, R. R., and Uhlmann, D. R. (1971). *J. Non-Cryst. Solids* **5**, 237.

Shchavelev, O. S., Kasymova, S. S., and Petrovskii, G. T. (1974). *J. Appl. Chem. USSR* (*Engl. Transl.*) **47**, 13.

Shelby, J. E. (1979). *J. Appl. Phys.* **50**, 8010.

Shelby, J. E., and Day, D. E. (1969). *J. Am. Ceram. Soc.* **52**, 169.

Sheludyakov, L. N. (1967). *Vestn. Akad. Nauk Kaz. SSR* **23**, 43.

Sheybany, H. A. (1948). *Verres Refract.* **2**, 127.

Shmidt, Y. A., and Alekseeva, Z. D. (1964). *J. Appl. Chem. USSR* (*Engl. Transl.*) **37**, 2266.

Shnypikov, A. D. (1967). *Steklo* No. 1, p. 85.

Shul'ts, M. M., Peshekhonova, N. V., Parfenov, A. I., Ivanova, E. A., and Petrova, V. N. (1963). *Vestn. Leningr. Univ., Fiz, Khim.* **18**(4), 104.

Simpson, H. E. (1959). *Glass Ind.* **40**, 454.

Simpson, H. E. (1961). *Glass Ind.* **42**, 222.

Soga, N., Ota, R., and Kunugi, M. (1972). *Proc. Int. Conf. Mech. Behav. Mater., 4th, Kyoto* p. 366.

Soga, N., Yamanaka, H., Misamoto, C., and Kunugi, M. (1976). *J. Non-Cryst. Solids* **22**, 67.

Sosman, R. B. (1927). "The Properties of Silica." Reinhold, New York.

Sun, K.-H., and Silverman, A. (1941). *J. Am. Ceram. Soc.* **24**, 160.

Syritskaya, Z. M., Rogozhin, Y. V., and Shakhova, R. I. (1971). *Stekloobraznye Sist. Nov. Stekla Ikh Osn., Mater. Vses. Soveshch., Minsk, 1967* p. 55.

Takahashi, K. (1962). *Adv. Glass Technol., Tech. Pap. Int. Congr. Glass, 6th, Washington, D.C.* 7, 366.

Takahashi, K., and Osaka, A. (1983). *Yogyo Kyokaishi* **91**, 116.

Takahashi, K., Mochida, N., and Hatta, G. (1975). *Yogyo Kyokaishi* **83**, 103.

Takahashi, K., Mochida, N., Matsui, H., Takeuchi, S., and Gohshi, Y. (1976). *Yogyo Kyokaishi* **84**, 482.

Takahashi, K., Mochida, N., and Yoshida, Y. (1977). *Yogyo Kyokaishi* **85**, 330.

Tarlakov, Y. P., Pronkin, A. A., Kekeliya, D. I., Verulashvili, R. D., and Kogan, V. E. (1980). *Izv. Akad. Nauk Gruz. SSR, Ser. Khim.* **6**, 166.

Terai, R. (1969). *Yogyo Kyokaishi* **77**, 318.

Terai, R. (1970). *Osaka Kogyo Gijutsu Shikensho Kiho* **21**, 64.

Terai, R. (1971a). *J. Non-Cryst. Solids* **6**, 121.

Terai, R. (1971b). *Osaka Kogyo Gijutsu Shikensho Kiho* **22**, 73.

Terai, R. (1972). *Osaka Kogyo Gijutsu Shikensho Kiho* **23**, 36.

Thiele, A. (1955). *Glastech. Ber.* **28**, 384.

Thompson, C. L., and Parmelee, C. W. (1937). *J. Am. Ceram. Soc.* **20**, 305.

Tiwari, A. N., and Das, A. R. (1972). *Am. Ceram. Soc. Bull.* **51**, 695.

Topping, J. A., Fuchs, P., and Murthy, M. K. (1974). *J. Am. Ceram. Soc.* **57**, 205.

Tsekhomskaya, T. S., and Dubrovo, S. K. (1968). *J. Appl. Chem. USSR (Engl. Transl.)* **41**, 468.

Tsekhomskii, V. A. (1966). Polyconducting glasses based on the oxides of iron and titanium. Candidate's Dissertation, Leningrad.

Urusovskaya, L. N. (1960). *J. Appl. Chem. USSR (Engl. Transl.)* **33**, 1971.

Vakhrameev, V. I. (1968). *Steklo* No. 3, p. 84.

Vakhrameev, V. I., and Evstrop'ev, K. S. (1969). *Inorg. Mater. (Engl. Transl.)* **5**, 82.

Vargin, V. V., Zasolotskaya, M. V., Kind, N. E., Kondratyev, Y. N., Milyukov, E. M., and Tudorovskaya, N. A. (1971). "Kataliziro-vannaya Kristalliatsiya Stekol Litiyevoalyumocili-Katnoy Sistemi." Leningrad.

Vargin, V. V., Dzhavuktsyan, S. G., Mishel', V. E., and Pevzner, B. Z. (1972). *J. Appl. Chem. USSR (Engl. Transl.)* **45**, 1228.

Varshal, B. G. (1972). *Inorg. Mater. (Engl. Transl.)* **8**, 812.

Watanabe, K., Sumiyoshi, Y., and Anbo, E. (1970). *Yogyo Kyokaishi* **78**, 165.

Weberbauer, A. (1932a). *Glastech. Ber.* **10**, 361.

Weberbauer, A. (1932b). *Glastech. Ber.* **10**, 426.

Weir, C. E., and Shartsis, L. (1955). *J. Am. Ceram. Soc.* **38**, 299.

Winks, F., and Turner, W. E. S. (1931). *J. Soc. Glass Technol.* **15**, 185.

Wulff, P., and Majumdar, S. K. (1936). *Z. Phys. Chem. B* **31**, 1936.

Yamane, M., and Okuyama, M. (1982). *J. Non-Cryst. Solids* **52**, 217.

Young, J. C., Glaze, F. W., Faick, C. A., and Finn, A. N. (1939). *J. Res. Natl. Bur. Stand. (U.S.)* **22**, 453.

TABLE 4.1

Densities of Single-Component Oxide Glasses

Glass	Density $(g\ cm^{-3})$	References
SiO_2	2.20	Sosman (1927)
GeO_2	3.65	Mackenzie (1958); Rao (1965); Guckelsberger and Lasjaunias (1970)
B_2O_3	1.84	Cousen and Turner (1928c); Wulff and Majumdar (1936); Shaw and Uhlmann (1969); Weir and Shartsis (1955)
As_2O_3	3.70	Morey (1954, p. 230)
Sb_2O_3	≥ 5.18	Kordes (1939a)
P_2O_5	≥ 2.23	Kordes and Becker (1949)

TABLE 4.2

Densities of Alkali Metal Oxide (M_2O)–Silica Glass Systems

Composition (mole % M_2O)	Density $(g\ cm^{-3})$				
	Li_2O^a	Na_2O^b	K_2O^c	Rb_2O^d	Cs_2O^e
2			2.225	2.284	2.35
4		2.230	2.246	2.348	2.47
6		2.250	2.266	2.413	2.585
8		2.270	2.286	2.475	2.69
10	2.235	2.289	2.305	2.541	2.79
12	2.245	2.308	2.323	2.605	2.89
14	2.254	2.328	2.341	2.670	2.985
16	2.263	2.347	2.357	2.733	3.075
18	2.273	2.365	2.373	2.798	3.155
20	2.283	2.383	2.389	2.862	3.24
22	2.292	2.400	2.403	2.926	3.315
24	2.301	2.418	2.417	2.990	3.395
26	2.311	2.435	2.430	3.055	3.475
28	2.320	2.451	2.443	3.118	3.555
30	2.330	2.466	2.453	3.183	3.634
32	2.340	2.481	2.463	3.248	3.71
34	2.349	2.496	2.472		3.79
36	2.348	2.509	2.479		3.864
38	2.347	2.521	2.484		3.935

(*continues*)

TABLE 4.2 (*continued*)

Composition (mole % M_2O)	Density (g cm^{-3})				
	Li_2O^a	Na_2O^b	K_2O^c	Rb_2O^d	Cs_2O^e
40	2.346	2.532	2.489		4.010
42	2.345	2.541	2.491		4.075
44	2.344	2.550			4.14
46	2.343	2.555			4.205
48		2.558			
50		2.560			

[a] Values derived from a data base consisting of 16 data sets of different workers. The useful studies are Kondrat'ev and Smirnova (1970), Nemilov (1973), Dubrovo and Shmidt (1959), Vargin *et al.* (1971), Shaw and Uhlmann (1971), and Takahashi and Osaka (1983).

[b] Values derived from a data base consisting of 15 data sets of different workers. The useful studies are Young *et al.* (1939), Peddle (1920), Imaoka *et al.* (1971), Winks and Turner (1931), Urusovskaya (1960), and Takahashi and Osaka (1983).

[c] Values derived from a data base consisting of 15 data sets of different workers. The useful studies are Young *et al.* (1939), Charles (1966), Urusovskaya (1960), and Shmidt and Alekseeva (1964).

[d] Values derived from a data base consisting of 10 data sets of different workers. The useful studies are Brosset (1962), Sasek and Lisy (1972a,b), Evstropev and Pavlovskii (1967), Alekseeva (1964), and Shmidt and Alekseeva (1964).

[e] Values derived from a data base consisting of 12 data sets of different workers. The useful studies are Otto and Milberg (1967), Brosset (1962), Shmidt and Alekseeva (1964), Terai (1969, 1970), and Takahashi and Osaka (1983).

TABLE 4.3

Densities of Tl_2O-SiO_2 Glass System

Composition (mole % Tl_2O)	Density[a] (g cm^{-3})	Composition (mole % Tl_2O)	Density[a] (g cm^{-3})
5	2.80	35	5.97
10	3.40	40	6.30
15	4.00	45	6.64
20	4.59	50	6.98
25	5.18	55	7.32
30	5.78		

[a] Values derived from a data base consisting of Otto and Millberg (1967) and Nemilov (1973).

TABLE 4.4

Densities of Some Binary Silicate Glass Systems

Composition (mole % MO)	Density (g cm^{-3})				
	CaO[a]	BaO[b]	CdO[c]	SnO[d]	PbO[e]
22					3.725
24					3.890
26					4.058
28		3.455			4.225
30		3.557	3.78		4.400
32		3.660		3.290	4.560
34		3.755		3.305	4.723
36		3.837		3.385	4.880
38		3.917		3.465	5.045
40	2.765	3.998	4.32	3.540	5.200
42	2.798			3.615	5.360
44	2.828			3.687	5.513
46	2.855			3.757	5.670
48	2.880			3.823	5.825
50	2.901		4.75	3.883	5.980
52	2.920			3.940	6.130
54	2.937			3.993	6.275
56	2.950			4.040	6.425
58	2.961				6.565
60			4.89		6.700
62					6.838
64					6.968
66					7.095
68					7.213
70			5.00		7.325
72					7.425
74					7.525

[a] Data from Schwartz and Mackenzie (1966), Morey and Merwin (1932), and Larsen (1909).

[b] Data from Hamilton et al. (1958), MacDowell (1965), and Graham and Rindone (1967).

[c] Data from Tarlakov et al. (1980).

[d] Data from Ishikawa and Akagi (1978).

[e] Values derived from a data base consisting of 12 data sets of different workers. The useful studies are Kordes (1939b), Abou-El-Azm and Hussein (1962), Demkina (1958), Nemilov (1973), and Bogatyreva et al. (1973).

TABLE 4.5

Densities of SrO–SiO$_2$ Glasses

Composition (mole % SrO)	Density[a] (g cm^{-3})
20	2.80
25	2.96
30	3.14
35	3.31
40	3.51

[a] Values from graphs in Shelby (1979) and Escola (1922).

TABLE 4.6

Densities of Some Binary Silicate Glass Systems

B$_2$O$_3$–SiO$_2$ system[a]		Al$_2$O$_3$–SiO$_2$ system[b]		Bi$_2$O$_3$–SiO$_2$ system[c]	
Composition (mole % B$_2$O$_3$)	Density (g cm^{-3})	Composition (mole % Al$_2$O$_3$)	Density (g cm^{-3})	Composition (mole % Bi$_2$O$_3$)	Density (g cm^{-3})
5	2.182				
10	2.143				
15	2.134				
20	2.110				
30	2.055				
40	2.020	2	2.220	25	4.87
45	2.002	4	2.240	27.5	5.14
50	1.985	6	2.260	30	5.35
55	1.968	8	2.281	32.5	5.52
60	1.951	10	2.301	35	5.63
65	1.935	12	2.321	37.5	5.67
70	1.920	14	2.342	40	5.76
75	1.905	16	2.362	42.5	5.96
80	1.890	18	2.382	45	6.15
85	1.877	20	2.403	47.5	6.34
90	1.864	22	2.423	50	6.52
95	1.853	24	2.443	52.5	6.71
100	1.843			55	6.87
				57.5	7.05
				60	7.20
				62.5	7.35
				65	7.49

[a] Data from Imaoka et al. (1971), Jabra et al. (1980), and Cousen and Turner (1928a,b).
[b] Data from Thompson and Parmelee (1937) and Nassau et al. (1975).
[c] Data from Shabanova (1967) and Tiwari and Das (1972).

TABLE 4.7

Densities of GeO$_2$–SiO$_2$ System[a]

Composition (mole % GeO$_2$)	Density (g cm^{-3})	Composition (mole % GeO$_2$)	Density (g cm^{-3})
5	2.270	45	2.850
10	2.344	50	2.920
15	2.415	55	2.995
20	2.488	60	3.067
25	2.560	65	3.140
30	2.633	70	3.212
35	2.705	100	3.647
40	2.776		

[a] Values derived from a data base consisting of Riebling (1968) and Huang et al. (1978).

TABLE 4.8

Densities of Some MO$_2$–SiO$_2$ Glasses

Composition (mole % MO$_2$)	Density (g cm^{-3})	Reference(s)
M = Ti		
2.4	2.199	Schroeder (1980); Schultz and Smyth (1970)
7.4	2.197	Schroeder (1980); Schultz and Smyth (1970)
9.0	2.196	Schroeder (1980); Schultz and Smyth (1970)
M = Zr[a]		
0.15	2.208	Bihuniak and Condrate (1981)
0.43	2.210	Bihuniak and Condrate (1981)
0.62	2.213	Bihuniak and Condrate (1981)
0.93	2.215	Bihuniak and Condrate (1981)
M = Hf[a]		
0.08	2.207	Bihuniak and Condrate (1981)
0.70	2.230	Bihuniak and Condrate (1981)
0.84	2.239	Bihuniak and Condrate (1981)

[a] Glasses prepared by the Sol–gel technique.

TABLE 4.9

Densities of P$_2$O$_5$–SiO$_2$ System[a]

Composition (mole % PO$_{2.5}$)	Density (g cm^{-3})	Composition (mole % PO$_{2.5}$)	Density (g cm^{-3})
17.4	2.204	49.6	2.584
28.5	2.323	50.2	2.587
32.8	2.295	57.9	2.649
42.8	2.473	59.8	2.686
45.8	2.422	64.2	2.702

[a] Data from Takahashi et al. (1976).

TABLE 4.10

Densities of Some Binary Silicate Glasses

Modifier oxide	Mole % modifier oxide	Density (g cm^{-3})	Reference
MgO	50	2.75	Schwartz and Mackenzie (1966)
V$_2$O$_5$	1	2.204	Goral'nik et al. (1972)

TABLE 4.11
Densities of Li_2O–Na_2O–SiO_2 Glasses

Composition (mole %)			Density[a] (g cm⁻³)	Reference[a]	Composition (mole %)			Density[b] (g cm⁻³)	Reference[a]
Li_2O	Na_2O	SiO_2			Li_2O	Na_2O	SiO_2		
0.5	24.5	75	2.430	1	4	11	85	2.337	2
1.25	23.75	75	2.431		7.5	7.5	85	2.323	
3.75	21.25	75	2.423		11	4	85	2.306	
8.75	16.25	75	2.409		6	19	75	2.446	
12.5	12.5	75	2.391						
16.25	8.75	75	2.367		19	6	75	2.373	
21.25	3.75	75	2.337		9	26	65	2.496	
23.75	1.25	75	2.316		17.5	17.5	65	2.459	
24.5	0.5	75	2.313		26	9	65	2.414	
24.875	0.125	75	2.309						

[a] Data from (1) Shelby and Day (1969), and (2) Sheybany (1948).

TABLE 4.12

Densities of Li_2O–K_2O–SiO_2 Glasses

Composition (mole %)			Density ($g\ cm^{-3}$)	Reference[a]	Composition (mole %)			Density ($g\ cm^{-3}$)	Reference[a]
Li_2O	K_2O	SiO_2			Li_2O	K_2O	SiO_2		
0.125	24.875	75	2.410	1	3.7	13	83.3	2.343	2
0.5	24.5	75	2.411		6.7	10	83.3	2.334	
1.25	23.75	75	2.414		9.7	7	83.3	2.315	
3.75	21.25	75	2.420		12.7	4	83.3	2.298	
8.75	16.25	75	2.405		4	11	85	2.334	3
12.5	12.5	75	2.387		7.5	7.5	85	2.321	
16.25	8.75	75	2.366		11	4	85	2.310	
21.25	3.75	75	2.331		6	19	75	2.430	
23.75	1.25	75	2.318		19	6	75	2.368	
24.5	0.5	75	2.310		9	26	65	2.468	
					17.5	17.5	65	2.440	
					26	9	65	2.400	

[a] Data from (1) Shelby and Day (1969), (2) Aleyhikov *et al.* (1967), and (3) Sheybany (1948).

TABLE 4.13

Densities of Li$_2$O–Cs$_2$O–SiO$_2$ Glasses

Composition (mole %)			Density	Reference[a]	Composition (mole %)			Density	Reference[a]
Li$_2$O	Cs$_2$O	SiO$_2$	(g cm^{-3})		Li$_2$O	Cs$_2$O	SiO$_2$	(g cm^{-3})	
12.5	12.5	75	2.762	1	3	24	73	3.425	4
23.75	1.25	75	2.461		6	21	73	3.381	
24.5	0.5	75	2.440		12	15	73	3.121	
					15	12	73	2.978	
5	15	80	3.073	2	18	9	73	2.812	
10	10	80	2.848		21	6	73	2.677	
15	5	80	2.579		24	3	73	2.506	
18.2	1.8	80	2.399		24	6	70	2.679	
19.1	0.9	80	2.341		24	9	67	2.858	
					27	3	70	2.533	
5	10	85	2.63	3	27	6	67	2.719	
7.5	7.5	85	2.58		30	3	67	2.523	
10	5	85	2.56		30	6	64	2.707	
					30	9	61	2.881	

[a] Data from (1) Shelby and Day (1969), (2) Alekseeva and Polozok (1972), (3) Hakim and Uhlmann (1967), and (4) Parfenov et al. (1959).

TABLE 4.14

Densities of Na_2O–K_2O–SiO_2 Glasses

Composition (mole %)			Density (g cm^{-3})	Reference[a]
Na_2O	K_2O	SiO_2		
9	26	65	2.510	1
17.5	17.5	65	2.515	
26	9	65	2.513	
0.5	24.5	75	2.417	2
1.25	23.75	75	2.419	
2.5	22.5	75	2.420	
5	20	75	2.433	
10	15	75	2.435	
12.5	12.5	75	2.439	
15	10	75	2.443	
20	5	75	2.436	
22.5	2.5	75	2.432	
23.75	1.25	75	2.432	
24.5	0.5	75	2.432	
24.875	0.125	75	2.428	3
8	16	76	2.432	
20	4	76	2.432	

Composition (mole %)			Density (g cm^{-3})	Reference[a]
Na_2O	K_2O	SiO_2		
4	19	77	2.418	3
8	15	77	2.424	
16	7	77	2.419	
20	3	77	2.419	
4	18	78	2.411	
16	6	78	2.416	
20	2	78	2.409	
4	17	79	2.406	
8	13	79	2.412	
12	9	79	2.412	
16	5	79	2.407	
18	3	79	2.403	
20	1	79	2.401	
4	16	80	2.3998	
8	12	80	2.398	
12	8	80	2.399	
16	4	80	2.396	
4	15	81	2.389	
8	11	81	2.389	

Composition (mole %)			Density (g cm^{-3})	Reference[a]
Na_2O	K_2O	SiO_2		
12	7	81	2.386	3
16	3	81	2.380	
4	14	82	2.381	
8	10	82	2.379	
12	6	82	2.375	
16	2	82	2.372	
4	13	83	2.371	
8	9	83	2.370	
5	12	83	2.365	
16	1	83	2.360	
8	8	84	2.363	
12	4	84	2.356	
4	11	85	2.357	
12	3	85	2.345	
4	10	86	2.345	
12	2	86	2.332	

[a] Data from (1) Sheybany (1948), (2) Shelby and Day (1969), and (3) Urusovskaya (1960).

TABLE 4.15

Densities of Na_2O–Rb_2O–SiO_2 Glasses

Composition (mole %)			Density	
Na_2O	Rb_2O	SiO_2	(g cm^{-3})	Reference[a]
6.25	18.75	75	2.901	1
12.5	12.5	75	2.769	1
18.75	6.25	75	2.617	1
23.75	1.25	75	2.461	2
24.5	0.5	75	2.440	2

[a] Data from (1) McVay and Day (1970) and (2) Shelby and Day (1969).

TABLE 4.16

Densities of Na_2O–Cs_2O–SiO_2 Glasses

Composition (mole %)			Density		Composition (mole %)			Density	
Na_2O	Cs_2O	SiO_2	(g cm^{-3})	Reference[a]	Na_2O	Cs_2O	SiO_2	(g cm^{-3})	Reference[a]
16.25	8.75	75	2.837	1	16	4	80	2.492	2
21.25	3.75	75	2.602		18.2	1.8	80	2.401	
22.5	2.5	75	2.541						
23.75	1.25	75	2.496		5	10	85	2.66	3
24.5	0.5	75	2.449		7.5	7.5	85	2.63	
					10	5	85	2.53	
1.5	18.5	80	3.087	2	1.78	14.31	83.91	3.08	4
5	15	80	2.987		3.78	13.54	82.68	3.01	
8	12	80	2.903		4.44	12.93	82.63	2.96	
10	10	80	2.839		5.60	12.02	82.38	2.94	
11.4	8.6	80	2.753		7.23	9.57	83.20	2.87	
13.3	6.7	80	2.607		9.26	7.79	82.95	2.79	
					10.68	6.73	82.59	2.71	
					12.32	4.12	83.56	2.58	

[a] Data from (1) Shelby and Day (1969), (2) Alekseeva and Polozok (1972), (3) Hakim and Uhlmann (1967), and (4) Terai (1971a, 1972).

TABLE 4.17

Densities of Na_2O–Tl_2O–SiO_2 Glasses with Na_2O:SiO_2 Ratio of 0.25[a]

Composition (wt. % Tl_2O)	Density (g cm^{-3})
10	2.569
12	2.592
21	2.773
39	3.194
51	3.544
60	4.247

[a] Data from Polukhin (1960).

TABLE 4.18

Densities of K_2O–Rb_2O–SiO_2 Glasses[a]

Composition (mole %)			Density
K_2O	Rb_2O	SiO_2	(g cm^{-3})
12.5	12.5	75.0	2.748
21.25	3.75	75.0	2.523
23.75	1.25	75.0	2.446
24.5	0.5	75.0	2.441

[a] Data from Shelby and Day (1969).

	TABLE 4.19
	Densities of $K_2O-Cs_2O-SiO_2$ Glasses[a]

Composition (mole %)			Density
K_2O	Cs_2O	SiO_2	(g cm^{-3})
5	10	85	2.62
7.5	7.5	85	2.55
10	5	85	2.53

[a] Data from Hakim and Uhlmann (1967).

	TABLE 4.20
	Densities of $Rb_2O-Cs_2O-SiO_2$ Glasses[a]

Composition (mole %)			Density
Rb_2O	Cs_2O	SiO_2	(g cm^{-3})
5	10	85	2.84
7.5	7.5	85	2.78
10	5	85	2.73

[a] Data from Hakim and Uhlmann (1967).

TABLE 4.21

Densities of $Li_2O-BeO-SiO_2$ Glasses

Composition (mole %)			Density	
Li_2O	BeO	SiO_2	(g cm^{-3})	Reference[a]
27	2.5	70.5	2.344	1
27	5	68	2.357	
27	10	63	2.433	
27	15	58	2.458	
27	20	53	2.471	
30	5	65	2.316	2
30	10	60	2.402	

[a] Data from (1) Shul'ts et al. (1963) and (2) Moore and McMillan (1956).

TABLE 4.22

Densities of $Li_2O-MgO-SiO_2$ Glasses

Composition (mole %)			Density	
Li_2O	MgO	SiO_2	(g cm^{-3})	Reference[a]
27	5	68	2.426	1
27	10	63	2.453	
27	15	58	2.480	
27	20	53	2.503	
20	30	50	2.601	2
20	25	55	2.550	
30	10	60	2.413	

[a] Data from (1) Shul'ts et al. (1963) and (2) Moore and McMillan (1956).

TABLE 4.23

Densities of Li$_2$O–CaO–SiO$_2$ Glasses[a]

Composition (mole %)			Density
Li$_2$O	CaO	SiO$_2$	(g cm^{-3})
27	5	68	2.394
27	10	63	2.454
27	15	58	2.500
27	20	53	2.545

[a] Data from Shul'ts et al. (1963).

TABLE 4.24

Densities of Li$_2$O–BaO–SiO$_2$ Glasses[a]

Composition (mole %)			Density
Li$_2$O	BaO	SiO$_2$	(g cm^{-3})
27	5	68	2.614
27	10	63	2.822
27	15	58	3.077
27	20	53	3.391

[a] Data from Shul'ts et al. (1963).

TABLE 4.25

Densities of Li$_2$O–ZnO–SiO$_2$ Glasses[a]

Composition (mole %)			Density	Composition (mole %)			Density
Li$_2$O	ZnO	SiO$_2$	(g cm^{-3})	Li$_2$O	ZnO	SiO$_2$	(g cm^{-3})
15	20	65	2.74	25	10	65	2.55
15	25	60	2.867	25	15	40	2.66
15	30	55	2.99	30	5	65	2.47
15	35	50	3.115	30	10	60	2.574
20	5	75	2.405	30	15	55	2.675
20	10	70	2.51	35	5	60	2.49
20	15	65	2.636	35	10	55	2.597
20	25	55	2.885	35	15	50	2.705
25	5	70	2.439				

[a] Values from graph in Vargin et al. (1972).

TABLE 4.26

Densities of Na$_2$O–BeO–SiO$_2$ Glasses[a]

Composition (mole %)			Density	Composition (mole %)			Density
Na$_2$O	BeO	SiO$_2$	(g cm^{-3})	Na$_2$O	BeO	SiO$_2$	(g cm^{-3})
15.4	19.2	65.4	2.456	17.4	17.4	65.2	2.467
16	16	68	2.444	17.4	21.7	60.9	2.489
16	20	64	2.461	18.2	9.1	72.7	2.427
16.7	12.5	70.8	2.432	18.2	13.6	68.2	2.455
16.7	16.7	66.6	2.450	18.2	18.2	63.6	2.479
16.7	20.8	62.5	2.478	19	14.3	66.7	2.464
17.4	8.7	73.9	2.417	19.1	9.5	71.4	2.439
17.4	13	69.6	2.440	20	10	70	2.449

[a] Data from Lai and Silverman (1928, 1930).

TABLE 4.27

Densities of Na_2O–MgO–SiO_2 Glasses[a]

Composition (mole %)			Density (g cm^{-3})	Composition (mole %)			Density (g cm^{-3})	Composition (mole %)			Density (g cm^{-3})
Na_2O	MgO	SiO_2		Na_2O	MgO	SiO_2		Na_2O	MgO	SiO_2	
7.1	22.9	70	2.363	11.8	18.2	70	2.406	18.8	11.2	70	2.432
10	20	70	2.385	13.2	21.8	65	2.425	21.5	21.5	57	2.470
10.5	14.5	75	2.390	15	15	70	2.425	21.9	13.1	65	2.453
10.6	32.4	57	2.412	15.9	19.1	65	2.431	22	13.1	64.9	2.453
11	11	78	2.386	16	14.8	69.2	2.425	22	15.3	62.7	2.457
11	11.5	77.5	2.389	16	18.5	65.5	2.432	22	17.7	60.3	2.463
11	12.8	76.2	2.388	16	19	65	2.430	22	19.9	58.1	2.471
11	17.2	71.8	2.391	16	22.5	61.5	2.441	22	21.4	56.6	2.478
11	23.5	65.5	2.394	16	24.4	59.6	2.453	22	23.4	54.6	2.486
11	24	65	2.414	16	25.7	58.3	2.464	22	24.8	53.2	2.491
11	27.3	61.7	2.397	16	26.3	57.7	2.462	22.2	20.8	57	2.471
11	29.6	59.4	2.402	16	28	56	2.474	23.2	19.8	57	2.489
11	32.3	56.7	2.427	16	30.5	53.5	2.487	25.8	9.2	65	2.465
11	34.3	54.7	2.439	16.6	18.4	65	2.436	27	16	57	2.505
11.3	23.7	65	2.403	17.2	25.8	57	2.462				
11.7	23.3	65	2.409	17.9	25.1	57	2.464				

[a] Data from Botvinkin and Yaroker (1966, 1967).

TABLE 4.28
Densities of Na$_2$O–CaO–SiO$_2$ Glasses[a]

Composition (wt. %)			Density (g cm^{-3})	Composition (wt. %)			Density (g cm^{-3})	Composition (wt. %)			Density (g cm^{-3})
Na$_2$O	CaO	SiO$_2$		Na$_2$O	CaO	SiO$_2$		Na$_2$O	CaO	SiO$_2$	
3.90	5.70	90.40	2.310	14.77	15.00	70.23	2.557	12.82	12.98	74.20	2.509
6.00	4.00	90.00	2.311	23.26	6.52	70.22	2.504	18.48	7.34	74.18	2.484
10.89	3.01	86.10	2.345	24.86	4.94	70.20	2.499	20.73	5.12	74.15	2.469
8.00	6.00	84.00	2.355	15.91	14.08	70.01	2.548	13.88	12.02	74.10	2.505
9.21	7.58	83.21	2.389	24.71	5.31	69.98	2.496	15.04	11.03	73.93	2.505
6.55	11.65	81.80	2.421	25.55	4.61	69.84	2.503	7.32	19.04	73.64	2.541
16.98	5.01	78.01	2.432	17.23	13.00	69.77	2.540	12.95	13.44	73.61	2.518
12.00	10.00	78.00	2.455	25.79	5.01	69.20	2.504	13.30	13.13	73.57	2.519
14.93	8.01	77.06	2.456	5.56	25.40	69.04	2.609	16.42	10.02	73.56	2.501
8.90	14.99	76.11	2.497	20.54	10.78	68.68	2.544	23.48	3.07	73.45	2.461
12.85	11.03	76.12	2.485	20.90	10.84	68.26	2.543	21.71	5.00	73.29	2.474
18.89	5.01	76.10	2.455	13.83	18.04	68.13	2.586	13.81	13.02	73.17	2.519
6.18	17.96	75.86	2.505	24.95	7.00	68.05	2.520	22.84	4.01	73.15	2.470
18.64	5.76	75.60	2.468	22.10	10.14	67.76	2.543	17.91	9.01	73.08	2.500
4.48	20.23	75.29	2.535	19.64	13.03	67.33	2.561	11.82	16.03	72.15	2.546
17.25	7.54	75.21	2.475	3.87	28.84	67.29	2.655	8.92	19.02	72.06	2.560
17.50	7.50	75.00	2.477	17.45	15.57	66.98	2.582	21.66	6.54	71.80	2.495
15.04	9.98	74.98	2.489	5.98	27.24	66.78	2.654	18.52	10.10	71.38	2.518
13.40	11.64	74.96	2.495	12.44	20.79	66.77	2.613	21.35	7.54	71.11	2.499
10.08	14.97	74.95	2.506	23.45	10.07	66.48	2.547	10.91	18.02	71.07	2.561
15.20	10.10	74.70	2.489	18.69	14.91	66.40	2.581	5.46	23.66	70.88	2.594
15.28	10.03	74.69	2.492	17.80	16.04	66.16	2.591	9.11	20.19	70.70	2.585
5.37	20.25	74.38	2.532	18.95	15.06	65.99	2.584	24.02	5.32	70.66	2.495
5.46	20.25	74.29	2.536	19.15	15.00	65.85	2.586	13.78	15.66	70.56	2.553
5.73	20.02	74.25	2.522	9.07	25.28	65.65	2.656	15.68	14.04	70.28	2.549

(continues)

TABLE 4.28 (Continued)

Composition (wt. %)				Composition (wt. %)				Composition (wt. %)			Density
Na$_2$O	CaO	SiO$_2$		Na$_2$O	CaO	SiO$_2$		Na$_2$O	CaO	SiO$_2$	(g cm^{-3})
23.50	21.14	55.36	2.683	16.64	17.83	65.53	2.595	37.44	11.68	50.88	2.641
9.97	34.72	55.31	2.773	21.80	13.09	65.11	2.573	17.52	31.61	50.87	2.761
16.41	28.31	55.28	2.733	10.53	24.38	65.09	2.649	9.60	39.64	50.76	2.820
33.87	10.87	55.26	2.610	32.82	2.40	64.78	2.509	21.40	27.90	50.70	2.741
6.13	38.75	55.12	2.808	32.43	3.02	64.55	2.517	33.47	15.98	50.55	2.670
4.56	40.42	55.02	2.816	27.20	8.33	64.47	2.546	29.77	20.18	50.05	2.696
41.94	3.33	54.73	2.568	14.15	21.47	64.38	2.638	34.42	15.54	50.04	2.663
38.32	7.50	54.18	2.587	8.22	28.51	63.27	2.691	26.81	23.57	49.62	2.710
31.33	14.71	53.96	2.630	5.11	32.59	62.30	2.719	5.30	45.44	49.26	2.867
21.06	24.97	53.97	2.705	9.71	28.18	62.11	2.694	28.42	22.46	49.12	2.717
4.44	42.04	53.52	2.838	28.60	10.04	61.36	2.575	30.00	21.00	49.00	2.697
5.18	41.50	53.32	2.832	10.42	28.46	61.12	2.702	7.40	44.30	48.30	2.859
28.44	18.29	53.27	2.664	7.44	32.17	60.39	2.729	16.98	35.54	47.48	2.794
39.91	7.52	52.57	2.601	16.81	23.04	60.15	2.672	40.76	12.32	46.92	2.654
20.80	27.10	52.10	2.730	19.99	20.00	60.01	2.657	7.98	45.16	46.86	2.863
31.87	16.08	52.05	2.665	5.01	35.70	59.29	2.761	36.50	17.27	46.23	2.681
10.23	37.98	51.79	2.807	34.08	6.66	59.26	2.562	44.77	9.05	46.18	2.630
10.61	37.63	51.76	2.802	9.97	31.14	58.89	2.733	44.56	10.07	45.37	2.640
23.49	24.95	51.56	2.712	26.22	15.00	58.78	2.618	9.27	45.38	45.35	2.874
18.00	30.65	51.35	2.761	17.14	24.24	58.62	2.692	14.47	40.24	45.29	2.837
14.68	34.13	51.19	2.777	28.47	14.31	57.22	2.626	4.79	50.06	45.15	2.902
9.88	39.03	51.09	2.821	15.79	27.07	57.14	2.712	34.91	20.02	45.07	2.703
17.38	31.65	50.97	2.764	33.91	9.04	57.05	2.588	22.60	32.50	44.90	2.784
36.37	12.66	50.97	2.648	13.80	29.40	56.80	2.735	26.86	32.24	40.90	2.800
23.99	25.10	50.91	2.719	23.06	20.15	56.79	2.665	38.91	23.43	37.66	2.742

[a] Data from Morey and Merwin (1932).

TABLE 4.29

Densities of $Na_2O-SrO-SiO_2$ Glasses[a]

Composition (mole %)			Density	Composition (mole %)			Density
Na_2O	SrO	SiO_2	(g cm^{-3})	Na_2O	SrO	SiO_2	(g cm^{-3})
10	10	80	2.595	10	30	60	3.212
15	5	80	2.474	15	25	60	3.085
10	15	75	2.757	17	23	60	3.050
12.5	12.5	75	2.698	20	20	60	2.971
20	5	75	2.537	22	18	60	2.944
10	20	70	2.912	23.4	16.6	60	2.901
15	15	70	2.800	24.3	15.7	60	2.883
20	10	70	2.689	25	15	60	2.870
25	5	70	2.576	10	40	50	3.517
25	10	65	2.718	25	25	50	3.158
5	35	60	3.356	40	10	50	2.776

[a] Data from Dubrovo and Kasimova (1964) and Kasimova (1964a,b).

TABLE 4.30

Densities of $Na_2O-BaO-SiO_2$ Glasses

Composition (mole %)			Density	
Na_2O	BaO	SiO_2	(g cm^{-3})	Reference[a]
14.3	7.2	78.5	2.66	1
14.3	10.7	75.0	2.83	
14.3	14.2	71.5	3.00	
14.3	17.9	6	3.17	
15	15	70	3.010	2
20	10	70	2.837	
20	20	60	3.225	

[a] Data from (1) graph in Din and Hennicke (1974) and (2) Kasimova (1964a,b).

TABLE 4.31

Densities of Na$_2$O–ZnO–SiO$_2$ Glasses[a]

Composition (mole %)			Density (g cm^{-3})	Composition (mole %)			Density (g cm^{-3})
Na$_2$O	ZnO	SiO$_2$		Na$_2$O	ZnO	SiO$_2$	
15	35	50	3.215	15	20	65	2.901
20	30	50	3.091	20	15	65	2.796
25	25	50	2.979	25	10	65	2.694
30	20	50	2.870	30	5	65	2.591
35	15	50	2.776	10	20	70	2.836
40	10	50	2.851	15	15	70	2.748
10	30	60	3.173	20	10	70	2.634
15	25	60	3.057	25	5	70	2.561
20	20	60	2.930	10	15	75	2.705
25	15	60	2.818	12.5	12.5	75	2.657
30	10	60	2.717	15	10	75	2.597
35	5	60	2.624	20	5	75	2.507
10	25	65	3.000				

[a] Data from Hurt and Phillips (1970).

TABLE 4.32

Density of Na$_2$O–CdO–SiO$_2$ Glass[a]

Composition (mole %)			Density (g cm^{-3})
Na$_2$O	CdO	SiO$_2$	
20	10	70	3.004
16	20	64	3.172

[a] Data from Terai (1971b) and Appen (1953).

TABLE 4.33

Densities of Na$_2$O–PbO–SiO$_2$ Glasses[a]

Composition (wt. %)				Density (g cm^{-3})	Composition (wt. %)				Density (g cm^{-3})
Na$_2$O	PbO	SiO$_2$	Al$_2$O$_3$		Na$_2$O	PbO	SiO$_2$	Al$_2$O$_3$	
9.6	19.7	69.4	1.0	2.720	18.19	2.40	78.91	0.50	2.408
10.12	19.88	69.64	0.36	2.821	18.20	14.46	66.82	0.48	2.711
14.80	19.68	65.06	0.46	2.854	18.24	3.96	77.82	0.18	2.445
15.0	20.1	64.2	0.8	2.832	18.36	10.54	70.44	0.66	2.647
17.06	28.64	54.00	0.52	3.092	20.0	20.0	59.6	0.6	2.897
17.50	19.16	62.76	0.38	2.861	20.82	20.44	58.56	0.42	2.895
17.50	49.30	32.50	0.76	3.932	25.69	20.38	53.39	0.54	2.926
17.66	39.88	42.20	0.48	3.524	29.91	18.70	50.91	0.48	2.953
17.67	23.80	58.01	0.52	2.981	35.25	19.55	44.86	0.28	2.991

[a] Data from Weberbauer (1932a,b).

TABLE 4.34

Densities of K_2O–BeO–SiO_2 Glasses[a]

Composition (mole %)			Density	Composition (mole %)			Density
K_2O	BeO	SiO_2	(g cm^{-3})	K_2O	BeO	SiO_2	(g cm^{-3})
15.4	26.9	57.7	2.455	17.4	21.7	60.9	2.468
16	24	60	2.459	17.4	26.1	56.5	2.472
16	28	56	2.458	17.4	30.4	52.2	2.473
16.7	20.8	62.5	2.464	18.2	22.7	59.1	2.476
16.7	25	58.3	2.463	18.2	27.3	54.5	2.478
16.7	29.1	54.2	2.468	19	23.8	57.2	2.483

[a] Data from Lai and Silverman (1928, 1930).

TABLE 4.35

Densities of K_2O–MgO–SiO_2 Glasses[a]

Composition (wt. %)			Density	Composition (wt. %)			Density
K_2O	MgO	SiO_2	(g cm^{-3})	K_2O	MgO	SiO_2	(g cm^{-3})
5.5	12.5	82	2.34	24	2.5	73.5	2.36
7	15	78	2.41	26.53	22.71	50.76	2.54
8	23	69	2.50	29.93	12.81	57.26	2.40
13.17	11.27	75.56	2.39	34	3	63	2.38
17.6	15.07	67.33	2.44	36.99	15.83	47.18	2.48
21.67	9.27	69.06	2.38	40.12	5	54.88	2.44

[a] Data from Roedder (1951).

TABLE 4.36
Densities of K_2O–CaO–SiO_2 Glasses[a]

Composition (wt. %)			Density (g cm⁻³)	Composition (wt. %)			Density (g cm⁻³)	Composition (wt. %)			Density (g cm⁻³)
K_2O	CaO	SiO_2		K_2O	CaO	SiO_2		K_2O	CaO	SiO_2	
3.9	30.4	65.7	2.659	16.6	19.8	63.6	2.569	27.7	24.2	48.1	2.665
5.1	9.8	85.1	2.360	17.3	10.2	72.5	2.448	28.3	8.4	63.3	2.498
5.3	39.9	54.8	2.783	18.3	11.0	70.7	2.473	28.3	17.0	54.7	2.596
5.4	22.4	72.2	2.546	19.8	5.0	75.2	2.412	30.6	19.7	49.7	2.635
8.4	19.8	71.8	2.528	20.0	22.3	57.7	2.613	30.7	7.7	61.6	2.502
8.6	31.7	59.7	2.189	20.4	9.9	69.7	2.473	31.4	9.3	59.3	2.519
9.6	40.6	49.8	2.793	22.1	15.4	62.5	2.544	32.2	7.2	60.6	2.500
10.5	16.9	72.6	2.501	22.4	8.1	69.5	2.461	32.6	4.9	62.5	2.481
10.5	24.4	65.1	2.593	24.3	7.6	68.1	2.470	32.8	23.5	43.7	2.672
11.6	9.1	79.3	2.412	24.3	29.0	46.7	2.701	34.0	15.6	50.4	2.606
12.6	15.0	72.4	2.492	24.7	15.4	59.9	2.554	34.3	7.2	58.5	2.509
12.7	22.6	64.7	2.584	25.1	26.5	48.4	2.686	34.5	20.7	44.8	2.658
15.0	5.0	80.0	2.371	25.4	14.1	60.5	2.546	38.3	4.7	57.0	2.505
15.2	29.7	55.1	2.683	27.3	1.0	71.7	2.405	46.0	5.8	48.2	2.542
16.5	41.4	42.1	2.800	27.7	14.6	57.7	2.560				

[a] Data from Morey (1954).

TABLE 4.37

Densities of K_2O–SrO–SiO_2 Glasses[a]

Composition (mole %)			Density	Composition (mole %)			Density	Composition (mole %)			Density
K_2O	SrO	SiO_2	(g cm^{-3})	K_2O	SrO	SiO_2	(g cm^{-3})	K_2O	SrO	SiO_2	(g cm^{-3})
5.4	10.1	84.5	2.528	13.4	30.0	56.6	3.139	18.6	22.4	59.0	2.947
6.9	8.3	84.8	2.510	13.5	9.6	76.9	2.558	19.6	18.3	62.1	2.831
7.6	14.2	78.2	2.675	14.0	26.4	59.6	3.052	20.5	14.4	65.1	2.758
8.7	6.1	85.2	2.487	14.2	4.5	81.3	2.464	21.3	26.4	52.3	3.065
9.9	18.7	71.4	2.821	14.7	4.6	80.7	2.482	21.4	10.6	68.0	2.646
10.8	7.4	81.8	2.524	15.5	18.7	65.8	2.824	22.3	6.9	70.8	2.580
10.9	27.1	62.0	3.053	16.8	11.8	71.4	2.660	23.1	3.4	73.5	2.481
11.6	3.6	84.8	2.435	16.8	30.0	53.2	3.136	24.0	17.2	58.8	2.831
11.6	22.1	66.3	2.896	17.6	26.6	55.8	3.057	26.5	8.5	65.0	2.640
12.6	15.4	72.0	2.699	18.0	5.6	76.4	2.521	34.7	10.8	54.5	2.727

[a] Data from Shchavelev et al. (1974).

TABLE 4.38

Densities of Other K_2O–SrO–SiO_2 Glasses[a]

Composition (mole %)			Density	Composition (mole %)			Density	Composition (mole %)			Density
K_2O	SrO	SiO_2	(g cm^{-3})	K_2O	SrO	SiO_2	(g cm^{-3})	K_2O	SrO	SiO_2	(g cm^{-3})
5	15	80	2.665	15	15	70	2.746	25	25	50	3.010
5	25	70	2.965	15	25	60	2.988	30	10	60	2.682
5	35	60	3.205	15	35	50	3.300	30	30	40	3.195
10	10	80	2.590	16.7	16.7	66.6	2.811	35	5	60	2.451
11.1	11.1	77.8	2.614	20	20	60	2.912	35	15	50	2.834
12.5	12.5	75.0	2.657	25	5	70	2.477	45	5	50	2.635
14.3	14.3	71.4	2.730	25	15	60	2.790	50	10	40	2.700
								55	10	35	2.780

[a] Data from Kas'imova and Makhkamova (1970).

TABLE 4.39

Densities of $K_2O-BaO-SiO_2$ Glasses[a]

| Composition (mole %) | | | Density |
K_2O	BaO	SiO_2	(g cm^{-3})
28.67	5.27	66.50	2.638
31.83	4.96	63.20	2.638
32.82	3.62	63.55	2.588
33.50	2.21	64.46	2.543
34.60	0.62	64.70	2.501

[a] Data from Antonova (1975).

TABLE 4.40

Densities of $K_2O-ZnO-SiO_2$ Glasses[a]

| Composition (mole %) | | | Density | Composition (mole %) | | | Density |
K_2O	ZnO	SiO_2	(g cm^{-3})	K_2O	ZnO	SiO_2	(g cm^{-3})
10	5	85	2.395	20	15	65	2.699
10	10	80	2.497	20	20	60	2.809
10	15	75	2.61	20	25	55	2.914
10	20	70	2.70	20	30	50	3.028
10	25	65	2.836	20	35	45	3.135
10	30	60	2.94	25	5	70	2.517
10	35	55	3.037	25	10	65	2.631
10	40	50	3.16	25	20	55	2.836
10	45	45	3.265	25	25	50	2.94
15	5	80	2.458	25	30	45	3.06
15	10	75	2.563	25	35	40	3.177
15	15	70	2.67	30	5	65	2.536
15	20	65	2.758	30	10	60	2.647
15	25	60	2.885	30	15	55	2.75
15	30	55	2.992	30	20	50	2.855
15	35	50	3.093	30	25	45	2.976
15	40	45	3.207	35	5	60	2.563
20	5	75	2.488	35	10	55	2.67
20	10	70	2.598	35	15	50	2.77
				35	20	45	2.878

[a] Values from a graph in Vargin et al. (1972).

TABLE 4.41

Densities of $K_2O-PbO-SiO_2$ Glasses[a]

| Composition (wt. %) | | | | Density |
K_2O	PbO	SiO_2	Al_2O_3	(g cm^{-3})
9.7	19.6	69.5	1.0	2.71
15.5	20.4	63.4	0.6	2.755
20.5	20.1	58.6	0.4	2.818

[a] Data from Weberbauer (1932a,b).

TABLE 4.42

Densities of Rb₂O–CaO–SiO₂ Glasses

Composition (mole %)			Density	Composition (mole %)			Density
Rb_2O	CaO	SiO_2	(g cm^{-3})	Rb_2O	CaO	SiO_2	(g cm^{-3})
3.3	44.6	52.1	3.02	11.7	39.3	49.0	2.91
3.3	55.3	41.4	2.86	11.8	26.4	61.8	2.90
3.4	22.6	74.0	2.60	12	13.4	74.6	2.72
3.4	33.6	63.0	2.73	17.2	42.8	40.0	2.97
7.3	24.4	68.3	2.72	17.3	28.8	53.9	2.95
7.3	36.2	56.5	2.84	17.4	14.6	68.0	2.92
7.3	48	44.7	2.97	23.8	31.8	44.4	2.94
7.4	12.3	80.3	2.58	24	16	60	2.92
11.6	52	36.4	2.88	32.1	17.9	50.0	2.86

[a] Data from Simpson (1959, 1961).

TABLE 4.43

Densities of Rb₂O–BaO–SiO₂ Glasses[a]

Composition (mole %)			Density	Composition (mole %)			Density
Rb_2O	BaO	SiO_2	(g cm^{-3})	Rb_2O	BaO	SiO_2	(g cm^{-3})
4.2	15.7	80.1	2.95	14.2	11.7	74.1	3.09
4.6	22.8	72.6	3.26	17.4	28.3	54.3	3.32
5.1	31.2	63.7	3.27	19.3	5.8	74.9	2.94
8.6	10.6	80.8	2.89	21.2	12.9	65.9	3.28
9.4	17.3	73.3	3.06	23.6	21.5	54.9	3.58
11.5	35	53.5	3.38	26.7	6.5	66.8	3.29
13.1	5.3	81.6	2.84				

[a] Data from Simpson (1959, 1961).

TABLE 4.44

Densities of Rb₂O–PbO–SiO₂ Glasses[a]

Composition (mole %)			Density	Composition (mole %)			Density
Rb_2O	PbO	SiO_2	(g cm^{-3})	Rb_2O	PbO	SiO_2	(g cm^{-3})
5	16.8	78.2	3.52	19.1	21.3	59.6	3.91
5.6	23.7	70.7	4.39	19.6	4.1	76.3	3.11
6.5	32.7	60.8	4.73	22	9.3	68.7	3.42
10	12.5	77.5	3.27	22.3	31.3	46.4	4.22
11.2	18.8	70.0	3.86	25.2	15.9	58.9	3.85
12.8	27	60.2	4.43	27.3	4.6	68.1	3.30
14.8	8.2	77.0	3.14	29.5	24.7	45.8	4.16
15.1	37.9	47.0	4.78	31.2	10.5	58.3	3.16
16.6	14	69.4	3.63				

[a] Data from Simpson (1959, 1961).

TABLE 4.45
Densities of Lithium Borosilicate Glasses[a]

Composition (mole %)			Density	Composition (mole %)			Density	Composition (mole %)			Density
Li_2O	B_2O_3	SiO_2	(g cm^{-3})	Li_2O	B_2O_3	SiO_2	(g cm^{-3})	Li_2O	B_2O_3	SiO_2	(g cm^{-3})
10	33.3	56.7	2.194	16.5	17.5	66	2.287	24	30	46.7	2.335
10	50	40	2.130	16.7	16.7	66.6	2.294	24.2	24	52	2.353
10.4	30	59.6	2.200	17	15	68	2.294	25	27.5	48.3	2.360
10.5	29.4	60.1	2.215	17.1	14.3	68.6	2.296	25.5	25	50	2.372
10.8	27.5	61.7	2.209	17.2	17.2	65.6	2.298	26.1	23.5	51	2.362
11.2	25	63.8	2.234	17.5	12.5	70	2.288	26.7	21.7	52.2	2.372
11.3	11.3	77.4	2.247	17.5	17.5	65	2.299	27.3	20	53.3	2.376
11.7	21.6	66.7	2.238	17.6	11.8	70.6	2.310	27.5	18.2	54.5	2.388
12	20	68	2.245	18	10	72	2.304	28	27.5	45	2.400
12	40	48	2.187	18.2	9.1	72.7	2.303	28.3	16	56	2.377
12.3	17.7	70	2.246	18.2	18.2	63.6	2.301	30	15	56.7	2.384
12.5	12.5	75	2.248	18.7	6.3	75	2.301	30	10	60	2.390
12.6	15.8	71.6	2.246	19	5	76	2.301	31.7	30	40	2.401
12.7	15	72.3	2.247	19	19	62	2.326	33.3	5	63.3	2.373
13	13	74	2.259	19.4	3.1	77.5	2.297	35	33.3	33.4	2.382
13.3	33.3	53.4	2.254	20	20	60	2.321	40	35	30	2.360
14	30	56	2.258	20	40	40	2.288	40	40	20	2.316
14.3	28.6	57.1	2.258	21	21	58	2.317				
15	15	70	2.279	22	22	56	2.346				
15	25	60	2.277	22.2	33.3	44.5	2.327				
15.4	23.1	61.5	2.285	22.5	22.5	55	2.356				
15.7	21.5	62.8	2.278	23	23	54	2.348				
16	20	64	2.284	23.3	30	46.7	2.335				

[a] Data from Akimov (1973).

77

TABLE 4.46

Densities of Lithium Aluminosilicate Glasses[a]

Composition (mole %)			Density (g cm^{-3})	Composition (mole %)			Density (g cm^{-3})
Li_2O	Al_2O_3	SiO_2		Li_2O	Al_2O_3	SiO_2	
25	30	45	2.452	31	4	65	2.368
25	25	50	2.363	33	2	65	2.361
29	14	57	2.410	13	21	66	2.350
20	20	60	2.410	25	9	66	2.381
19	20	61	2.414	13	20	67	2.432
30	9	61	2.395	15	18	67	2.396
17	21	62	2.414	11	20	69	2.408
15	22	63	2.419	15	16	69	2.377
17	20	63	2.411	15	15	70	2.369
32	5	63	2.376	17	13	70	2.369
13	22	65	2.430	23	7	70	2.363
15	20	65	2.423	25	5	70	2.355
17	18	65	2.398	15	12	73	2.369
17.5	17.5	65	2.400	15	10	75	2.353
19	16	65	2.393	15	8	77	2.334
21	14	65	2.399	15	6	79	2.317
23	12	65	2.392	15	4	81	2.299
25	10	65	2.385				
27	8	65	2.381				
29	6	65	2.372				

[a] Data from Karapetyan et al. (1979, 1980).

TABLE 4.47

Densities of $Li_2O-Ga_2O_3-SiO_2$ Glasses[a]

Composition (mole %)			Density (g cm^{-3})	Composition (mole %)			Density (g cm^{-3})
Li_2O	Ga_2O_3	SiO_2		Li_2O	Ga_2O_3	SiO_2	
12	22	66	3.088	20	25	55	3.224
12.5	12.5	75	2.743	20	27.5	52.5	3.314
12.5	17.5	70	2.876	25	5	70	2.540
13.3	21.7	65	3.050	25	10.7	64.3	2.780
16	12	72	2.758	25	15	60	2.921
16	21	63	3.074	25	25	50	3.245
16	28	56	3.317	27.5	10.4	62.1	2.760
16.7	16.7	66.6	2.908	27.5	18.1	54.4	3.032
20	2.5	77.5	2.382	27.5	27.5	45.0	3.317
20	5	75	2.500	30	5	65	2.557
20	10	70	2.706	30	10	60	2.767
20	12.5	67.5	2.801	30	14	56	2.907
20	15	65	2.886	30	20	50	3.108
20	20	60	3.056	35	5	60	2.584
20	22.5	57.5	3.128				

[a] Data from Dubrovo et al. (1969) and Tsekhomskaya and Dubrovo (1968).

TABLE 4.48

Densities of Some $Li_2O-M_2O_3-SiO_2$ Glass Systems[a]

Composition (mole %)			Density (g cm^{-3})
Li_2O	M_2O_3	SiO_2	
M = Y			
27	1	72	2.393
27	3	70	2.536
27	6	67	2.787
M = La			
25	3	72	2.582
25	6	69	2.859
27	3	70	2.615
27	6	67	2.905
M = Nd			
27	3	70	2.620
27	9	64	3.330

[a] Data from Parfenov *et al.* (1963).

TABLE 4.49

Densities of $Li_2O-Fe_2O_3-SiO_2$ Glasses[a,b]

Composition (mole %)			Density (g cm^{-3})	Composition (mole %)			Density (g cm^{-3})
Li_2O	Fe_2O_3[c]	SiO_2		Li_2O	Fe_2O_3[c]	SiO_2	
24	1	75	2.34	24	16	60	2.94
24	2	74	2.42	24	17	59	2.97
24	3	73	2.43	24	18	58	3.01
24	7	69	2.59	24	19	57	3.05
24	10	66	2.71	24	20	56	3.12
24	14	62	2.86	24	21	55	3.15
24	15	61	2.90				

[a] Data from Belyustin and Volkov (1967).
[b] Glasses prepared under oxidizing conditions.
[c] Calculated as Fe_2O_3; contains $FeO + Fe_2O_3$.

TABLE 4.50
Densities of Sodium Borosilicate Glasses

Composition (mole %)			Density (g cm⁻³)	Reference[a]	Composition (mole %)			Density (g cm⁻³)	Reference[a]	Composition (mole %)			Density (g cm⁻³)	Reference[a]
Na$_2$O	B$_2$O$_3$	SiO$_2$			Na$_2$O	B$_2$O$_3$	SiO$_2$			Na$_2$O	B$_2$O$_3$	SiO$_2$		
3	82	15	1.834	1	10	45	45	2.182		20	65	15	2.252	
5	80	15	1.883		10	50	40	2.164		20	70	10	2.216	
7	78	15	1.942		10	55	35	2.145		20	75	5	2.202	
10	75	15	2.002		10	60	30	2.133		25	50	25	2.416	
14.5	70.5	15	2.085		10	65	25	2.118		25	55	20	2.378	
17	68	15	2.148		10	70	20	2.094		25	60	15	2.351	
20	65	15	2.244		10	75	15	2.065		25	65	10	2.319	
30	55	15	2.441		10	80	10	2.052		25	70	5	2.283	
1	34	65	2.035		10	85	5	2.039		30	5	65	2.520	
5	30	65	2.105		20	5	75	2.447		30	10	60	2.540	
15	20	65	2.345		20	10	70	2.502		30	15	55	2.539	
20	15	65	2.462		20	15	65	2.521		30	20	50	2.543	
25	10	65	2.507		20	20	60	2.525		30	25	45	2.553	
30	5	65	2.537		20	25	55	2.512		30	30	40	2.529	
10	10	80	2.370	2	20	30	50	2.478		30	35	35	2.513	
10	15	75	2.357		20	35	45	2.445		30	40	30	2.492	
10	20	70	2.326		20	40	40	2.419		30	45	25	2.473	
10	25	65	2.292		20	45	35	2.389		30	50	20	2.448	
10	30	60	2.264		20	50	30	2.331		30	55	15	2.421	
10	35	55	2.239	2	20	55	25	2.330	2	30	60	10	2.393	
10	40	50	2.211		20	60	20	2.282		30	65	5	2.356	

[a] Data from (1) Konijnendijk (1975) and (2) Imaoka et al. (1971).

TABLE 4.51
Densities of Sodium Aluminosilicate Glasses

Composition (wt. %)			Density	Reference[a]
Na_2O	Al_2O_3	SiO_2	(g cm^{-3})	
20.8	40.4	38.8	2.505	1
21.8	38.3	39.9	2.503	
21.9	35.5	42.6	2.498	
22.1	33.0	44.9	2.503	
25.3	28.0	46.7	2.502	
26.9	21.6	51.5	2.503	
29.6	14.1	56.3	2.497	
32.8	5.1	62.1	2.497	
21.36	0.93	77.71	2.401	2
21.54	2.85	75.61	2.412	
19.70	4.82	75.48	2.402	
19.67	5.12	75.21	2.403	
23.02	2.00	74.98	2.419	
23.98	1.07	74.95	2.423	
22.12	3.00	74.88	2.416	
24.27	0.98	74.75	2.428	
21.50	4.69	73.81	2.418	

Composition (wt. %)			Density	Reference[a]
Na_2O	Al_2O_3	SiO_2	(g cm^{-3})	
25.87	0.98	73.15	2.440	2
19.60	9.32	71.08	2.421	
22.63	6.49	70.88	2.434	
26.31	2.85	70.84	2.451	
24.45	4.92	70.63	2.443	
24.84	4.86	70.30	2.445	
27.86	2.03	70.11	2.457	
29.04	0.97	69.99	2.461	
25.45	5.77	68.78	2.453	
25.15	6.75	68.10	2.454	
29.30	2.95	67.75	2.468	
28.34	4.45	67.21	2.465	
31.61	2.03	66.36	2.481	
31.67	2.41	65.92	2.481	
31.23	2.87	65.90	2.483	
24.69	9.43	65.88	2.458	
29.55	4.75	65.70	2.475	
24.94	9.96	65.10	2.462	

Composition (wt. %)			Density	Reference[a]
Na_2O	Al_2O_3	SiO_2	(g cm^{-3})	
34.20	1.02	64.78	2.492	2
32.46	6.57	60.97	2.498	
29.53	9.59	60.88	2.489	
38.13	1.09	60.78	2.517	
36.42	2.82	60.76	2.511	
34.46	4.86	60.68	2.503	
29.71	9.84	60.45	2.490	
39.23	4.83	55.94	2.530	
41.32	2.88	55.80	2.536	
34.66	9.68	55.66	2.516	
37.50	6.94	55.56	2.526	
34.92	9.83	55.25	2.518	
44.17	4.88	50.95	2.553	
39.54	9.57	50.89	2.541	
42.43	6.71	50.86	2.548	
42.43	7.00	50.57	2.550	
46.82	2.86	50.32	2.561	

[a] Data from (1) Terai (1969, 1971b) and (2) Faick et al. (1935).

TABLE 4.52

Densities of Na$_2$O–Ga$_2$O$_3$–SiO$_2$ Glasses

Composition (mole %)			Density		Composition (mole %)			Density	
Na$_2$O	Ga$_2$O$_3$	SiO$_2$	(g cm^{-3})	Reference[a]	Na$_2$O	Ga$_2$O$_3$	SiO$_2$	(g cm^{-3})	Reference[a]
12.5	12.5	75	2.717	1	18.3	11.7	70	2.790	1
12.5	17.5	70	2.913	1	20	10	70	2.744	1
13	1.5	85.5	2.372	2	20	20	60	3.039	1
13	3	84	2.423	2	20	28	52	3.266	1
13	4.5	82.5	2.472	2	21	9	70	2.783	3
13	6	81	2.539	2	21	11.3	67.7	2.792	1
					23.3	30	46.7	3.323	4
13	9	78	2.64	2	24.2	10.8	65	2.809	1
13	13	74	2.77	2	25	15	60	2.943	1
13	17	70	2.93	2	25	25	50	3.203	1
					25	50	25	3.766	4
					26.7	20	53.3	3.066	4
13.3	21.7	65	3.070	1	28.6	14.3	57.1	2.923	1
16	21	63	3.016	1	30	5	65	2.658	3
16.66	8.33	75.01	2.682	1	30	10	60	2.802	3
					30	40	30	3.558	4
16.66	12.5	70.84	2.804	1	30.3	9.1	60.6	2.775	1
16.66	16.66	66.68	2.900	1	31.7	4.8	63.5	2.644	1
					32.5	35	32.5	3.451	4
					35	30	35	3.338	4
					40	20	40	3.110	4
					45	10	45	2.865	4

[a] Data from (1) Dubrovo et al. (1965), (2) Galant (1961) and Ivanov and Galant (1963), (3) Takahashi et al. (1975), and (4) Hanada et al. (1981).

TABLE 4.53

Densities of Some Na$_2$O–M$_2$O$_3$–SiO$_2$ Glasses[a]

Composition (mole %)			Density
Na$_2$O	M$_2$O$_3$	SiO$_2$	(g cm^{-3})
M = In			
21	9	70	3.201
30	5	65	2.899
30	10	60	3.257
M = Sc			
21	9	70	2.699
30	5	65	2.611
30	10	60	2.731
M = Y			
21	9	70	2.963
30	5	65	2.754
30	10	60	2.990
M = Sb			
21	9	70	2.988
30	5	65	2.825
30	10	60	2.977

[a] Data from Takahashi et al. (1975).

TABLE 4.54

Densities of $Na_2O-La_2O_3-SiO_2$ Glasses

Composition (mole %)			Density $(g\ cm^{-3})$	Reference[a]	Composition (mole %)			Density $(g\ cm^{-3})$	Reference[a]
Na_2O	La_2O_3	SiO_2			Na_2O	La_2O_3	SiO_2		
5	7	88	2.87	1	16	11	73	3.27	1
10	10	80	3.16	1	17	5	78	2.83	1
11	16	73	3.67	1	18	2	80	2.56	1
12	1	87	2.40	1	20	5	75	2.85	1
12	5	83	2.77	1	21	5	74	2.86	1
12	8	80	3.03	1	25	5	70	2.90	1
12.5	12.5	75	3.46	1					
15	5	80	2.80	1	21	9	70	3.273	2
15	10	75	3.24	1	30	5	65	2.946	2
15	15	70	3.66	1	30	10	60	3.283	2

[a] Data from (1) Dubrovo and Shnypikov (1966) and Shnypikov (1967) and (2) Takahashi *et al.* (1975).

TABLE 4.55

Densities of Some $3Na_2O-xM_2O_3-7SiO_2$ Glass Systems[a]

x (mole %)	Density $(g\ cm^{-3})$
M = Nd	
1.25	2.68
2.5	2.83
5.0	3.06
7.0	3.26
M = Yb	
1.23	2.63
2.4	2.73
3.25	2.83
4.86	2.92
6.0	3.01
7.0	3.09
9.05	3.19

[a] Values from a graph in Komiyama (1974).

TABLE 4.56
Densities of $Na_2O-Bi_2O_3-SiO_2$ Glasses

Composition (wt. %)			Density	Reference[a]	Composition (wt. %)			Density	Reference[a]
Na_2O	Bi_2O_3	SiO_2	(g cm^{-3})		Na_2O	Bi_2O_3	SiO_2	(g cm^{-3})	
5	60	35	4.367	1	20.83	29.52	49.65	3.171	2
10	50	40	3.931	1	21.87	22.00	56.13	2.956	2
10	60	30	4.481	1	22.73	20.67	56.60	2.941	2
10	65	25	4.756	1	23.78	12.40	63.82	2.7166	2
13.55	52.87	33.58	4.110	2	24.30	12.71	62.99	2.725	2
14.24	44.92	40.84	3.7629	2	25	28.6	46.4	3.171	1
14.48	45.75	39.77	3.730	2	25	40	35	3.590	1
14.99	34.51	50.50	3.3699	2	30	30	40	3.253	1
15	40	45	3.564	1	32	9	59	2.681	1
18.23	34.34	47.43	3.333	2					
19.18	29.66	51.16	3.180	2					
20	30	50	3.215	1					
20	50	30	4.004	1					

[a] Data from (1) Watanabe *et al.* (1970) and (2) Riegel and Sharp (1934).

TABLE 4.57
Densities of $Na_2O-Fe_2O_3-SiO_2$ Glasses

Composition (mole %)			Density	Reference[a]	Composition (mole %)			Density	Reference[a]
Na_2O	Fe_2O_3	SiO_2	(g cm^{-3})		Na_2O	Fe_2O_3	SiO_2	(g cm^{-3})	
13	0.5	86.5	2.33	1	13	9	78	2.57	1
13	1	86	2.35	1	13	11	76	2.61	1
13	2	85	2.39	1	13	14.5	72.5	2.61	1
13	3	84	2.42	1	13	17	70	2.76	1
13	4	83	2.45	1					
13	5	82	2.47	1	21	9	70	2.699	2
13	7	80	2.52	1	30	5	65	2.631	2
					30	10	60	2.723	2

[a] Data from (1) Galant (1961) and Ivanov and Galant (1963) and (2) Takahashi *et al.* (1975).

TABLE 4.58
Densities of Potassium Borosilicate Glasses

Composition (mole %)			Density	Reference[a]	Composition (mole %)			Density	Reference[a]
K_2O	B_2O_3	SiO_2	(g cm^{-3})		K_2O	B_2O_3	SiO_2	(g cm^{-3})	
1	34	65	2.038	2	16	12	72	2.473	1
3	82	15	1.842	2	16	16	68	2.477	1
4	28	68	2.121	1	16	20	64	2.463	1
5	30	65	2.130	2	16	24	60	2.444	1
8	24	68	2.217	1	16	32	52	2.387	1
10	75	15	2.014	2	20	15	65	2.468	2
14.5	70.5	15	2.083	2	20	65	15	2.223	2
16	4	80	2.403	1	24	8	68	2.459	1
16	8	76	2.445	1	28	4	68	2.468	1
					30	55	15	2.448	2

[a] Data from (1) Appen and Gan'Fu-Si (1959a,b) and (2) Konijnendijk (1975).

<table>

| | TABLE 4.59 | | | | | TABLE 4.60 | | |

</table>

TABLE 4.59

Density of Potassium Aluminosilicate Glass[a]

Composition (mole %)			Density
K_2O	Al_2O_3	SiO_2	(g cm^{-3})
20.5	9.5	70.0	2.368

[a] Data from Rothermel (1967).

TABLE 4.60

Densities of $K_2O-Bi_2O_3-SiO_2$ Glasses[a]

Composition (wt. %)			Density
K_2O	Bi_2O_3	SiO_2	(g cm^{-3})
20.15	47.72	32.13	3.884
22.87	40.66	36.47	3.584
26.46	31.35	42.19	3.264
28.71	25.50	45.79	3.051
31.38	18.58	50.04	2.824
34.59	10.25	55.16	2.607

[a] Data from Riegel and Sharp (1934).

TABLE 4.61

Densities of Rubidium Borosilicate Glasses

Composition (mole %)			Density	Composition (mole %)			Density
Rb_2O	B_2O_3	SiO_2	(g cm^{-3})	Rb_2O	B_2O_3	SiO_2	(g cm^{-3})
3.5	61	35.5	2.04	13.3	59.2	27.5	2.59
3.6	72.5	23.9	2.09	13.4	72.3	14.3	2.42
7.9	20.7	71.4	2.47	15.2	33.3	51.5	2.08
8	31.4	60.6	2.33	17.8	11.9	70.3	3.00
8.1	42.2	49.7	2.22	17.9	24.8	57.3	3.02
8.3	53.8	37.9	2.12	18.4	37.7	43.9	2.47
8.5	66.1	25.4	2.24	18.9	51.4	29.7	2.50
8.6	78.1	13.3	2.25	19.3	65.1	15.6	2.42
12.4	10.8	76.8	3.16	25	13	62	2.88
12.5	22.7	64.8	2.71	25.5	27.4	47.1	2.89
12.7	34.1	53.2	2.41	26.2	41.8	32.0	2.46
13	46.3	40.7	2.57	26.7	56.5	16.8	2.33

[a] Data from Simpson (1961).

TABLE 4.62

Densities of Cesium Aluminosilicate Glasses[a]

Composition (mole %)			Density
Cs_2O	Al_2O_3	SiO_2	(g cm^{-3})
5	5	90	2.454
5	10	85	2.610
7.5	12.5	80	2.698
10	5	85	2.773
10	25	65	2.923
20	5	75	3.036

[a] Data from Bollin (1972).

85

TABLE 4.63

Densities of Copper Aluminosilicate Glasses

Composition (mole %)			Density
$Cu_2O + (CuO)_2$	Al_2O_3	SiO_2	(g cm^{-3})
12.5	12.5	75	2.81
17.5	17.5	65	2.98
25	10	65	3.06

[a] Data from Matusita and Mackenzie (1979).

TABLE 4.64

Densities of Thallium Borosilicate Glasses[a]

Composition (mole %)			Density
Tl_2O	B_2O_3	SiO_2	(g cm^{-3})
1.4	45.8	52.8	2.142
3.9	44.2	51.9	2.358
6.1	43.5	50.4	2.702
8.8	42.2	49.0	3.168
13.3	40	46.7	3.578
18.4	38.2	43.4	4.097
25.8	33.9	40.3	4.886
38	28	34	5.906
58.3	19.5	22.2	6.825

[a] Data from Kee (1968).

TABLE 4.65

Densities of Some $Li_2O-MO_2-SiO_2$ Glasses

Composition (mole %)			Density	Reference[a]
Li_2O	MO_2	SiO_2	(g cm^{-3})	
	M = Sn			
32.5	1.5	66.0	2.43	1
32.5	3	64.5	2.49	1
32.5	4.5	63.0	2.53	1
	M = Zr			
20	2.5	77.5	2.38	2
20	5	75	2.46	2
20	7.5	72.5	2.55	2
20	10	70	2.62	2
20	12.5	67.5	2.68	2

[a] Data from (1) Vakhrameev (1968) and Vakhrameev and Evstrop'ev (1969) and (2) from a graph in Ellern and Pavlushkin (1969).

TABLE 4.66

Densities of $Na_2O-GeO_2-SiO_2$ Glasses[a]

Composition (mole %)			Density	Composition (mole %)			Density
Na_2O	GeO_2	SiO_2	(g cm^{-3})	Na_2O	GeO_2	SiO_2	(g cm^{-3})
20	10	70	2.617	30	10	60	2.654
20	20	60	2.833	30	20	50	2.835
20	30	50	3.064	30	30	40	3.017
20	40	40	3.275	30	40	30	3.212
20	50	30	3.471	30	50	20	3.378
20	60	20	3.665	30	60	10	3.562
20	70	10	3.852				

[a] Data from Takahashi et al. (1977).

TABLE 4.67
Densities of Na_2O–SnO_2–SiO_2 Glasses[a]

Composition (mole %)			Density		Composition (mole %)			Density	
Na_2O	SnO_2	SiO_2	(g cm^{-3})	Reference[a]	Na_2O	SnO_2	SiO_2	(g cm^{-3})	Reference[a]
13	1	86	2.354	1	20	2.5	77.5	2.50	2
13	2	85	2.451	1	20	5	75	2.60	2
13	5	82	2.540	1	20	7.5	72.5	2.72	2
13	7	80	2.630	1	25	2.5	72.5	2.52	2
13	8.5	78.5	2.700	1	25	5	70	2.68	2
13	10	77	2.761	1	25	7.5	67.5	2.71	2
					30	2.5	67.5	2.57	2
15	2.5	82.5	2.47	2	30	5	65	2.66	2
15	5	80	2.57	2	30	7.5	62.5	2.76	2
15	7.5	77.5	2.66	2	30	10	60	2.84	2

[a] Data from (1) Galant et al. (1966) and (2) Vakhrameev (1968) and Vakhrameev and Evstrop'ev (1969).

TABLE 4.68
Densities of Na_2O–TiO_2–SiO_2 Glasses[a]

Composition (mole %)			Density	Composition (mole %)			Density
Na_2O	TiO_2	SiO_2	(g cm^{-3})	Na_2O	TiO_2	SiO_2	(g cm^{-3})
10	10	80	2.439	25	30	45	2.869
10	15	75	2.520	25	35	40	2.934
10	20	70	2.570	30	5	65	2.533
15	5	80	2.429	30	10	60	2.615
15	10	75	2.517	30	15	55	2.684
15	15	70	2.601	30	20	50	2.749
15	20	65	2.678	30	25	45	2.815
15	25	60	2.744	30	30	40	2.871
15	30	55	2.813	30	35	35	2.932
20	5	75	2.480	35	5	60	2.567
20	10	70	2.561	35	10	55	2.635
20	15	65	2.641	35	15	50	2.696
20	20	60	2.712	35	20	45	2.746
20	25	55	2.789	35	25	40	2.817
20	30	50	2.855	35	30	35	2.869
23	17	60	2.692	35	35	30	2.926
25	5	70	2.512	40	5	55	2.590
25	10	65	2.595	40	10	50	2.649
25	15	60	2.668	40	15	45	2.706
25	20	55	2.740	40	20	40	2.750
25	25	50	2.806	45	5	50	2.607

[a] Data from Hamilton and Cleek (1958).

TABLE 4.69

Densities of Na$_2$O–ZrO$_2$–SiO$_2$ Glasses[a]

Composition (mole %)			Density (g cm^{-3})	Composition (mole %)			Density (g cm^{-3})
Na$_2$O	ZrO$_2$	SiO$_2$		Na$_2$O	ZrO$_2$	SiO$_2$	
14.8	1.2	84.0	2.340	23.2	5.1	71.7	2.607
15.0	2.5	82.5	2.454	23.8	8.0	68.2	2.718
15.2	3.3	81.5	2.518	25.4	3.8	70.8	2.592
15.4	5.2	79.4	2.562	26.4	8.0	65.6	2.736
15.6	6.6	77.8	2.594	27.6	2.5	69.9	2.571
15.8	8.0	76.2	2.625	29.1	8.0	62.9	2.752
16.0	9.4	74.6	2.660	29.7	1.3	69.0	2.539
16.3	10.9	72.8	2.680	31.8	8.0	60.2	2.768
16.5	12.5	71.0	2.70	34.5	8.0	57.5	2.786
16.6	7.2	76.2	2.671	37.2	8.0	54.8	2.796
18.5	8.0	73.5	2.626	39.8	8.0	52.2	2.805
21.1	8.0	70.9	2.693				

[a] Data from Botvinkin and Demichev (1964).

TABLE 4.70

Densities of Some Na$_2$O–MO$_2$–SiO$_2$ Glasses

Composition (mole %)			Density (g cm^{-3})	Reference[a]
Na$_2$O	MO$_2$	SiO$_2$		
	M = Hf			
18	12	70	3.096	1
	M = Th			
20	5	75	2.838	2
30	5	65	2.893	

[a] Data from (1) Brekhovskikh and Sesorova (1960) and (2) Takahashi et al. (1977).

TABLE 4.71

Densities of Na$_2$O–TeO$_2$–SiO$_2$ Glasses[a]

Composition (mole %)			Density (g cm^{-3})	Composition (mole %)			Density (g cm^{-3})
Na$_2$O	TeO$_2$	SiO$_2$		Na$_2$O	TeO$_2$	SiO$_2$	
25	4.8	70.2	2.592	30	20.2	49.8	3.107
25	9.8	65.2	2.757	30	24.9	45.1	3.240
25	19.6	55.4	3.070	35	5.0	60	2.656
25	24.9	50.1	3.212	35	10	55	2.813
30	4.8	65.2	2.631	40	5	55	2.677
30	10.0	60.0	2.793	45	5	50	2.692
30	15.0	55.0	2.950	50	4.8	45.2	2.707

[a] Data from Mochida et al. (1980).

TABLE 4.72
Densities of $K_2O-TiO_2-SiO_2$ Glasses[a]

Composition (wt. %)			Density	Composition (wt. %)			Density	Composition (wt. %)			Density
K_2O	TiO_2	SiO_2	(g cm^{-3})	K_2O	TiO_2	SiO_2	(g cm^{-3})	K_2O	TiO_2	SiO_2	(g cm^{-3})
15	5	80	2.351	30	10	60	2.501	40	50	10	2.773
15	12.5	72.5	2.454	30	15	55	2.549	40	55	5	2.802
15	20	65	2.485	30	20	50	2.582	45	5	50	2.513
15	35	50	2.657	30	30	40	2.655	45	10	45	2.550
15	40	45	2.720	30	40	30	2.744	45	20	35	2.612
20	5	75	2.409	30	45	25	2.784	45	30	25	2.663
20	10	70	2.469	40	5	55	2.498	45	40	15	2.701
20	15	65	2.521	40	10	50	2.531	45	45	10	2.727
20	20	60	2.566	40	20	40	2.607	45	50	5	2.733
20	30	50	2.673	40	30	30	2.675	60	10	30	2.524
20	40	40	2.775	40	40	20	2.727	60	20	20	2.571
20	45	35	2.823	40	45	15	2.759				

[a] Data from Rao (1963a,b, 1964).

TABLE 4.73
Densities of $Na_2O-P_2O_5-SiO_2$ Glasses with $Na_2O:P_2O_5 = 1:1.5$[a]

Mole % of SiO_2 in the glass	Density (g cm^{-3})
7	2.502
14	2.510
17	2.515
25	2.522

[a] Values from a graph in Takahashi (1962).

TABLE 4.74
Densities of $Na_2O-Nb_2O_5-SiO_2$ Glasses

Composition (mole %)			Density		Composition (mole %)			Density	
Na_2O	Nb_2O_5	SiO_2	(g cm^{-3})	Reference[a]	Na_2O	Nb_2O_5	SiO_2	(g cm^{-3})	Reference[a]
12.6	12.5	74.9	2.908	1	26	20	54	3.321	1
15	7.5	77.5	2.742	1	27.3	27.3	45.4	3.571	2
15	15	70	3.044	1	29	9	62	2.908	1
19.1	19.1	61.8	3.251	1	29	29	42	3.622	2
19.8	26	54.2	3.468	1	29.9	5	65.1	2.735	1
19.9	5	75.1	2.673	1	29.9	16	54.1	3.203	1
20	10	70	2.914	1	30	20	50	3.333	1
21.9	7	71.1	2.776	1	30.3	30.3	39.4	3.655	2
22.1	16	61.9	3.155	1	32.2	32.2	35.6	3.721	2
22.9	23	54.1	3.387	1	33.9	19.1	47.0	3.313	1
24.9	13	62.1	3.072	1	34.5	34.5	31.0	3.782	2
25	5	70	2.711	1	35.1	11	53.9	3.015	1
25	25	50	3.454	1	38.1	12	49.9	3.067	1
25.1	25.1	49.8	3.501	2	40.9	13.6	45.5	3.207	1

[a] Data from (1) Hirayama and Berg (1963) and (2) Layton and Herczog (1967).

TABLE 4.75

Densities of $K_2O-Nb_2O_5-SiO_2$ Glasses[a]

Composition (mole %)			Density $(g\ cm^{-3})$	Composition (mole %)			Density $(g\ cm^{-3})$
K_2O	Nb_2O_5	SiO_2		K_2O	Nb_2O_5	SiO_2	
16.1	19.4	64.5	3.159	21.7	13.1	65.2	2.946
16.7	16.7	66.6	3.067	22.7	9.1	68.2	2.811
17.2	13.8	69.0	2.967	23.8	4.8	71.4	2.639
17.9	10.7	71.4	2.866	23.8	28.6	47.6	3.403
18.5	7.4	74.1	2.738	25	25	50	3.304
19.2	3.8	77.0	2.572	26.3	21.1	52.6	3.196
19.2	23.1	57.7	3.257	27.8	16.7	55.5	3.069
20	20	60	3.161	29.4	11.8	58.8	2.917
20.8	16.7	62.5	3.059	31.3	6.2	62.5	2.736

[a] Data from Rao (1962b).

TABLE 4.76

Densities of $K_2O-Ta_2O_5-SiO_2$ Glasses[a]

Composition (wt. %)			Density $(g\ cm^{-3})$	Composition (wt. %)			Density $(g\ cm^{-3})$
K_2O	Ta_2O_5	SiO_2		K_2O	Ta_2O_5	SiO_2	
21	12.3	66.7	2.575	28.6	19.2	52.2	2.785
22.3	20.9	56.8	2.770	28.7	25.2	46.1	2.956
22.4	6.4	71.2	2.477	29.2	8.6	62.2	2.562
23.2	27.1	49.7	2.915	29.3	17.2	53.5	2.741
23.6	16.6	59.8	2.679	30.1	5.9	64.0	2.515
24.3	13.9	61.8	2.636	30.1	15.1	54.8	2.702
24.4	23.8	51.8	2.869	30.1	22	47.9	2.866
24.5	35.9	39.6	3.215	31.2	10.5	58.3	2.619
25	11.4	63.6	2.585	31.4	18.1	50.5	2.789
25	22	53	2.829	32.3	28.5	39.2	3.079
25.6	20	54.4	2.783	32.9	14.2	52.9	2.707
26.1	26.3	47.6	2.949	33.4	5.6	61.0	2.522
26.6	6.1	67.3	2.493	34.5	10	55.5	2.620
26.6	15.5	57.9	2.690	34.8	20.5	44.7	2.885
26.6	30.5	42.9	3.106	36.1	5.3	58.6	2.533
27.3	22.9	49.8	2.867	37.5	22.2	40.3	2.959
28.5	10.9	60.6	2.610				

[a] Data from Sun and Silverman (1941).

TABLE 4.77

Densities of $Na_2O-U_3O_8-SiO_2$ Glasses[a]

Composition (wt. %)			Density	Composition (wt. %)			Density
Na_2O	U_3O_8	SiO_2	(g cm^{-3})	Na_2O	U_3O_8	SiO_2	(g cm^{-3})
11.0	50	39	4.174	23.5	30	46.5	3.397
11.2	52	36.8	3.729	27	20	53	3.040
14	41	45	3.328	30.5	10	59.5	2.786
16.8	29	54.2	2.899	35.5	30	34.5	3.260
20	15	65	2.665	40.5	20	39.5	3.065
20	40	40	3.699	45.5	10	44.5	2.813

[a] Data from Chakraborty (1969).

TABLE 4.78

Densities of $CuO-PbO-SiO_2$ Glasses

Composition (mole %)			Density	Composition (mole %)			Density
CuO	PbO	SiO_2	(g cm^{-3})	CuO	PbO	SiO_2	(g cm^{-3})
1.5	53.5	45	6.25	1	44	55	5.41
3	52	45	6.10	2.3	42.7	55	5.26
4.2	50.8	45	6.02	3	42	55	5.18
5.4	49.6	45	5.94	5	40	55	5.14
8	47	45	5.83	8	37	55	5.05
10	45	45	5.75	10	35	55	4.99
12	43	45	5.68	12	33	55	4.97
15	40	45	5.64	15.2	29.8	55	4.91
20	35	45	5.49	20.3	24.7	55	4.82

[a] Values of densities and mole % CuO from a graph in Ershov et al. (1972).

TABLE 4.79

Densities of $MgO-CaO-SiO_2$ Glasses[a]

Composition (mole %)			Density	Composition (mole %)			Density
MgO	CaO	SiO_2	(g cm^{-3})	MgO	CaO	SiO_2	(g cm^{-3})
2.8	47.2	50.0	2.899	25	25	50	2.854
4.9	50.9	44.2	2.953	31.7	18.3	50.0	2.835
8.4	41.6	50.0	2.892	32.7	23.3	44.0	2.920
14.4	35.6	50.0	2.881	36.5	13.5	50.0	2.820
19.7	30.3	50.0	2.872	45.6	4.4	50.0	2.879
20.3	38.9	40.8	2.967	47.8	2.2	50.0	2.777
21.7	28.3	50.0	2.858				

[a] Data from Larsen (1909).

TABLE 4.80

Densities of Some $MO-M'O-SiO_2$ Glasses

Composition (mole %)			Density		Composition (mole %)			Density	
MO	M'O	SiO$_2$	(g cm^{-3})	Reference[a]	MO	M'O	SiO$_2$	(g cm^{-3})	Reference[a]
M = Mg, M' = Sr					M = Ca, M' = Ba				
12.5	37.5	50	3.529	1	16.67	16.67	66.66	3.230	1
25	25	50	3.302	1	25	25	50	3.650	1
M = Mg, M' = Ba					M = Sr, M' = Ba				
12.5	37.5	50	4.032	1	16.67	16.67	66.66	3.504	1
25	25	50	3.658	1					
					M = Ba, M' = Pb				
M = Mg, M' = Fe					17.5	17.5	65.0	4.214	3
25	25	50	3.068	2	20	20	60	4.504	3
					22.5	22.5	55.0	5.001	3
M = Ca, M' = Fe					23.3	20	56.7	4.709	3
25	25	50	3.162	2	25	25	50	5.164	3

[a] Data from (1) Soga *et al.* (1976), (2) Soga *et al.* (1972), and (3) Sheludyakov (1967).

TABLE 4.81

Densities of Some $MO-MnO-SiO_2$ Glasses[a]

Composition (mole %)			Density	Composition (mole %)			Density
MO	MnO	SiO$_2$	(g cm^{-3})	MO	MnO	SiO$_2$	(g cm^{-3})
M = Ca				M = Ba			
10	40	50	2.99	10	40	50	4.24
20	30	50	3.11	20	30	50	4.03
30	20	50	3.22	30	20	50	3.82
40	10	50	3.31	40	10	50	3.61
				M = Zn			
M = Sr				5	45	50	3.56
10	40	50	3.69	10	40	50	3.53
20	30	50	3.62	20	30	50	3.51
30	20	50	3.51	30	20	50	3.46
40	10	50	3.46	40	10	50	3.43
				M = Cd			
				10	40	50	4.34
				20	30	50	4.11
				30	20	50	3.87
				40	10	50	3.64

[a] Data from Kuznetsova (1972).

TABLE 4.82

Densities of Magnesium Aluminosilicate Glasses[a]

Composition (wt. %)			Density	Composition (wt. %)			Density
MgO	Al_2O_3	SiO_2	(g cm^{-3})	MgO	Al_2O_3	SiO_2	(g cm^{-3})
7.0	20.0	73.0	2.410	10.0	25.0	65.0	2.500
7.0	25.0	68.0	2.456	10.4	23.3	66.3	2.487
7.0	30.0	63.0	2.498	12.0	20.0	68.0	2.481
8.0	23.0	69.0	2.452	12.0	22.5	65.5	2.497
8.5	24.0	67.5	2.465	12.0	25.0	63.0	2.519
9.5	22.5	68.0	2.469	17.0	20.0	63.0	2.558
9.5	25.0	65.5	2.487				

[a] Data from Aslanova *et al.* (1980).

TABLE 4.83

Densities of Calcium Aluminosilicate Glasses[a]

Composition (mole %)			Density	Composition (mole %)			Density
CaO	Al_2O_3	SiO_2	(g cm^{-3})	CaO	Al_2O_3	SiO_2	(g cm^{-3})
19.2	19.2	61.6	2.605	26.3	31.6	42.1	2.748
20	20	60	2.624	27.2	23.4	49.4	2.720
20.5	28.2	51.3	2.688	27.8	27.8	44.4	2.735
20.8	20.8	58.4	2.640	29.3	21.9	48.8	2.735
22.2	11.1	66.7	2.593	29.4	29.4	41.2	2.754
22.7	13.6	63.7	2.623	32.1	2.9	65.0	2.661
22.7	22.7	54.6	2.671	32.8	6.0	61.2	2.728
22.8	26.6	50.6	2.700	33.3	19.1	47.6	2.772
23.2	16.3	60.5	2.641	34.2	12.5	53.2	2.763
23.8	19.1	57.1	2.661	35.8	19.7	44.5	2.760
24.4	22	53.6	2.686	37.2	16.3	46.5	2.800
25	25	50	2.704	37.5	27.5	35.0	2.805
25.3	26.6	48.1	2.714	39.4	36.1	24.5	2.827
25.6	28.2	46.2	2.731	40.4	40.6	18.9	2.847
26.3	26.3	47.4	2.722				

[a] Data from Chang and Ying (1965) and Yamane and Okuyama (1982).

TABLE 4.84

Densities of CaO–Ga$_2$O$_3$–SiO$_2$ Glasses[a]

Composition (mole %)			Density (g cm^{-3})	Composition (mole %)			Density (g cm^{-3})
CaO	Ga$_2$O$_3$	SiO$_2$		CaO	Ga$_2$O$_3$	SiO$_2$	
12.5	12.5	75	2.846	25	35	40	3.833
15	10	75	2.776	30	10	60	3.051
15	15	70	2.985	30	30	40	4.728
15	25	60	3.365	35	35	30	3.912
19	19	62	3.202	38	38	24	3.998
20	10	70	2.858	40	10	50	3.190
20	15	65	3.073	40	20	40	3.533
20	20	60	3.262	40	30	30	3.826
20	25	55	3.442	40	40	20	4.079
20	30	50	3.614	40	45	15	4.206
25	15	60	3.154	45	45	10	4.209
25	25	50	3.509	50	10	40	3.312
				50	40	10	4.157

[a] Data from Danilova (1967), Danilova and Dubrovo (1967), and Lisenenkov and Vasilev (1979).

TABLE 4.85

Densities of Strontium Aluminosilicate Glasses

Composition (mole %)			Density (g cm^{-3})	Composition (mole %)			Density (g cm^{-3})
SrO	Al$_2$O$_3$	SiO$_2$		SrO	Al$_2$O$_3$	SiO$_2$	
25	10	65	2.898	40	20	40	3.352
30	5	65	2.917	40	25	35	3.339
30	10	60	3.061	45	5	50	3.321
35	5	60	3.229	45	10	45	3.383
35	10	55	3.238	45	15	40	3.397
35	15	50	3.265	45	20	35	3.333
35	20	45	3.183	50	5	45	3.357
35	25	40	3.212	50	10	40	3.418
35	30	35	3.172	50	15	35	3.351
40	5	55	3.281	55	5	40	3.370
40	10	50	3.336	55	10	35	3.326
40	15	45	3.353	60	5	35	3.350

[a] Data from Iskhakov (1971).

TABLE 4.86

Densities of Barium Borosilicate Glasses[a]

Composition (mole %)			Density (g cm^{-3})	Composition (mole %)			Density (g cm^{-3})	Composition (mole %)			Density (g cm^{-3})
BaO	B$_2$O$_3$	SiO$_2$		BaO	B$_2$O$_3$	SiO$_2$		BaO	B$_2$O$_3$	SiO$_2$	
16.7	75.4	7.9	2.692	32.3	53.6	14.1	3.597	37.2	27.7	35.1	3.912
20.6	71.4	8.0	2.908	32.3	59.4	8.3	3.556	37.7	37.1	25.2	3.879
20.7	63.6	15.7	2.970	32.4	16.1	51.5	3.747	38.2	51.9	9.9	3.817
24.0	39.9	36.1	3.352	32.4	33.8	33.8	3.722	39.8	55.1	5.1	3.818
24.3	42.9	32.8	3.332	32.4	45.7	21.9	3.658	40.0	16.5	43.5	4.031
24.4	51.4	24.2	3.268	32.4	49.5	18.1	3.628	40.1	26.0	33.9	4.005
24.4	55.5	20.1	3.241	32.6	25.5	41.9	3.748	40.3	31.7	28.0	3.985
24.5	63.5	12.0	3.170	32.6	41.4	26.0	3.682	40.4	39.8	19.8	3.947
24.5	71.4	4.1	3.101	32.6	64.2	3.2	3.508	40.5	47.6	11.9	3.900
28.1	12.5	59.4	3.572	34.2	32.6	33.2	3.775	40.7	55.5	3.8	3.840
28.2	23.9	47.9	3.586	34.3	35.8	29.9	3.785	44.0	20.4	35.6	4.135
28.2	32.0	39.8	3.570	35.9	16.7	47.4	3.899	44.1	32.1	23.8	4.078
28.2	36.2	35.6	3.550	36.3	28.0	35.7	3.879	44.2	28.1	27.7	4.103
28.6	40.0	31.4	3.544	36.4	35.5	28.1	3.851	44.3	24.1	31.6	4.125
28.6	55.8	15.6	3.427	36.4	39.7	23.9	3.832	44.7	39.7	15.6	4.045
28.8	63.2	8.0	3.359	36.4	43.5	20.1	3.808	47.2	22.9	29.9	4.196
30.1	20.3	49.6	3.671	36.6	51.5	11.9	3.771	47.5	30.7	21.8	4.158
32.1	22.6	45.3	3.758	36.6	59.4	4.0	3.699	50.0	26.5	23.5	4.249
32.1	50.4	17.5	3.623	36.9	48.2	14.9	3.798				
32.3	6.5	61.2	3.728								
32.3	30.1	37.6	3.740								

[a] Data from Hamilton et al. (1958).

TABLE 4.87

Densities of Some MO–Al$_2$O$_3$–SiO$_2$ Glasses

Composition (mole %)			Density (g cm^{-3})	Reference[a]
MO	Al$_2$O$_3$	SiO$_2$		
M = Ba				
15.1	7.6	77.3	2.78	1
21.9	8.2	69.9	3.05	1
29.9	9.0	61.1	3.15	1
M = Pb				
31.5	3.5	65.0	4.454	2
M = Zn				
14.3	14.3	71.4	2.653	3
16.7	16.7	66.6	2.700	3
20	20	60	2.824	3

[a] Data from (1) Haydenov (1963), (2) Buchanan and Zuegel (1968), and (3) Varshal (1972).

TABLE 4.88

Densities of BaO–Ga$_2$O$_3$–SiO$_2$ Glasses[a]

Composition (mole %)			Density (g cm^{-3})	Composition (mole %)			Density (g cm^{-3})
BaO	Ga$_2$O$_3$	SiO$_2$		BaO	Ga$_2$O$_3$	SiO$_2$	
20	10	70	3.362	30	30	40	4.276
20	15	65	3.525	40	10	50	4.181
20	20	60	3.666	40	20	40	4.391
20	25	55	3.836	40	30	30	4.571
20	30	50	3.994	40	40	20	4.742
30	10	60	3.796	40	45	15	4.843

[a] Data from Danilova and Dubrovo (1971).

TABLE 4.89

Densities of BaO–La$_2$O$_3$–SiO$_2$ Glasses[a]

Composition (mole %)			Density (g cm^{-3})	Composition (mole %)			Density (g cm^{-3})
BaO	La$_2$O$_3$	SiO$_2$		BaO	La$_2$O$_3$	SiO$_2$	
23	5	72	3.663	30	7	63	4.121
23	7	70	3.849	32	2	66	3.819
23	10	67	4.085	33	5	62	4.076
25	5	70	3.768	35	2	63	3.918
25	7	68	3.923	35	5	60	4.132
25	10	65	4.169	35	7	58	4.276
28	2	70	3.623	37	5	58	4.243
28	5	67	3.885	38	2	60	3.997
30	5	65	3.964	38	7	55	4.387

[a] Data from Cleek and Babcock (1973).

TABLE 4.90

Densities of Some MO–Fe₂O₃–SiO₂ Glasses[a]

Composition (mole %)			Density
MO	Fe₂O₃	SiO₂	(g cm⁻³)
	M = Ba		
21	10	69	3.38
35	10	55	3.83
37.5	5	57.5	3.90
	M = Pb		
40	10	50	5.25
45	10	45	5.53
47.5	5	47.5	5.82

[a] Data from Tsekhomskii (1966).

TABLE 4.91

Densities of CdO–Bi₂O₃–SiO₂ Glasses[a]

Composition (wt. %)			Density	Composition (wt. %)			Density
CdO	Bi₂O₃	SiO₂	(g cm⁻³)	CdO	Bi₂O₃	SiO₂	(g cm⁻³)
4.8	92.2	3.0	8.18	19.5	78.7	1.8	8.14
7.5	85	7.5	7.92	19.6	78.4	2.0	8.07
9.8	88.2	2.0	8.25	19.9	79.6	0.5	8.12
19.4	77.7	2.9	7.97	37.3	36.5	26.2	6.46

[a] Data from Rao (1962a).

TABLE 4.92

Densities of Lead Borosilicate Glasses

Composition (mole %)			Density	
PbO	B₂O₃	SiO₂	(g cm⁻³)	Reference[a]
31.5	12.6	55.9	4.482	1
39.7	18.2	42.1	5.075	2
41.5	47.5	11.0	4.936	2
53.6	21.5	24.9	6.00	2
54.5	32.8	12.7	5.931	2
62.7	23.6	13.7	6.45	2
69.2	7.7	23.1	7.181	3
71.4	14.3	14.3	7.177	3
73.3	20	6.7	7.226	3

[a] Data from (1) Buchanan and Zuegel (1968), (2) Thiele (1955), and (3) Kordes (1939c).

TABLE 4.93

Densities of BaO–TiO$_2$–SiO$_2$ Glasses[a]

Composition (mole %)			Density (g cm^{-3})	Composition (mole %)			Density (g cm^{-3})
BaO	TiO$_2$	SiO$_2$		BaO	TiO$_2$	SiO$_2$	
25	20	55	3.699	30	20	50	3.945
25	25	50	3.804	30	25	45	4.004
25	30	45	3.887	30	30	40	4.079
25	35	40	3.954	35	5	60	3.634
30	10	60	3.743	35	10	55	3.968
30	15	55	3.859	35	15	50	4.047

[a] Data from Cleek and Babcock (1973).

TABLE 4.94

Densities of PbO–GeO$_2$–SiO$_2$ Glasses[a]

Composition (mole %)			Density (g cm^{-3})	Composition (mole %)			Density (g cm^{-3})
PbO	GeO$_2$	SiO$_2$		PbO	GeO$_2$	SiO$_2$	
10	20	70	3.178	30	60	10	5.520
10	40	50	3.382	50	10	40	6.085
10	60	30	3.750	50	20	30	6.263
10	80	10	4.199	50	30	20	6.402
30	20	50	4.650	50	40	10	6.564
30	40	30	5.009				

[a] Data from Topping et al. (1974).

TABLE 4.95

Densities of BaO–Ta$_2$O$_5$–SiO$_2$ Glasses[a]

Composition (mole %)			Density (g cm^{-3})	Composition (mole %)			Density (g cm^{-3})
BaO	Ta$_2$O$_5$	SiO$_2$		BaO	Ta$_2$O$_5$	SiO$_2$	
27	5	68	3.952	38	7	55	4.646
28	2	70	3.672	40	2	58	4.192
30	2	68	3.770	40	5	55	4.481
30	5	65	4.105	40	7	53	4.703
32	5	63	4.140	43	7	50	4.816
33	2	65	3.929	45	5	50	4.676
33	7	60	4.480	45	7	48	4.771
35	2	63	4.005	45	10	45	5.105
35	5	60	4.317	46	7	47	4.787
35	7	58	4.508	47	5	48	4.696
37	5	58	4.418	48	7	45	4.947
38	2	60	4.151				

[a] Data from Cleek and Babcock (1973).

TABLE 4.96

Densities of Some Al_2O_3–M_2O_3–SiO_2 Glasses

Composition (mole %)			Density $(g\ cm^{-3})$	Reference[a]
Al_2O_3	M_2O_3	SiO_2		
	M = La			
20	25	55	4.17	1
22	23	55	4.08	
23.5	21.5	55	4.05	
25	20	55	3.94	
26.5	18.5	55	3.86	
28.5	16.5	55	3.77	
'33	12	55	3.58	
35.5	9.5	55	3.46	
37	8	55	3.38	
19	12	69	3.52	
20.5	10.5	69	3.40	
21	10	69	3.34	
23	8	69	3.22	
25	6	69	3.03	
26	5	69	2.89	
	M = Nd			
29.6	12.1	58.3	3.63	2

[a] Data from (1) graph in Aleksandrov *et al.* (1980) and (2) Karlsson (1970).

TABLE 4.97

Densities of Al_2O_3–P_2O_5–SiO_2 Glasses[a]

Composition (wt. %)			Density $(g\ cm^{-3})$
Al_2O_3	P_2O_5	SiO_2	
19.9	78.9	0.7	2.584
21.2	78.6	0.9	2.571
23.9	74.5	0.8	2.554

[a] Data from Moriya *et al.* (1960).

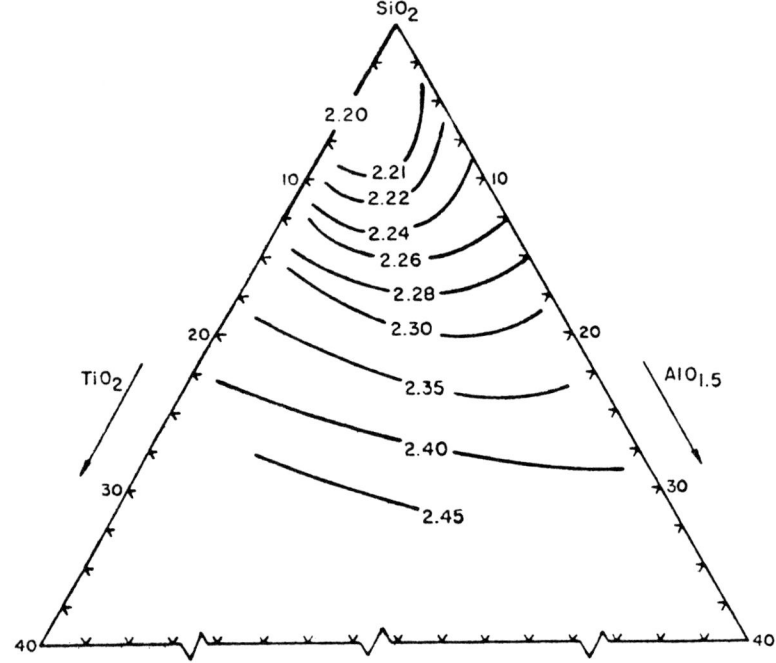

Fig. 4.1. Densities of $Al_2O_3-TiO_2-SiO_2$ glasses. The number on each curve is the density (grams per cubic centimeter); composition is in cation %. Data from Schultz and Dumbaugh (1980).

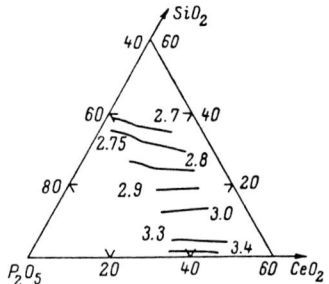

Fig. 4.2. Densities of $CeO_2-P_2O_5-SiO_2$ glasses. The number on each curve is the density (grams per cubic centimeter); composition is in wt.%. Data from Syritskaya *et al.* (1971).

Chapter 5

Surface Tension

The surface energy γ at temperature T and pressure p can be defined as the reversible work dW_r to form a surface of area dA:

$$\gamma = -(dW_r/dA)_{T,p}. \tag{1}$$

For liquid, surface energy and tension are equal. The units of surface tension are joules per square meter; $10^3 \, \text{dyn cm}^{-1} = 1 \, \text{J m}^{-2}$.

A number of measurements of surface tensions of glass melts have been made by a dipping cylinder method developed by Shartsis and co-workers at the United States National Bureau of Standards. In this method the force F required to pull a small thin-walled platinum cylinder of radius R out of the melt is measured, and

$$\gamma = F/4\pi R. \tag{2}$$

A correction factor is necessary for the most accurate work.

The surface tension decreases as the temperature increases; for most silicate glasses this change is quite small (less than 10%) from 1200 to 1500°C.

The influence of composition on surface tension of silicate glass melts is not large. For most sodium silicate melts the surface tension is about 0.3 J m^{-2} at 1200°C (Dunken, 1981), and this value is a reasonable approximation for the surface tension of most sodium-containing silicate glasses at this temperature.

Boric oxide (B_2O_3) has the low surface tension of about 0.085 J m^{-2} at 800°C, and addition of alkali oxides to it raises the surface tension sharply.

II   TABLES AND FIGURES

Data on surface tension for binary glasses are given in Tables 5.1–5.14 and Figs. 5.1–5.3. Tables 5.15–5.42 and Figs. 5.4 and 5.5 give data for ternary glasses.

BIBLIOGRAPHY

Adachi, A., Ogino, K., and Toritani, H. (1960). *Tech. Rep. Osaka Univ.* **10**, 149.
Anisimov, Yu. S., Grits, E. F., and Mitin, B. S. (1977). *Inorg. Mater. (Engl. Transl.)* **13**, 1168.
Appen, A. A. (1936a). *Opt.-Mekh. Prom.* (6), 5.
Appen, A. A. (1936b). *Opt.-Mekh. Prom.* (12), 17.
Appen, A. A., and Kayalova, S. S. (1962). *Dokl. Akad. Nauk SSSR* **145**, 592.
Appen, A. A., and Kayalova, S. S. (1963). *In* "Advances in Glass Technology," Part 2, p. 61. New York.
Appen, A. A., Shishov, K. A., and Kayalova, S. S. (1952). *Zh. Fiz. Khim.* **26**, 1131.
Appen, A. A., Shishov, K. A., and Kayalova, S. S. (1953). *Silikattechnik* **4**, 104.
Badger, A. E., Parmelee, C. W., and Williams, A. E. (1937). *J. Am. Ceram. Soc.* **20**, 325.
Cooper, C. F., and Kitchener, J. A. (1959). *J. Iron Steel Inst.* **196**, 48.
Dunken, H. H. (1981). "Physikalische Chemie der Glasoberflache." VEB Dtsch. Verlag Grundstoffind., Leipzig.
Ejima, A., and Shimoji, M. (1970). *Trans. Faraday Soc.* **66**, 99.
Ermolaeva, E. V. (1955). *Ogneupory* **20**, 221.
Frischat, G. H., and Beier, W. (1979). *Glastech. Ber.* **52**, 116.
Hino, M., Ejima, T., and Kameda, M. (1967). *Nippon Kinzoku Gakkaishi* **31**, 113.
Imaoka, M., Hasegawa, H., Hamaguchi, Y., and Kurotaki, Y. (1971). *Yogyo Kyokaishi* **79**, 164.
Ito, H., Yanagase, T., and Suginohara, Y. (1961). *Nippon Kogyo Kaishi* **77**, 895.
Kargin, Y. F., Zhereb, V. P., Skorikov, V. M., Kosov, A. V., Kutvitskii, V. A., and Nuriev, E. I. (1977). *Inorg. Mater. (Engl. Transl.)* **13**, 114.
King, T. B. (1951). *J. Soc. Glass Technol.* **35**, 241.
Kozakevitch, P. (1949). *Rev. Metall. (Paris)* **46**, 572.
Morey, G. W. (1954). "The Properties of Glass." Reinhold, New York.
Mukai, K., and Ishikawa, I. (1981). *Nippon Kinzoku Gakkaishi* **45**, 147.
Ono, K., Gunji, K., and Araki, T. (1967). *Nippon Kinzoku Gakkaishi* **31**, 102.
Parikh, N. M. (1958). *J. Am. Ceram. Soc.* **41**, 18.
Passerone, A., Sangiorgi, R., and Valbusa, G. (1979). *Ceramurgia Int.* **5**, 18.
Perminov, A. A., Mirova, T. V., and Popel, S. I. (1971). *J. Appl. Chem. USSR (Engl. Transl.)* **44**, 1041.
Popel, S. I., and Esin, O. A. (1957a). *Zh. Neorg. Khim.* **2**, 632.
Popel, S. I., and Esin, O. A. (1957b). *Fiz.-Khim. Osn. Proizvod. Stav., Akad. Nauk SSSR, Inst. Met. im. A. A. Baikova Tr. 3-ei Konf. 1955* p. 497.
Popel, S. I., Sokolov, V. I., and Esin, O. A. (1969). *Russ. J. Phys. Chem. (Engl. Transl.)* **43**, 1782.
Prabhu, A., Fuller, G. L., Reed, R. L., and Vest, R. W. (1975). *J. Am. Ceram. Soc.* **58**, 144.
Sharma, S. K., and Philbrook, W. O. (1970). *Scr. Metall.* **4**, 107.
Shartsis, L., and Smock, A. W. (1947). *J. Am. Ceram. Soc.* **30**, 130.
Shartsis, L., and Spinner, S. (1951). *J. Res. Natl. Bur. Stand. (U.S.)* **46**, 385.
Shartsis, L., Spinner, S., and Smock, A. W. (1948). *J. Am. Ceram. Soc.* **31**, 23.
Shermer, H. F. (1956). *J. Res. Natl. Bur. Stand. (U.S.)* **56**, 155.

Sokolov, L. N., Baidov, V. V., and Kunin, L. L. (1969). *In* "Physical Chemistry and Electrochemistry of Fused Salts and Slags", Vol. IV, All Union Conference, Part 1, Thermodynamics and the Structure of Fused Salts," p. 299. Kiev.

Sugai, M., and Somiya, S. (1982). *Yogyo Kyokaishi* **90,** 56.

Toropov, N. A., and Bryantsev, B. A. (1965). *In* "Structural Transformations at High Temperatures," p. 205. Moscow.

TABLE 5.1

Surface Tension of Lithium Silicate Glass Melts[a]

Composition (mole % Li_2O)	γ (mJ m^{-2}) at temperature given (°C)					$d\gamma/dt$ at 1300°C (mJ m^{-2} °C^{-1})
	1000	1100	1200	1300	1400	
21.5					312.1	
22.9				310.8	316.3	
28.6		311.9	313.5	315.1	316.5	+0.015
30.3	312.8	314.7	316.2	316.7	316.8	+0.003
32.6	314.8	316.9	317.8	318.4	318.8	+0.005
33.4		319.4	320.0	320.2	320.1	0
35.9		323.3	323.9	323.5	322.7	−0.006
38.7		328.3	329.1	328.4	327.9	−0.006
41.3		334.1	333.1	332.1	331.1	−0.010
43.6		338.4	336.3	334.2	332.2	−0.020
45.9			343.4	340.8	337.1	−0.032
50.3			355.5	352.3	349.1	−0.032
60.4			373.9	369.1	364.4	−0.048
64.0			379.1	374.9	370.0	−0.046
66.2					373.6	−0.08

[a] Data from Shartsis and Spinner (1951).

TABLE 5.2

Surface Tension of Sodium Silicate Glass Melts[a]

Composition (mole % Na_2O)	γ (mJ m^{-2}) at temperature given (°C)						$d\gamma/dt$ (mJ m^{-2} °C^{-1})
	900	1000	1100	1200	1300	1400	
19.5	—	276.6	276.2	275.6	274.6	273.1	−0.008
30.1	286.0	284.3	282.1	279.7	277.0	273.7	−0.026
32.9	289.0	286.4	283.8	280.9	277.9	274.4	−0.030
36.2	—	288.5	285.6	282.8	279.9	275.6	−0.035
49.2	—	—	300.0	294.6	289.0	284.3	−0.052

[a] Data from Shartsis and Spinner (1951) and Morey (1954).

TABLE 5.3

Surface Tension of Potassium Silicate Glass Melts[a]

Composition (mole % K_2O)	γ (mJ m^{-2}) at temperatures given (°C)					$d\gamma/dt$ at 1300°C (mJ m^{-2} °C^{-1})
	1000	1100	1200	1300	1400	
16.7	—	224.8	222.4	221.0	218.9	−0.018
18.8	227.2	224.8	223.4	220.3	218.5	−0.024
21.4	226.6	222.3	220.2	218.8	216.4	−0.019
23.8	226.9	222.0	219.2	216.3	213.6	−0.028
26.9	225.2	221.2	216.9	213.0	210.2	−0.034
28.7	225.4	221.3	217.6	213.0	210.0	−0.038
31.2	225.5	221.2	217.2	213.0	208.9	−0.045
33.0	223.0	220.3	215.4	210.8	206.2	−0.046

[a] Data from Shartsis and Spinner (1951) and Morey (1954).

TABLE 5.4

Surface Tension of Rubidium Silicate Glass Melts[a]

Composition (mole % Rb_2O)	γ (mJ m^{-2}) at temperature given (°C)			$d\gamma/dt$ (mJ m^{-2} °C^{-1})
	1200	1300	1400	
17	—	200.1	197.1	−0.03
20	—	192.7	188.8	−0.039
24	—	188.0	183.5	−0.045
32	175.1	173.4	170.9	−0.025
40	155.0	146.3	—	−0.087

[a] Data from Appen and Kayalova (1962).

TABLE 5.5

Temperature Dependence of Surface Tension of Rb_2O–SiO_2 System[a]

Composition (mole % Rb_2O)	Temperature range (°C)	γ (mJ m^{-2})
25	978–1396	$\gamma = 326.36 - 0.09844T$ (°C)

[a] Data from Frischat and Beier (1979).

TABLE 5.6

Surface Tension of Cesium Silicate Glass Melts[a]

Composition (mole % Cs_2O)	γ (mJ m^{-2}) at temperature given (°C)		$d\gamma/dt$ (mJ m^{-2} °C^{-1})
	1300	1400	
16	166.1	163.8	−0.023
20	165.1	162.5	−0.026
32	144.3	—	
44	120.5	—	

[a] Data from Appen and Kayalova (1962).

TABLE 5.7

Surface Tension of Magnesium Silicate Glass Melts[a]

Composition (mole % MgO)	γ at 1570°C (mJ m^{-2})	$d\gamma/dt$ (1540–1620°C) (mJ m^{-2} °C^{-1})
45.6	365	0.110
47.7	371	0.098
50.9	378	0.090

[a] Data from King (1951).

TABLE 5.8

Surface Tension of Calcium Silicate Glass Melts[a]

Composition (mole % CaO)	γ (mJ m^{-2}) at temperature given (°C)		
	1500	1550	1600
36.5	403.1	404.1	407.8
41.6	413.2	413.2	412.9
42.3	430.1	433.7	431.5
43.7	436.3	435.8	436.8
46.2	444 ± 0.4[b]	442.6 ± 1[c]	446.5 ± 2
49.3	—	452 ± 2.5[c]	453 ± 2
52.5	462.8[d]	460.5[e]	461.9
55.1	491.8	490.3	491.8

[a] Data from Mukai and Ishikawa (1981).
[b] 1520°C.
[c] 1610°C.
[d] 1540°C.
[e] 1570°C.

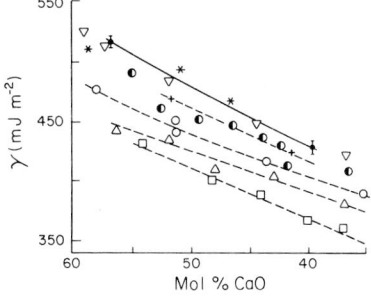

Fig. 5.1. Comparison of surface tension data of CaO–SiO$_2$ glass melts at 1600°C reported by various workers; ⊥, Ejima and Shimoji (1970); ▽, Sharma and Philbrook (1970); +, Popel and Esin (1957a,b); ○, Cooper and Kitchener (1959); △, Ono et al. (1967); □, King (1951); *, Popel' et al. (1969) at 1550°C; ◑, Mukai and Ishikawa (1981).

TABLE 5.9

Surface Tension of Lead Silicate Glass Melts[a]

Composition	γ (mJ m^{-2}) at temperature given (°C)							
(mole % PbO)	700	800	900	1000	1100	1200	1300	1400
33.4				233.0	233.7	234.0	235.0	235.7
35.6				233.7	235.0	235.6	234.5	235.1
38.4				230.0	230.8	230.8	231.7	232.2
38.5				231.3	232.0	232.7	230.5	229.1
40.6				227.8	228.4	227.5	228.5	
42.0			224.5	225.6	225.7	226.5	227.2	
45.6			217.1	219.1	221.0	222.6	223.4	
56.0	193.2	196.8	199.1	201.6	204.4	208.2	209.4	
57.2				199.4	202.2	205.7	210.2	
59.9		187.3	192.4	194.6	196.1			
64.0			183.7	186.5	189.6	193.9	197.3	
72.3			173.6	176.5	179.2	182.0	183.3	
83.2			147.8	157.7	152.1	158.0	164.1	162.7
89.1			134.4	142.0	145.4	154.2	158.0	161.3

[a] Data from Shartsis *et al.* (1948).

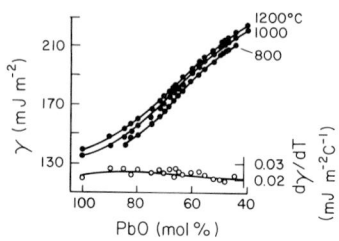

Fig. 5.2. Surface tension and its temperature coefficient for lead silicate glass melts. From Ito *et al.* (1961).

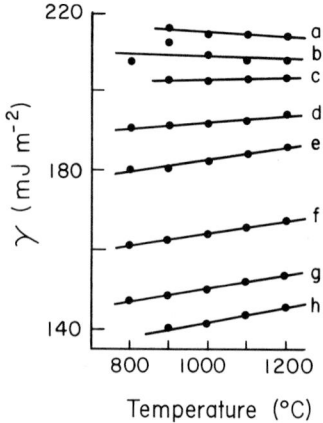

Fig. 5.3. Surface tension of lead silicate glass melts of different compositions. Mole % PbO: (a) 47.7, (b) 51.6, (c) 55.6, (d) 56.1, (e) 59.7, (f) 67.4, (g) 75.9, (h) 85.7. From Hino *et al.* (1967).

TABLE 5.10

Surface Tension of Manganese Silicate Glass Melts[a]

Composition (mole % MnO)	γ at 1570°C (mJ m^{-2})	Temperature coefficient	
		Temperature range (°C)	$d\gamma/dt$ (mJ m^{-2} °C^{-1})
47.8	409	1490–1610	0.086
50.1	415	1450–1570	0.086
53.9	432	1420–1570	0.064
58.0	460	1300–1580	0.047
66.4	492	1400–1600	0.015

[a] Data from King (1951).

TABLE 5.11

Surface Tension of MnO–SiO$_2$ System[a]

Composition (mole %)			γ (mJ m^{-2}) at temperature given (°C)			$d\gamma/dt$ (mJ m^{-2} °C^{-1})
MnO	FeO	SiO$_2$	1400	1510	1630	
52.7	1.5	42.8	420	420	430	0.04
60.6	3.2	36.2	440	460	470	0.13
67.5	3.1	32.4	460	480	480	0.09
69.7	3.8	26.5	490	500	490	—

[a] Data from Popel et al. (1969).

TABLE 5.12

Surface Tension of FeO–SiO$_2$ Glass Melts[a]

Composition (mole % FeO)	γ at 1300°C (mJ m^{-2})	Temperature Coefficient	
		Temperature range (°C)	$d\gamma/dt$ (mJ m^{-2} °C^{-1})
60.1	414	1300–1475	0.034
71.5	450	1300–1475	−0.006
77.0	474	1300–1475	−0.012

[a] Data from King (1951).

TABLE 5.13

Surface Tension of Al$_2$O$_3$–SiO$_2$ Glass Melts[a]

Composition (wt. % Al$_2$O$_3$)	Temperature range (K)	γ (mJ m^{-2}) (T in K)
20	1970–2570	405
30	2070–2570	$410 + 0.02(T - 2040)$
50	2100–2570	$415 + 0.02(T - 2070)$
72	2210–2570	$438 + 0.07(T - 2180)$
80	2250–2570	$475 + 0.09(T - 2220)$
100	2325–2625	$570 - 0.187(T - 2325)$

[a] Data from Anisimov et al. (1977).

TABLE 5.14

Surface Tension of Bismuth Silicate Glass Melts[a]

Composition (mole % Bi_2O_3[b])	γ[b] (mJ m^{-2}) at temperature given (°C)					
	900	920	940	1030	1040	1060
36				126	123	
38				132	127.5	121
40				137	132.5	126
42				141	137	130
44				142.5	138	132
50				144	140	136
80	190	185	175			
84	198	192	184			
86	202	196	187.5			
88	203	197	191			
90	205	200	194			
98	209	206	203			

[a] Data from Kargin et al. (1977).
[b] Values read from graph.

TABLE 5.15

Surface Tension of Some M_2O–M'_2O–SiO_2 Glass Melts[a]

Composition (mole %)			Temperature (°C)	γ (mJ m^{-2})
M_2O	M'_2O	SiO_2		
Li_2O	Na_2O			
20	16	64	1300	320
Na_2O	K_2O			
16	20	64	1300	236

[a] Data from Appen et al. (1952).

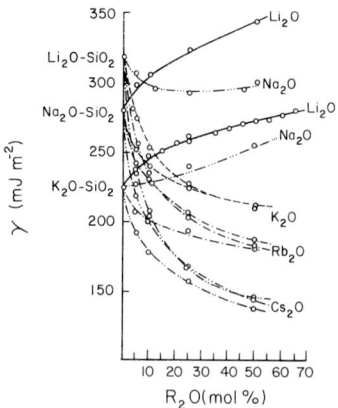

Fig. 5.4. Change in surface tension from the additions of R_2O (R = Li, Na, K, Rb, Cs) to the silicate glass melts of composition (mole %) $16.7R'_2O$–$83.3SiO_2$ (R' = Li, Na, K). Temperature 1300°C. From Appen and Kayalova (1962).

TABLE 5.16

Surface Tension of $Na_2O-Rb_2O-SiO_2$ Glass Melts[a]

Composition (mole %)			Temperature range (°C)	γ (mJ m^{-2}) = $a - bt$ (°C)	
Na_2O	Rb_2O	SiO_2		a	b
25.0	—	75	1020–1412	298.5	0.018
22.5	2.5	75	978–1396	322.36	0.04276
20.0	5.0	75	1038–1412	371.80	0.08965
17.5	7.5	75	1020–1396	363.35	0.09170
15.0	10.0	75	894–1396	394.97	0.12386
12.5	12.5	75	1020–1396	330.02	0.07590
7.5	17.5	75	1020–1396	338.25	0.08838
5.0	20.0	75	1078–1412	325.42	0.08455
2.5	22.5	75	996–1412	361.23	0.11475
—	25.0	75	978–1396	326.36	0.09844

[a] Data from Frischat and Beier (1979).

TABLE 5.17

Surface Tension of $Li_2O-MO-SiO_2$ Glass Melts[a]

Composition (mole %)			
Li_2O	MO	SiO_2	γ (mJ m^{-2}) at 1300°C
	M = Be		
17.4	17.4	65.2	337
	M = Mg		
17.4	17.4	65.2	359
	M = Ca		
17.4	17.4	65.2	357
	M = Sr		
17.4	17.4	65.2	360
	M = Ba		
5.9	19.8	74.3	345
11.2	18.7	70.1	345
17.4	17.4	65.2	353
22.1	16.4	61.5	350
	M = Zn		
17.4	17.4	65.2	342
	M = Cd		
17.4	17.4	65.2	340
	M = Pb		
16	20	64	258

[a] Data from Appen et al. (1952, 1953) and Appen and Kayalova (1963).

TABLE 5.18

Surface Tension of Some $Na_2O-MO-SiO_2$ Glass Melts at 1300°C[a]

Composition (mole %)			γ (mJ m^{-2})	Composition (mole %)			γ (mJ m^{-2})
Na_2O	MO	SiO_2		Na_2O	MO	SiO_2	
M = Be				**M = Sr**			
16	20	64	312	16	20	64	334
17.4	17.4	65.2	276	17.4	17.4	65.2	323
17.9	25	57.1	313	17.9	25	57.1	341
20	13.3	66.7	304				
				M = Zn			
M = Mg				16	20	64	327
16	20	64	348	17.4	17.4	65.2	322
17.4	17.4	65.2	334				
17.9	25	57.1	353	**M = Cd**			
20	13.3	66.7	322	16	20	64	318
38.3	13.3	48.4	320	17.4	17.4	65.2	306
40.4	8.8	50.8	315				
41.9	5.2	52.9	306	**M = Mn**			
				38.1	14.0	47.9	345
M = Ca				40.2	9.2	50.6	341
16	20	64	335	42.0	5.0	53.0	328
17.4	17.4	65.2	324				
17.9	25	57.1	338	**M = Co**			
38.0	14.2	47.8	330	32.68	1.96	65.36	296[b]
40.0	9.6	50.4	319	32.68	1.96	65.36	288[c]
41.9	5.1	53.0	310				

[a] Data from Appen *et al.* (1952, 1953), Appen and Kayalova (1963), Adachi *et al.* (1960), and Appen (1936a,b).
[b] At 1070°C.
[c] At 1200°C.

TABLE 5.19

Effect of Various Added Oxides on the Surface Tension of the Base Glass of Molecular Composition $1.4Na_2O-0.9CaO-6SiO_2$[a]

Composition (mole %)				γ (mJ m^{-2}) at temperature (°C)	
Na$_2$O	CaO	SiO$_2$	Added substance	1200	1350
17.4	10.1	72.5	—	304	302
17.2	9.9	71.6	1.3Li$_2$O	306	305
16.9	9.8	70.6	2.7Na$_2$O	300	298
16.7	9.6	69.6	4.1K$_2$O	292	286
16.8	9.7	70.0	3.5MgO	316	314
16.6	9.6	69.0	4.8CaO	312	310
15.3	8.9	63.7	12.1BaO	312	311
16.2	9.4	67.6	6.8ZnO	315	312
16.9	9.8	70.6	2.7B$_2$O$_3$	298	295
16.9	9.7	70.3	3.1B$_2$O$_3$	294	290
16.6	9.6	69.4	4.4Al$_2$O$_3$	315	313
16.5	9.6	68.8	5.1SiO$_2$	307	303
14.5	8.4	60.4	16.7PbO	273	276
16.2	9.4	67.7	6.7Fe$_2$O$_3$	314	311
16.3	9.4	68.0	6.3CoO	314	313
16.3	9.4	68.0	6.3NiO	313	311
16.3	9.4	67.7	6.6Mn$_2$O$_3$	312	310
15.8	9.1	66.0	9.1Mn$_2$O$_3$	320	319
16.2	9.4	67.7	6.7TiO$_2$	301	300
16.1	9.3	67.1	7.5V$_2$O$_5$	235	236
15.7	9.0	65.3	10.0ZrO$_2$	320	314
15.1	8.7	62.8	13.4CeO$_2$	318	315
16.3	9.4	67.8	6.5CaF$_2$	308	304

[a] Data from Badger *et al.* (1937).

TABLE 5.20

Surface Tension of $Na_2O-BaO-SiO_2$ Glass Melts[a]

Composition (mole %)			Temperature (°C)	γ (mJ m^{-2})
Na$_2$O	BaO	SiO$_2$		
5.9	19.8	74.3	1300	333
10	10	80		309
11.2	18.7	70.1		322
12.5	12.5	75		310
15	15	70		312
16	20	64		329
16.4	22.1	61.5		329
17.4	17.4	65.2		320
18.7	11.2	70.1		307
19.8	5.9	74.3		296
20	20	60		321
22.1	16.4	61.5		306
32.68	1.96	65.36	1070	295
32.68	1.96	65.36	1200	290

[a] Data from Appen *et al.* (1952, 1953), Appen (1963a,b), and Appen and Kayalova (1963).

TABLE 5.21

Surface Tension of Na$_2$O–PbO–SiO$_2$ Glass Melts[a]

Composition (mole %)			γ (mJ m^{-2}) at temperature given (°C)	
Na$_2$O	PbO	SiO$_2$	1000	1200
12.8	44.3	42.9	186.0	195.7
14.8	35.5	49.7	201.4	209.6
16.4	28.4	55.2	212.7	216.4
17.7	22.6	59.7	220.7	222.7
18.7	17.1	64.2	227.3	230.1
20.1	12.3	67.6	236.9	241.2
20.8	9.2	70.0	243.3	248.0
21.7	5.2	73.1	254.1	257.0
22.2	2.9	74.9	263.7[b]	264.4
22.6	1.4	76.0	270.3[b]	269.9
19.7	41.0	39.3	186.8	195.8
23.1	30.6	46.3	202.0	212.2
26.3	21.0	52.7	223.6	229.7
30.2	9.3	60.5	248.6	253.2
31.6	5.3	63.1	262.8	264.8
32.2	3.3	64.5	268.3	269.3
32.7	2.0	65.3	276.3[b]	275.4

[a] Data from Appen (1936a,b).
[b] At 1070°C.

TABLE 5.22

Surface Tension for Some K$_2$O–MO–SiO$_2$ Glass Melts[a]

Composition (mole %)				γ
K$_2$O	MO	SiO$_2$	Temperature (°C)	(mJ m^{-2})
	M = Be			
17.4	17.4	65.2	1300	262
	M = Mg			
17.4	17.4	65.2	1300	290
	M = Ca			
17.4	17.4	65.2	1300	262
	M = Sr			
17.4	17.4	65.2	1300	259
	M = Ba			
5.9	19.8	74.3	1300	297
11.2	18.7	70.1		267
17.4	17.4	65.2		261
22.1	16.4	61.5		235
	M = Zn			
17.4	17.4	65.2	1300	272
	M = Cd			
17.4	17.4	65.2	1300	247

[a] Data from Appen et al. (1952, 1953) and Appen and Kayalova (1963).

TABLE 5.23

Surface Tension of $K_2O-PbO-SiO_2$ Glass Melts[a]

Composition (wt. %)			γ (mJ m^{-2}) at temperature given (°C)			
K_2O	PbO	SiO_2[b]	1000	1100	1200	1300
2.3	66.2	31.5	219.0	219.2	218.8	218.7
3.2	62.4	34.4	218.4	217.4	217.8	218.4
5.5	55.4	39.1	214.3	214.1	215.0	216.6
6.0	54.4	39.6	213.6	213.7	214.6	216.2

[a] Data from Shartsis and Smock (1947).
[b] Includes 0.3% As_2O_3.

TABLE 5.24

Surface Tension of Lithium Borosilicate Glass Melts[a]

Composition (mole %)			Temperature (°C)	γ[b] (mJ m^{-2})
Li_2O	B_2O_3[b]	SiO_2		
16	—	84	1300	325
16	8	76		306
16	16	68		225
16	20	64		233
16	24	60		210
16	32	52		201

[a] Data from Appen *et al.* (1952, 1953) and Appen and Kayalova (1963).
[b] Values read from graph.

TABLE 5.25

Surface Tension of Sodium Borosilicate Glass Melts at 1300°C[a]

Composition (mole %)			γ[b] (mJ m^{-2})
Na_2O	B_2O_3[b]	SiO_2	
16	—	84	288
16	4	80	284
16	8	76	287
16	12	72	274
16	18	66	231
16	20	64	226
16	24	60	221
16	32	52	207

[a] Data from Appen *et al.* (1952, 1953) and Appen and Kayalova (1963).
[b] Values read from graph.

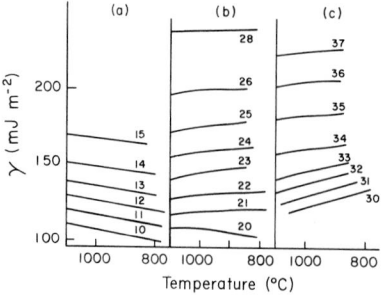

Fig. 5.5. Temperature dependence of surface tension of sodium borosilicate glass melts of various compositions containing Na_2O mole %: (a) 10, (b) 20, (c) 30. The number on each curve is mole % of SiO_2. From Imaoka et al. (1971).

TABLE 5.26

Surface Tension of Some $Na_2O-M_2O_3-SiO_2$ Glass Melts

Composition (mole %)			Temperature (°C)	γ (mJ m^{-2})	Reference[a]
Na_2O	M_2O_3	SiO_2			
	M = Al				
39.9	9.6	50.5	1300	324	1
42.0	4.9	53.1	1300	312	
	M = Cr				
32.68	1.96	65.36	1070	181	2
			1200	188	
	M = Mn				
32.68	1.96	65.36	1070	294	2
			1200	287	
	M = Fe				
32.68	1.96	65.36	1070	295	2
			1200	289	
	M = Ce				
32.68	1.96	65.36	1070	278	2
			1200	283	
	M = As				
32.68	1.96	65.36	1200	253	2

[a] Data from (1) Adachi et al. (1960), or (2) Appen (1936a).

TABLE 5.27

Surface Tension of $La_2O_3-Na_2O-2SiO_2$ Glass Melts at Various Temperatures[a]

Mole % La_2O_3	γ^b (mJ m^{-2}) at temperature given (°C)			
	900	1100	1300	1500
—	294	289	283	276
1	310	306	304	300
5	342	340	338	334
10	367	355	342	329

[a] Data from Passerone et al. (1979).
[b] Values read from graph.

TABLE 5.28

**Surface Tension of Potassium Borosilicate
Glass Melts at 1300°C[a]**

Composition (mole %)			γ^b
K_2O	$B_2O_3{}^b$	SiO_2	$(mJ\ m^{-2})$
16	—	84	228
16	4	80	229
16	8	76	231
16	12	72	232
16	16	68	231
16	20	64	230
16	24	60	223
16	32	52	182

[a] Data from Appen *et al.* (1952, 1953) and
Appen and Kayalova (1963).
[b] Values read from graph.

TABLE 5.29

Surface Tension of Some $M_2O-TiO_2-SiO_2$ Glass Melts[a]

Composition (mole %)				γ
M_2O	TiO_2	SiO_2	Temperature (°C)	$(mJ\ m^{-2})$
M = Li				
16	20	64	1300	309
M = Na				
16	20	64	1300	289
M = K				
16	20	64	1300	235

[a] Data from Appen *et al.* (1952).

TABLE 5.30

Surface Tension of $Na_2O-P_2O_5-SiO_2$ Glass Melt[a]

Composition (mole %)			$\gamma\ (mJ\ m^{-2})$ at given temperature (°C)	
Na_2O	P_2O_5	SiO_2	1070	1200
32.68	1.96	65.36	283	280

[a] Data from Appen (1936a) and Appen and Kayalova
(1963).

TABLE 5.31

Surface Tension of Na$_2$O–MoO$_3$–SiO$_2$ Glass Melts[a]

Composition (mole %)			γ at 900°C (mJ m^{-2})	$d\gamma/dt$ in range 900–1200°C (mJ m^{-2} °C^{-1})
Na$_2$O	MoO$_3$	SiO$_2$		
31	5	64	197	−0.063
32.5	3.5	64	200	−0.070
34	2	64	230	−0.083
35	1	64	256	−0.070
35.4	0.6	64	271	−0.050
36	—	64	292	−0.030
36	0.6	63.4	274	−0.046
36	1	63	261	−0.050
36	2	62	235	−0.083
36	3.5	60.5	201	−0.074
36	5	59	198	−0.070

[a] Data from Perminov *et al.* (1971).

TABLE 5.32

Surface Tension of Some Na$_2$O–RO$_3$–SiO$_2$ Glass Melts[a]

Composition (mole %)			Temperature (°C)	γ (mJ m^{-2})
Na$_2$O	RO$_3$	SiO$_2$		
	R = W			
32.68	1.96	65.36	1070	249
			1200	246
	R = U			
32.68	1.96	65.36	1070	291
			1200	285

[a] Data from Appen (1936a) and Appen and Kayalova (1963).

TABLE 5.33

Surface Tension of MgO–CaO–SiO$_2$ Glass Melts[a]

Composition (%)			Temperature (°C)	γ (mJ m^{-2})
MgO	CaO	SiO$_2$		
22.3	37.8	39.9	1498	346
26.45	32.2	41.35	1502	348

[a] Data from Ermolaeva (1955).

TABLE 5.34

Surface Tension of FeO–MgO–SiO₂ Glass Melts[a]

Composition (wt. %)			γ (mJ m^{-2}) at temperature given (°C)				
FeO	MgO	SiO₂	1350	1400	1450	1500	1550
20	25	55	391	397	405	414	420
25	20	55	380	391	398	407	413
30	15	55	373	384	393	400	409
35	10	55	367	378	389	395	409
40	5	55	350	363	380	389	399
25	25	50	404	409	415	425	430
30	20	50	391	400	406	410	421
35	15	50	387	390	401	407	411
40	10	50	382	386	392	403	410
30	25	45	412	419	421	432	—
35	20	45	400	409	415	424	426
40	15	45	388	396	408	409	420
35	25	40	424	429	434	441	446
40	20	40	403	415	423	427	429
40	25	35	428	434	440	442	450

[a] Data from Toropov and Bryantsev (1965).

TABLE 5.35

Surface Tension of FeO–MnO–SiO₂ Glass Melts at 1400°C[a]

Composition (wt. %)			γ
FeO	MnO	SiO₂	(mJ m^{-2})
11.2	51.4	37.4	435
12.0	52.0	36.0	444
12.5	45.0	42.5	400
12.8	58.2	29.0	476
13.2	50.0	36.8	460
14.5	65.0	20.5	500
15.0	51.1	33.9	474
17.6	54.6	27.8	475
21.3	58.9	19.8	517
21.5	41.7	36.8	400
23.7	30.9	45.4	403

[a] Data from Popel and Esin (1957a,b).

TABLE 5.36

Surface Tension of MgO–Al$_2$O$_3$–SiO$_2$ Glass Melts^a

Composition (%)			Temperature	γ
MgO	Al$_2$O$_3$	SiO$_2$	(°C)	(mJ m^{-2})
8.3	17.7	74	1475	240
10.0	23.5	66.5	1425	320
14.0	44.0	42.0	1600	362
15.2	42.0	42.8	1575	362
16.1	34.8	49.1	1460	358
25.7	22.8	51.5	1370	349
28.8	22.7	48.5	1450	364

^{a} Data from Ermolaeva (1955).

TABLE 5.37

Surface Tension and Its Temperature Coefficient for Calcium Aluminosilicate Glass Melts

Composition (wt. %)			Temperature	γ	$\gamma = a + bT$ (°C)		$d\gamma/dT$
CaO	Al$_2$O$_3$	SiO$_2$	(°C)	(mJ m^{-2})	a	b	(mJ m^{-2} °C^{-1})
15	20	65	1512	441.7	184.3	0.171	0.19
			1551	449.3			
			1604	458.5			
			1649	464.9			
23	15	62	1325	417.2	217.1	0.152	0.16
			1427	432.9			
			1477	437.2			
			1540	451.0			
			1608	461.7			
25	10	65	1439	415.1	59.0	0.248	0.16
			1495	426.6			
			1559	446.3			
			1623	456.3			
25	20	55	1445	445.0	308.6	0.096	0.15
			1516	454.0			
			1577	459.4			
			1674	467.2			
29	40	31	1502	468.1	316.7	0.100	0.09
			1574	475.1			
			1628	480.9			
30˙	10	60	1358	433.6	266.0	0.123	0.14
			1460	446.6			
			1515	453.0			
			1517	455.6			
			1596	460.9			

TABLE 5.37 (*Continued*)

Composition (wt. %)			Temperature (°C)	γ (mJ m^{-2})	$\gamma = a+bT$ (°C)		$d\gamma/dT$ (mJ m^{-2} °C^{-1})
CaO	Al$_2$O$_3$	SiO$_2$			a	b	
30	25	45	1491	454.4	330.1	0.085	0.10
			1518	459.6			
			1562	461.2			
			1640	469.0			
34	30	36	1450	458.8	329.6	0.089	0.08
			1538	465.7			
			1582	470.9			
			1625	473.1			
35	10	55	1457	452.2	338.0	0.078	0.11
			1515	456.1			
			1578	460.2			
			1632	466.7			
39	19	42	1431	453.6	352.6	0.0715	0.09
			1433	454.1			
			1477	460.2			
			1482	460.9			
			1537	463.7			
			1609	467.5			

[a] Data from Barrett and Thomas (1959).

TABLE 5.38
Surface Tension of Calcium Aluminosilicate Glass Melts[a]

Composition (mole %)			γ (mJ m^{-2}) at temperature (°C)				Composition (mole %)			γ (mJ m^{-2}) at temperature (°C)			
CaO	Al$_2$O$_3$	SiO$_2$	1450	1500	1550	1600	CaO	Al$_2$O$_3$	SiO$_2$	1450	1500	1550	1600
17.0	11.3	71.7	387.9	391.9	394.4	398.6	39.2	15.1	45.7	462.0	460.9	466.6	467.3
22.6	8.0	69.4	392.5	398.2	400.9	404.4	51.8	3.1	45.1	—	480.7	480.9	480.7
20.8	15.3	63.9	—	400.3	403.1	405.2	39.7	15.5	44.8	470.9	473.5	474.3	473.9
29.6	7.0	63.4	402.3	407.8	410.6	411.2	35.6	20.0	44.4	450.6	452.8	452.8	455.3
22.9	13.8	63.3	400.7	398.5	400.9	403.4	46.9	10.0	43.1	486.5	484.1	484.9	487.8
32.2	10.8	57.0	423.1	424.3	423.9	428.0	30.0	29.0	41.0	—	—	459.6	460.3
17.5	27.1	55.4	—	—	—	411.9	43.6	16.9	39.5	490.6	487.7	488.9	487.1
19.0	27.0	54.0	—	—	—	415.6	41.5	20.9	37.6	470.7	470.7	471.9	469.6
30.4	16.3	53.3	—	411.7	411.8	416.9	46.4	26.5	27.1	—	—	—	522.5
35.9	12.9	51.2	430.3	431.7	434.7	432.5	49.7	27.4	22.9	—	—	—	536.2
44.7	4.3	51.0	—	442.5	443.7	448.8	55.4	29.5	15.1	—	—	558.7	558.2
44.4	8.8	46.8	458.1	458.5	459.5	457.6	58.8	31.9	9.3	590.5	590.3	587.5	585.2
27.0	27.0	46.0	—	—	434.5	436.1							

[a] Data from Mukai and Ishikawa (1981).

TABLE 5.39

Surface Tension of Barium Borosilicate Glass Melts[a]

Composition[b] (wt. %)			γ (mJ m^{-2})[c] at temperature (°C)				
BaO	B$_2$O$_3$	SiO$_2$	1300	1200	1100	1000	900
2.1	92.9	5.0	99.3	93.3	—	87.1	84.9
4.9	84.1	11.0	107.1	103.3	97.9	92.8	93.9
10.4[d]	77.2	12.4	113.8	108.2	103.8	100.5	97.0
10.4[d]	65.0	24.6	133.5	128.7	121.5	117.9	116.0
10.6[d]	52.9	36.5	152.5	148.6	145.7	143.8	—
22.8[d]	65.0	12.2	133.4	126.7	118.8	114.8	114.0
23.5[d]	52.9	23.6	146.5	140.4	141.0	140.2	137.3
22.0[d]	41.4	36.6	174.1	175.2	—	—	—
29.8	67.2	3.0	120.7	119.6	113.8	110.9	110.1
29.6	61.2	9.2	132.8	128.9	123.5	118.2	124.9
30.7	57.7	11.6	142.0	138.9	135.7	132.6	135.4
31.2	63.2	5.6	132.4	128.9	126.5	125.5	124.9
33.8	20.4	45.8	241.9	—	—	—	—
35.5[d]	40.2	24.3	185.2	182.9	178.1	—	—
36.9	57.4	5.7	153.2	155.2	155.2	158.8	167.2
37.0	45.9	17.1	178.7	179.8	181.3	182.6	187.0
37.3	51.6	11.1	164.0	165.6	167.0	170.2	179.8
37.4	39.2	23.4	193.4	194.2	193.5	193.7	—
42.1	55.3	2.6	172.4	175.6	178.7	184.0	—
42.2	52.5	5.3	178.5	181.7	185.9	191.1	197.6
42.3	49.6	8.1	182.5	186.2	187.5	193.8	200.9
42.3	46.8	10.9	188.8	193.7	197.7	202.5	212.0
42.7	43.6	13.7	196.1	199.6	202.7	207.7	216.4
42.6	43.6	13.8	196.6	199.0	202.2	206.1	216.8
42.7	40.7	16.6	203.6	206.6	209.7	214.3	221.3
42.8	37.8	19.4	211.6	212.5	215.5	219.8	226.2
43.0	34.3	22.7	219.2	220.5	221.8	225.2	231.2
43.0	31.9	25.1	224.0	225.7	226.0	229.8	235.1
43.2	28.9	27.9	233.7	235.6	236.6	241.5	—
43.1	28.9	28.0	232.9	234.3	235.1	238.5	—
43.2	26.0	30.8	241.4	241.6	241.1	—	—
47.6	47.2	5.2	202.7	205.7	210.2	214.4	221.1
47.6	44.6	7.8	206.2	211.0	214.8	219.6	227.6
47.8	41.7	10.5	213.2	216.9	220.6	225.9	234.5
48.1	36.0	15.9	227.1	229.6	232.2	236.9	243.6
48.5	30.5	21.0	241.2	243.8	246.9	251.6	260.3
48.1	30.4	21.5	240.4	243.1	246.6	252.4	260.5
48.3	27.7	24.0	245.8	248.7	252.6	257.5	264.6
48.4	27.4	24.2	246.4	249.8	252.7	257.5	264.2
48.5	24.7	26.8	254.1	257.1	255.8	262.8	272.8
48.7	21.6	29.7	261.9	264.1	265.3	269.9	274.9
48.9	18.7	32.4	270.0	272.0	273.4	277.3	282.6

(*continues*)

TABLE 5.39 (*Continued*)

Barium Borosilicate Glass Melts

Composition[b] (wt. %)			γ (mJ m^{-2})[c] at temperature (°C)				
BaO	B$_2$O$_3$	SiO$_2$	1300	1200	1100	1000	900
49.5	12.5	38.0	291.9	291.8	291.4	294.6	—
49.3	9.7	41.0	302.1	301.9	302.0	—	—
49.7	6.2	44.1	318.4	—	—	—	—
51.4	15.3	33.3	284.0	285.5	286.2	290.5	—
51.7	46.4	1.9	212.1	215.7	219.6	225.4	229.3
51.8	44.7	3.5	215.9	219.4	223.2	228.2	233.6
50.5	43.2	5.3	219.6	223.3	227.0	231.4	238.7
52.0	42.3	5.7	220.3	224.0	227.6	232.7	239.6
52.0	39.1	8.9	226.6	230.7	234.3	239.7	247.7
52.2	37.8	10.0	228.1	231.9	235.8	241.2	248.3
52.0	36.9	11.1	230.9	234.3	238.2	242.6	249.5
52.3	36.6	11.1	228.8	232.1	236.5	240.6	247.6
52.2	36.4	11.4	231.4	235.0	238.7	243.3	246.3
52.5	35.3	12.2	229.0	231.9	235.8	240.5	245.9
52.4	33.7	13.9	235.3	240.7	244.3	249.3	256.9
52.7	31.9	15.4	242.5	246.1	250.3	254.6	260.6
52.8	30.7	16.5	243.3	248.1	251.7	256.4	264.2
53.1	26.5	20.4	255.2	258.7	262.2	266.5	271.5
53.1	25.0	21.9	257.1	260.5	263.9	269.0	274.6
52.7	24.1	23.2	261.3	264.2	267.1	272.4	280.9
53.3	22.4	24.3	265.0	267.4	270.3	275.4	279.5
53.5	19.4	27.1	272.9	275.6	278.8	282.8	294.6
53.4	18.3	28.3	276.9	279.2	282.2	284.9	289.1
53.6	16.9	29.5	270.3	273.0	275.7	279.0	282.3
53.6	15.5	30.9	280.9	283.3	285.1	287.1	297.3
53.7	12.6	33.7	292.6	—	—	—	—
54.1	8.2	37.7	313.0	—	—	—	—
54.7	24.3	21.0	265.1	265.5	268.9	273.0	278.1
55.2	25.9	18.9	258.3	261.8	265.7	270.6	276.0
56.1	41.6	2.3	227.6	231.4	236.1	242.1	247.8
56.4	38.7	4.9	232.4	236.5	241.9	246.8	253.5
56.9	35.9	7.2	238.8	243.2	247.3	252.0	257.7
56.9	33.2	9.9	241.5	246.2	251.1	256.7	264.3
57.3	33.7	9.0	244.8	249.6	254.3	259.6	268.0
56.9	30.8	12.3	248.3	253.0	258.3	263.0	269.0
57.1	30.7	12.2	248.5	252.6	257.5	261.8	267.3
57.1	28.1	14.8	254.2	258.9	263.4	268.2	275.1
57.3	28.0	14.7	254.4	259.3	264.1	269.0	275.1
57.4	25.3	17.3	261.7	266.5	270.3	274.8	282.0
57.5	22.8	19.7	268.0	271.5	274.7	279.5	286.7
58.6	19.9	21.5	277.1	281.2	284.9	288.9	297.3
57.5	20.1	22.4	272.6	277.2	280.5	285.5	294.0
58.8	17.4	23.8	287.8	290.1	—	—	—

(*continues*)

TABLE 5.39 (*Continued*)

Barium Borosilicate Glass Melts

Composition[b] (wt. %)			γ (mJ m^{-2})[c] at temperature (°C)				
BaO	B$_2$O$_3$	SiO$_2$	1300	1200	1100	1000	900
57.7	17.4	24.9	277.1	282.2	285.4	288.8	296.1
58.5	26.3	15.2	260.9	265.7	270.6	276.0	283.4
58.0	12.2	29.8	297.8	301.6	304.7	—	—
58.5	6.2	35.3	325.6	—	—	—	—
58.4	35.7	5.9	242.7	247.7	253.0	258.8	265.6
59.6	37.5	2.9	238.3	243.9	249.6	255.4	—
60.4	37.4	2.2	239.0	246.6	252.2	257.6	—
60.5	34.9	4.6	244.0	250.1	255.9	261.3	268.3
60.7	32.4	6.9	249.5	255.3	261.0	267.1	272.6
60.8	29.8	9.4	254.2	260.5	266.8	272.3	277.8
61.1	27.0	11.9	260.8	266.5	271.9	277.5	284.1
61.2	24.4	14.4	265.7	271.7	277.5	282.5	290.0
61.5	21.6	16.9	272.8	278.7	283.9	287.4	295.4
61.7	19.2	19.1	279.5	284.9	289.5	293.2	304.0
61.5	18.3	20.2	284.2	289.4	292.5	297.5	304.0
61.8	16.7	21.5	287.6	292.6	296.9	299.5	—
62.0	14.2	23.8	296.2	300.9	304.3	—	—
62.0	11.6	26.4	306.6	311.1	313.5	—	—
62.4	9.0	28.6	318.8	322.1	—	—	—
62.6	5.8	31.6	335.5	—	—	—	—
64.1	31.3	4.6	252.8	259.3	265.5	271.7	—
64.3	28.9	6.8	256.2	263.5	269.8	270.1	—
64.9	26.1	9.0	262.9	269.9	276.3	281.6	288.1
65.2	21.1	13.7	275.7	282.2	288.1	293.0	299.4
64.9	21.2	13.9	275.3	282.1	287.9	293.0	299.6
65.2	18.8	16.0	280.6	286.6	292.6	296.4	307.9
65.2	18.7	16.1	283.0	288.7	293.3	297.9	311.4
65.5	16.1	18.4	289.0	295.3	300.3	305.4	—
65.6	16.1	18.3	290.2	296.1	300.5	305.6	—
65.5	13.7	20.8	299.9	305.3	309.0	—	—
65.8	10.9	23.3	309.7	314.3	324.4	—	—
67.8	19.9	12.3	280.2	287.0	293.0	298.2	302.7
68.0	17.3	14.7	287.7	294.2	299.8	304.4	—
67.8	15.4	16.8	295.8	302.0	307.6	311.9	—
69.8	19.3	10.9	282.9	290.1	296.3	302.5	—
70.0	17.0	13.0	289.9	297.4	303.6	309.2	—
70.6	14.1	15.3	298.3	305.8	311.6	316.9	—

[a] Data from Shermer (1956).

[b] Average composition of before and after analyses except for two-liquid compositions, which are batch compositions except as noted by footnote d.

[c] Interpolated values. The temperature of any measurement may have been ±4°C from that given here.

[d] Average composition of two-liquid glasses. Analyzed values before measurement.

TABLE 5.40

Surface Tension of Lead Borosilicate Glass Melt[a,b]

Temperature (°C)	γ^c (mJ m^{-2})	Temperature (°C)	γ^c (mJ m^{-2})
635	225	768	163
652	207	776	162
657	205	810	160
668	192	824	158
673	188	868	156
695	177	888	155
733	168	913	153
749	165	920	152.5

[a] Composition by wt. % is 63PbO, 25B$_2$O$_3$, 12SiO$_2$.
[b] Data from Prabhu et al. (1975).
[c] Values from graph in Prabhu et al. (1975).

TABLE 5.41

Surface Tension of FeO–Fe$_2$O$_3$–SiO$_2$ Glass Melts[a]

Composition (mole %)			γ (mJ m^{-2}) at temperature (°C)			$d\gamma/dt$ (mJ m^{-2} °C^{-1})
FeO	Fe$_2$O$_3$	SiO$_2$	1290	1400	1540	
58.0	0.7	41.5	—	380[b]	390[c]	0.07
59.4	1.2	39.4	360	380	390	0.12
62.6	0.4	37.0	380	390	410	0.12
65.2	0.8	34.0	400	400	400	0.0
66.5	0.3	33.2	390	420	440	−0.20
71.5	0.5	28.0	470	465	425	−0.18
74.0	0.8	25.2	485	470	445	−0.16
73.9	1.1	25.0	—	475	455	−0.14
82.8	1.2	16.0	—	540	—	—
90.2	1.3	8.5	—	565	—	—
98.5	1.5	—	—	630	—	—

[a] Data from Popel et al. (1969).
[b] At 1440°C.
[c] At 1660°C.

TABLE 5.42

Surface Tension of Al$_2$O$_3$–TiO$_2$–SiO$_2$ Glass Melts at 1600°C[a]

Al$_2$O$_3$ (wt. %)[b]	γ^b (mJ m^{-2})	Al$_2$O$_3$ (wt. %)[b]	γ^b (mJ m^{-2})
0	424	12.5	380
2.5	393	14	396
5	388	15	423
10	384	20	429

[a] Data from Sugai and Somiya (1982). The TiO$_2$/SiO$_2$ ratio was maintained at the eutectic composition (wt. %) 10.5TiO$_2$–89.5SiO$_2$.
[b] Values read from graph.

Chapter 6

Coefficient of Thermal Expansion

Materials expand when they are heated. Thermal shock resistance and the fracture of seals between different materials depend on thermal expansion. The coefficient of linear thermal expansion α is defined as

$$\alpha = \Delta L/L_0 \Delta T, \tag{1}$$

where ΔL is the change of length of a material of initial length L_0 heated to a temperature difference ΔT. The units of α are reciprocal degrees. The coefficient usually depends upon temperature, as shown in Fig. 6.1, so it is necessary to specify the temperature range over which ΔL is measured. The usual range is from 0 to 300°C, or from room temperature to a temperature just below the glass transition range, where volume changes become greater.

A volume expansion coefficient can be defined as $\Delta V/V_0 \Delta T$ for a volume change ΔV and initial volume V_0. If ΔV is small, as is almost always the case, the volume coefficient equals 3α. The density ρ and specific volume V of the glass at temperatures up to the strain temperature can be calculated from the density ρ_0 at 25°C and the coefficient of thermal expansion:

$$V = \frac{1}{\rho_0}[1 + 3\alpha(T - 25)], \tag{2}$$

$$\rho = \frac{\rho_0}{1 + 3\alpha(T - 25)}, \tag{3}$$

where T is the temperature in degrees Celsius.

A thermal contraction coefficient can be defined as $\Delta L/L_0 \Delta T$ for cooling from the annealing temperature (see Chapter 9) to 25°C. Some contraction coefficients for commercial glasses are given in Table 3.5 and are always larger than α.

The expansion of glass as the temperature is increased is an important measure of its resistance to thermal shock. The higher the expansion coefficient, the more likely is fracture during rapid heating and cooling. Nonuniform volume changes, resulting from temperature gradients during heating or cooling, cause stresses that are larger the greater the volume changes. A vitreous silica beaker heated to 1000°C can be dashed into water without breaking it, because it has the low expansion coefficient of about $5 \times 10^{-7}\,°C^{-1}$. On the other hand, a plate of soda-lime glass 0.6 cm thick, with a coefficient of about $90 \times 10^{-7}\,°C^{-1}$, can be cracked with a temperature difference between faces of about 50°C. The temperature differences that lead to fracture for different commercial glasses are given in Table 3.5.

Matching of expansion coefficients of two materials being sealed together is essential to prevent fracture in the seal during heating or cooling. A difference in coefficients of less than about $2 \times 10^{-7}\,°C^{-1}$ is optimum for glass sealing, although acceptable seals can often be made with differences up to about $5 \times 10^{-7}\,°C^{-1}$. The greater the difference, the more likely is failure of the seal. Other factors that influence the strength of a seal are its geometry, the completeness of removal of residual stresses by annealing, and differences in strain temperatures in sealing two glasses. Special seal designs allow materials with large differences in coefficients of thermal expansion to be sealed together (Partridge, 1949; Varshneya, 1982).

The change in length of a glass rod can be measured with a dilatometer, in which the increase in length of a glass rod is compared to that of a vitreous

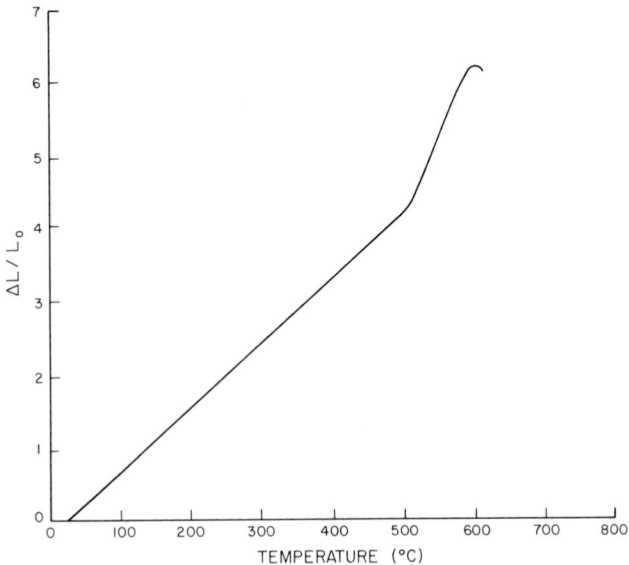

Fig. 6.1. Fractional change in length plotted against temperature for an annealed glass.

silica rod. Both rods are held in the same furnace and heated together at a constant rate, and the difference in their lengths is measured with a differential sensor.

A plot of length as a function of temperature is given in Fig. 6.1, for an annealed glass. If the glass has been rapidly cooled, it shows a different behavior in the glass transition range. The expanded structure of the rapidly cooled glass relaxes to a lower volume when the glass structure can rearrange.

TABLES AND FIGURES II

The thermal expansion coefficients for binary glasses are given in Tables 6.1–6.19. Tables 6.20–6.104 and Figs. 6.2–6.8 show the coefficients for ternary glasses.

BIBLIOGRAPHY

Aleksandrov, V. I., Borik, M. A., Dechev, G. K., Markov, N. I., Myzina, V. A., Osiko, V. V., Tatarintsev, V. M., and Khodakovskaya, R. Y. (1980). *Sov. J. Glass Phys. Chem.* (*Engl. Transl.*) **6**, 117.

Alekseeva, Z. D. (1974). *Inorg. Mater.* (*Engl. Transl.*) **10**, 772.

Alekseeva, Z. D., and Polozok, N. V. (1972). *Inorg. Mater.* (*Engl. Transl.*) **8**, 135.

Aleynikov, F. K., Vaytkus, Y. P., and Zhitkyavichyute, I. T. (1967). *Liet. Mokslu Akad. Darb., Ser. B* **2**, 75.

Amrhein, E. M. (1963). *Glastech. Ber.* **36**, 425.

Appen, A. A. (1953). *Zh. Prikl. Khim.* **24**, 1122.

Bezborodov, M. A., and Bobkova, N. M. (1957). *Dokl. Akad. Nauk BSSR* No. 1, 13.

Bezborodov, M. A., and Bobkova, N. M. (1959). *Silikattechnik* **10**, 584.

Bezborodov, M. A., and Mazurenko, V. D. (1960). *Dokl. Akad. Nauk BSSR* No. 4, 58.

Bezborodov, M. A., and Mazurenko, V. D. (1961a). *Izv. Vuzov* **4**, 261.

Bezborodov, M. A., and Mazurenkov, V. D. (1961b). *Szklo Ceram.* **12**, 161.

Bezborodov, M. A., and Mel'nik, M. T. (1959). *Dokl. Akad. Nauk SSR* No. 3, 338.

Bezborodov, M. A., and Rzhevuskaya, T. L. (1959). *Dokl. Akad. Nauk SSR* No. 13, 488.

Bezborodov, M. A., and Ulazovskii, V. A. (1957). "Issledvaniye Sistemy $Li_2O-Al_2O_3-B_2O_3-SiO_2$ Stekloobraznov Sostoyanii." Minsk.

Bezborodov, M. A., Mazo, E. E., Grishina, N. P., and Kaminskaya, V. S. (1959). *Vestsi Akad. Nauk BSSR, Ser. Fiz.-Tekh. Navuk* No. 1, p. 53.

Bogatyreva, V. V., Bogatyrev, Y. Z., and Solov'yeva, T. I. (1973). *Sov. J. Opt. Technol.* (*Engl. Transl.*) **40**, 495.

Bollin, P. L. (1972). *J. Am. Ceram. Soc.* **55**, 483.

Botvinkin, O. K., and Yaroker, K. G. (1967a). *Inorg. Mater.* (*Engl. Transl.*) **3**, 1427.

Botvinkin, O. K., and Yaroker, K. G. (1967b). *Steklo* No. 1, 1.

Brackbill, C. E., Mckinstry, H. A., and Hummel, F. A. (1951). *J. Am. Ceram. Soc.* **34**, 107.

Brekhovskikh, S. M., and Sesorova, V. N. (1960). *In* "Stekloobraznaye Sostoyaniye," p. 444. Moscow.

Bruckner, R., and Navarro, J. F. (1966). *Glastech. Ber.* **39**, 283.

Cleek, G. W. and Babcock, C. L. (1973). Properties of glasses in some ternary systems containing BaO and SiO_2. *NBS Monogr.* (*U.S.*) No. 135.

Cousen, A., and Turner, W. E. S. (1928). *J. Soc. Glass Technol.* **12**, 169.

Dale, A. E., Pegg, E. F., and Stanworth, J. E. (1951). *J. Soc. Glass Technol.* **35,** 136.
Danilova, N. P. (1967). *Steklo* **1,** 89.
Danilova, N. P., and Dubrovo, S. K. (1965). *In* "Issledovaniya v Oblasti Khimii i Okislov," p. 18. Moscow.
Danilova, N. P., and Dubrovo, S. K. (1967). *J. Appl. Chem. USSR (Engl. Transl.)* **40,** 959.
Danilova, N. P., and Dubrovo, S. K. (1971). "Steklovbraznye Silikaty Bariya," 3509-71 Dep. VINITI.
Deeg, E. (1958a). *Glastech. Ber.* **31,** 85.
Deeg, E. (1958b). *Glastech. Ber.* **31,** 124.
Dietzel, A., and Sheybany, H.-A. (1948). *Verres Refract.* **2,** 63.
Din, A. (1968). Untersuchungen zum strukturellen Aufbau der Gläser $Na_2O \cdot xMe_nO \cdot (6-x)SiO_2$ aus der Temperaturabhangigkeit der Dichte, der Viskositat und davon abgeleiteter Groben. Dissertation, Clausthal.
Dubrovo, S. K. (1965). *Inorg. Mater. (Engl. Transl.)* **1,** 893.
Dubrovo, S. K., and Kasymova, S. S. (1964). *Uzb. Khim. Zh.* **8,** 14.
Dubrovo, S. K., and Shnypikov, A. D. (1966). *Inorg. Mater. (Engl. Transl.)* **2,** 1417.
Dubrovo, S. K., and Shnypikov, A. D. (1968). *Proc. Conf. Silic. Ind., 9th, Budapest* p. 459.
Dubrovo, S. K., and Tsekhomskaya, T. S. (1964a). *J. Appl. Chem. USSR (Engl. Transl.)* **37,** 1224.
Dubrovo, S. K., and Tsekhomskaya, T. S. (1964b). *In* "Khimiya Redkikh Elementov," p. 91. Leningrad.
Dubrovo, S. K., Danilova, N. P., and Tsekhomskaya, T. S. (1965a). *In* "Issledovaniya v. Oblasti Khimii Silikatov i Okislov," p. 11. Moscow.
Dubrovo, S. K., Danilova, N. P., and Tsekhomskaya, T. S. (1965b). *Proc. Int. Congr. Glass, 7th, Versailles* **2,** 314.
Dubrovo, S. K., Zasolotskaya, M. V., and Tsekhomskaya, T. S. (1969). *J. Appl. Chem. USSR (Engl. Transl.)* **42,** 2295.
Geller, R. F., Bunting, E. N., and Creamer, A. S. (1938). *J. Res. Natl. Bur. Stand. (U.S.)* **20,** 57.
Gorashenko, N. G., Fomcenkov, L. P., Mayer, A. A., and Mishenkova, T. N. (1973). *Tr.-Mosk. Khim.-Tekhnol. Inst.* No. 76, p. 75.
Hamilton, E. H., Waxler, R. M., and Nivert, J. M. (1959). *J. Res. Natl. Bur. Stand. (U.S.)* **62,** 59.
Hasegawa, Y. (1980). *Glastech. Ber.* **53,** 277.
Hirao, K., and Soga, N. (1982). *Yogyo Kyokaishi* **90,** 476.
Hirayama, C., and Berg. D. (1961). *Phys. Chem. Glass* **2,** 145.
Hirayama, C., and Berg, D. (1963). *J. Am. Ceram. Soc.* **46,** 85.
Horn, W. F., and Hummel, F. A. (1955). *J. Soc. Glass. Technol.* **39,** 113.
Hoshikawa, T., and Akagi, S. (1974). *Yogyo Kyokaishi* **82,** 185.
Huang, Y. Y., Sarkar, A., and Schultz, P. C. (1978). *J. Non-Cryst. Solids* **27,** 29.
Hummel, F. A., and Reid, H. W. (1951). *J. Am. Ceram. Soc.* **34,** 319.
Hurt, I. C., and Phillips, C. J. (1970). *J. Am. Ceram. Soc.* **53,** 269.
Iskhakov, K. S. (1971). *Uzb. Khim. Zh.* **2,** 79.
Kaneko, T. (1943). *Yogyo Kyokaishi* **51,** 487.
Karkhanavala, M. D. (1952). *Glass Ind.* **33,** 480.
Karkhanavala, M. D., and Hummel, F. A. (1952). *J. Am. Ceram. Soc.* **35,** 215.
Karlsson, K. (1970). *Suom. Kemistil* **43,** 479.
Karpechenko, V. G. (1965). *In* "Issledovanie v Oblasti Silipatov i Okislov," p. 84. Moscow.
Kasymova, S. S. (1964). "Physico-chemical properties of sodium–strontium–silicate glasses. Candidate's Dissertation, Tashkent.
Kasymova, S. S., and Dubrovo, S. K. (1965). *Proc. Int. Congr. Glass, 7th, Versailles* **2,** 364.
Kheifets, V. S., Shevyakov, A. M., and Tarlakov, Y. P. (1972). *In* "Mekhaniceskie i Teplovye Svoystva i Stroyniye Neorg. Stekol," p. 280. Moscow.
Kim, K. H. (1968). *Taehan Hwahakhoe Chi* **12,** 65.
Komiyama, T. (1974). *Osaka Kogyo Gijutsu Shikensho Kiho* **25,** 84.
Kuznetsova, M. G. (1972). Investigation of the physico-chemical properties of the glass systems $RO-MnO-SiO_2$ and $RO-MnO-B_2O_3$. Candidate's Dissertation, Leningrad.
MacDowell, J. F., and Beall, G. H. (1969). *J. Am. Ceram. Soc.* **52,** 17.

McMillan, P. W. (1962). *Adv. Glass Technol., Tech. Pap. Int. Congr. Glass, 6th, Washington, D.C.*, Part 1, 333.

Matusita, K., and Mackenzie, J. D. (1979). *J. Non-Cryst. Solids* **30**, 285.

Matveev, M. A., Rzhevusskaya, T. L., and Milevskaya, R. N. (1970). *In* "Steklovbraznaye Sostoyaniye," p. 276. Erevan.

Mazurenko, V. D. (1962). *In* "Steklo i Siliketniye Materialy," p. 90. Minsk.

Mazurenko, V. D. (1963). *In* "Sintez. Steklo i Silikatnykh Materialov," p. 14. Minsk.

Mazurin, O. V., Tatesh, A. S., and Strel'tsina, M. V. (1969). *Sklar Keram.* **19**, 100.

Milyukov, E. M., and Kheifets, V. S. (1968). *J. Appl. Chem. USSR (Engl. Transl.)* **41**, 2465.

Mochida, N., Takahashi, K., and Shibusawa, S. (1980). *Yogyo Kyokaishi* **88**, 583.

Moore, H., and McMillan, P. W. (1956). *J. Soc. Glass Technol.* **40**, 139T.

Moriya, Y., and Ueno, T. (1964). *Osaka Kogyo Gijutsu Shikensho Kiho* **15**, 66.

Nagaoka, K., Hara, M., and Tanaka, H. (1962). *Osaka Kogyo Gijutsu Shikensho Kiho* **13**, 105.

Nassau, K., Shiever, J. W., and Krause, J. T. (1975). *J. Am. Ceram. Soc.* **58**, 461.

Nemkovich, I. K., Yasinskii, L. G., and Levchenye, A. A. (1974). *In* "Steklo, Silally i Sibikatnye Materialy," p. 60. **3**, Minsk.

Ogura, T., Hayami, R., and Kadota, M. (1968). *Yogyo Kyokaishi* **76**, 277.

Ogura, T., Hayami, R. and Kadota, M. (1969). *Osaka Kogyo Gijutsu Shikensho Kiho* **20**, 48.

Partridge, J. H. (1949). "Glass-to-Metal Seals," Soc. Glass Technol., Sheffield, England.

Pavlova, G. A., and Amatuni, A. N. (1974). *In* "Fiziko-Khim. Issledov. Strukt. i Svoistv Kvartsevogo Stekla," p. 130. Moscow.

Pavlova, G. A., and Amatuni, A. N. (1975). *Inorg. Mater. (Engl. Transl.)* **11**, 1443.

Pavlushkin, N. M., Gastev, Y. A., Beletskii, B. I., and Udovenko, N. G. (1974). "Elektronnya Tekhnika, Materialy," p. 68. **2**.

Rao, B. V. J. (1962). *J. Sci. Ind. Res. Sect. B* **21**, 108.

Rao, B. V. J. (1963a). *Phys. Chem. Glasses* **4**, 22.

Rao, B. V. J. (1963b). *J. Am. Ceram. Soc.* **46**, 107.

Rencker, E. (1933). *C. R. Hebd. Seances Acad. Sci.* **197**, 840.

Rencker, E. (1934). *C. R. Hebd. Seances Acad. Sci.* **199**, 1114.

Rencker, E. (1935). *Bull. Soc. Chim. Fr.* **2**, 1389.

Riebling, E. F. (1968). *J. Am. Ceram. Soc.* **51**, 406.

Rothermel, D. L. (1967). *J. Am. Ceram. Soc.* **50**, 574.

Sack, W., Scheidler, H., and Petzoldt, J. (1968). *Glastech. Ber.* **41**, 138.

Schmid, B. C., Finn, A. N., and Young, J. C. (1934a). *J. Res. Natl. Bur. Stand. (U.S.)* **12**, 421.

Schmid, B. C., Finn, A. N., and Young, J. C. (1934b). *Glass Ind.* **15**, 65.

Scholze, H. (1959). *Glastech. Ber.* **32**, 81.

Schultz, P. C., and Dumbaugh, W. H. (1980). *J. Non-Cryst. Solids* **38/39**, 33.

Schultz, P. C., and Smyth, H. T. (1970). *Res. Dev. Lab., Corning Glass Works, New York*, **14**, 830.

Shabanova, E. B. (1967). *Tr. Gor'k. Politekh. Inst. im. A. A. Zhdanova* **23**, 38.

Shchavelev, O. S., Kasymova, S. S., and Sipovskaya, R. V. (1973). *Sov. J. Opt. Technol. (Engl. Transl.)* **40**, 308.

Shelby, J. E. (1979). *J. Appl. Phys.* **50**, 8010.

Shelby, J. E. (1983). *Glastech. Ber.* **56**, 1057.

Sheludyakov, L. N. (1967). *Vestn. Akad. Nauk Kaz. SSR* **23**, 43.

Shermer, H. F. (1956). *J. Res. Natl. Bur. Stand. (U.S.)* **57**, 97.

Sheybany, H.-A. (1948). *Verres. Refract.* **2**, 127.

Shmidt, Y. A. (1964). *In* "Khimiya Redkikh Elementov," p. 116. Leningrad.

Shmidt, Y. A., and Alekseeva, Z. D. (1964). *J. Appl. Chem. USSR (Engl. Transl.)* **37**, 2266.

Shnypikov, A. D. (1967). *Steklo* No. 1, p. 85.

Simpson, H. E. (1959). *Glass Ind.* **40**, 454.

Simpson, H. E. (1961). *Glass Ind.* **42**, 222.

Strel'tsina, M. V. (1967). *Steklo* No. 1, p. 82.

Strel'tsina, M. V. (1970). Vzaimosvyaz' strukturi i nekotorikb-svoystv likviruyuscikh stekol. Candidate's dissertation, Leningrad.

Syritskaya, Z. M., and Feikner, S. Y. (1966). *Latv. PSR Zinat. Akad. Vestis, Kim. Ser.* **5**, 515.

Syritskaya, Z. M., Rogozhin, Y. V., and Shakhova, R. I. (1971). *In* "Steklovbraznye Sistemy i Novye Stekla na ikh Osnove," p. 55. Moscow.
Takahashi, K. (1962). *Adv. Glass Technol., Tech. Pap. Int. Congr. Glass, 6th, Washington, D.C.*, Part 1, 366.
Takahashi, K., Mochida, N., and Hatta, G. (1975). *Yogyo Kyokaishi* **83**, 103.
Takahashi, K., Mochida, N., Matsui, H., Takeuchi, S., and Gohshi, Y. (1976). *Yogyo Kyokaishi* **84**, 482.
Takahashi, K., Mochida, N., and Yoshida, Y. (1977). *Yogyo Kyokaishi* **85**, 330.
Tarlakov, Y. P., Pronkin, A. A., Vekeliya, D. I., Verulashvili, R. D., and Kogan, V. E. (1980). *Izv. Akad. Nauk Gruz. SSR, Ser. Khim.* **6**, 2.
Terai, R. (1968a). *Yogyo Kyokaishi* **76**, 189.
Terai, R. (1968b). *Osaka Kogyo Gijutsu Shikensho Kiho* **19**, 283.
Terai, R. (1969). *Phys. Chem. Glasses* **10**, 146.
Terai, R. (1971). *J. Non-Cryst. Solids* **6**, 121.
Thompson, C. L., and Parmelee, C. W. (1937). *J. Am. Ceram. Soc.* **20**, 305.
Topping, J. A., Fuchs, P., and Murthy, M. K. (1974). *J. Am. Ceram. Soc.* **57**, 205.
Tsekhomskaya, T. S., and Dubrovo, S. K. (1968). *J. Appl. Chem. USSR (Engl. Transl.)* **41**, 468.
Turner, W. E. S., and Winks, F. (1930). *J. Soc. Glass Technol.* **14**, 84.
Vakhrameev, V. I. (1968). *Steklo* No. 3, 84.
Vakhrameev, V. I., and Evstrop'ev, K. S. (1969). *Inorg. Mater. (Engl. Transl.)* **5**, 82.
Vargin, V. V., and Kheifets, V. S. (1964). *J. Appl. Chem. USSR (Engl. Transl.)* **37**, 1362.
Vargin, V. V., Zasolotskaya, M. V., Kind, N. E., Kondrat'ev, Y. N., Milyukov, E. M., and Tudorovskaya, N. A. (1971). "Katalizirovannaya Kristalliatskiya Stekol Litiyevaalyumocili Katnoy Sistem," p. 204. Leningrad.
Vargin, V. V., Dzhavuktsyan, S. G., Mishel', V. E., and Pevzner, B. Z. (1972). *J. Appl. Chem. USSR (Engl. Transl.)* **45**, 1228.
Varshal, B. G., Rabinovich, E. M., Levitina, A. V., and Vaisfeld, N. M. (1971). *In* "Steklovbraznye Sistemy i Novye Stekla na ikh Osnove," p. 86. Moscow.
Varshneya, A. K. (1982). Stresses in glass-to-metal seals. *In* "Glass III" (M. Tomozawa and R. H. Doremus, eds.), Treatise on Materials Science and Technology, Vol. 22, p. 241. Academic Press, New York.
Watanabe, K., Sumiyoshi, Y., and Anbo E. (1970). *Yogyo Kyokaishi* **78**, 165.
Waterton, S. C., and Turner, W. E. S. (1934). *J. Soc. Glass Technol.* **18**, 268.
White, G. K., Birch, J. A., and Manghnani, M. H. (1977). *J. Non-Cryst. Solids* **23**, 99.
Yasuhara, N. (1942). *Yogyo Kyokaishi* **50**, 61.

TABLE 6.1

Thermal Expansion Coefficients of Lithium Silicate Glasses[a]

Composition (mole % Li$_2$O)	$\alpha \times 10^7$ (°C^{-1})				
	RT[b]–100°C	100–200°C	200–300°C	300–400°C	RT[b]–~450°C
16.1					54
22.2					73
25.4					81.5
26.4					86
28.3					95
32.0	95.2	102.1	108.9	120.8	
33.2					102
34.6	99.6	111.2	120.0	127.1	
36.3					117
37.9	106.2	117.0	126.8	141.4	
39.6	109.5	121.6	131.4	147.7	
40.1					120

[a] Data from Shermer (1956), Ogura *et al.* (1968, 1969), and Dietzel and Sheybany (1948).
[b] RT means room temperature.

TABLE 6.2

Thermal Expansion Coefficients of Sodium Silicate Glasses[a]

Composition (mole % Na$_2$O)	$\alpha \times 10^7$ (°C^{-1})		
	0–130°C	130–250°C	250–350°C
14.6		73.1[b]	
17.2		83.3[b]	
19.6	90.7	98.7	
22.2	101.5	108.6	112.4
24.8	106.0	117.6	124.7
25.8	109.4	120.6	128.8
26.6	114.6	125.3	135.7
29.1	122.7	135.1	145.6
32.2	134.5	147.8	159.4
35.8	141.9	165.3	176.9
39.0	155.0	177.3	192.6
43.8	165.7	191.9	212.7
48.1	176.5	196.8	224.3

[a] Data from Turner and Winks (1930).
[b] In the temperature interval 0–250°C.

TABLE 6.3

Thermal Expansion Coefficients of Na_2O–SiO_2 Glasses at Low Temperatures[a]

Temperature (K)	Values of $\alpha \times 10^8$ (K^{-1}) for various compositions (mole % Na_2O)				
	10	15	20	30	40
2	− 0.30	− 0.15$_5$	− 0.10		
3	− 0.53	− 0.36	− 0.11	0.10	0.19
4	− 1.10	− 0.65	− 0.28	0.14	0.38
5	− 2.2$_5$	− 1.4	− 0.50	0.3$_5$	0.63
6	− 3.7	− 2.5	− 0.9$_5$	0.4$_5$	1.05
8	− 7.3	− 4.6	− 1.7$_5$	1.1	2.40
10	− 10.9	− 6.6	− 2.2	2.5	4.8
12	− 13.2	− 7.4	− 1.4	4.9	8.5
14	− 14.5	− 7.0	− 0.6	8.7	13.7
16	− 13.6	− 4.7	4.3	14.1	20.5
18	− 11.6	− 0.8	9.8	21.2	29.0
20	− 7.8$_5$	4.7	16.8	30.0	39.3
22	− 2.9	11.5	25.4	39.8	51.2
24	2.5	19.2	34.6	51.5	64.4
26	9.1	27.8	45.0	64.0	78.3
28	16.9	37.7	56.9	77.3	94.0
30	24.1	48.1	68.7	92.4	111.5
65	179		315	416	469
75	214		380	487	538
85	252		430	569	677
283	500	720	900	1225	1490

[a] Data from White *et al.* (1977).

TABLE 6.4

Thermal Expansion Coefficients of Potassium Silicate Glasses

Composition (mole % K_2O)	$\alpha \times 10^7$ ($°C^{-1}$) 25–325°C[a]	Composition (mole % K_2O)	$\alpha \times 10^7$ ($°C^{-1}$) 20–400°C[b]
10.1	62	5.0	32.9
13.8	88	10.0	63.2
17.5	107	12.0	74.7
21.5	127	14.2	87.5
25.6	155	20.0	116.6
29.9	177	25.9	143.3
34.3	196	32.6	178.3
48.9	230		

[a] Data from Rao (1963a).
[b] Data from Shmidt (1964) and Shmidt and Alekseeva (1964).

TABLE 6.5

**Thermal Expansion Coefficients of
Rubidium Silicate Glasses**[a]

Composition (mole % Rb_2O)	$\alpha \times 10^7$ ($°C^{-1}$) 20–400°C
8.4	54.9
11.8	75.8
13.5	82.4
15.3	95.5
20.0	120.1
26.6	154.2

[a] Data from Shmidt and Alekseeva (1964).

TABLE 6.6

**Thermal Expansion Coefficients of
Cesium Silicate Glasses**[a]

Composition (mole % Cs_2O)	$\alpha \times 10^7$ ($°C^{-1}$) 20–400°C
4.0	26.6
5.0	33.7
5.8	37.9
8.0	52.3
12.2	74.4
20.0	118.4
26.7	152.6

[a] Data from Shmidt and Alekseeva (1964).

TABLE 6.7

**Thermal Expansion Coefficients of
Thallium Silicate Glasses**[a]

Composition (mole % Tl_2O)	$\alpha \times 10^7$ ($°C^{-1}$) 20–170°C
40	83
50	118
60	137
70	150
80	178

[a] Data from graph in Karpechenko (1965).

TABLE 6.8
Thermal Expansion Coefficients of Some MO–SiO₂ Glasses

Composition (mole % CaO)	$\alpha \times 10^7$ (°C⁻¹) 300–400°C[a]	Composition (mole % SrO)	$\alpha \times 10^7$ (°C⁻¹) 300–400°C[a]	Composition (mole % BaO)	$\alpha \times 10^7$ (°C⁻¹) 300–400°C[a]	$\alpha \times 10^7$ (°C⁻¹) 20–500°C[b]
35	76.5	23	62	10	40	
37	82	25	67	15	56.5	
40	88	27	72.5	20	70	
45	96	30	76.5	21	72	
48	102	35	86	22	76	
52	108	40	98	25	85	
55	114	45	108	30	98	84
57	119			32	—	91
				33	103	—
				34	—	94
				36	—	96
				38	106	99
				40	112	101

[a] Data from graph in Shelby (1979).
[b] Data from graph in Varshal et al. (1971).

TABLE 6.9

Thermal Expansion Coefficients of Cadmium Silicate Glasses[a]

Composition (mole % CdO)	$\alpha \times 10^7$ ($^\circ$C^{-1}) 200–600°C
30	52.5
50	71.4
70	69.0

[a] Data from Tarlakov *et al.* (1980).

TABLE 6.10

Thermal Expansion Coefficients of Lead Silicate Glasses[a]

Composition (mole % PbO)	$\alpha \times 10^7$ ($^\circ$C^{-1}) 200–300°C	Composition (mole % PbO)	$\alpha \times 10^7$ ($^\circ$C^{-1}) 200–300°C
15	31	39.5	75
17.5	37	44.5	83
20	40	49.5	98
22.5	46	55	98
25	48	60	114
27.5	54	64.5	122
30	60	70	128
35	63		

[a] Data from graph in Shelby (1983).

TABLE 6.11

Thermal Expansion Coefficients of Lead Silicate Glasses[a]

Composition (mole % PbO)	$\alpha \times 10^7$ ($^\circ$C^{-1}) at given temperature			
	20°C	70°C	120°C	170°C
25.7	51.4	51.7	52.0	52.2
30.0	57.7	58.1	58.6	59.1
32.5	60.6	61.2	61.8	62.3
33.2	61.6	62.2	62.8	63.3
35.0	64.0	64.7	65.4	66.2
37.5	68.8	69.6	70.6	71.4
42.6	75.2	76.3	77.4	78.6
45.8	78.8	80.1	81.4	82.6
47.8	83.0	84.4	85.7	87.0
49.8	85.6	87.0	88.4	89.8
53.8	90.6	92.2	93.7	95.2
57.5	95.6	97.2	98.8	100.4
59.0	97.0	98.6	100.3	101.9
61.0	100.7	102.3	104.0	105.6
61.75	101.4	103.0	104.6	106.3
67.7	110.4	112.1	113.8	115.5

[a] Data from Bogatyreva *et al.* (1973).

TABLE 6.12

Thermal Expansion Coefficients of $B_2O_3-SiO_2$ Glasses[a]

Composition (mole % B_2O_3)	$\alpha \times 10^7$ (°C^{-1})	
	0–100°C	100–200°C
39.76	47.5	44.9
44.17	49.8	50.8
50.81	57.6	54.8
58.44	71.9	70.1
72.72	87.0	89.7
83.23	111.4	116.6
88.60	118.1	126.0
92.59	131.7	141.9

[a] Data from Cousen and Turner (1928).

TABLE 6.13

Thermal Expansion Coefficients of $B_2O_3-SiO_2$ Glasses[a]

Composition (mole %)			$\alpha \times 10^7$ (°C^{-1})	
B_2O_3	SiO_2	H_2O	0–100°C	0–200°C
39.26	60.69	0.044	44	43
44.55	55.40	0.048	47	49
51.58	48.36	0.059	58	59
56.38	43.57	0.053	59	62
64.61	35.35	0.041	77	79
70.56	29.39	0.049	91	90
77.65	22.27	0.082	106	107
84.07	15.84	0.091	124	123
89.19	10.72	0.089	133	136
90.38	9.55	0.074	135	134
93.97	5.95	0.082	140	144
95.31	4.59	0.097	142	147
97.69	2.17	0.139	151	156

[a] Data from Bruckner and Navarro (1966).

TABLE 6.14

Thermal Expansion Coefficients of Al_2O_3–SiO_2 Glasses[a]

Composition (mole % Al_2O_3)	$\alpha \times 10^7$ ($°C^{-1}$)	
	0–300°C	400–700°C
6.13	—	10[c]
15	13.4[b]	

[a] For effect of heat treatment on the thermal expansion behavior of some Al_2O_3–SiO_2 glasses, see Thompson and Parmelee (1937).
[b] Data from MacDowell and Beall (1969).
[c] Data from Nassau et al. (1975).

TABLE 6.15

Thermal Expansion Coefficients of Bi_2O_3–SiO_2 Glasses[a]

Composition (mole % Bi_2O_3)	$\alpha \times 10^7$ ($°C^{-1}$)	Composition (mole % Bi_2O_3)	$\alpha \times 10^7$ ($°C^{-1}$)
25	67.6	50	82.6
30	73.1	55	87.6
35	75.2	60	94.5
40	76.1	63	97.3
45	80.7	65	102.8

[a] Data from Shabanova (1967).

TABLE 6.16

Thermal Expansion Coefficients of GeO_2–SiO_2 Glasses

Composition (mole % GeO_2)	$\alpha \times 10^7$ ($°C^{-1}$) 25–300°C	Composition (mole % GeO_2)	$\alpha \times 10^7$ ($°C^{-1}$) 25–300°C
		14	21
2	7	20	27
4	10	42.2	43.7[a]
6	12	50.6	51[a]
10	17	100	76

[a] From Riebling (1968); rest data points read from graph of Huang et al. (1978), who prepared glasses using flame deposition technique.

TABLE 6.17

Thermal Expansion Coefficients of TiO_2–SiO_2 Glasses in the Temperature Range 20 to $t°C$[a]

t (°C)	$\alpha \times 10^7$ (°C^{-1}) at TiO_2 composition (mole %)							
	1.8	4.5	5.4	6.1	6.5	7.3	8.3	10.0
−60	—	−1.0	—	—	−2.5	−2.7	−4.1	−5.7
−40	—	−0.8	—	—	−2.3	−2.4	−3.8	−5.5
−20	—	−0.4	—	—	−1.8	−1.9	−3.6	−4.9
0	—	0	—	—	−1.8	—	−3.2	−4.5
75	3.6	0.8	0.5	−0.3	−0.4	−1.1	−2.7	−4.0
100	4.0	1.2	0.6	−0.2	−0.4	−1.0	−2.4	−3.8
125	4.2	1.3	0.8	0.2	−0.3	−0.8	−2.3	−3.8
150	4.3	1.6	1.0	0.2	−0.3	−0.7	−2.0	−3.7
175	4.4	1.8	1.1	0.3	−0.2	−0.5	−1.9	−3.4
200	4.5	1.8	1.2	0.2	0	−0.6	−1.9	−3.4
225	4.6	2.0	1.3	0.3	−0.1	−0.5	−1.8	−3.3
250	4.7	2.0	1.4	0.3	−0.1	−0.4	−1.8	−3.3
275	4.7	2.1	1.4	0.3	0	−0.4	−1.8	−3.2
300	4.7	2.2	1.4	0.3	0	−0.4	−1.8	−3.2
325	4.8	2.2	1.4	0.4	0	−0.4	−1.7	−3.2
350	4.8	2.2	1.4	0.4	−0.1	−0.4	−1.7	−3.2

[a] Data from Pavlova and Amatuni (1974, 1975).

TABLE 6.18

Thermal Expansion Coefficients of TiO_2–SiO_2 Glasses[a,b]

Composition (mole % TiO_2)	$\alpha \times 10^7$ (°C^{-1})	
	5–35°C	25–700°C
2.6	3.10	2.312
4.6	1.25	0.139
5.5	0.0	−0.741
5.7	0.36	−0.326
6.4	—	−1.156
7.3	−1.80	−1.926

[a] Data from Schultz and Smyth (1970).
[b] Glasses made using melting by flame hydrolysis technique.

TABLE 6.19

Thermal Expansion Coefficients of P_2O_5–SiO_2 Glasses[a]

Composition (mole % $PO_{5/2}$)	$\alpha \times 10^7$ (°C^{-1}) 50–350°C
19	15
29	32
43	47
50	49
59	54
95	103
100	137

[a] Data from graph in Takahashi et al. (1976).

TABLE 6.20

Thermal Expansion Coefficients of $Li_2O-Na_2O-SiO_2$ Glasses

Composition (mole %)				$\alpha \times 10^7$ (°C^{-1})		
Li_2O	Na_2O	SiO_2	t_g (°C)	10–300°C	20–150°C	Reference[a]
9	26	65	392	143		1
17.5	17.5	65	387	133		
26	9	65	400	124		
6	19	75	416	123		
12.5	12.5	75	409	113		
19	6	75	419	98		
4	11	85	—	71		
7.5	7.5	85	414	66		
11	4	85	418	59		
7	14.3	78.7			94	2
10.5	14.3	75.2			105	
14	14.3	71.7			109	
18	14.3	67.7			117	
21.5	14.3	64.2			126	

[a] Data from (1) Sheybany (1948) and (2) from graph in Din (1968).

TABLE 6.21

Thermal Expansion Coefficients of $Li_2O-K_2O-SiO_2$ Glasses

Composition (mole %)					
Li_2O	K_2O	SiO_2	t_g (°C)	$\alpha \times 10^7$ (°C^{-1})	Reference[a]
9	26	65	—	151	1[b]
17.5	17.5	65	420	142	
26	9	65	430	126	
6	19	75	444	119	
12.5	12.5	75	447	107	
19	6	75	444	95	
—	16.7	83.3	—	83.8,[c] 88.6[d]	2
6.7	10	83.3	—	70.7,[c] 77.5[d]	
9.7	7	83.3	—	63.5,[c] 72.8[d]	
12.7	4	83.3	—	58.5,[c] 63.6[d]	
16.7	—	83.3	—	46.2,[c] 50.5[d]	
4	11	85	465	75	1[b]
7.5	7.5	85	519	68	
11	4	85	474	59	

[a] Data from (1) Sheybany (1948) and (2) Aleynikov *et al.* (1967).
[b] Temperature range 10–300°C.
[c] 200°C.
[d] 400°C.

TABLE 6.22
Thermal Expansion Coefficients of $Li_2O-Cs_2O-SiO_2$ Glasses[a]

Composition (mole %)				$\alpha \times 10^7$ (°C^{-1})
Li_2O	Cs_2O	SiO_2	t_g (°C)	100–300°C
—	20	80	520	118.5
5	15	80	500	97
10	10	80	515	87.5
15	5	80	500	85
18.2	1.8	80	445	81
19.1	0.9	80	440	77
2.0	—	80	455	75
—	35	65	525	185
9	26	65	460	159
17.5	17.5	65	480	135.5
25	10	65	480	127.5
30	5	65	475	125
33	2	65	465	125
34	1	65	455	125
34.5	0.5	65	470	125
35	—	65	465	125

[a] Data from Alekseeva (1974) and Alekseeva and Polozok (1972).

TABLE 6.23
Thermal Expansion Coefficients of $Na_2O-K_2O-SiO_2$ Glasses

Composition (mole %)					
Na_2O	K_2O	SiO_2	t_g (°C)	$\alpha \times 10^7$ (°C^{-1})	Reference[a]
10	30	60	380	197	1[b]
20	20	60	350	204	
30	10	60	360	191	
9	26	65	—	170	2[c]
17.5	17.5	65	—	174	
26	9	65	392	167	
5	25	70	435	176	1[b]
10	20	70	400	174	
15	15	70	380	168	
20	10	70	410	161	
25	5	70	420	150	
6	19	75	—	141	2[c]
12.5	12.5	75	—	135	
19	6	75	424	130	
6	18	76	410	$150(<t_g); 210(>t_g)$	3
12	12	76	410	$170(<t_g); 370(>t_g)$	
18	6	76	425	$160(<t_g); 330(>t_g)$	
5	15	80	450	108	1[b]
10	10	80	430	118	
15	5	80	440	106	
4	11	85	461	87	2[c]
7.5	7.5	85	457	85	
11	4	85	458	80	

[a] Data from (1) Deeg (1958a,b), (2) Sheybany (1948), and (3) Amrhein (1963).
[b] Temperature 20–t_g °C.
[c] Temperature 10–300°C.

TABLE 6.24

Thermal Expansion Coefficients of $Na_2O-Rb_2O-SiO_2$ Glasses[a]

Composition (mole %)				$\alpha \times 10^7 \, (°C^{-1})$	
Na_2O	Rb_2O	SiO_2	$t_g \, (°C)$	$<t_g$	$>t_g$
18	6	76	400	130	210
6	18	76	425	125	320

[a] Data from Amrhein (1963).

TABLE 6.25

Thermal Expansion Coefficients of $Na_2O-Cs_2O-SiO_2$ Glasses

Composition (mole %)					
Na_2O	Cs_2O	SiO_2	$t_g \, (°C)$	$\alpha \times 10^7 \, (°C^{-1})$	Reference[a]
—	16.67	83.33	555	98.1	1[b]
1.67	15.0	83.33	524	95.0	
3.33	13.34	83.33	508	94.3	
4.17	12.5	83.33	—	89.2	
5.0	11.67	83.33	498	93.0	
6.67	10.0	83.33	493	87.9	
10	6.67	83.33	495	86.3	
8.33	8.34	83.33	484	86.4	
12.5	4.17	83.33	460	91.3	
16.67	—	83.33	—	95.4	
1.5	18.5	80	485	111	2[c]
5	15	80	485	109	
8	12	80	485	108	
10	10	80	480	108.5	
11.4	8.6	80	490	109	
13.3	6.7	80	485	110	
16	4	80	470	111	
18.2	1.8	80	465	111	

[a] Data from (1) Terai (1971) and (2) Alekseeva (1974) and Alekseeva and Polozok (1972).

[b] Temperature interval of measurement not given.

[c] Temperature range 100–300°C; heating rate 2.5°C min^{-1}.

TABLE 6.26

Thermal Expansion Coefficients of $Li_2O-BeO-SiO_2$ Glasses

Composition (mole %)			Temperature range (°C)	$\alpha \times 10^7$ (°C^{-1})	Reference[a]
Li_2O	BeO	SiO_2			
30	5	65	20–150	99.2	1
30	10	60	20–150	95.9	1
20	20	60	20–400	85	2

[a] Data from (1) Moore and McMillan (1956) and (2) Scholze (1959).

TABLE 6.27

**Thermal Expansion Coefficients of
$Li_2O-MgO-SiO_2$ Glasses[a]**

Composition (mole %)			$\alpha \times 10^7$ (°C^{-1}) 20–150°C
Li_2O	MgO	SiO_2	
20	10	70	78.7
22.5	10	67.5	84.1
20	15	65	84.5
25	10	65	88.8
20	20	60	89.6
30	10	60	99.6
20	25	55	92.1
20	30	50	99.2

[a] Data from Moore and McMillan (1956).

TABLE 6.28

Thermal Expansion Coefficients of $Li_2O-CaO-SiO_2$ Glasses[a]

Composition (wt. %)			t_g (°C)	Temperature range (°C)	$\alpha \times 10^7$ (°C^{-1})
Li_2O	CaO	SiO_2			
15.09	10.10	74.81	435	0–165	93.2
				165–265	106.0
				265–355	117.5
				355–435	135.0

[a] Data from Waterton and Turner (1934).

TABLE 6.29

Thermal Expansion Coefficients of $Li_2O-BaO-SiO_2$ Glasses[a]

Composition (mole %)			$\alpha \times 10^7 \, (^{\circ}C^{-1})$	Composition (mole %)			$\alpha \times 10^7 \, (^{\circ}C^{-1})$
Li_2O	BaO	SiO_2	20–400°C	Li_2O	BaO	SiO_2	20–400°C
35	5	60	126.3	25	30	45	143.8
40	5	55	137.6	35	30	35	166.5
45	5	50	149.7	5	35	60	106.8
35	10	55	134.7	10	35	55	116.9
45	10	45	157.0	15	35	50	128.6
25	15	60	119.2	20	35	45	140.6
30	15	55	131.0	25	35	40	151.4
35	15	50	142.4	30	35	35	163.3
40	15	45	154.2	35	35	30	174.8
45	15	40	165.4	5	40	55	113.8
25	20	55	127.4	15	40	45	136.4
35	20	45	150.1	25	40	35	161.0
45	20	35	173.8	5	45	50	121.4
15	25	60	112.7	10	45	45	133.0
20	25	55	123.9	15	45	40	144.2
25	25	50	135.6	20	45	35	156.4
30	25	45	147.0	25	45	30	167.6
35	25	40	159.2	5	50	45	128.9
40	25	35	170.4	15	50	35	152.1
				5	55	40	137.8
45	25	30	181.9	10	55	35	149.1
15	30	55	120.5	15	55	30	161.0

[a] Data from Bezborodov and Mazurenko (1960, 1961a, b) and Mazurenko (1962).

TABLE 6.30

Thermal Expansion Coefficients of $Li_2O-ZnO-SiO_2$ Glasses[a]

Composition (mole %)			$\alpha \times 10^7 \, (^{\circ}C^{-1})$	Composition (mole %)			$\alpha \times 10^7 \, (^{\circ}C^{-1})$
Li_2O	ZnO	SiO_2	100–300°C	Li_2O	ZnO	SiO_2	100–300°C
15	20	65	64	25	15	60	106
15	25	60	68	25	20	55	110
15	30	55	73	25	25	50	112
15	35	50	77	25	30	45	116
20	5	75	80	30	5	65	116
20	10	70	83	30	10	60	117
20	15	65	88	30	15	55	119
20	20	60	91	30	20	50	121
20	25	55	95	35	5	60	130
20	30	50	97	35	10	55	131
25	5	70	99	35	15	50	132
25	10	65	101				

[a] Data from graph in Vargin et al. (1972).

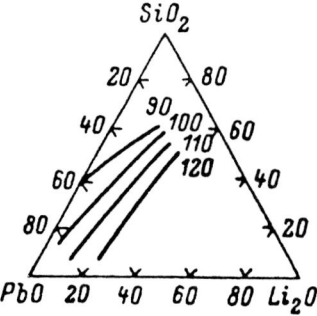

Fig. 6.2. Thermal expansion coefficients ($\alpha \times 10^7 \ ^\circ\mathrm{C}^{-1}$) in the temperature range 20–220°C of $Li_2O-PbO-SiO_2$ glasses; compositions in mole %. Data from Bezborodov and Rzhevuskaya (1959).

TABLE 6.31

Thermal Expansion Coefficients of $Na_2O-CuO-SiO_2$ Glasses[a]

Composition (mole %)			$\alpha \times 10^7 \ (^\circ\mathrm{C}^{-1})$			
Na_2O	CuO	SiO_2	RT–150°C	150–300°C	300–450°C	450–474°C
14.3	14.3	71.4	78.0	89.0	134.3	176.2

[a] Data from Karkhanavala and Hummel (1952).

TABLE 6.32

Thermal Expansion Coefficients of $Na_2O-BeO-SiO_2$ Glasses[a]

Composition (wt. %)			$\alpha \times 10^7 \ (^\circ\mathrm{C}^{-1})$ 20–100°C	Composition (wt. %)			$\alpha \times 10^7 \ (^\circ\mathrm{C}^{-1})$ 20–400°C
Na_2O	BeO	SiO_2		Na_2O	BeO	SiO_2	
7.6	6.4	86	45	7.5	6.5	86.0	50
9	4	87	49	9.5	4.5	86.0	54
13.2	6.8	80	67	12.5	7.5	80.0	74
14	11.5	74.5	66	12.5	12.5	75.0	73
15	11	74	76	15.0	11.0	74.0	84
16	12	72	80	16.0	12.0	72.0	88
18	10	72	87	18.0	10.0	72.0	96
19	4	77	90	19.0	5.0	76.0	100
20	14	66	91	20.0	14.0	66.0	100
22.4	3.2	74.4	99	22.5	3.5	74.0	109
23.5	7.5	69	102	23.5	7.5	69.0	112
25.9	2.7	71.4	107	26.0	3.0	71.0	118
26.8	11	62.2	111	26.5	11.0	62.5	122
28.7	6.6	64.7	116	28.5	7.0	64.5	128
32	11	57	121	31.0	12.0	57.0	133

[a] Data from Rencker (1933, 1935) and Karkhanavala (1952).

TABLE 6.33

Thermal Expansion Coefficients of $Na_2O-MgO-SiO_2$ Glasses[a]

Composition (mole %)			t_g (°C)	$\alpha \times 10^7$ (°C^{-1}) $20-t_g$ °C	Composition (mole %)			t_g (°C)	$\alpha \times 10^7$ (°C^{-1}) $20-t_g$ °C
Na_2O	MgO	SiO_2			Na_2O	MgO	SiO_2		
10	15	75	613	69.3	23.2	19.8	57	500	105
15	10	75	520	82.5	25.2	17.8	57	476	108.7
20	5	75	476	104.5	27	16	57	450	121.8
5	25	70	644	58.3	30	13	57	415	144.5
15	15	70	516	90.5	20.5	29.5	50	480	129.5
20	10	70	490	107	25	25	50	466	134.5
25	5	70	460	112.6	40	10	50	370	165.5
5	30	65	710	63					
7.7	27.3	65	672	73	10	23.5	66.5	620	74.5
11.7	23.3	65	614	77.5	11	32.8	56.2	624	86
13.3	21.7	65	590	84	15	14.3	70.7	520	91.2
16.6	18.4	65	514	96.6	15	18.6	66.4	537	93.2
20	15	65	436	104.3	15	25	60	558	94.3
24.1	10.5	65.4	432	121	20	7	73	472	101.5
25.8	9.2	65	416	131	20	19.5	60.5	480	112.2
26.5	8.5	65	396	135	20	27.2	52.8	475	121.4
28.5	6.5	65	370	142	25	10	65	416	118
10.6	32.4	57	610	87	25	15	60	440	126.5
22.2	20.8	57	540	103	25	20	55	456	129

[a] Data from Botivinkin and Yaroker (1967a,b).

TABLE 6.34

Thermal Expansion Coefficients of Na_2O–CaO–SiO_2 Glasses[a]

| Composition (wt. %) | | | t_g (°C) | $\alpha \times 10^7$ (°C^{-1}) | | Composition (wt. %) | | | t_g (°C) | $\alpha \times 10^7$ (°C^{-1}) | |
Na_2O	CaO	SiO_2		25–400°C	25–t_g °C	Na_2O	CaO	SiO_2		25–400°C	25–t_g °C
27.97	19.83	52.20	498	157.3	169.1	21.74	11.79	66.47	522	128.0	144.9
32.85	12.78	54.37	477	165.3	174.8	11.62	21.15	67.23	601	98.7	111.1
34.00	9.80	56.20	470	178.7	188.8	29.43	3.27	67.30	480	141.3	151.6
23.91	19.75	56.34	517	136.0	148.4	22.50	9.52	67.98	518	125.3	137.9
37.48	5.76	56.76	444	178.7	183.8	15.03	14.85	70.12	567	104.0	116.2
38.54	3.05	58.41	437	181.3	186.9	23.00	6.50	70.50	506	120.0	128.9
19.79	20.16	60.05	548	130.7	141.6	12.10	16.20	71.70	598	93.3	104.7
24.50	15.18	60.32	520	138.7	149.5	14.21	13.71	72.08	571	96.0	108.1
17.39	19.68	62.93	568	114.7	127.1	24.24	3.15	72.61	490	122.7	131.2
24.39	12.27	63.34	518	138.7	150.1	15.23	10.68	74.09	555	96.0	107.5
21.22	14.64	64.14	530	125.3	136.6	12.28	13.03	74.69	589	88.0	101.1
26.84	8.46	64.70	493	138.7	151.7	15.26	9.26	75.48	553	96.0	102.3
28.79	5.50	65.71	493	141.3	153.8	18.45	2.99	78.56	504	101.3	108.6
17.92	16.03	66.05	563	117.3	128.3	16.33	4.90	78.77	523	90.7	102.4
13.97	19.91	66.12	587	106.7	119.2	16.17	3.24	80.59	505	85.3	89.6

[a] Data from Schmid et al. (1934a,b).

TABLE 6.35
Thermal Expansion Coefficients of $Na_2O-SrO-SiO_2$ Glasses[a]

Composition (mole %)			$\alpha \times 10^7 (°C^{-1})$	Composition (mole %)			$\alpha \times 10^7 (°C^{-1})$
Na_2O	SrO	SiO_2	20–400°C	Na_2O	SrO	SiO_2	20–400°C
10	10	80	76.7	5	35	60	104.2
15	5	80	85.5	10	30	60	110.6
10	15	75	89.9	15	25	60	121.0
12.5	12.5	75	95.1	17	23	60	123.0
20	5	75	110.8	20	20	60	131.3
10	20	70	96.2	22	18	60	133.2
15	15	70	106.4	23.4	16.6	60	135.8
20	10	70	117.5	24.3	15.7	60	138.4
25	5	70	132.6	25	15	60	140.9
25	10	65	138.8				

[a] Data from Kasymova and Dubrovo (1965), Kasymova (1964), and Dubrovo and Kasymova (1964).

TABLE 6.36
Thermal Expansion Coefficients of $Na_2O-BaO-SiO_2$ Glasses[a]

Composition (wt. %)			$\alpha \times 10^7 (°C^{-1})$	Composition (wt. %)			$\alpha \times 10^7 (°C^{-1})$
Na_2O	BaO	SiO_2	50–300°C	Na_2O	BaO	SiO_2	50–300°C
15	10	75	91	20	20	60	116
20	5	75	90	25	15	60	117
10	20	70	90	30	10	60	147
15	15	70	98	35	5	60	151
20	10	70	99	5	40	55	97
25	5	70	99	10	35	55	107
5	30	65	76	15	30	55	116
10	25	65	91	20	25	55	123
15	20	65	100	25	20	55	134
20	15	65	111	30	15	55	111
25	10	65	106	5	45	50	102
30	5	65	112	10	40	50	118
5	35	60	87	15	35	50	129
10	30	60	99	20	30	50	131
15	25	60	112				

[a] Data from Yasuhara (1942).

TABLE 6.37

Thermal Expansion Coefficients of $Na_2O-ZnO-SiO_2$ Glasses[a]

Composition (mole %)			$\alpha \times 10^7$ ($^\circ C^{-1}$) 20–300°C	Composition (mole %)			$\alpha \times 10^7$ ($^\circ C^{-1}$) 20–300°C
Na_2O	ZnO	SiO_2		Na_2O	ZnO	SiO_2	
10	15	75	67	10	30	60	67
12.5	12.5	75	78	15	25	60	86
15	10	75	93	20	20	60	111
20	5	75	106	25	15	60	127
10	20	70	67	30	10	60	141
15	15	70	92	35	5	60	154
20	10	70	110	15	35	50	88
25	5	70	128	19	31	50	100
10	25	65	67	25	25	50	128
15	20	65	91	30	20	50	148
20	15	65	111	35	15	50	162
25	10	65	131	40	10	50	173
30	5	65	147				

[a] Data from graph in Hurt and Phillips (1970).

TABLE 6.38

Thermal Expansion Coefficients of
$Na_2O-CdO-SiO_2$ Glass[a]

Composition (mole %)			$\alpha \times 10^7$ ($^\circ C^{-1}$) 20–400°C
Na_2O	CdO	SiO_2	
16	20	64	112.4

[a] Data from Appen (1953).

TABLE 6.39

Thermal Expansion Coefficients of $Na_2O-PbO-SiO_2$ Glasses[a]

Composition (wt. %)			$\alpha \times 10^7$ ($^\circ C^{-1}$) 30–100°C	Composition (wt. %)			$\alpha \times 10^7$ ($^\circ C^{-1}$) 30–100°C
Na_2O	PbO	SiO_2		Na_2O	PbO	SiO_2	
4.61	44.76	49.82	66.6	11.35	31.57	55.87	88.0
5.61	41.83	51.21	70.5	12.40	28.01	58.52	90.3
6.90	39.24	52.61	72.9	13.83	24.33	60.30	93.4
7.88	37.45	53.58	76.1	14.06	22.94	61.67	92.6
8.27	36.60	54.10	78.1	16.34	16.35	65.95	96.6
10.02	33.96	54.84	84.4				

[a] Data from Karkhanavala (1952).

TABLE 6.40

Thermal Expansion Coefficients of $K_2O-MgO-SiO_2$ Glasses[a]

Composition (mole %)				$\alpha \times 10^7$ ($°C^{-1}$)
K_2O	MgO	SiO_2	t_g ($°C$)	200–400°C
12.5	12.5	75	590	91
14.63	12.20	73.17	620	95.5
16.67	11.90	71.43	560	105
17.65	11.76	70.59	580	120.5

[a] Data from Kaneko (1943).

TABLE 6.41

Thermal Expansion Coefficients of $K_2O-CaO-SiO_2$ Glass[a]

Composition (wt. %)				Temperature range	
K_2O	CaO	SiO_2	t_g ($°C$)	($°C$)	$\alpha \times 10^7$ ($°C^{-1}$)
14.95	10.04	75.01	610	0–170	72.4
				170–260	76.5
				260–360	80.0
				360–610	83.0

[a] Data from Waterton and Turner (1934).

TABLE 6.42

Thermal Expansion Coefficients of $K_2O-SrO-SiO_2$ Glasses[a,b]

Composition (mole %)			$\alpha \times 10^7$ ($°C^{-1}$) at		$(d\alpha/dT) \times 10^7$
K_2O	SrO	SiO_2	70°C	170°C	($°C^{-2}$)
14.9	—	84.7	91	—	—
14.2	4.5	80.9	92.5	96	0.04
13.4	9.5	76.6	95	100	0.05
12.6	15.3	71.7	98.5	102.5	0.04
11.6	22.0	66.1	104	108.5	0.05
10.8	27.0	61.8	106	111.5	0.05
18.0	5.6	76.4	114	121	0.065
16.8	11.8	71.3	114.5	121	0.065
15.5	18.7	65.8	115	121.5	0.065
14.0	26.4	59.6	116	122	0.06
13.4	30.0	56.6	117	123.5	0.07
23.1	3.4	76.5	126	133	0.07
22.3	6.9	70.8	125.5	132	0.07
21.4	10.6	68.0	125	132	0.07

(*continues*)

TABLE 6.42 (*Continued*)

Composition (mole %)			$\alpha \times 10^7$ ($°C^{-1}$) at		$(d\alpha/dT) \times 10^7$
K_2O	SrO	SiO_2	70°C	170°C	($°C^{-2}$)
20.5	14.4	65.1	125	132	0.07
19.6	18.3	62.1	125	132	0.07
18.6	22.4	59.0	124	131	0.07
17.6	26.6	55.8	124	131	0.07
16.5	30.9	52.2	124	130	0.065
26.5	8.5	65.0	142	149	0.07
24.0	17.2	58.8	135	—	—
26.5	8.5	65.0	142	149	0.07
22.3	6.9	70.8	125.5	132	0.07
18.0	5.6	76.4	114	121	0.065
14.7	4.6	80.7	93.	97.5	0.045
14.2	4.5	80.9	92.5	96	0.04
24.0	17.2	58.8	135	—	—
20.5	14.4	65.1	125	132	0.07
16.8	11.8	71.2	114.5	121	0.065
13.4	9.5	76.6	95	100	0.05
10.8	7.6	81.6	85.5	86.5	0.01
8.7	6.1	85.2	68	69	0.01
18.6	22.4	59.0	124	131	0.07
15.5	18.7	65.8	115	121.5	0.065
12.6	15.3	71.7	98.5	102.5	0.04
16.5	30.9	52.5	124	130	0.065
14.0	26.4	59.5	116	122	0.06
11.6	22.0	66.1	104	108	50.05
9.9	18.7	71.4	97	101	0.04
7.6	14.2	78.2	81	83	0.025
5.4	10.1	84.5	63	64 5	0.015
10.8	27.0	61.8	106.5	111.5	0.05
14.0	26.4	59.6	116	122	0.065
17.6	26.6	55.8	124	131	0.07
15.5	18.7	65.8	115	121.5	0.065
19.6	18.3	62.1	125	132	0.07
24.0	17.2	58.8	135	—	—

[a] Data from Shchavelev *et al.* (1973).

[b] For some glasses, the combined concentration of K_2O, SrO, and SiO_2 is less than 100% since minor impurities are not listed here.

TABLE 6.43

Thermal Expansion Coefficients of $K_2O-ZnO-SiO_2$ Glasses[a]

Composition (mole %)			$\alpha \times 10^7\ (°C^{-1})$	Composition (mole %)			$\alpha \times 10^7\ (°C^{-1})$
K_2O	ZnO	SiO_2	100–300°C	K_2O	ZnO	SiO_2	100–300°C
10	5	85	71	20	20	60	116
10	10	80	72	20	25	55	117
10	15	75	73	20	30	50	117
10	20	70	74	20	35	45	119
10	25	65	75	25	5	70	138
10	30	60	75	25	10	65	140
10	35	55	76	25	15	60	137
10	40	50	75	25	20	55	140
10	45	45	78	25	25	50	141
15	5	80	88	25	30	45	140
15	10	75	89	25	35	40	141
15	15	70	90	30	5	65	155
15	20	65	91	30	10	60	155
15	25	60	91	30	15	55	156
15	30	55	92	30	20	50	156
15	35	50	94	30	25	45	155
15	40	45	94	35	5	60	168
20	5	75	117	35	10	55	168
20	10	70	117	35	15	50	168
20	15	65	116	35	20	45	170

[a] Data from graph in Vargin *et al.* (1972).

TABLE 6.44

Thermal Expansion Coefficients of $K_2O-PbO-SiO_2$ Glasses

Composition (mole %)			Temperature range (°C)	$\alpha \times 10^7\ (°C^{-1})$	Reference[a]
K_2O	PbO	SiO_2			
9.5	16	74.5	20–200	90	1
12.2	30.8	57.0		112	
14	42	44		152	
18.5	11.5	70		117	
20	17	63		137	
22.7	23.8	53.5		173	
25.8	7.2	67.0		176	
28.4	12	59.6		178	
18.27	1.54	80.19	25–325	111	2
19.07	3.22	77.71		115.4	
25.84	17.45	56.71		160	

[a] Data from (1) Bezborodov *et al.* (1959) and (2) Rao (1963b).

TABLE 6.45

Thermal Expansion Coefficients of Rb_2O–CaO–SiO_2 Glasses

Composition (mole %)			Temperature range (°C)	$\alpha \times 10^7$ (°C^{-1})	Reference[a]
Rb_2O	CaO	SiO_2			
3.3	55.3	41.4	0–300	163	1
11.6	52	36.4		206	
7.3	48	44.7		170	
3.3	44.6	52.1		130	
17.2	42.8	40		222	
11.7	39.3	49		185	
7.3	36.2	56.5		145	
3.4	33.6	63.0		109	
23.8	31.8	44.4		224	
17.3	28.8	53.9		214	
11.8	26.4	61.8		163	
7.3	24.4	68.3		125	
3.4	22.6	74		87.5	
32.1	17.9	50		242	
24	16	60		204	
17.4	14.6	68		180	
12	13.4	74.6		141	
7.4	12.3	80.3		104	
5.36	11.85	82.79	0–190	51.3	2
			190–335	55.0	
			335–715[b]	58.3	

[a] Data from (1) Simpson (1959, 1961) and (2) Waterton and Turner (1934).
[b] $t_g = 715°C$.

<table>
<tr><td colspan="4" align="center">TABLE 6.46</td><td colspan="4" align="center">TABLE 6.47</td></tr>
</table>

Thermal Expansion Coefficients of Rb_2O–BaO–SiO_2 Glasses[a]			**Thermal Expansion Coefficients of Rb_2O–PbO–SiO_2 Glasses**		

Composition (mole %)			$\alpha \times 10^7$ (°C^{-1}) 0–300°C	Composition (mole %)			$\alpha \times 10^7$ (°C^{-1}) 0–300°C
Rb_2O	BaO	SiO_2		Rb_2O	PbO	SiO_2	
13.1	5.3	81.6	139	22	9.3	68.7	143
19.3	5.8	74.9	110	16.6	14	69.4	128
9.4	17.3	73.3	100	11.2	18.8	70	103
21.2	12.9	65.9	135	19.1	21.3	59.6	146
15.6	19.2	65.2	123	12.8	27	60.2	116
10.3	25.3	64.4	116	22.3	31.3	46.4	154

[a] Data from Simpson (1959, 1961).   [a] Data from Simpson (1959, 1961).

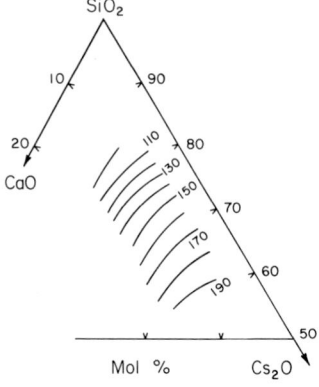

Fig. 6.3. Thermal expansion coefficients ($\alpha \times 10^7 \,^\circ\text{C}^{-1}$) in the temperature range 20–400°C of Cs_2O–CaO–SiO_2 glasses. Data from Bezborodov and Bobkova (1957, 1959).

TABLE 6.48

Thermal Expansion Coefficients of Lithium Borosilicate Glasses

Composition (mole %)			Temperature range (°C)	$\alpha \times 10^7 \,(^\circ\text{C}^{-1})$	Reference[a]
Li_2O	B_2O_3	SiO_2			
33.6	3.6	62.8	20–400	104.5	1
33.8	7.3	58.9		105.2	
34.0	10.9	55.1		110.0	
34.2	14.7	51.1		103.8	
34.4	18.5	47.1		100.3	
34.6	22.3	43.1		109.0	
40.3	3.5	56.2		117.0	
40.5	7.0	52.5		113.0	
40.7	10.5	48.8		127.7	
41.0	14.1	44.9		116.0	
41.2	17.7	41.1		119.4	
46.4	3.3	50.3		144.0	
46.7	6.8	46.5		142.3	
47.0	10.1	42.9		127.5	
47.2	13.5	39.3		124.2	
30.70	1.46	67.84	0–300	105	2
31.43	16.49	52.08		98	
32.19	32.25	35.56		95	
32.99	48.78	18.23		95	
33.40	57.37	9.23		95	

[a] Data from (1) Bezborodov and Ulazovskii (1957) and (2) Dale *et al.* (1951).

TABLE 6.49

Thermal Expansion Coefficients of Lithium Aluminosilicate Glasses

Composition (mole %)			Temperature range (°C)	$\alpha \times 10^7$ (°C^{-1})	Reference[a]	Composition (mole %)			Temperature range (°C)	$\alpha \times 10^7$ (°C^{-1})	Reference[a]
Li$_2$O	Al$_2$O$_3$	SiO$_2$				Li$_2$O	Al$_2$O$_3$	SiO$_2$			
30	5	65	50–150	93.7	1	13	13	74	20–420	50.8	3
35	5	60		106.0		18	18	64		67	
27.5	10	62.5		83.8		20	20	60		70.8	
30	10	60		90.3		17	—	83		61	
32.5	10	57.5		95.0		17	5	78		62	
35	10	55		98.9		17	9	74		63	
37.5	10	52.5		104.5		17	13	70		62	
40	10	50		107.4		17	17	66		63	
						17	21	62		66	
8.3	8.3	83.4	30–500	39.3	2	25	—	75		88	
10	10	80		43.6		25	9	66		81	
12.5	12.5	75		52.5		25	25	50		78	
16.7	16.7	66.6		66.6							
25	25	50		76.0							

[a] Data from (1) Moore and McMillan (1956), (2) Brackbill et al. (1951), and (3) Vargin et al. (1971).

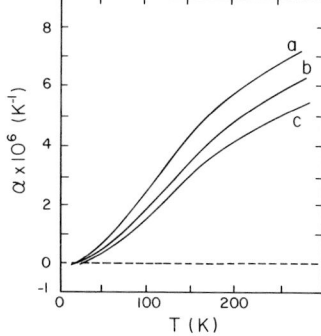

Fig. 6.4. Low-temperature thermal expansion coefficients of three lithium aluminosilicate glasses. (a) LiAlSiO$_4$, (b) LiAlSi$_2$O$_6$, (c) LiAlSi$_3$O$_8$. Data from Hirao and Soga (1982).

TABLE 6.50

Thermal Expansion Coefficients of Lithium Gallosilicate Glasses[a]

Composition (mole %)			$\alpha \times 10^7 \, (°C^{-1})$	Composition (mole %)			$\alpha \times 10^7 \, (°C^{-1})$
Li$_2$O	Ga$_2$O$_3$	SiO$_2$	20–400°C	Li$_2$O	Ga$_2$O$_3$	SiO$_2$	20–400°C
12	22	66	54.1	20	12.5	67.5	73.2
12.5	12.5	75	54.7	20	15	65	73.8
12.5	17.5	70	55.2	20	20	60	71.6
13.3	21.7	65	56.7	20	25	55	68.5
16	12	72	62.3	20	27.5	52.5	68.6
16	21	63	62.8	25	15	60	84.8
16	28	56	63.9	25	25	50	80.0
16.7	16.7	66.6	65.3	27.5	10.4	62.1	93.0
20	5	75	71.0	27.5	18.1	54.4	88.7
20	10	70	73.5	30	10	60	99.6

[a] Data from Dubrovo and Tsekhomskaya (1964a,b), Dubrovo et al. (1969), and Tsekhomskaya and Dubrovo (1968).

TABLE 6.51

Thermal Expansion Coefficients of Sodium Borosilicate Glasses[a]

Composition (mole %)			$\alpha \times 10^7 \, (°C^{-1})$	Composition (mole %)			$\alpha \times 10^7 \, (°C^{-1})$
Na$_2$O	B$_2$O$_3$	SiO$_2$	20–300°C	Na$_2$O	B$_2$O$_3$	SiO$_2$	20–300°C
1	39	60	45	8	22	70	47
2	28	70	36	8	27	65	49
2	38	60	46	8	32	60	51
3	37	60	45	8	37	55	57
4	21	75	32	10	20	70	56
4	26	70	33	10	30	60	57
4	31	65	40	12	28	60	64
4	36	60	43	15	15	70	75
4	41	55	50	15	25	60	72
5	35	60	45	20	5	75	113
6	24	70	41	20	10	70	95
6	34	60	46	20	20	60	91
8	17	75	42	25	15	60	102

[a] Data from Strel'tsina (1967, 1970) and Mazurin et al. (1969).

TABLE 6.52
Thermal Expansion Coefficients of Sodium Aluminosilicate Glasses[a]

Composition (wt. %)			t_g (°C)	$\alpha \times 10^7$ (°C^{-1}) 20–100°C	Composition (wt. %)			t_g (°C)	$\alpha \times 10^7$ (°C^{-1}) 20–100°C	Reference
Na$_2$O	Al$_2$O$_3$	SiO$_2$			Na$_2$O	Al$_2$O$_3$	SiO$_2$			
12.9	22.1	65.0	660	75	20	37	43	810	77	(1)
14.4	14.5	71.1	690	64	20.3	23.5	56.2	660	93	
14.6	9.5	75.9	570	68	21	5.9	73.1	520	85	
15.9	21.1	63.0	640	78	22.5	12.2	65.3	540	91	
16	16	68.0	600	75	22.6	17	60.4	560	90	
16.8	26.2	57.0	690	84	23.9	25.8	50.3	720	90	
17	15	68.0	600	76	24.2	13.3	62.5	540	95	
17.8	36.7	45.5	790	80	25.7	8.3	66.0	530	102	
17.9	10.2	71.9	570	80	26.5	18.5	55.0	570	103	
18	16	66	580	82	26.6	4.5	68.9	520	107	
18	28.6	53.4	710	90	29	5	66.0	610	117	
18.2	10.9	70.9	570	81	30.1	14.3	55.6	540	115	
18.9	34.3	46.8	790	80	33.8	7.8	58.4	490	136	
19.2	19.8	61.0	590	85	36.2	14.1	49.7	500	146	
19.9	23.8	56.3	660	93						
20.8	40.4	38.8	760	100^b	25.3	28.0	46.7	710	105^b	(2)
21.8	38.3	39.9	740	105^b	26.9	21.6	51.5	535	123^b	
21.9	35.5	42.6	730	101^b	29.6	14.1	56.3	425	123^b	
22.1	33.0	44.9	740	112^b	32.8	5.1	62.1	430	131^b	

[a] Data from (1) Rencker (1934, 1935) and (2) Terai (1968a,b, 1969).
[b] Temperature range not known.

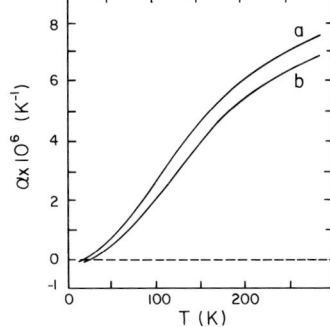

Fig. 6.5. Low-temperature thermal expansion coefficients of two sodium aluminosilicate glasses. (a) $NaAlSi_2O_6$, (b) $NaAlSi_3O_8$. Data from Hirao and Soga (1982).

TABLE 6.53

Thermal Expansion Coefficients of Sodium Gallosilicate Glasses

Composition (mole %)			Temperature range	$\alpha \times 10^7$ (°C^{-1})	Reference[a]
Na_2O	Ga_2O_3	SiO_2			
15	5	80	RT–400°C	81.9	1
15	10	75		82.8	
15	20	65		82.5	
20	10	70		101.2	
20	14	66		101.2	
20	16	64		102.0	
20	18	62		102.0	
20	20	60		103.0	
20	28	52		88.7	
21	9	70	30–350°C	113.1	2
30	5	65		144.0	
30	10	60		135.0	

[a] Data from (1) Dubrovo (1965), Danilova and Dubrovo (1965), Dubrovo *et al.* (1965a,b), and (2) Takahashi *et al.* (1975).

TABLE 6.54

Thermal Expansion Coefficients of Some $Na_2O-M_2O_3-SiO_2$ Glasses[a]

Composition (mole %)			$\alpha \times 10^7$ (°C^{-1}) 50–350°C	Composition (mole %)			$\alpha \times 10^7$ (°C^{-1}) 50–350°C
Na_2O	M_2O_3	SiO_2		Na_2O	M_2O_3	SiO_2	
	M = Sc				M = Sb		
21	9	70	96.8	21	9	70	114.8
30	5	65	130.7	30	5	65	134.6
30	10	60	114.7	30	10	40	121.6
	M = Y				M = Fe		
21	9	70	104.5	21	9	70	116.4
30	5	65	138.1	30	5	65	145.6
30	10	60	122.1	30	10	60	138.6
	M = In						
21	9	70	102.3				
30	5	65	133.6				
30	10	60	121.2				

[a] Data from Takahashi *et al.* (1975).

TABLE 6.55

Thermal Expansion Coefficients of
$3Na_2O-7SiO_2$ after the Addition of
Various M_2O_3 Oxides[a]

M_2O_3 (mole %)	$\alpha \times 10^7$ ($°C^{-1}$)	M_2O_3 (mole %)	$\alpha \times 10^7$ ($°C^{-1}$)
M = Y		M = Nd	
1	136	2.4	137
2.4	133	3.6	134
4.8	123	4.8	129
7	122	6.0	128
M = Yb		7.0	125
1.2	139	9.0	125
2.4	130		
3.6	125		
6.0	120.5		
7.0	119		

[a] Data from graph in Komiyama (1974).

TABLE 6.56

Thermal Expansion Coefficients of $Na_2O-La_2O_3-SiO_2$ Glasses[a]

Composition (mole %)			$\alpha \times 10^7$ ($°C^{-1}$) 20–400°C	Composition (mole %)			$\alpha \times 10^7$ ($°C^{-1}$) 20–400°C
Na_2O	La_2O_3	SiO_2		Na_2O	La_2O_3	SiO_2	
5	7	88	49.0	17	5	78	90.9
6	7	87	49.7	18	2	80	92.3
10	10	80	74.3	20	5	75	103.5
12	1	87	63.6	20	10	70	105.5
12	5	83	67.8	20	14	66	106.4
12	8	80	76.1	21	5	74	118.6
12	12	76	83.2	27	13	60	120.5
12	16	72	88.5	30	5	65	125.8
12.5	12.5	75	83.8	30	10	60	126.5
15	5	80	84.6	35	5	60	147
15	10	75	89.7	39	1	60	170
15	15	70	89.2				
16	11	73	91.7				

[a] Data from Dubrovo and Shnypikov (1966, 1968) and Shnypikov (1967).

TABLE 6.57

Thermal Expansion Coefficients of $Na_2O-Bi_2O_3-SiO_2$ Glasses[a]

Composition (wt. %)				$\alpha \times 10^7 \ (°C^{-1})$
Na_2O	Bi_2O_3	SiO_2	$t_g \ (°C)$	20–400°C
10.1	16.2	73.5	465	93
17.2	11.4	71.2	460	112
20.4	16.2	63.2	444	126
22.5	19.4	58.0	426	135
22.5	8	69.5	458	132
26.4	5.3	68.2	449	142
35	11.5	53.3	405	161
32.5	4.95	62.4	445	145
37.6	8	54.4	395	180[b]
39.8	5.3	54.8	380	176[b]
34	1.27	64.7	420	169

[a] Data from Watanabe et al. (1970).
[b] $200-t_g$ °C.

TABLE 6.58

Thermal Expansion Coefficients of Potassium Borosilicate Glasses[a]

Composition (mole %)			$\alpha \times 10^7 \ (°C^{-1})$
K_2O	B_2O_3	SiO_2	20–300°C
2	38	60	47
4	36	60	50
6	34	60	56
8	32	60	65
10	30	60	74
8	37	55	65
8	27	65	63
8	22	70	60
8	17	75	56
4	26	70	58

[a] Data from Strel'tsina (1967, 1970) and Mazurin et al. (1969).

TABLE 6.59

Thermal Expansion Coefficients of Potassium Aluminosilicate Glass[a]

Composition (mole %)			$\alpha \times 10^7 \ (°C^{-1})$
K_2O	Al_2O_3	SiO_2	0–300°C
20.5	9.5	70	64.6

[a] Data from Rothermel (1967).

TABLE 6.60

Thermal Expansion Coefficients of K$_2$O–Fe$_2$O$_3$–SiO$_2$ Glasses[a]

Composition (mole %)			$\alpha \times 10^7 \, (°C^{-1})$
K$_2$O	Fe$_2$O$_3$	SiO$_2$	RT–430°C
16.67	16.67	66.66	103
17.24	13.79	68.97	108
17.70	11.50	70.80	110
18.18	9.09	72.73	108

[a] Data from Hoshikawa and Akagi (1974).

TABLE 6.61

Thermal Expansion Coefficients of Rubidium Borosilicate Glasses[a]

Composition (mole %)			$\alpha \times 10^7 \, (°C^-)$
Rb$_2$O	B$_2$O$_3$	SiO$_2$	0–300°C
7.9	20.7	71.4	66.8
8.1	42.2	49.7	74.5
8.3	53.8	37.9	76.6
12.4	10.8	76.8	70
12.5	22.7	64.8	71
12.7	34.1	53.2	74.7
13.0	46.3	40.7	75.8
17.8	11.9	70.3	71
17.9	24.8	57.3	75.7
18.4	37.7	43.9	82.2
26.2	41.8	32.0	84.2

[a] Data from Simpson (1961).

TABLE 6.62

Thermal Expansion Coefficients of Cesium Aluminosilicate Glasses[a]

Composition (mole %)			$\alpha \times 10^7 \, (°C^{-1})$
Cs$_2$O	Al$_2$O$_3$	SiO$_2$	0–300°C
5	5	90	22.8
5	10	85	28.7
10	5	85	47.8
7.5	12.5	80	33.6
10	10	80	34.5
20	5	75	68.4
10	20	70	39.7
10	25	65	51.0

[a] Data from Bollin (1972).

TABLE 6.63

Thermal Expansion Coefficients of $Cu_2O-Al_2O_3-SiO_2$ Glasses[a]

Composition (mole %)			$\alpha \times 10^7 \, (°C^{-1})$
$Cu_2O + (CuO)_2$	Al_2O_3	SiO_2	20–300°C
12.5	12.5	75	5
17.5	17.5	65	7
25	10	65	10

[a] Data from Matusita and Mackenzie (1979).

TABLE 6.64

Thermal Expansion Coefficients of Thallium Borosilicate Glasses[a]

Composition (mole %)			$\alpha \times 10^7 \, (°C^{-1})$
Tl_2O	B_2O_3	SiO_2	RT–250°C
1.4	45.8	52.8	46
3.9	44.2	51.9	54
6.1	43.5	50.4	58
8.8	42.2	49.0	62
13.3	40	46.7	72
18.4	38.2	43.4	78
25.8	33.9	40.3	96
38	28	34.0	122

[a] Data from Kim (1968).

TABLE 6.65

Thermal Expansion Coefficients of $Li_2O-SnO_2-SiO_2$ Glasses[a]

Composition (mole %)			$\alpha \times 10^7 \, (°C^{-1})$
Li_2O	SnO_2	SiO_2	20–400°C
32.5	—	67.5	110
32.5	1.5	66	107
32.5	4.5	63	99
32.5	6.0	61.5	95.5

[a] Data from graphs in Vakhrameev (1968) and Vakhrameev and Evstrop'ev (1969).

TABLE 6.66

**Thermal Expansion Coefficients of
$Li_2O-ZrO_2-SiO_2$ Glasses[a]**

Composition (mole %)			$\alpha \times 10^7 \ (°C^{-1})$
Li_2O	ZrO_2	SiO_2	$20-300°C$
20	2.5	77.5	64
20	5	75	62
20	7.5	72.5	64
20	10	70	65
25	2.5	72.5	82
25	5	70	79
25	7.5	67.5	77
25	10	65	75
30	5	65	88
30	10	60	86

[a] Data from Kheifets *et al.* (1972) and Milyukov and
Kheifets (1968).

TABLE 6.67

Thermal Expansion Coefficients of $Na_2O-GeO_2-SiO_2$ Glasses[a]

Composition (mole %)			$\alpha \times 10^7 \ (°C^{-1})$	Composition (mole %)			$\alpha \times 10^7 \ (°C^{-1})$
Na_2O	GeO_2	SiO_2	$50-350°C$	Na_2O	GeO_2	SiO_2	$50-350°C$
20	—	80	103	30	—	70	144
20	10	70	101	30	10	60	143
20	20	60	100	30	20	50	143.5
20	30	50	101	30	30	40	143.5
20	40	40	102	30	40	30	142
20	50	30	103	30	50	20	143
20	60	20	105	30	60	10	143
20	70	10	106.5	30	70	—	142
20	80	—	107.5				

[a] Data from graph in Takahashi *et al.* (1977).

TABLE 6.68

Thermal Expansion Coefficients of $Na_2O-SnO_2-SiO_2$ Glasses[a]

Composition (mole %)			$\alpha \times 10^7$ (°C^{-1}) 20–400°C	Composition (mole %)			$\alpha \times 10^7$ (°C^{-1}) 20–400°C
Na_2O	SnO_2	SiO_2		Na_2O	SnO_2	SiO_2	
15	2.5	82.5	72.8	25	7.5	67.5	99.0
15	5	80	70.2				
15	7.5	77.5	68.5	30	2.5	67.5	123.0
20	2.5	77.5	95[b]				
20	5	75	91[b]	30	5	65	118.0
20	7.5	72.5	87[b]				
25	2.5	72.5	105.0	30	7.5	62.5	114.0
25	5	70	102.0	30	10	60	112.0

[a] Data from Vakhrameev (1968) and Vakhrameev and Evstrop'ev (1969).
[b] Values read from graph.

TABLE 6.69

Thermal Expansion Coefficients of
$Na_2O-TiO_2-SiO_2$ Glasses[a]

Composition (mole %)			$\alpha \times 10^7$ (°C^{-1}) 50–300°C
Na_2O	TiO_2	SiO_2	
15.5	14.0	70.5	84
16.5	26.8	56.7	110
21.7	24.9	53.4	112
22.6	37.4	40.3	108
23.0	30.2	46.8	116
23.3	5.4	71.5	110
23.6	21.8	54.8	107
25	40	35	115
26.2	12.9	61.0	118
27.6	25.8	46.6	116
27.7	15.8	56.5	127
28.0	35.8	36.3	142
29.1	31.4	39.7	130
30	20	50	131
30	40	30	124
32.2	34.1	33.7	161
37.8	23.2	39.0	162

[a] Data from Hirayama and Berg (1961).

TABLE 6.70

Thermal Expansion Coefficients of Na_2O–ZrO_2–SiO_2 Glasses[a]

Composition (mole %)			$\alpha \times 10^7$ ($°C^{-1}$) 20–400°C	Composition (mole %)			$\alpha \times 10^7$ ($°C^{-1}$) 20–400°C
Na_2O	ZrO_2	SiO_2		Na_2O	ZrO_2	SiO_2	
10	10	80	49	30	10	60	106
10	15	75	49	30	15	55	100
10	20	70	49	30	20	50	89
15	5	80	70	35	5	60	137
15	10	75	69	35	10	55	122
15	15	70	65	35	15	50	110
15	20	65	68	35	20	45	94
20	5	75	86	40	5	55	144
20	10	70	79	40	10	50	130
20	15	65	84	40	15	45	112
25	5	70	105	40	20	40	105
25	10	65	94	45	5	50	148
25	15	60	96	45	10	45	144
30	5	65	120	45	15	40	115

[a] Data from Kheifets et al. (1972) and Milyukov and Kheifets (1968).

TABLE 6.71

Thermal Expansion Coefficient of Na_2O–HfO_2–SiO_2 Glass[a]

Composition (mole %)			$\alpha \times 10^7$ ($°C^{-1}$) RT–450°C
Na_2O	HfO_2	SiO_2	
18	12	70	64.0

[a] Data from Brekhovskikh and Sesorova (1960).

TABLE 6.72

Thermal Expansion Coefficients of Na_2O–ThO_2–SiO_2 Glasses[a]

Composition (mole %)			$\alpha \times 10^7$ ($°C^{-1}$) 50–350°C
Na_2O	ThO_2	SiO_2	
20	5	75	97.5
30	5	65	136.1

[a] Data from graph in Takahashi et al. (1977).

TABLE 6.73
Thermal Expansion Coefficients of $Na_2O-TeO_2-SiO_2$ Glasses[a]

Composition (mole %)			t_g	$\alpha \times 10^7 \, (°C^{-1})$	Composition (mole %)			t_g	$\alpha \times 10^7 \, (°C^{-1})$
Na_2O	TeO_2	SiO_2	(°C)	50–350°C	Na_2O	TeO_2	SiO_2	(°C)	50–350°C
25	4.8	70.2	439	128.7	30	20.2	49.8	344	168.8[b]
25	9.8	65.2	412	136.3	30	24.9	45.1	320	181.7[b]
25	19.6	55.4	358	150.4[b]	35	5	60	417	165.0
25	24.9	50.1	336	159.6[b]	35	10	55	382	168.6[b]
30	4.8	65.2	418	148.0	40	5	55	395	182.3
30	10.0	60.0	395	155.1	45	5	50	390	207.0[b]
30	15.0	55.0	370	162.7[b]	50	4.8	45.2	380	219.9[b]

[a] Data from Mochida et al. (1980).
[b] 50–300°C.

TABLE 6.74
Thermal Expansion Coefficients of $K_2O-SnO_2-SiO_2$ Glasses[a]

Composition (mole %)			$\alpha \times 10^7 \, (°C^{-1})$
K_2O	SnO_2	SiO_2	20–400°C
20	—	80	102.5
20	2.5	77.5	97.5
20	5	75	93
20	7.5	72.5	88.5

[a] Data from graph in Vakhrameev (1968) and Vakhrameev and Evstrop'ev (1969).

TABLE 6.75
Thermal Expansion Coefficients of $K_2O-TiO_2-SiO_2$ Glasses[a]

Composition (wt. %)			$\alpha \times 10^7 \, (°C^{-1})$	Composition (wt. %)			$\alpha \times 10^7 \, (°C^{-1})$
K_2O	TiO_2	SiO_2	25–325°C	K_2O	TiO_2	SiO_2	25–325°C
15	5	80	54	30	45	25	136
15	12.5	72.5	62	40	5	55	171
15	20	65	67	40	10	50	176
15	35	50	74	40	20	40	179
15	40	45	77	40	40	20	188
20	5	75	84	45	5	50	199
20	15	65	81	45	10	45	201
20	20	60	86	45	20	35	205
20	40	40	96	45	30	25	208
20	45	35	94	45	40	15	210
30	10	60	123	45	45	10	205
30	15	55	120	60	10	30	265
30	30	40	128				
30	40	30	138				

[a] Data from Rao (1963a,b).

TABLE 6.76

Thermal Expansion Coefficient of Li$_2$O–P$_2$O$_5$–SiO$_2$ Glass[a]

Composition (wt. %)			$\alpha \times 10^7$ ($°C^{-1}$)
Li$_2$O	P$_2$O$_5$	SiO$_2$	50–325°C
19.2	3.9	76.9	125

[a] Data from Nagaoka *et al.* (1962).

TABLE 6.77

Thermal Expansion Coefficients of Na$_2$O–P$_2$O$_5$–SiO$_2$ Glasses with P$_2$O$_5$:Na$_2$O Ratio of 1.5[a]

Mole % SiO$_2$	$\alpha \times 10^7$ ($°C^{-1}$) 20–100°C
0	190
4	184
7.5	156
10.7	146
14	142
16.7	141
25	114

[a] Data from graph in Takahashi (1962).

TABLE 6.78

Thermal Expansion Coefficients of Na$_2$O–Nb$_2$O$_5$–SiO$_2$ Glasses[a]

Composition (mole %)			$\alpha \times 10^7$ ($°C^{-1}$)	Composition (mole %)			$\alpha \times 10^7$ ($°C^{-1}$)
Na$_2$O	Nb$_2$O$_5$	SiO$_2$	50–300°C	Na$_2$O	Nb$_2$O$_5$	SiO$_2$	50–300°C
15	7.5	77.5	62	19.1	19.1	61.8	72
19.9	5	75.1	92	19.8	26	54.2	66
12.6	12.5	74.9	45	22.9	23	54.1	70
21.9	7	71.1	92	29.9	16	54.1	103
15	15	70	60	26	20	54	84
20	10	70	75	35.1	11	53.9	125
25	5	70	99	25	25	50	76
29.9	5	65.1	127	30	20	50	95
24.9	13	62.1	92	38.1	12	49.9	135
29	9	62	87	33.9	19.1	47.0	112
22.1	16	61.9	90	40.9	13.6	45.5	143

[a] Data from Hirayama and Berg (1963).

TABLE 6.79

Thermal Expansion Coefficients of $K_2O-Nb_2O_5-SiO_2$ Glasses[a]

Composition (mole %)			$\alpha \times 10^7 \, (°C^{-1})$	Composition (mole %)			$\alpha \times 10^7 \, (°C^{-1})$
K_2O	Nb_2O_5	SiO_2	25–325°C	K_2O	Nb_2O_5	SiO_2	25–325°C
16.1	19.4	64.5	68	23.8	28.6	47.6	88
16.7	16.7	66.6	71	25	25	50	94
17.2	13.8	69.0	72.5	26.3	21.1	52.6	98
17.9	10.7	71.4	76	27.8	16.7	55.5	103
18.5	7.4	74.1	81	29.4	11.8	58.8	118
19.2	3.8	77.0	95	31.3	6.2	62.5	137

[a] Data from Rao (1962).

TABLE 6.80

Thermal Expansion Coefficients of $RO-MnO-SiO_2$ Glasses[a]

Composition (mole %)			$\alpha \times 10^7 \, (°C^{-1})$	Composition (mole %)			$\alpha \times 10^7 \, (°C^{-1})$
RO	MnO	SiO_2	200–400°C	RO	MnO	SiO_2	200–400°C
R = Ca				R = Zn			
10	40	50	67.5	5	45	50	52.5
20	30	50	76.5	10	40	50	57.0
30	20	50	81.0	20	30	50	62.0
40	10	50	74.5	30	20	50	50.5
R = Sr				40	10	50	42.5
10	40	50	70.0	R = Cd			
20	30	50	81.8	10	40	50	48.2
30	20	50	89.6	20	30	50	71.0
40	10	50	84.6	30	20	50	67.5
R = Ba				40	10	50	61.3
10	40	50	78.0				
20	30	50	91.7				
30	20	50	101.0				
40	10	50	99.3				

[a] Data from Kuznetsova (1972).

TABLE 6.81

**Thermal Expansion Coefficients of
BaO–ZnO–SiO$_2$ Glasses[a]**

Composition (mole %)			
ZnO	BaO	SiO$_2$	$\alpha \times 10^7$ (°C^{-1})
2	28	70	86
10	30	60	92
12	28	60	92
14	36	50	105
24	16	60	69
24	26	50	93
24	28	48	95
30	20	50	79
38	12	50	72
42	8	50	57

[a] Data from Cleek and Babcock (1973).

TABLE 6.82

**Thermal Expansion Coefficients of
BaO–PbO–SiO$_2$ Glasses[a]**

Composition (mole %)			$\alpha \times 10^7$ (°C^{-1})
BaO	PbO	SiO$_2$	20–500°C
17.5	17.5	65	82.4
20	20	60	89.2
23.3	20	56.7	93.6
22.5	22.5	55	95.6
25	25	50	102.1

[a] Data from Sheludyakov (1967).

TABLE 6.83

**Thermal Expansion Coefficient of ZnO–PbO–SiO$_2$
Glass[a]**

Composition (wt. %)			$\alpha \times 10^7$ (°C^{-1})
PbO	ZnO	SiO$_2$	20–300°C
60.33	19.74	19.97	64.6

[a] Data from Sack *et al.* (1968).

TABLE 6.84
Thermal Expansion Coefficients of Magnesium Aluminosilicate Glasses

Composition (wt. %)			$\alpha \times 10^7 \ (^\circ C^{-1})$ (25–600°C)	Reference[a]	Composition (wt. %)			$\alpha \times 10^7 \ (^\circ C^{-1})$ (20–300°C)	Reference[a]
MgO	Al$_2$O$_3$	SiO$_2$			MgO	Al$_2$O$_3$	SiO$_2$		
5.42	13.72	80.86	20.2	1	25	10	65	38.5	2
8.02	20.28	71.70	27.1		25	15	60	37	
10.54	26.65	62.81	33.1		25	20	55	37	
20.3	18.3	61.4	43.1		30	10	60	42.5	
					30	15	55	41.5	
					30	20	50	39	
					35	10	55	47	
					35	15	50	45	
					40	10	50	50.5	
					45	10	45	55	

[a] Data from (1) Hummel and Reid (1951) and (2) Hasegawa (1980).

TABLE 6.85

Thermal Expansion Coefficients of Calcium Aluminosilicate Glasses[a]

Composition (mole %)			$\alpha \times 10^7 \, (^\circ C^{-1})$
CaO	Al_2O_3	SiO_2	$20-300^\circ C$
20	10	70	45.6
20	20	60	43.8
20	25	55	39.9
25	15	60	51.7
30	10	60	67.2
40	10	50	73.7
40	20	40	67.5
40	30	30	61.7
50	10	40	83.2

[a] Data from Danilova (1967) and Danilova and Dubrovo (1967).

TABLE 6.86

Thermal Expansion Coefficients of Calcium Gallosilicate Glasses[a]

Composition (mole %)			$\alpha \times 10^7 \, (^\circ C^{-1})$	Composition (mole %)			$\alpha \times 10^7 \, (^\circ C^{-1})$
CaO	Ga_2O_3	SiO_2	$20-400^\circ C$	CaO	Ga_2O_3	SiO_2	$20-400^\circ C$
15	10	65	39.5	25	25	50	54.7
15	15	70	43.4	25	35	40	57.0
15	25	60	45.5	30	10	60	64.0
20	10	70	44.9	30	30	40	59.0
20	15	65	49.7	40	10	50	74.0
20	20	60	50.4	40	20	40	68.3
20	25	55	50.8	40	30	30	67.1
20	30	50	50.2	40	40	20	67.4
25	15	60	56.5	50	10	40	88.6

[a] Data from Danilova (1967) and Danilova and Dubrovo (1967).

TABLE 6.87

Thermal Expansion Coefficients of Strontium Aluminosilicate Glasses[a]

| Composition (mole %) | | | t_g (°C) | $\alpha \times 10^7$ (°C^{-1}) | | Composition (mole %) | | | t_g (°C) | $\alpha \times 10^7$ °C^{-1}) | |
SrO	Al$_2$O$_3$	SiO$_2$		20–400°C	20–t_g	SrO	Al$_2$O$_3$	SiO$_2$		20–400°C	20–t_g
25	10	65	800	65.7	63.7	50	5	45	800	93.9	96.5
30	5	65	765	74.2	74.1	35	25	40	860	73.3	79.1
30	10	60	785	74.4	76.2	40	20	40	840	81.5	83.3
35	5	60	750	85.4	87.4	45	15	40	820	87.8	84.5
35	10	55	785	81.3	80.1	50	10	40	815	88.8	89.0
40	5	55	770	90.0	90.4	55	5	40	800	90.3	97.0
35	15	50	800	80.6	82.3	35	30	35	865	71.9	78.0
40	10	50	795	88.3	86.8	40	25	35	860	79.6	76.4
45	5	50	775	89.9	94.8	45	20	35	850	83.7	80.3
35	20	45	850	74.9	82.0	50	15	35	840	88.6	86.1
40	15	45	810	86.2	84.8	55	10	35	830	88.6	91.1
45	10	45	800	91.3	89.7	60	5	35	810	90.4	98.3

[a] Data from Iskhakov (1971).

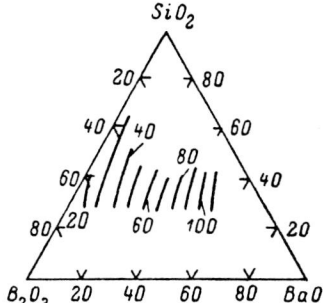

Fig. 6.6. Thermal expansion coefficients ($\alpha \times 10^7$ °C^{-1}) in the temperature range 20–400°C of barium borosilicate glasses; compositions in mole %. Data from Mazurenko (1963).

TABLE 6.88

**Thermal Expansion Coefficients of
Barium Aluminosilicate Glasses**[a]

Composition (mole %)			$\alpha \times 10^7 \, (°C^{-1})$
BaO	Al_2O_3	SiO_2	20–300°C
10.5	19.5	70	31
13	17	70	38
16	14	70	44
18	12	70	50
16	19	65	38
18	17	65	44
21	14	65	51
23	12	65	56
23.5	16.5	60	52
25.5	14.5	60	57
28	12	60	63
28	17	55	57
31	14	55	64
33.5	11.5	55	70

[a] Data from graph in Pavlushkin *et al.* (1974).

TABLE 6.89

**Thermal Expansion Coefficients of
Barium Gallosilicate Glasses**[a]

Composition (mole %)			$\alpha \times 10^7 \, (°C^{-1})$
BaO	Ga_2O_3	SiO_2	20–400°C
20	10	70	57.7
20	15	65	56.5
20	20	40	55.6
20	25	55	55.5
20	30	50	54.3
30	10	60	78.3
30	30	40	67.0
40	10	50	92.0
40	20	40	85.9
40	30	30	79.7
40	40	20	73.5
40	45	15	71.7

[a] Data from Danilova and Dubrovo (1971).

TABLE 6.90

Thermal Expansion Coefficients of BaO–La$_2$O$_3$–SiO$_2$ Glasses[a]

Composition (mole %)			
BaO	La$_2$O$_3$	SiO$_2$	$\alpha \times 10^7$ (°C^{-1})
28	2	70	89
33	2	65	100
38	2	60	104
23	5	72	84
28	5	67	94
33	5	62	108
37	5	58	114
40	5	55	108
23	7	70	86
30	7	63	98
35	7	58	113
38	7	55	109
25	10	65	93

[a] Data from Cleek and Babcock (1973).

TABLE 6.91

Thermal Expansion Coefficients of Zinc Borosilicate Glasses[a]

Composition (mole %)			$\alpha \times 10^7$ (°C^{-1})	
ZnO	B$_2$O$_3$	SiO$_2$	25–100°C	100–400°C
50	45	5	34.9	51.8
55.5	34.5	10	35.4	50.1
60	20	20	33.7	49.6
60	30	10	36.5	50.9
60	32	8	32.1	50.4

[a] Data from Hamilton et al. (1959).

TABLE 6.92

Thermal Expansion Coefficient of Zinc Aluminosilicate Glass[a]

Composition (mole %)			$\alpha \times 10^7$ (°C^{-1}) 20–300°C
(ZnO)$_2$	Al$_2$O$_3$	SiO$_2$	
25	10	65	41

[a] Data from Matusita and Mackenzie (1979).

TABLE 6.93

Thermal Expansion Coefficients of Lead Borosilicate Glasses[a]

Composition (wt. %)			$\alpha \times 10^7 \, (°C^{-1})$	Composition (wt. %)			$\alpha \times 10^7 \, (°C^{-1})$
PbO	B_2O_3	SiO_2	RT–300°C	PbO	B_2O_3	SiO_2	RT–300°C
10.0	9.7	80.3	21	40.2	29.1	30.7	54
9.9	19.5	70.6	28	50.2	19.9	29.9	53
19.8	10.0	70.2	25	60.1	10.3	29.6	79
8.9	29.1	62.0	34	10.0	68.9	21.1	78
19.8	19.2	61.0	32	20.5	59.6	19.9	71
29.5	9.9	60.6	31	30.6	49.1	20.3	63
8.7	39.7	51.6	47	40.2	39.6	20.2	60
19.4	28.9	51.7	37	50.5	29.2	20.3	62
24.8	24.0	51.2	40	60.4	19.4	20.2	63
30.1	19.2	50.7	39	69.9	10.2	19.9	79
39.9	10.0	50.1	42	20.2	69.9	9.9	89
10.1	48.9	41.0	54	30.1	59.9	10.0	82
20.0	39.7	40.3	48	40.2	49.9	9.9	69
30.2	29.3	40.5	46	50.2	39.8	10.0	64
40.4	19.7	39.9	47	60.1	29.9	10.0	70
50.2	10.2	39.6	53	70.1	19.9	10.0	76
20.6	49.0	30.4	59	80.0	10.1	9.9	98
30.4	39.0	30.6	55				

[a] Data from Geller *et al.* (1938).

TABLE 6.94

Thermal Expansion Coefficients of Manganese Aluminosilicate Glasses[a]

Composition (mole %)			t_g (°C)	$\alpha \times 10^7 \, (°C^{-1})$ 20–500°C
MnO	Al_2O_3	SiO_2		
36	8	56	650	48.1
40	5	55	640	48.8
40	10	50	620	48.5
40	15	45	645	50.6
40	20	40	660	46.1
45	10	45	620	53.0
50	10	40	600	60.9
55	10	35	600	66.4

[a] Data from McMillan (1962).

TABLE 6.95

Thermal Expansion Coefficients of $CoO-Al_2O_3-SiO_2$ Glasses[a]

Composition (mole %)				$\alpha \times 10^7$ ($°C^{-1}$)
CoO	Al_2O_3	SiO_2	t_g (°C)	20–500°C
40	10	50	625	44.7
50	10	40	625	51.1

[a] Data from McMillan (1962).

TABLE 6.96

Thermal Expansion Coefficients of $BaO-TiO_2-SiO_2$ Glasses

Composition (mole %)			Temperature range (°C)	$\alpha \times 10^7$ ($°C^{-1}$)	Reference[a]
BaO	TiO_2	SiO_2			
27.5	27.5	45	100–500	90	1
65	5	30	20–220	98.1	2
60	10	30	20–220	96.2	2

[a] Data from (1) Moriya and Ueno (1964) and (2) Matveev et al. (1970).

TABLE 6.97

Thermal Expansion Coefficients of $PbO-GeO_2-SiO_2$ Glasses[a]

Composition (mole %)			$\alpha \times 10^7$ ($°C^{-1}$) 50–350°C	Composition (mole %)			$\alpha \times 10^7$ ($°C^{-1}$) 50–350°C
PbO	GeO_2	SiO_2		PbO	GeO_2	SiO_2	
30	—	70	61	30	20	50	70
40	—	60	76	30	40	30	77
50	—	50	95	30	60	10	84
60	—	40	110	50	10	40	97
10	20	70	39	50	20	30	101
10	40	50	48	50	30	20	105
10	60	30	61	50	40	10	108
10	80	10	73				

[a] Data from Topping et al. (1974).

TABLE 6.98

Thermal Expansion Coefficients of $BaO-Ta_2O_5-SiO_2$ Glasses[a]

Composition (mole %)			$\alpha \times 10^7$ ($°C^{-1}$)
BaO	Ta_2O_5	SiO_2	
30	5	65	83
45	10	45	95

[a] Data from Cleek and Babcock (1973).

TABLE 6.99

Thermal Expansion Coefficients of $Al_2O_3-La_2O_3-SiO_2$ Glasses[a]

Composition (wt. %)				$\alpha \times 10^7$ ($°C^{-1}$)
Al_2O_3	La_2O_3	SiO_2	t_g (°C)	$20-t_g°C$
15	40	45	825	41
22.5	51	26.5	610	77
28.5	38.5	33	755	37

[a] Data from Karlsson (1970).

TABLE 6.100

Thermal Expansion Coefficients of $Al_2O_3-La_2O_3-SiO_2$ Glasses[a]

Composition (mole %)				$\alpha \times 10^7$ ($°C^{-1}$)
Al_2O_3	La_2O_3	SiO_2	t_g (°C)	$20-800°C$
21.5	6.3	72.0	865	38.9
25.6	6.1	68.3	860	39.1
26.4	2.0	71.6	900	33.7
28.4	16.6	55.0	825	62.5
32.8	17.0	50.1	820	64.0
34.7	6.3	59.1	850	43.2

[a] Data from Aleksandrov et al. (1980).

TABLE 6.101

Thermal Expansion Coefficients of
$Nd_2O_3-Bi_2O_3-SiO_2$ Glasses[a]

Composition (wt. %)			$\alpha \times 10^7$ ($°C^{-1}$)
Nd_2O_3	Bi_2O_3	SiO_2	$100-400°C$
2	35	63	40
2	42	56	47
2	50	48	52

[a] Data from Gorashenko et al. (1973).

TABLE 6.102

Thermal Expansion Coefficients of
Al_2O_3–CeO_2–SiO_2 Glasses[a]

Composition (mole %)			$\alpha \times 10^7 (°C^{-1})$
Al_2O_3	CeO_2	SiO_2	20–$420°C$
5	35	60	50
10	30	60	49
15	25	60	46
20	20	60	41
25	15	60	33
20	15	65	35
20	25	55	43
20	30	50	46

[a] Data from graph in Nemkovich *et al.* (1974).

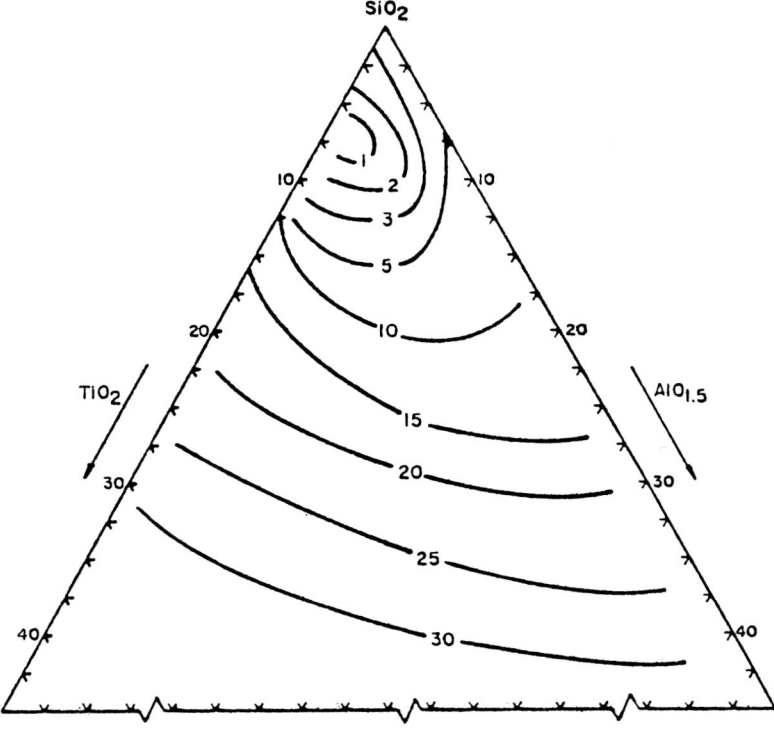

Fig. 6.7. Values of $\alpha \times 10^7 (°C^{-1})$ $(0$–$600°C)$ for Al_2O_3–TiO_2–SiO_2 glasses; composition in cation %. Data from Schultz and Dumbaugh (1980).

TABLE 6.103

**Thermal Expansion Coefficients of
B_2O_3–P_2O_5–SiO_2 Glasses[a]**

Composition (wt. %)			$\alpha \times 10^7 \,(°C^{-1})$
B_2O_3	P_2O_3	SiO_2	75–375°C
1.36	1.02	97.62	6.8
5.55	1.41	93.04	8.5
7.35	9.57	83.08	19.4
10.41	18.07	71.52	40.0
17.20	23.81	58.99	45.5
25.41	32.74	41.85	57.5

[a] Data from Horn and Hummel (1955).

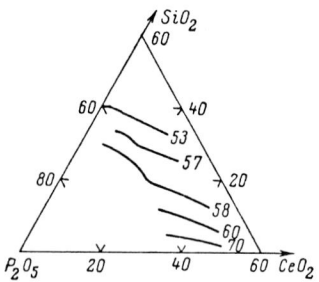

Fig. 6.8. Thermal expansion coefficients ($\alpha \times 10^{-7}$, $°C^{-1}$) in the temperature range 20–300°C of CeO_2–P_2O_5–SiO_2 glasses; compositions in wt. %. Data from Syritskaya *et al.* (1971).

TABLE 6.104

**Thermal Expansion Coefficients of
TiO_2–P_2O_5–SiO_2 Glasses[a]**

Composition (mole %)			$\alpha \times 10^7 \,(°C^{-1})$
TiO_2	P_2O_5	SiO_2	20–300°C
61	35	4	55.0
68	30	2	47.0
73	25	2	45.3
75	20	5	39.8

[a] Data from Syritskaya and Feikner (1966).

Chapter 7

Heat Capacity

The heat capacity C of a material is the quantity of heat dQ required to raise its temperature dT:

$$C = dQ/dT.$$

The heat capacity depends on the constraints imposed during heating; in other words, C is a path property. Thus it is necessary to state the constraints; in this work the heat capacity at constant pressure is used. The units of C are joules per gram per degree celsius. Often a mean heat capacity over a temperature range is reported.

The heat capacity of a solid is small near absolute zero temperature and increases sharply at lower temperatures. From room temperature to higher temperatures, the change in heat capacity is small. For example, the heat capacity of vitreous silica is about 0.73 J g^{-1} °C^{-1} at 25°C, 1.12 at 500°C, and 1.21 at 1000°C (see Chapter 2). The temperature dependence for most silicate glasses is similar.

The heat capacities of silicate glasses containing more than about 60% silica are not much dependent on glass composition, all being about 0.8 ± 0.1 J g^{-1} °C^{-1} at 25°C. Even other oxide glasses such as borate and phosphate have similar values of C_p at 25°C.

In this work the terms heat capacity and specific heat are used interchangeably.

At constant pressure the heat change $dQ = dH$, where H is the thermodynamic enthalpy. Thus for two temperatures T_1 and T_2

$$H_2 - H_1 = \int_{T_1}^{T_2} C_p \, dT.$$

The value of $H_2 - H_1$ found in this way is called the heat content and is usually referred to a reference temperature T_1 of 25°C or 298 K.

The units of heat capacity are energy/mass kelvin. Energy is in joules or calories; 4.1840 joules = 1 calorie. Mass is in grams or moles. To convert these units, the molecular weight M of the glass is needed. If the mole fraction of an oxide component i (Na_2O, CaO, SiO_2) is X_i and its molecular weight M_i, then

$$M = \sum_i x_i M_i$$

summing over all component oxides.

II TABLES, FIGURES, AND EQUATIONS

Tables 7.1–7.16, Figs. 7.1–7.5, and Eqs. 7.1–7.6 give data for the heat capacity of binary glasses; Tables 7.17–7.38, Figs. 7.6–7.16, and Eqs. 7.7–7.17 for ternary glasses.

BIBLIOGRAPHY

Baboian, R. (1969). *J. Chem. Eng. Data* **14**, 63.
Bansal, N. P., and Doremus, R. H. (1984). Unpublished observations.
Haggerty, J. S., Cooper, A. R., and Heasley, J. H. (1968). *Phys. Chem. Glasses* **9**, 47.
Haselton, H. T., Hemingway, B. S., and Robie, R. A. (1984). *Am. Mineral.* **69**, 481.
Hirao, K., and Soga, N. (1982). *Yogyo Kyokaishi* **90**, 476.
Hirao, K., Soga, N., and Kunugi, M. (1979). *J. Am. Ceram. Soc.* **62**, 570.
Krupka, K. M., Robie, R. A., and Hemingway, B. S. (1979). *Am. Mineral.* **64**, 86.
Moore, J., and Sharp, D. E. (1958). *J. Am. Ceram. Soc.* **41**, 461.
Moynihan, C. T., Easteal, A. J., Tran, D. C., Wilder, J. A., and Donovan, E. P. (1976). *J. Am. Ceram. Soc.* **59**, 137.
Muratov, A. V. (1978). *Fiz. Khim. Stekla* **2**, 219.
Neiman, T. S., Yinnon, H., and Uhlmann, D. R. (1982). *J. Non-Cryst. Solids* **48**, 393.
Nemilov, S. V., and Muratov, A. V. (1983). *Sov. J. Glass Phys. Chem. (Engl. Transl.)* **9**, 413.
Primenko, V. I., and Gudovich, O. D. (1978). *Sov. J. Glass Phys. Chem. (Engl. Transl.)* **4**, 69.
Primenko, V. I., and Gudovich, O. D. (1979). *Sov. J. Glass Phys. Chem. (Engl. Transl.)* **5**, 218.
Richet, P., and Bottinga, Y. (1980). *Geochim. Cosmochim. Acta* **44**, 1535.
Richet, P., and Bottinga, Y. (1984). *Geochim. Cosmochim. Acta* **48**, 453.
Robie, R. A., Hemingway, B. S., and Wilson, W. H. (1978). *Am. Mineral.* **63**, 109.
Schwiete, H. E., and Ziegler, G. (1955). *Glastech. Ber.* **28**, 137.
Senapati, H., and Rao, K. J. (1982). *Rev. Chim. Miner.* **19**, 187.
Soga, N. (1982). *J. Phys., Colloq. (Orsay, Fr.)* **43**(C9), 557.
Stebbins, J. F., Weill, D. F., Carmichael, I. S. E., and Moret, L. K. (1982). *Contrib. Mineral. Petrol.* **80**, 276.
Stebbins, J. F., Carmichael, I. S. E., and Weill, D. E. (1983). *Am. Mineral.* **68**, 717.
Stebbins, J. F., Carmichael, I. S. E., and Moret, L. K. (1984). *Contrib. Mineral. Petrol.* **86**, 131.
Stephens, R. B. (1973). *Phys. Rev. B* **8**, 2896.
Stephens, R. B. (1976). *Phys. Rev. B* **13**, 852.
Takahashi, K., and Yoshio, T. (1973). *Yogyo Kyokaishi* **81**, 524.

Takahashi, K., and Yoshio, T. (1978). *Zairyo* **27**, 196.
Tarasov, V. V., and Stroganov, E. F. (1956). *Tr.-Mosk. Khim.-Tekhnol. Inst.* 21, p. 26.
Terai, R., Hori, M., and Yamanaka, H. (1979). *Am. Ceram. Soc. Bull.* **58**, 1125.
Tydlitat, V., Blazek, A., Endrys, J., and Stanek, J. (1972). *Glastech. Ber.* **45**, 352.
White, G. K., Birch, J. A., and Manghnani, M. H. (1977). *J. Non-Cryst. Solids* **23**, 99.
White, W. P. (1919). *Am. J. Sci.* **47**, 1.
Yageman, V. D., and Matveev, G. M. (1982). *Sov. J. Glass Phys. Chem.* (*Engl. Transl.*) **8**, 168.

TABLE 7.1

Heat Contents of $Li_2O-2SiO_2$ Glass[a]

t (°C)	H_t^{25} [kcal (g mole SiO_2)$^{-1}$]	t (°C)	H_t^{25} [kcal (g mole SiO_2)$^{-1}$]
100	1.65	600	13.80
150	2.78	650	15.15
200	3.98	700	16.50
250	5.10	750	17.85
300	6.30	800	19.21
350	7.50	850	20.63
400	8.70	900	22.06
450	9.90	1000	24.75
500	11.10	1033	25.60
550	12.45		

[a] Data from Takahashi and Yoshio (1973, 1978).

TABLE 7.2

Mean Heat Capacities of $Li_2O-2SiO_2$ Glass[a]

Temperature range (°C)	Mean heat Capacity C_m (cal g^{-1} °C^{-1})	Temperature range (°C)	Mean heat capacity C_m (cal g^{-1} °C^{-1})
25–100	0.2932	25–650	0.3230
25–150	0.2963	25–700	0.3257
25–200	0.3031	25–750	0.3281
25–250	0.3020	25–800	0.3303
25–300	0.3053	25–850	0.3332
25–350	0.3075	25–900	0.336
25–400	0.3092	25–1000	0.3383
25–450	0.3104	25–1033	0.3384
25–500	0.3114		
25–550	0.3160		
25–600	0.3198		

[a] Calculated from the heat content data of Takahashi and Yoshio (1973, 1978).

TABLE 7.3

Heat Contents of $Na_2O-3.3SiO_2$ Glass[a]

t (°C)	$H_t - H_{25°C}$ (cal mole^{-1})	t (°C)	$H_t - H_{25°C}$ (cal mole^{-1})
100	4,271	600	40,779
200	10,535	700	49,395
300	17,377	800	58,321
400	24,697	900	67,319
500	32,491	1000	76,579

[a] Data from Schwiete and Ziegler (1955).

TABLE 7.4

Mean Heat Capacities of $Na_2O-3.3SiO_2$ Glass[a]

Temperature range (°C)	Mean heat capacity C_m (cal g^{-1} °C^{-1})	Temperature range (°C)	Mean heat capacity C_m (cal g^{-1} °C^{-1})
25–100	0.2187	25–600	0.2724
25–200	0.2312	25–700	0.2811
25–300	0.2427	25–800	0.2891
25–400	0.2530	25–900	0.2955
25–500	0.2628	25–1000	0.3017

[a] Calculated from the heat content data of Schwiete and Ziegler (1955).

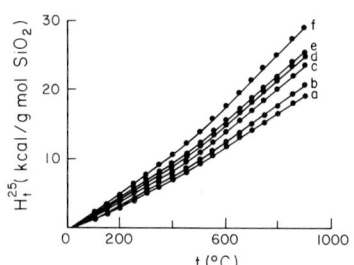

Fig. 7.1. Temperature dependence of heat contents of sodium silicate glasses. (a) $Na_2O-4SiO_2$; (b) $Na_2O-3SiO_2$; (c) $3Na_2O-7SiO_2$; (d) $Na_2O-2SiO_2$; (e) $7Na_2O-13SiO_2$; (f) $2Na_2O-3SiO_2$. From Takahashi and Yoshio (1973).

TABLE 7.5

Heat Capacities of Na₂O–SiO₂ Glasses at Low Temperatures[a]

T (K)	$10^3 C_p$ (J g⁻¹ K⁻¹) for glasses with given mole % Na₂O		
	15	25	40
2	0.0191	0.0190	0.0202
3	0.0624	0.0597	0.0572
4	0.168	0.142	0.137
5	0.374	0.299	0.277
6	0.726	0.574	0.510
7	1.28	0.991	0.876
8	2.04	1.58	1.40
10	4.14	3.38	2.94
12	7.18	5.85	5.17
14	10.82	9.65	8.12
16	15.06	14.0	12.4
18	—	18.8	16.7
20	—	24.4	—

[a] Data from White *et al.* (1977).

TABLE 7.6

Heat Capacities of Na₂O–3SiO₂ Glass[a]

T (K)	$10^3 C_p$ (J g⁻¹ K⁻¹)	T (K)	$10^3 C_p$ (J g⁻¹ K⁻¹)
10	4.50	160	482.2
20	23.96	170	507.1
30	47.00	180	530.1
40	80.87	190	554.6
50	101.1	200	584.0
60	141.8	210	606.0
70	176.5	220	629.1
80	220.6	230	653.8
90	257.9	240	676.9
100	299.3	250	698.3
110	331.7	260	721.5
120	364.4	270	741.8
130	397.7	280	761.7
140	427.8	290	775.7
150	455.1	300	790.7

[a] Data from Muratov (1978).

EQ. 7.1

Specific Heat of Na₂O–3SiO₂ Glass in the Temperature Range 0.05–2 K[a]

$$C \ (\text{erg g}^{-1} \ \text{K}^{-1}) = 20T + 34T^3$$

[a] Data from Stephens (1973, 1976).

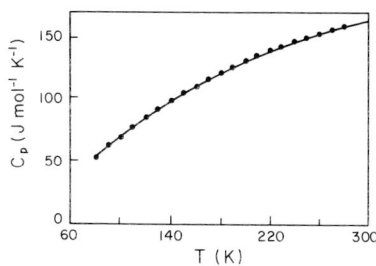

Fig. 7.2. Temperature dependence of low-temperature heat capacity of Na₂O–SiO₂ glass. From Soga (1982).

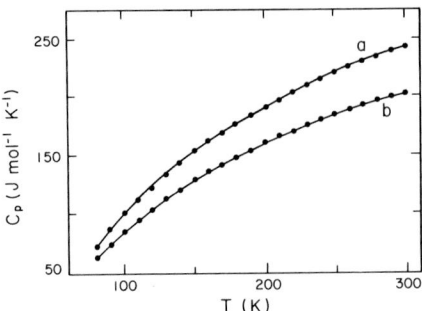

Fig. 7.3. Temperature dependence of low-temperature heat capacity of (a) $Na_2O-4SiO_2$ and (b) $Na_2O-3SiO_2$ glasses. For low-temperature heat capacities of $Na_2O-2SiO_2$ glass, see Fig. 7.5. From Hirao *et al.* (1979).

TABLE 7.7

Heat Capacities of Sodium Silicate Glasses[a]

$T(K)$	C_p (J mol^{-1} K^{-1}) for glass compositions (mole %)		
	$15Na_2O-85SiO_2$	$27.6Na_2O-72.4SiO_2$	$33.3Na_2O-66.7SiO_2$
346	51.5	54.7	55.9
374	53.7	57.2	58.3
401	55.6	59.4	59.8
426	57.4	61.1	61.1
451	59.0	62.8	62.3
475	60.4	64.0	63.6
498	61.7	65.3	64.7
521	62.8	66.4	65.7
544	63.7	67.5	66.6
566	64.5	68.4	67.2
588	65.2	69.2	67.8
609	65.9	69.9	68.6
631	66.5	70.7	69.3
652	67.1	71.3	69.9
673	67.5	71.6	70.8
694	68.0	71.9	72.2
704	68.2	72.2	73.2
715	68.4	72.8	75.9
725	68.9	73.7	80.2
735	69.4	76.0	88.2
746	70.1	80.3	95.4
756	71.6	88.2	95.6
766	74.5	92.3	91.3
776	77.9	89.7	91.0
787	80.0	88.2	91.1
797	80.0	87.7	91.1
807	79.2	87.4	91.2
827	78.7	87.3	91.3
846	78.9	87.3	—
866	79.4	87.3	89.1
886	80.2	87.4	76.1
905	81.0	87.4	

(continues)

TABLE 7.7 (*Continued*)

	C_p (J mol^{-1} K^{-1}) for glass compositions (mole %)		
$T(K)$	$15Na_2O-85SiO_2$	$27.6Na_2O-72.4SiO_2$	$33.3Na_2O-66.7SiO_2$
925	81.4	87.3	
944	82.0	87.0	
963	82.2	86.7	
982	82.9	86.5	
1001	83.4	86.5	
1019	83.8	86.6	
1039	84.0	87.0	
1057	84.1	87.1	
1075	84.6	87.1	
1093	—	87.4	

[a] Data from Yageman and Matveev (1982).

TABLE 7.8

Heat Capacity C_p (cal K^{-1} mole^{-1}) of a Sodium Silicate Glass of Composition $1.0061Na_2O-1.9937SiO_2$[a]

$T(K)$	C_p	$T(K)$	C_p	$T(K)$	C_p
10.0	0.1340	63.0	9.574	180.1	28.39
11.2	0.1934	65.8	10.25	185.1	28.84
12.1	0.2392	68.7	10.93	190.1	29.26
13.0	0.3000	71.2	11.33	195.2	29.73
14.1	0.3853	74.1	12.11	200.0	30.18
15.0	0.4708	77.8	12.93	205.4	30.64
16.2	0.5852	80.1	13.29	210.2	31.03
17.3	0.6900	83.3	14.01	215.1	31.49
18.1	0.7717	86.1	14.50	220.1	31.94
19.0	0.8751	80.0	15.03	225.2	32.35
20.1	1.003	92.2	15.66	230.4	32.82
21.5	1.184	95.0	16.33	235.1	33.15
22.8	1.307	98.4	16.98	240.0	33.49
24.1	1.463	101.2	17.58	245.2	33.87
26.2	1.740	106.5	18.67	250.3	34.25
28.4	2.038	111.2	19.31	255.1	34.62
30.1	2.299	116.4	20.25	260.4	35.01
32.6	2.808	121.3	21.00	265.2	35.26
34.8	3.257	125.1	21.34	270.0	35.48
36.0	3.511	130.4	22.12	272.1	35.63
38.1	3.934	135.1	22.77	273.2	35.69
40.2	4.296	140.2	23.69	275.1	35.84
43.1	5.005	145.1	24.41	280.0	36.21
46.3	5.741	150.0	25.09	285.1	36.64
49.2	6.444	155.2	25.85	290.2	37.04
52.4	7.179	160.1	26.50	295.4	37.41
55.2	7.788	165.3	27.09	297.0	37.53
57.9	8.409	170.1	27.43	298.15	37.61
59.4	8.871	175.0	27.92	300.0	37.75

[a] Data from Nemilov and Muratov (1983).

TABLE 7.9

Heat Contents of $K_2O-2.5SiO_2$ Glass[a]

t (°C)	$H_t - H_{25°C}$ (cal mole^{-1})	t (°C)	$H_t - H_{25°C}$ (cal mole^{-1})
100	3527	900	56000
300	14436	1100	69760
500	27285	1300	83797
700	41505		

[a] Data from Schwiete and Ziegler (1955).

TABLE 7.10

Mean Heat Capacities of $K_2O-2.5SiO_2$ Glass

Temperature range (°C)	Mean heat capacity C_m (cal g^{-1} °C^{-1})	Temperature range (°C)	Mean heat capacity C_m (cal g^{-1} °C^{-1})
25–100	0.1924	25–900	0.2619
25–300	0.2148	25–1100	0.2659
25–500	0.2351	25–1300	0.2690
25–700	0.2516		

[a] Calculated by Moore and Sharp (1958) from the heat content data of Schwiete and Ziegler (1955).

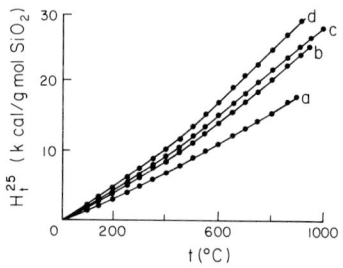

Fig. 7.4. Temperature dependence of heat contents of potassium silicate glasses. (a) $K_2O-4SiO_2$; (b) $3K_2O-7SiO_2$; (c) $K_2O-2SiO_2$; (d) $2K_2O-3SiO_2$. From Takahashi and Yoshio (1973).

EQ. 7.2

Temperature Dependence of the Heat Capacity of $K_2O-4SiO_2$ Glass in the Range 298–752 K[a,b]

$$C_p \text{ (J mole}^{-1}\text{ K}^{-1}) = 450.60 - (116.10 \times 10^{-3}T) - (166.79 \times 10^5 T^{-2})$$

[a] Data from Richet and Bottinga (1980).
[b] For low-temperature heat capacities of $K_2O-2SiO_2$ glass, see Fig. 7.5.

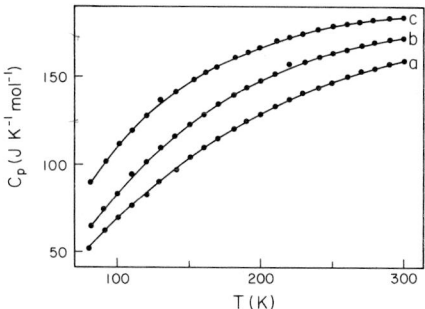

Fig. 7.5. Temperature dependence of low-temperature heat capacities of $M_2O-2SiO_2$ glasses (M = Na, K, and Cs). (a) $Na_2Si_2O_5$; (b) $K_2Si_2O_5$; (c) $Cs_2Si_2O_5$. From Hirao *et al.* (1979).

TABLE 7.11

Heat Contents of MgO–SiO$_2$ Glass

t (°C)	$H_t - H_{25°C}$ (cal/100 g)	t (°C)	$H_t - H_{25°C}$ (cal/100 g)
100	1470	500	11,800
300	6336	700	17,616

[a] Data from Schwiete and Ziegler (1955).

TABLE 7.12

Mean Heat Capacities of MgO–SiO$_2$ Glass[a]

Temperature range (°C)	Mean heat capacity C_m (cal g^{-1} °C^{-1})	Temperature range (°C)	Mean heat capacity C_m (cal g^{-1} °C^{-1})
25–100	0.196	25–500	0.2484
25–300	0.2304	25–700	0.2610

[a] Calculated from the heat content data of Schwiete and Ziegler (1955).

TABLE 7.13

Heat Contents of CaO–SiO$_2$ Glass

t (°C)	$H_t - H_{25°C}$ (cal/100 g)	t (°C)	$H_t - H_{25°C}$ (cal/100 g)
100	1419	500	10,607
300	5801	700	16,052

[a] Data from Schwiete and Ziegler (1955).

TABLE 7.14

Mean Heat Capacities of CaO–SiO$_2$ Glass

Temperature range (°C)	Mean heat capacity C_m (cal g^{-1} °C^{-1})	Temperature range (°C)	Mean heat capacity C_m (cal g^{-1} °C^{-1})
25–100	0.1892	25–500	0.2233
25–300	0.2109	25–700	0.2378

[a] Calculated from the heat content data of Schwiete and Ziegler (1955).

EQ. 7.3

Temperature Dependence of the Heat Capacity of 50.02CaO–49.97SiO$_2$ (mole %) Glass up to the Temperature of 995 K

$$C_p \ (\text{J K}^{-1} \text{ gfw}^{-1}) = 11.651 + 2.0219 \times 10^{-2}T - 30.626$$
$$\times \ 10^5 T^{-2} + 986.88 T^{-1/2}$$

[a] Data from Stebbins *et al.* (1984).

TABLE 7.15

Heat Contents of PbO–SiO$_2$ and 2PbO–SiO$_2$ Glasses[a]

Composition	T (K)	$H_T - H_{298.15}$ (cal mole^{-1})	Composition	T (K)	$H_T - H_{298.15}$ (cal mole^{-1})
PbO–SiO$_2$	387	2250	2PbO–SiO$_2$	388	3000
	433	3450		433	4790
	441	3560		510	8170
	508	6000		622	13010

[a] Data from Baboian (1969).

TABLE 7.16

Mean Heat Capacities of PbO–SiO$_2$ and 2PbO–SiO$_2$ Glasses[a]

Composition	Temperature range (K)	Mean heat capacity $10^2 C_m$ (cal g^{-1} K^{-1})	Composition	Temperature range (K)	Mean heat capacity $10^2 C_m$ (cal g^{-1} K^{-1})
PbO–SiO$_2$	298.15–387	8.92	2PbO–SiO$_2$	298.15–388	6.58
	298.15–433	9.02		298.15–433	7.00
	298.15–441	8.79		298.15–510	7.61
	298.15–508	10.08		298.15–622	7.93

[a] Calculated from the heat content data of Baboian (1969).

EQ. 7.4

Temperature Dependence of Heat Capacities of Two Lead Silicate Glasses[a]

PbO–SiO$_2$ glass: $\quad C_p$ (cal mole^{-1} K^{-1}) = 35.07 + (1.98 × 10^{-3})T
[387–508 K] $\qquad\qquad - 10.432T^{-2}$
2PbO–SiO$_2$ glass: $\quad C_p$ (cal mole^{-1} K^{-1}) = 37.81 + (15.32 × 10^{-3})T
[388–622 K] $\qquad\qquad\qquad -8.061 × 10^5 T^{-2}$

[a] Data from Baboian (1969).

EQ. 7.5

**Temperature Dependence of Heat Capacity of
PbO–SiO$_2$ Glass in the Range 340–690 K[a]**

$$C_p \text{ (cal mole}^{-1}\text{ K}^{-1}) = 15.60 + (1.88 × 10^{-2})T$$

[a] Data from Neiman et al. (1982).

TABLE 7.17

Heat Contents of Lithium–Sodium–Silicate Glasses[a]

t (°C)	H_t^{25} [kcal (g mole SiO$_2$)$^{-1}$]		
	0.3Li$_2$O–0.7Na$_2$O–2SiO$_2$	0.5Li$_2$O–0.5Na$_2$O–2SiO$_2$	0.7Li$_2$O–0.3Na$_2$O–2SiO$_2$
100	1.47	1.58	1.28
150	2.50	2.49	2.24
200	3.62	3.65	3.35
250	4.66	4.73	4.47
300	5.87	5.90	5.67
350	7.25	7.14	7.03
400	8.71	8.55	8.46
450	10.18	10.05	10.06
500	11.64	11.63	11.74
550	13.28	13.20	13.49
600	14.92	14.78	15.17
650	16.56	16.44	16.93
700	18.20	18.02	18.68
750	19.84	19.60	20.44
800	21.48	21.18	22.20
850	23.12	22.84	23.87
900	24.76	24.38	25.55

[a] Data from Takahashi and Yoshio (1978).

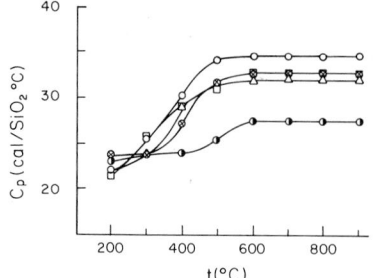

Fig. 7.6. Temperature dependence of the heat capacities of $(1 - x)Li_2O-xNa_2O$ $2SiO_2$ glasses. Value of x: ◑, 0; ⊗, 1; ◯, 0.3; △, 0.5; ▢, 0.7. From Takahashi and Yoshio (1978).

TABLE 7.18

Heat Contents of Lithium–Potassium–Silicate Glasses[a]

t (°C)	H_t^{25} [kcal (g mole SiO_2)$^{-1}$]		
	$0.3Li_2O-0.7K_2O-2SiO_2$	$0.5Li_2O-0.5K_2O-2SiO_2$	$0.7Li_2O-0.3K_2O-2SiO_2$
100	1.46	1.28	1.44
150	2.44	2.28	2.54
200	3.51	3.37	3.64
250	4.68	4.46	4.83
300	5.85	5.74	6.01
350	7.02	7.02	7.28
400	8.39	8.38	8.64
450	9.66	9.84	10.08
500	11.12	11.39	11.60
550	12.78	13.21	13.21
600	14.44	15.03	14.99
650	16.09	16.76	16.77
700	17.85	18.59	18.54
750	19.61	20.41	20.32
800	21.26	22.23	22.10
850	22.92	24.05	23.88
900	24.58	25.78	25.66

[a] Data from Takahashi and Yoshio (1978).

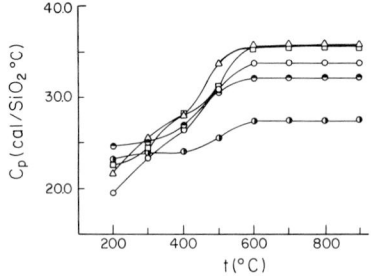

Fig. 7.7. Temperature dependence of the heat capacities of $(1 - x)Li_2O-xK_2O-2SiO_2$ glasses. Value of x: ◯, 1; ◯, 0; ◯, 0.7; △, 0.5; ▢, 0.3. From Takahashi and Yoshio (1978).

TABLE 7.19

Heat Contents of Sodium–Potassium Silicate Glasses[a]

t (°C)	H_t^{25} [kcal (g mole SiO_2)$^{-1}$]		
	$0.3Na_2O-0.7K_2O-2SiO_2$	$0.5Na_2O-0.5K_2O-2SiO_2$	$0.7Na_2O-0.3K_2O-2SiO_2$
100	1.43	1.49	1.63
150	2.46	2.68	2.78
200	3.48	3.77	3.93
250	4.71	4.86	5.08
300	6.04	6.25	6.33
350	7.47	7.53	7.67
400	8.80	8.82	9.11
450	10.34	10.31	10.55
500	11.98	11.80	12.18
550	13.61	13.38	13.81
600	15.30	14.97	15.54
650	16.99	16.55	17.17
700	18.63	18.14	18.80
750	20.37	19.83	20.43
800	22.01	21.51	22.06
850	23.75	23.10	23.69
900	25.38	24.68	25.32

[a] Data from Takahashi and Yoshio (1978).

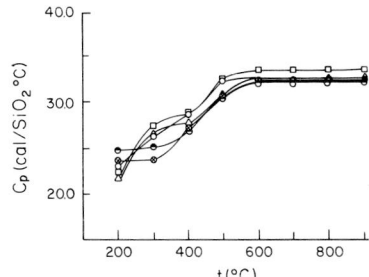

Fig. 7.8. Temperature dependence of the heat capacities of $(1-x)Na_2O-xK_2O-2SiO_2$ glasses. Value of x: ⊗, 0; ●, 1; ○, 0.3; △, 0.5; □, 0.7. From Takahashi and Yoshio (1978).

TABLE 7.20

Heat Capacity of Sodium–Potassium–Silicate Glasses[a,b]

Composition (mole %)			C_p at 700 K	
Na_2O	K_2O	SiO_2	cal g^{-1} K^{-1}	cal (g · atom)$^{-1}$ K^{-1}
—	24.2	75.8	0.245	5.59
2.52	21.5	75.98	0.245	5.52
6.14	18.5	75.36	0.249	5.53
12.20	12.2	75.60	0.266	5.71
18.20	6.03	75.77	0.273	5.70
24.9	—	75.1	0.273	5.51

[a] Data from Moynihan et al. (1976).
[b] See also Figs. 7.10 and 7.11.

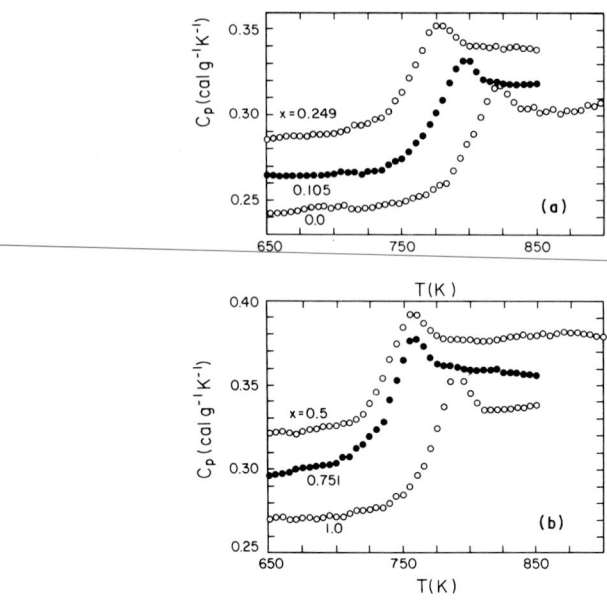

Fig. 7.9. Temperature dependence of heat capacity of $0.244x\mathrm{Na_2O}-0.244(1-x)\mathrm{K_2O}-0.756\mathrm{SiO_2}$ glasses. (a) C_p scale is correct for $x = 0.000$. Plots for $x = 0.105$ and 0.249 have been displaced upward by 0.02 and 0.02 cal g^{-1} K^{-1}, respectively. (b) C_p scale is correct for $x = 1.000$. Plots for $x = 0.751$ and 0.500 have been displaced upward by 0.03 and 0.06 cal g^{-1} K^{-1}, respectively. From Moynihan *et al.* (1976).

TABLE 7.21

Room Temperature Heat Capacities of $(1 - x)\mathrm{Na_2O}-x\mathrm{Cs_2O}-5\mathrm{SiO_2}$ Glasses[a,b]

x^c	$C_p^{\ c}$ (cal cm^{-3} $^\circ$C^{-1})
0	0.465
0.25	0.441
0.4	0.432
0.5	0.421
0.6	0.411
0.7	0.399
0.8	0.391
0.9	0.380
1.0	0.375

[a] Data from Terai *et al.* (1979).
[b] See also Figs. 7.10 and 7.11.
[c] Values read from graph.

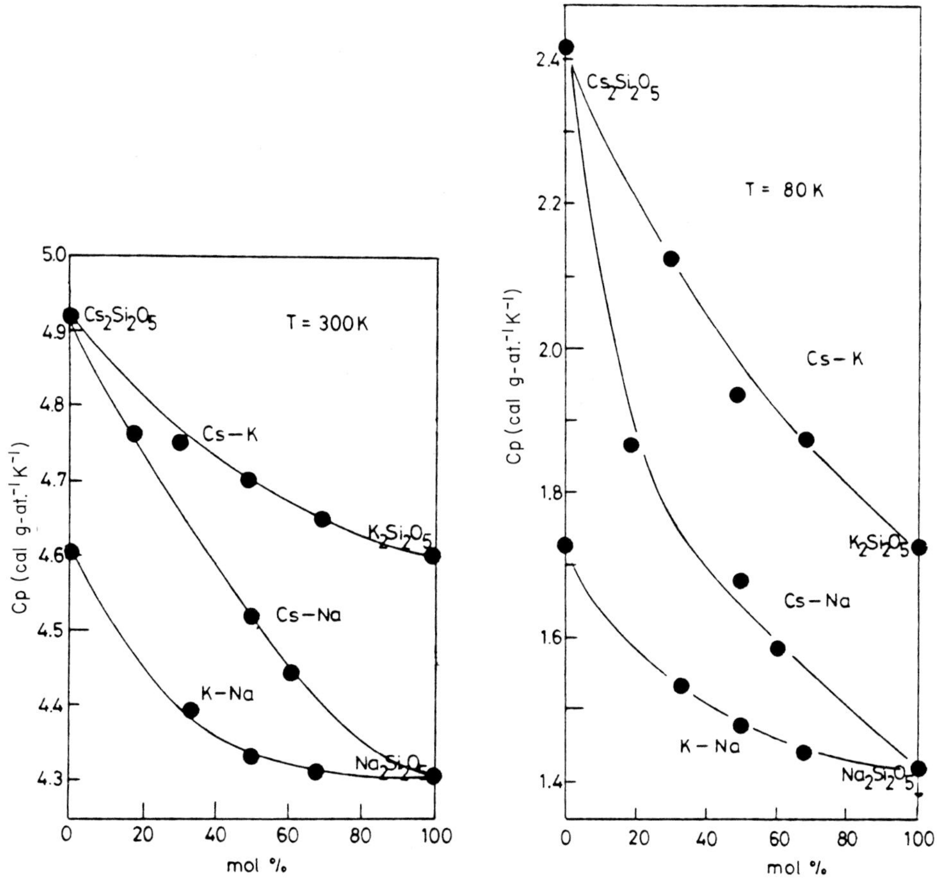

Fig. 7.10. Composition dependence of heat capacity of mixed-alkali silicate glasses. From Hirao *et al.* (1979).

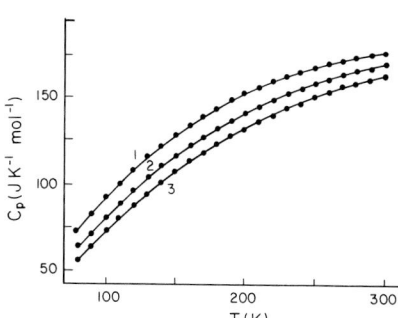

Fig. 7.11. Temperature dependence of low-temperature heat capacities of (1) $KCsSi_2O_5$, (2) $NaCsSi_2O_5$, and (3) $NaKSi_2O_5$ glasses. From Hirao *et al.* (1979).

EQ. 7.8

Temperature Dependence of Heat Content of NaAlSi$_3$O$_8$ Glass in the Temperature Interval 300–1248 K[a]

$\Delta H = H_T - H_{300}$ (kcal mole^{-1})

$\quad = 3.03 + 0.108224T + 3.233 \times 10^{-6}T^2 - 1.70 \times 10^{-9}T^3$

$\quad - 2.061T^{1/2} - 13T^{-1}$

$\Delta H = H_T - H_{300}$ (kcal mole^{-1})

$\quad = -23.56 + 6.0066 \times 10^{-2}T + 9.728 \times 10^{-6}T^2$

$\quad + 1.399 \times 10^3 T^{-1}$

[a] Data from Stebbins et al. (1982).

TABLE 7.26

Coefficients of the Equations C_p (J K^{-1} mole^{-1}) = $a + bT + cT^{-2} + dT^{-1/2}$ and
ΔH (J mole^{-1}) = $H_T - H_{273} = Z + aT + bT^2/2 - CT^{-1} + 2dT^{1/2}$ for
Sodium Aluminosilicate Glasses[a]

System	Temperature range (K)	Coefficients				
		z	a	$10^3 b$	$10^{-5}c$	d
NaAlSiO$_4$ (glass)	918–1013	−72,250	179.48	—	−63.442	
NaAlSiO$_4$ (liquid)	1033–1834	−80,259	182.84	19.826	—	
NaAlSi$_2$O$_6$ (glass)	903–1127	−112,721	261.58	—	−112.73	
NaAlSi$_2$O$_6$ (liquid)	1127–1810	−101,493	242.77	31.344	—	
NaAlSi$_3$O$_8$ (glass)	270–1141	78,737	616.32	−75.480	24.641	−7117.0
NaAlSi$_3$O$_8$ (liquid)	1120–1791	−119,716	300.67	42.620	—	

[a] Data from Richet and Bottinga (1984).

TABLE 7.27

Heat Capacities of NaAlSi$_3$O$_8$ Glass at Various Temperatures[a]

T (K)	C_p (J mole^{-1} K^{-1})	T (K)	C_p (J mole^{-1} K^{-1})	T (K)	C_p (J mole^{-1} K^{-1})
5	0.111	110	92.53	260	191.8
10	0.877	120	101.1	270	196.7
15	2.88	130	109.3	273.15	198.2
20	6.10	140	117.2	280	201.3
25	10.11	150	124.7	290	205.8
30	14.62	160	132.0	298.15	209.3
35	19.36	170	139.1	300	210.1
40	24.29	180	145.9	310	214.3
45	29.34	190	152.4	320	218.4
50	34.45	200	158.7	330	222.3
60	45.09	210	164.8	340	226.0
70	55.11	220	170.6	350	229.5
80	64.87	230	176.2	360	233.2
90	74.39	240	181.5	370	236.7
100	83.63	250	186.8	380	239.4

[a] Data from Robie *et al.* (1978).

TABLE 7.28

Heat Capacities of NaAlSi$_3$O$_8$ Glass (cal mole^{-1} K^{-1})

T (K)	C_p	T (K)	C_p	T (K)	C_p	T (K)	C_p	T (K)	C_p
404.9	59.13	514.4	65.18	624.2	69.35	734.2	72.51	833.5	74.24
409.9	59.49	519.4	65.38	629.2	69.50	739.2	72.67	835.0	74.48
414.8	59.84	524.3	65.62	634.1	69.59	744.1	72.82	838.5	74.31
419.8	60.19	529.3	65.87	639.2	69.81	749.1	73.02	839.9	74.56
424.8	60.45	534.3	66.07	644.1	69.92	754.3	73.09	843.5	74.50
429.8	60.82	539.3	66.28	649.1	70.07	759.3	73.16	844.9	74.70
434.8	61.10	544.3	66.53	654.1	69.95	764.2	73.31	848.5	74.54
439.7	61.42	549.2	66.74	659.1	70.09	769.2	73.36	849.0	74.76
444.7	61.73	554.2	66.95	664.1	70.33	774.2	73.52	854.8	74.85
449.7	62.00	559.2	67.15	669.0	70.36	779.1	73.51	859.8	74.92
454.7	62.28	564.2	67.30	674.0	70.59	784.1	73.56	864.7	74.95
459.7	62.60	569.2	67.45	679.0	70.72	789.1	73.64	869.7	75.10
464.7	62.83	574.1	67.65	683.9	70.94	794.1	73.74	874.6	75.19
469.6	63.11	579.1	67.83	688.9	71.05	799.0	73.72	879.6	75.24
474.6	63.32	584.1	67.97	694.2	71.32	804.0	73.72	884.6	75.44
479.6	63.58	589.1	68.20	699.2	71.56	809.0	73.77	889.5	75.45
484.6	63.81	594.3	68.45	704.2	71.70	813.9	73.54	894.5	75.52
489.6	64.05	599.3	68.60	709.1	71.88	818.9	73.80	899.4	75.60
494.5	64.31	604.3	68.74	714.1	72.12	823.5	74.00	904.4	75.75
499.5	64.55	609.3	68.87	719.1	72.29	825.1	74.37	909.4	75.75
504.1	64.73	614.2	69.04	724.2	72.31	828.5	74.05	914.3	75.95
509.4	64.97	619.2	69.16	729.2	72.47	830.0	74.45		

[a] Data from Stebbins *et al.* (1982).

EQ. 7.9

**Temperature Dependence of Heat Capacity of $NaAlSi_3O_8$ Glass
in the Range 300–968 K[a]**

$$C_p \text{ (cal mole}^{-1} \text{ K}^{-1}) = 108.22 + 6.465 \times 10^{-3}T + 1.3 \times 10^4 T^{-2}$$
$$- 1.0305 \times 10^3 T^{-1/2} - 5.10 \times 10^{-6}T^2$$
$$C_p \text{ (cal mole}^{-1} \text{ K}^{-1}) = 60.06 + 1.9456 \times 10^{-2}T - 1.399 \times 10^6 T^{-2}$$

[a] Data from Stebbins *et al.* (1982).

TABLE 7.29

**Parameters of the Equation C_p (J K^{-1} gfw^{-1}) = $a + bT + cT^{-2} + dT^{-1/2}$ for the
Heat Capacity of Sodium Aluminosilicate Glasses[a]**

Composition (mole %)			Equation parameters				Highest temperature $T(K)$
Na_2O	Al_2O_3	SiO_2	a	$10^2 b$	$10^{-5} c$	d	
6.25	6.25	87.50	63.739	1.1218	−28.677	122.40	995
25.0	25.0	50.0	−81.980	7.1363	−82.263	3490.62	995

[a] Data from Stebbins *et al.* (1984).

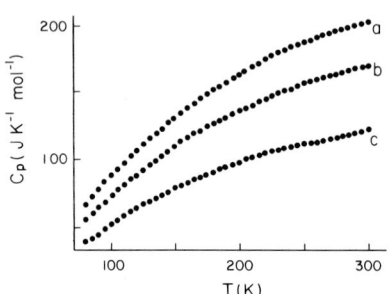

Fig. 7.14. Temperature dependence of heat capacities of soda aluminosilicate glasses. (a) $NaAlSi_3O_8$; (b) $NaAlSi_2O_6$; and (c) $NaAlSiO_4$. From Hirao and Soga (1982).

EQ. 7.10

**Temperature Dependence of Heat Capacity for
$NaAlSi_3O_8$ Glass in the Temperature Range 298–1200 K[a]**

$$C_p \text{ (J mole}^{-1} \text{ K}^{-1}) = 934.4 - 0.3891T + 5.594 \times 10^6 T^{-2}$$
$$- 11,820 T^{-1/2} + 1.476 \times 10^{-4}T^2$$

[a] Data from Krupka *et al.* (1979).

TABLE 7.30

Heat Capacities of KAlSi$_3$O$_8$ Glass at Various Temperatures[a]

T (K)	C_p (J mole^{-1} K^{-1})	T (K)	C_p (J mole^{-1} K^{-1})	T (K)	C_p (J mole^{-1} K^{-1})
5	0.183	110	96.16	260	192.3
10	1.40	120	104.4	270	197.0
15	4.27	130	112.4	273.15	198.5
20	8.51	140	120.0	280	201.6
25	13.45	150	127.4	290	205.9
30	18.66	160	134.5	298.15	209.4
35	24.10	170	141.3	300	210.1
40	29.27	180	147.8	310	214.1
45	34.81	190	154.1	320	218.0
50	40.07	200	160.1	330	221.7
60	50.27	210	166.0	340	225.3
70	60.04	220	171.7	350	228.8
80	69.47	230	177.1	360	232.3
90	78.65	240	182.4	370	235.7
100	87.56	250	187.4	380	238.4

[a] Data from Robie *et al.* (1978).

TABLE 7.31

Coefficients of the Equations C_p (J K^{-1} mole^{-1}) = $a + bT + cT^{-2} + dT^{-1/2}$ and ΔH (J mole^{-1}) = $H_T - H_{273} = z + aT + bT^2/2 - cT^{-1} + 2dT^{1/2}$ for KAlSi$_3$O$_8$[a]

Phase	Temperature range (K)	Coefficients				
		z	a	$10^3 b$	$10^{-5} c$	d
Glass	270–1192	−81,219	620.28	−82.255	26.793	−7193.3
Liquid	1224–1805	−93,461	261.84	61.872	—	—

[a] Data from Richet and Bottinga (1984).

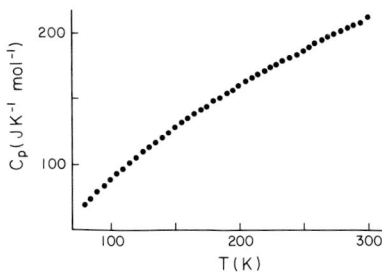

Fig. 7.15. Temperature dependence of the heat capacity of KAlSi$_3$O$_8$ glass. From Hirao and Soga (1982).

EQ. 7.11

Temperature Dependence of Heat capacity for KAlSi$_3$O$_8$ Glass, in the Temperature Range 298–1300 K[a]

$$C_p \,(\text{J mole}^{-1}\,\text{K}^{-1}) = 629.5 - 0.1084T + 2.496 \times 10^6 T^{-2}$$
$$- 7210T^{-1/2} + 1.928 \times 10^{-5}T^2$$

[a] Data from Krupka *et al.* (1979).

TABLE 7.32

Mean Heat Capacity of K$_2$O–Al$_2$O$_3$–6SiO$_2$ Glass[a]

Temperature range (°C)	Mean heat capacity C_m (cal g^{-1} °C^{-1})
0–100	0.1919
0–300	0.2163
0–500	0.2321
0–700	0.2431
0–900	0.2515
0–1100	0.2598

[a] Data from White (1919).

EQ. 7.12

Temperature Dependence of the Heat Capacity of 26.30Na$_2$O–16.03TiO$_2$–57.67SiO$_2$ (mole %) Glass up to a Temperature of 720 K[a]

$$C_p \,(\text{J K}^{-1}\,\text{gfw}^{-1}) = 63.820 + 1.6744 \times 10^{-2}T - 17.037 \times 10^5 T^{-2}$$

[a] Data from Stebbins *et al.* (1984).

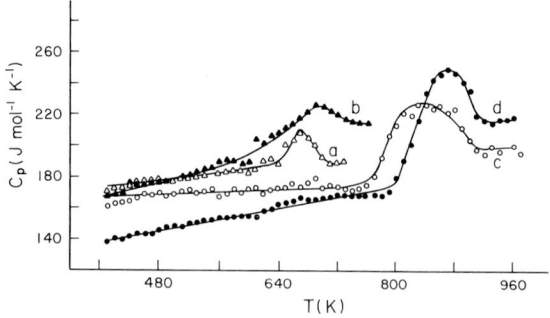

Fig. 7.16. Temperature dependence of heat capacities of glasses: (a) 5(Na$_2$O–2SiO$_2$)UO$_3$; (b) 4(Na$_2$O–2SiO$_2$)UO$_3$; (c) 3(Na$_2$O–2SiO$_2$)UO$_3$; (d) 2(Na$_2$O–2SiO$_2$)UO$_3$. From Senapati and Rao (1982).

TABLE 7.33

Heat Capacities (cal mole^{-1} K^{-1}) of MgCaSi$_2$O$_6$ Glass[a]

T (K)	C_p	T (K)	C_p	T (K)	C_p	T (K)	C_p	T (K)	C_p
405.0	46.81	544.4	52.29	683.9	55.59	794.2	57.29	914.0	60.10
409.9	47.07	549.4	52.39	688.8	55.72	799.1	57.43	915.4	60.13
414.9	47.29	554.4	52.54	693.8	55.86	804.1	57.58	918.9	60.20
419.9	47.53	559.4	52.66	694.9	55.71	809.1	57.81	920.3	60.16
424.9	47.78	564.3	52.77	698.8	55.92	814.0	57.72	924.0	60.38
429.9	48.01	569.3	52.88	699.9	55.80	814.7	57.66	925.3	60.28
434.8	48.23	574.3	53.04	703.7	56.02	819.0	57.83	928.9	60.42
439.8	48.45	579.3	53.13	704.8	55.92	819.6	57.74	930.2	60.32
444.8	48.68	584.3	53.27	708.7	56.11	824.7	57.81	933.9	60.60
449.8	48.86	589.2	53.41	709.8	55.90	829.6	57.94	935.2	60.35
454.8	49.03	594.3	53.53	713.6	56.24	834.6	58.04	938.8	60.69
459.7	49.28	599.3	53.68	714.7	55.83	839.5	58.11	940.2	60.31
464.7	49.45	604.3	53.74	718.6	56.32	844.5	58.22	943.8	60.78
469.7	49.69	609.3	53.83	719.7	55.94	849.5	58.37	945.1	60.40
474.7	49.85	614.2	54.00	724.7	56.01	854.4	58.48	948.7	60.85
479.7	50.03	619.2	54.14	729.6	56.14	859.4	58.58	950.1	60.43
484.6	50.21	624.2	54.26	734.6	56.26	864.3	58.68	953.7	60.93
489.6	50.42	629.2	54.37	739.6	56.27	868.3	58.80	955.0	60.38
494.6	50.60	634.1	54.48	744.5	56.23	874.2	58.92	958.6	61.02
499.6	50.82	639.1	54.61	749.5	56.41	879.2	59.05	960.0	60.49
504.6	51.00	644.1	54.70	754.5	56.63	884.2	59.15	963.6	61.19
509.6	51.16	649.1	54.78	759.4	56.71	889.1	59.29	965.0	60.45
514.5	51.29	654.1	54.87	764.4	56.75	894.1	59.47	968.5	61.20
519.5	51.49	659.0	55.07	769.4	56.74	899.0	59.60	969.9	60.57
524.5	51.65	664.0	55.19	774.3	57.03	904.0	59.77	973.5	61.35
529.5	51.77	669.0	55.38	779.3	57.06	905.5	59.96	974.9	60.66
534.5	51.92	673.9	55.41	784.3	57.04	909.0	59.96		
539.4	52.11	678.9	55.56	789.2	57.20	910.4	60.02		

[a] Data from Stebbins *et al.* (1983).

TABLE 7.34

Mean Heat Capacity of MgO–CaO–2SiO$_2$ Glass[a]

Temperature range (°C)	Mean heat capacity C_m (cal g^{-1} °C^{-1})
0–100	0.1938[b]
0–300	0.2189[b]
0–500	0.2333[b]
0–700	0.2439

[a] Data from White (1919).
[b] Preliminary determination.

EQ. 7.13

Temperature Dependence of the Heat Capacity of
$50.88MgO-0.13Al_2O_3-48.99SiO_2$ (mole %) Glass up to a
Temperature of 995 K[a]

$$C_p \text{ (J K}^{-1}\text{ gfw}^{-1}) = -24.283 + 3.9763 \times 10^{-2}T - 44.513 \times 10^5 T^{-2}$$
$$+ 1692.46T^{-1/2}$$

[a] Data from Stebbins *et al.* (1984).

EQ. 7.14

Temperature Dependence of Heat Content, $\Delta H = H_T - H_{300}$
(kcal mole^{-1}) of $CaAl_2Si_2O_8$ Glass in the Range 300–1413 K[a]

$$\Delta H \text{ (kcal mole}^{-1}) = -2.55 + 0.10089T + 3.254 \times 10^{-6}T^2$$
$$+ 3.22 \times 10^2 T^{-1} - 1.6776T^{1/2} - 1.056 \times 10^{-9}T^3$$
$$\Delta H \text{ (kcal mole}^{-1}) = -23.84 + 6.070 \times 10^{-2}T + 9.58 \times 10^{-6}T^2$$
$$+ 1.43 \times 10^3 T^{-1}$$

[a] Data from Stebbins *et al.* (1982).

TABLE 7.35

Mean Heat Capacity of $CaO-Al_2O_3-2SiO_2$ Glass[a]

Temperature range (°C)	Mean heat capacity C_m (cal g^{-1} °C^{-1})
0–100	0.1885
0–300	0.2152[b]
0–500	0.2304[b]
0–700	0.2405

[a] Data from White (1919).
[b] Preliminary determination.

EQ. 7.15

Temperature Dependence of Heat Capacity for $CaAl_2Si_2O_8$ Glass, in
the Temperature Range 298–1500 K[a]

$$C_p \text{ (J mole}^{-1}\text{ K}^{-1}) = 375.2 + 0.03197T - 2.815 \times 10^6 T^{-2} - 2459T^{-1/2}$$

[a] Data from Krupka *et al.* (1979).

TABLE 7.36

Heat Capacities of $CaAl_2Si_2O_8$ Glass at Various Temperatures[a]

T (K)	C_p (J mole^{-1} K^{-1})	T (K)	C_p (J mole^{-1} K^{-1})	T (K)	C_p (J mole^{-1} K^{-1})
5	0.255	110	85.82	260	192.2
10	0.333	120	94.83	270	197.5
15	1.401	130	103.6	273.15	199.1
20	3.346	140	112.1	280	202.5
25	6.442	150	120.3	290	207.1
30	9.386	160	128.4	298.15	210.6
35	13.20	170	136.1	300	211.4
40	17.43	180	143.4	310	215.5
45	21.97	190	150.4	320	219.6
50	26.74	200	157.1	330	223.8
60	36.68	210	163.5	340	227.8
70	46.77	220	169.7	350	231.5
80	56.83	230	175.6	360	235.0
90	66.77	240	181.3	370	238.4
100	76.46	250	186.8		

[a] Data from Robie et al. (1978).

TABLE 7.37

Heat Capacities of $CaAl_2Si_2O_8$ Glass (cal mole^{-1} K^{-1})[a]

T (K)	C_p	T (K)	C_p	T (K)	C_p	T (K)	C_p	T (K)	C_p
404.9	59.33	509.4	64.98	614.2	68.95	719.2	72.18	824.9	74.50
409.9	59.65	514.4	65.25	619.1	69.09	724.3	72.37	829.9	74.56
411.9	59.66	519.4	65.45	624.1	69.29	729.3	72.53	834.8	74.68
419.8	60.31	524.4	65.64	629.1	69.36	734.3	72.67	839.8	74.81
424.8	60.60	529.3	65.87	634.1	69.52	739.2	72.94	844.8	74.92
429.8	60.92	534.3	66.09	639.0	69.73	744.2	73.02	849.7	75.02
434.8	61.22	539.3	66.27	644.0	69.81	749.2	73.21	854.7	75.12
439.8	61.53	544.3	66.47	649.0	69.99	754.2	73.32	859.7	75.19
444.7	61.83	549.3	66.66	654.0	70.02	759.4	74.38	864.6	75.36
449.7	62.10	554.2	66.87	659.0	70.14	764.4	73.50	869.6	75.54
454.7	62.37	559.2	66.98	664.0	70.27	769.4	73.58	874.6	75.51
459.7	62.63	564.2	67.22	668.9	70.32	774.3	73.68	879.5	75.65
464.7	62.88	569.2	67.45	673.9	70.50	779.3	73.60	884.5	75.78
469.7	63.15	574.2	67.68	678.9	70.65	784.3	73.69	889.4	75.97
474.6	63.43	579.2	67.81	683.8	70.87	789.2	73.83	894.4	76.05
479.6	63.64	584.1	68.02	688.8	70.96	794.6	73.95	899.4	76.17
484.6	63.91	589.1	68.19	694.3	71.29	799.6	73.93	904.3	76.30
489.6	64.14	594.3	68.36	699.3	71.49	804.6	74.04	909.3	76.37
494.6	64.37	599.2	68.53	704.3	71.69	809.5	73.98	914.3	76.63
499.5	64.61	604.2	68.69	709.2	71.81	814.7	74.10	919.2	76.47
504.4	64.75	609.2	68.84	714.2	71.99	819.6	74.22		

[a] Data from Stebbins et al. (1982).

EQ. 7.16

Temperature Dependence of Heat Capacity of $CaAl_2Si_2O_8$ Glass in the Range 300–997 K[a]

$$C_p \, (\text{cal mole}^{-1} \, K^{-1}) = 100.89 + 6.509 \times 10^{-3}T - 3.22 \times 10^{5}T^{-2}$$
$$- 8.388 \times 10^{2}T^{-1/2} - 3.169 \times 10^{-6}T^{2}$$
$$C_p \, (\text{cal mole}^{-1} \, K^{-1}) = 60.70 + 1.916 \times 10^{-2}T - 1.43 \times 10^{6}T^{-2}$$

[a] Data from Stebbins et al. (1982).

TABLE 7.38

Heat Capacities of $CaAl_2SiO_6$ Glass at Various Temperatures[a]

T (K)	C_p (J mole^{-1} K^{-1})	T (K)	C_p (J mole^{-1} K^{-1})	T (K)	C_p (J mole^{-1} K^{-1})
5	0.026	110	67.53	260	151.6
10	0.224	120	74.85	270	155.7
15	0.872	130	81.93	273.15	157.2
20	2.190	140	88.72	280	159.5
25	4.063	150	95.25	290	163.1
30	6.504	160	101.4	298.15	166.1
35	9.482	170	107.5	300	166.5
40	12.87	180	113.4	310	169.8
45	16.32	190	118.8	320	173.1
50	19.84	200	124.1	330	176.2
60	27.64	210	129.1	340	179.1
70	35.97	220	133.9	350	181.8
80	44.28	230	138.6	360	184.5
90	52.29	240	143.2	370	187.1
100	60.02	250	147.5	380	189.9

[a] Data from Haselton et al. (1984).

EQ. 7.17

Temperature Dependence of the Heat Capacity of $12.50CaO-12.50Al_2O_3-75.00SiO_2$ (mole %) Glass up to a Temperature of 995 K[a]

$$C_p \, (\text{J K}^{-1} \, \text{gfw}^{-1}) = 113.335 - 0.5212 \times 10^{-2}T - 6.337 \times 10^{5}T^{-2}$$
$$- 1008.70T^{-1/2}$$

[a] Data from Stebbins et al. (1984).

Chapter **8**

Thermal Conductivity

Heat can be transported through and to solids by conduction and radiation. Conduction is the transport of heat through contacting phases by thermal gradients, and is the most important transport mechanism for temperatures below about $500°C$. At higher temperatures radiative transfer becomes important and is controlled by the emissivity of the material.

The thermal conductivity λ can be defined by considering a slab of material of thickness L and cross-sectional area A, with one face at temperature T_1 and the other at T_2. Then

$$\lambda = QLt/A(T_1 - T_2), \tag{1}$$

where Q is the heat flux that passes through the slab in time t. The units of λ are watts per meter degree Celsius. Another coefficient sometimes used is the thermal diffusivity K, with units of square meters per second, which is

$$K = \lambda/\rho C_p, \tag{2}$$

where ρ is the density and C_p the heat capacity at constant pressure.

The thermal conductivity of a glass can be measured by a linear heat flow technique. A heat source supplies a known flux through a cross-section A of a cylindrical sample of thickness L, and the other side of the sample is provided with a heat sink. Then the heat flux q is

$$q = \lambda A \Delta T/L, \tag{3}$$

from which λ can be calculated. This and a number of other methods are described by Eckert and Goldstein (1976).

The thermal conductivity of a silicate glass is low near absolute zero and increases with temperature. Above room temperature the thermal conductivity increases slowly as the temperature increases. The thermal conductivity of glasses is not influenced much by composition; most glasses have λ values of about 1 W m^{-1} °C^{-1} at room temperature.

II   TABLES AND FIGURES

Thermal conductivity data for binary glasses are given in Tables 8.1–8.6 and Fig. 8.1. Data for ternary glasses are given in Tables 8.7–8.32 and Figs. 8.2 and 8.3.

BIBLIOGRAPHY

Ammar, M. M., El-Badry, K., Moussa, M. R., Gharib, S., and Halawa, M. (1975). *Cent. Glass Ceram. Res. Inst. Bull.* **22,** 10.

Ammar, M. M., Gharib, S., El-Badry, K., Ghoneim, N. A., and El-Batal, H. A. (1982a). *Sprechsaal* **115,** 692.

Ammar, M. M., Gharib, S., Halawa, M. M., El-Badry, K., Ghoneim, N. A., and El-Batal, H. A. (1982b). *J. Non-Cryst. Solids* **53,** 165.

Ammar, M. M., Gharib, S. A., Halawa, M. M., El-Batal, H. A., and El-Badry, K. (1983). *J. Am. Ceram. Soc.* **66,** C76.

Eckert, E. R. G., and Goldstein, R. J., Eds. (1976). "Measurements in Heat Transfer." Hemisphere Publ. Co., Washington, D. C.

Ghoneim, N. A. (1980). *Sprechsaal* **113,** 610.

Ghoneim, N. A., El-Batal, H. A., and El-Badry, K. (1983). *Glastech. Ber.* **56,** 934.

Muratov, A. B., and Chernyshov, A. V. (1979). *Sov. J. Glass Phys. Chem.* (*Engl. Transl.*) **5,** 100.

Primenko, V. I. (1980). *Glass Ceram.* **37,** 240.

Ratcliffe, E. H. (1963). *Glass Technol.* **4,** 113.

Russ, A. (1928). *Sprechsaal* **15,** 907.

Stephens, R. B. (1973). *Phys. Rev. B* **8,** 2896.

Stephens, R. B. (1976). *Phys. Rev. B* **13,** 852.

Terai, R., Hori, M., and Yamanaka, H. (1979). *Am. Ceram. Soc. Bull.* **58,** 1125.

VanVelden, P. F. (1965). *Glass Technol.* **6,** 166.

Vavilov, Y. V., Komarov, V. E., and Tabunova, N. A. (1982). *Sov. J. Glass Phys. Chem.* (*Engl. Transl.*) **8,** 326.

TABLE 8.1

Thermal Conductivity of Lithium Silicate Glasses[a]

Composition (mole % Li_2O)	λ at 30°C (W m^{-1} deg^{-1})
29.97	0.95
31.97	0.90
33.37	0.88
36.10	0.85
40.14	0.77
46.29	0.65

[a] Data from Ammar *et al.* (1975, 1983).

TABLE 8.2

Low-Temperature Thermal Conductivity of Sodium Silicate Glass[a]

Composition (mole % Na_2O)	Temperature (K)	λ (W m^{-1} deg^{-1})
25	0.05–2	$0.017T^{1.92}$

[a] Data from Stephens (1973, 1976).

TABLE 8.3

Thermal Conductivity of Na_2O–$3SiO_2$ Glass[a]

T (K)	λ (W m^{-1} deg^{-1})	T (K)	λ (W m^{-1} deg^{-1})
32	0.18	154	0.65
53	0.28	193	0.77
76	0.37	239	0.88
106	0.48	276	0.94
133	0.58	301	0.98

[a] Data from graph in Muratov and Chernyshov (1979).

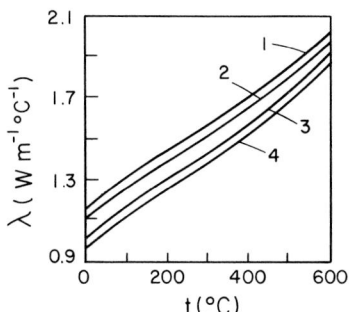

Fig. 8.1. Temperature dependence of thermal conductivity of K_2O–SiO_2 glasses with K_2O concentration (wt. %): (1) 12.64; (2) 17.85; (3) 27.04; (4) 30.88. From Primenko (1980).

TABLE 8.4

Room-Temperature Thermal Conductivity of Cesium Silicate Glass[a]

Composition (mole % Cs_2O)	λ (W m^{-1} deg^{-1})
16.67	0.637

[a] Data from graph in Terai et al. (1979).

TABLE 8.5

Thermal Conductivity of Lead Silicate Glasses[a]

Composition (mole % PbO)	Temperature (°C)	λ (W m^{-1} deg^{-1})
49.3	40	0.544
51.9	−150	0.373
	−100	0.419
	−50	0.465
	0	0.515
	50	0.561
	100	0.611
66.2	40	0.469

[a] Data from Ratcliffe (1963) and VanVelden (1965).

TABLE 8.6

Thermal Conductivity of M_2O_3–SiO_2 Glasses

Composition (mole % M_2O_3)	Temperature (°C)	λ (W m^{-1} deg^{-1})	Reference[a]
M = B			
12.3	27	1.12	1
M = Bi			
39.5	40	0.645	2

[a] Data from (1) Vavilov et al. (1982) and (2) VanVelden (1965).

TABLE 8.7

Thermal Conductivities of
$(1 - x)Na_2O-xCs_2O-5SiO_2$
Glasses at Room
Temperature[a]

x	λ (W m^{-1} deg^{-1})
0.25	0.928
0.4	0.871
0.5	0.833
0.6	0.798
0.7	0.735
0.8	0.697
0.9	0.645

[a] Data from graph in Terai *et al.* (1979).

TABLE 8.8

Thermal Conductivity of Li$_2$O–ZnO–SiO$_2$ Glasses[a]

Composition (mole %)			
Li$_2$O	ZnO	SiO$_2$	λ (W m^{-1} K^{-1}) at 30°C
26.80	6.27	66.63	0.913
33.84	3.03	63.15	0.855
34.19	6.28	59.53	0.828
40.14	—	59.86	0.771
40.59	2.91	56.50	0.789
46.29	—	53.71	0.653

[a] Data from Ammar *et al.* (1982b, 1983).

TABLE 8.9

Thermal Conductivity of Na$_2$O–MgO–SiO$_2$ Glasses

Composition (mole %)			λ (W m^{-1} deg^{-1}) at			
Na$_2$O	MgO	SiO$_2$	0°C	30°C	100°C	Reference[a]
18	5	77	0.992		1.094	1
18	8	74	1.013		1.107	
18	10	72	0.974		1.069	
18	12	70	0.956		1.030	
18	15	67	0.951		1.027	
18	18	64	0.951		1.025	
39.31	2.40	58.29		0.612		2
39.35	4.84	55.81		0.621		
39.40	7.26	53.34		0.628		
39.44	9.70	50.85		0.641		

[a] Data from (1) Russ (1928) and (2) Ammar *et al.* (1982b).

TABLE 8.10

Thermal Conductivity of $Na_2O-CaO-SiO_2$ Glasses

Composition (mole %)			λ (W m^{-1} deg^{-1}) at			
Na_2O	CaO	SiO_2	0°C	30°C	100°C	Reference[a]
18	5	77	0.933		1.043	1
18	10	72	0.910		1.016	
18	15	67	0.905		0.972	
39.27	2.50	58.23		0.644		2
39.51	4.52	55.98		0.634		
39.30	7.49	53.12		0.625		
39.29	10.02	50.69		0.657		

[a] Data from (1) Russ (1928) and (2) Ammar et al. (1982b, 1983).

TABLE 8.11

Low-Temperature Thermal Conductivity of $Na_2O-CaO-SiO_2$ Glass[a]

Composition (mole %)			Temperature	λ
Na_2O	CaO	SiO_2	(K)	(W m^{-1} deg^{-1})
15	10	75	23	0.12
			64	0.28
			107	0.43
			149	0.57
			186	0.69
			224	0.77
			265	0.86
			303	0.92

[a] Data from graph in Muratov and Chernyshov (1979).

TABLE 8.12

Thermal Conductivity of $Na_2O-SrO-SiO_2$ Glasses[a]

Composition (mole %)			
Na_2O	SrO	SiO_2	λ (W m^{-1} K^{-1}) at 30°C
39.33	2.51	58.16	0.588
39.02	4.91	56.08	0.616
39.26	7.57	53.16	0.542
39.41	10.14	50.45	0.573

[a] Data from Ammar et al. (1982b, 1983).

TABLE 8.13

Thermal Conductivity of $Na_2O-BaO-SiO_2$ Glasses

Composition (mole %)			λ (W m^{-1} deg^{-1}) at			
Na_2O	BaO	SiO_2	0°C	30°C	100°C	Reference[a]
18	5	77	0.946		1.029	1
18	10	72	0.922		1.004	
18	15	67	0.824		0.919	
18	25	57	0.799		0.884	
39.25	2.55	58.20		0.548		2
39.25	5.09	55.66		0.518		
39.25	7.62	53.13		0.540		
39.24	10.15	50.60		0.466		

[a] Data from (1) Russ (1928) and (2) Ammar *et al.* (1982b, 1983).

TABLE 8.14

Thermal Conductivity of $Na_2O-ZnO-SiO_2$ Glasses[a]

Composition (mole %)			λ (W m^{-1} deg^{-1}) at	
Na_2O	ZnO	SiO_2	0°C	100°C
18	5	77	0.999	1.091
18	10	72	0.927	0.974
18	15	67	0.900	0.972
18	20	62	0.848	0.925
18	30	52	0.807	0.877

[a] Data from Russ (1928).

TABLE 8.15

Thermal Conductivity of $Na_2O-CdO-SiO_2$ Glass[a]

Composition (mole %)			λ at 27°C
Na_2O	CdO	SiO_2	(W m^{-1} K^{-1})
18.2	8.8	73	0.93

[a] Data from Vavilov *et al.* (1982).

TABLE 8.16

Thermal Conductivity of Na$_2$O–PbO–SiO$_2$ Glasses[a]

Composition (mole %)			λ (W m^{-1} deg^{-1}) at	
Na$_2$O	PbO	SiO$_2$	0°C	100°C
10	20	70	0.948	1.028
15	20	65	0.887	0.985
18	5	77	0.902	0.986
18	10	72	0.901	0.945
18	20	62	0.875	0.962
18	25	57	0.845	0.930
20	20	60	0.846	0.954
25	20	55	0.829	0.864

[a] Data from Russ (1928).

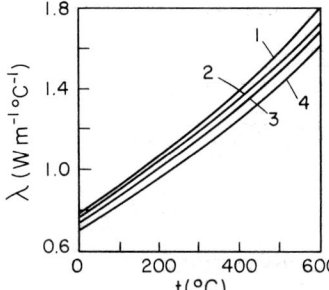

Fig. 8.2. Temperature dependence of thermal conductivities of glasses of compositions (wt. %):
(1) 14.6Na$_2$O–28.9PbO–56.5SiO$_2$;
(2) 9.56Na$_2$O–34.66PbO–55.78SiO$_2$;
(3) 10.59Na$_2$O–38.13PbO–51.28SiO$_2$;
(4) 8.3Na$_2$O–44.76PbO–46.94SiO$_2$.
From Primenko (1980).

TABLE 8.17

Thermal Conductivity of K$_2$O–CaO–SiO$_2$ Glass[a]

Composition (wt. %)			λ at 50°C (W m^{-1} deg^{-1})
K$_2$O	CaO	SiO$_2$	
33	8	59	0.879–0.921

[a] Data from Ratcliffe (1963).

TABLE 8.18

Thermal Conductivity of K$_2$O–PbO–SiO$_2$ Glasses[a]

Composition (mole %)			λ (W m^{-1} deg^{-1}) at	
K$_2$O	PbO	SiO$_2$	0°C	100°C
15	20	65	0.812	0.899
20	20	60	0.759	0.852

[a] Data from Russ (1928).

TABLE 8.19

Thermal Conductivity of K$_2$O–PbO–SiO$_2$ Glasses[a]

Composition (wt. %)			λ (W m^{-1} deg^{-1}) at given temperature					
K$_2$O	PbO	SiO$_2$	$-150°$C	$-100°$C	$-50°$C	$0°$C	$50°$C	$100°$C
9.2	44.8	46.0	0.486	0.574	0.649	0.729	0.796	0.863
4.4	59.7	35.6	0.431	0.519	0.599	0.662	0.708	0.737

[a] Data from Ratcliffe (1963).

TABLE 8.20

Thermal Conductivity of Na$_2$O–B$_2$O$_3$–SiO$_2$ Glasses[a]

Composition (mole %)			λ (W m^{-1} deg^{-1}) at	
Na$_2$O	B$_2$O$_3$	SiO$_2$	$0°$C	$100°$C
18	5	77	1.017	1.118
18	10	72	1.010	1.075
18	15	67	0.985	1.075
18	20	62	0.974	1.059
18	30	52	0.941	1.038

[a] Data from Russ (1928).

TABLE 8.21

Thermal Conductivity of Sodium Aluminosilicate Glasses[a]

Composition (mole %)			λ (W m^{-1} deg^{-1}) at	
Na$_2$O	Al$_2$O$_3$	SiO$_2$	$0°$C	$100°$C
18	3	79	0.972	1.081
18	6	76	0.964	1.072
18	9	73	0.960	1.031
18	12	70	0.947	1.026

[a] Data from Russ (1928).

TABLE 8.22

Thermal Conductivity of $Na_2O-Al_2O_3-SiO_2$ Glasses[a]

Composition (wt. %)			
Na_2O	Al_2O_3	SiO_2	λ (W m^{-1} deg^{-1}) at 30°C
37.43	15.82	46.77	0.672
38.03	12.06	49.91	0.651
38.44	8.39	53.17	0.625
39.32	4.16	56.52	0.594

[a] Data from Ammar *et al.* (1982a, 1983).

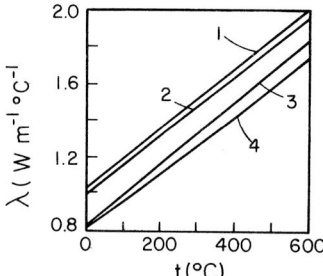

Fig. 8.3. Temperature dependence of thermal conductivity of the glass compositions (wt. %):
(1) $15.87Na_2O-6.12Al_2O_3-78.01SiO_2$;
(2) $16.12Na_2O-7.67Al_2O_3-76.21SiO_2$;
(3) $25.08Na_2O-15.45Al_2O_3-59.47SiO_2$;
(4) $25.12Na_2O-20.81Al_2O_3-54.07SiO_2$.
From Primenko (1980).

TABLE 8.23

Thermal Conductivity of Some $Na_2O-M_2O_3-SiO_2$ Glasses[a]

Composition (mole %)			λ at 27°C
Na_2O	M_2O_3	SiO_2	(W m^{-1} deg^{-1})
	M = In		
19.28	5.87	74.85	1.00
	M = La		
18.8	4.7	76.5	0.86

[a] Data from Vavilov *et al.* (1982).

TABLE 8.24

Thermal Conductivity of $Na_2O-Fe_2O_3-SiO_2$ Glasses[a]

Composition (mole %)			λ (W m^{-1} deg^{-1}) at	
Na_2O	Fe_2O_3	SiO_2	0°C	100°C
18	5	77	0.925	1.059
18	10	72	0.942	1.036
18	15	67	0.918	1.049
18	20	62	0.893	0.982

[a] Data from Russ (1928).

TABLE 8.25

Thermal Conductivity of Sodium Silicate Glass after Various Additions of Fe_2O_3[a]

Composition (wt. %)			
Na_2O	SiO_2	Fe_2O_3 added (g)	λ at 30°C ($W\ m^{-1}\ deg^{-1}$)
30	70	—	0.764
30	70	0.2	0.766
30	70	1	0.769
30	70	2	0.774
30	70	5	0.781
30	70	10	0.787

[a] Data from Ghoneim *et al.* (1983).

TABLE 8.26

Thermal Conductivity of $Li_2O-TiO_2-SiO_2$ Glasses[a]

Composition (wt. %)			
Li_2O	TiO_2	SiO_2	λ at 30°C ($W\ m^{-1}\ K^{-1}$)
16.17	8.10	75.72	0.61
16.47	6.40	77.11	0.64
16.78	4.67	78.53	0.75
17.06	3.04	79.89	0.85
17.26	1.92	80.82	0.88
17.46	0.78	81.75	0.903

[a] Data from Ammar *et al.* (1975, 1983).

TABLE 8.27

Thermal Conductivity of $K_2O-TiO_2-SiO_2$ Glasses[a]

Composition (wt. %)			
K_2O	TiO_2	SiO_2	λ ($W\ m^{-1}\ deg^{-1}$) at 30°C
20	45	35	2.34
25	40	35	2.49
25	45	30	2.11
30	45	25	1.86
35	45	20	1.78

[a] Data from Ghoneim (1980).

TABLE 8.28

Thermal Conductivity of BaO–PbO–SiO$_2$ Glasses[a]

Composition (wt. %)			λ (W m^{-1} deg^{-1}) at	
BaO	PbO	SiO$_2$	40°C	120°C
14.8	65.9	19.3	0.486	0.570
1.1	81.6	17.3	0.503	0.536

[a] Data from Van Velden (1965).

TABLE 8.29

Thermal Conductivity of BaO–Bi$_2$O$_3$–SiO$_2$ Glasses[a]

Composition (wt. %)			λ at 40°C
BaO	Bi$_2$O$_3$	SiO$_2$	(W m^{-1} deg^{-1})
21.2	66.6	12.2	0.532
6.7	81.6	11.7	0.616
34.8	42.9	22.3	0.553
22.8	65.0	13.0	0.561

[a] Data from Van Velden (1965).

TABLE 8.30

Thermal Conductivity of Lead Aluminosilicate Glasses[a]

Composition (wt. %)			λ (W m^{-1} deg^{-1}) at	
PbO	Al$_2$O$_3$	SiO$_2$	40°C	120°C
84.6	7.1	8.3	0.532	0.586
57.9	6.0	36.1	0.779	0.858
61.9	2.1	36.0	0.741	0.791

[a] Data from Van Velden (1965).

TABLE 8.31

Thermal Conductivity of
PbO–Bi$_2$O$_3$–SiO$_2$
Glasses[a]

Composition (wt. %)			λ (W m^{-1} deg^{-1}) at	
PbO	Bi$_2$O$_3$	SiO$_2$	40°C	120°C
55.3	38.8	5.9	0.503	
54.1	36.7	9.2	0.519	
28.5	60.5	11.0	0.561	
19.7	69.0	11.3	0.586	
32.6	55.5	11.9	0.570	
12.1	74.6	13.3	0.611	
27.5	57.9	14.6	0.595	
44.5	37.3	18.2	0.578	
23.3	49.8	26.9	0.657	
31.4	39.2	29.4	0.708	0.775
45.2	26.7	28.1	0.683	0.720
50.8	14.9	34.3	0.729	0.796

[a] Data from VanVelden (1965).

TABLE 8.32

Thermal conductivity of
Al$_2$O$_3$–Bi$_2$O$_3$–SiO$_2$
Glasses[a]

Composition (wt. %)			λ (W m^{-1} deg^{-1}) at	
Al$_2$O$_3$	Bi$_2$O$_3$	SiO$_2$	40°C	120°C
9.4	78.8	11.8	0.745	0.846
3.3	86.6	10.1	0.666	
5.7	75.4	18.9	0.724	
4.7	75.7	19.6	0.733	0.791
1.8	77.3	20.9	0.678	

[a] Data from VanVelden (1965).

Part III
Mechanical Properties

Chapter 9

Viscosity

The viscosity of a glass is one of its most important technological properties. It determines the melting conditions, the temperatures of working and annealing, fining behavior (removal of bubbles from the melt), upper temperature of use, and devitrification rate. The viscosities of different glasses vary enormously with composition and are strong functions of temperature.

When a shearing force is applied to a liquid, it flows, and the viscosity is a measure of the ratio between the force and rate of flow. If two parallel planes of area A, a distance d apart, are subjected to a tangential force difference F, the viscosity η is defined as

$$\eta = Fd/Av, \tag{1}$$

where v is the relative velocity of the two planes. The unit of viscosity is the dyne second per square centimeter which is called the poise (P). Almost all measurements of viscosities of glasses have been reported in this unit, and it is used in this volume. The SI unit for viscosity is the newton second per square meter, or pascal second; one of these units equals 10 P. The viscosity of most common fluids such as water and organic liquids is about one-hundredth of a poise at room temperature.

The working point of a glass is defined as the temperature at which it has a viscosity of 10^4 P. At this temperature the glass can be readily formed or sealed. The softening point of a glass is the temperature at which it has a viscosity of $10^{7.6}$ P. At this viscosity, a rod about 24 cm long and 0.7 mm in diameter elongates 1 mm/min under its own weight. The annealing point is the temperature at which the viscosity is $10^{13.4}$ P, and the strain point where it is $10^{14.6}$ P. The thermal expansion curve of an annealed glass begins to deviate considerably from linearity at a viscosity of about 10^{14} or 10^{15} P. At the

223

annealing temperature a rapidly cooled silicate glass becomes reasonably strain-free in about 15 min.

The viscosities of several commercially important silicate glasses are compared in Fig. 9.1, as a function of temperature. The working, softening, and annealing points are marked on the plot.

Two methods are most frequently used to measure viscosity in glass: one at low viscosity and the other at high viscosity. At low viscosity (up to 10^8 P), the rotating cylinder or crucible method appears to be the most reliable. In this method the relative rate of rotation of two concentric cylinders is measured at a constant torque, and the viscosity is inversely proportional to the rate of rotation. Usually the viscometer is calibrated with liquids of known viscosity, but with care absolute measurements can be made. Higher viscosities are

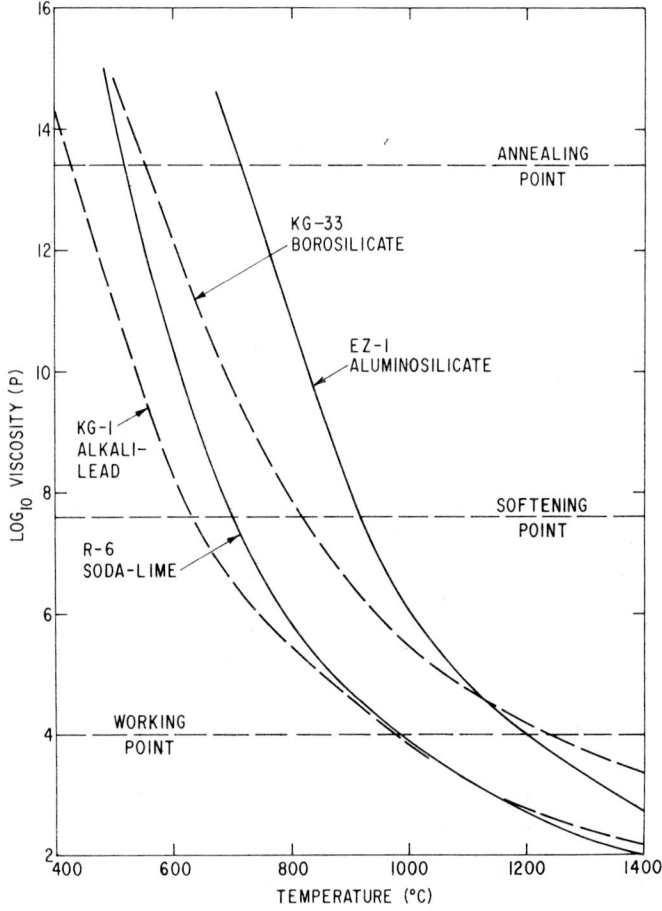

Fig. 9.1. Viscosities of some commercial silicate glasses. For compositions see Chapter 3 (Table 3.1) on commercial glasses.

measured by the rate at which a glass rod elongates under a fixed force. The viscosity is then given by the formula

$$\eta = Lmgf/3\pi R^2 v, \tag{2}$$

where L is the length of the glass rod, R is its radius, m is the mass hanging on it, g is the gravitational constant, v is an instrumental reading proportional to the rate of elongation, and f is a calibration factor for the instrument. A beam-bending method is also possible for high viscosities (Hagy, 1963).

Two problems arise in measuring the viscosity of viscous materials: one is the variation of viscosity with time, and the other is its variation with the force applied. Many investigators have found that the measured viscosity is not a function of applied force or velocity of flow, implying that glass is a Newtonian liquid in which the flow rate is directly proportional to the shearing stress.

Lillie (1933) studied the change of viscosity of glasses with time at the measuring temperature. He found that above a viscosity of about 10^{12} P the viscosity of samples cooled quickly from a higher temperature initially appeared to be low and asymptotically approached some higher value with time of holding at the measuring temperature. The same asymptoic "equilibrium" viscosity was found when a sample was held at a lower temperature for a long time and then raised to the measuring temperature. The deviation of the initially measured viscosity from the equilibrium viscosity became progressively greater at higher viscosities, until above about 10^{16} P it was not possible to measure the equilibrium viscosity even after very long times at the measurement temperature. Thus this viscosity of 10^{16} P should be considered as the upper limit of measureable values, and any viscosity measured higher than this cannot be considered an equilibrium value.

The viscosities η of glasses sometimes fit an Arrhenius-type equation,

$$\eta = \eta_0 \exp(E/RT) \tag{3}$$

where E and η_0 are temperature-independent coefficients called the activation energy and the preexponential factor, respectively.

Figure 9.2 shows the logarithm of viscosities of various simple glass-forming oxides as a function of reciprocal temperature. The activation energies for viscous flow of these glasses are given in Table 9.1. The data for silica and germania were carefully selected from those of many workers as being the most reliable. The viscosity is decreased by the presence of small amounts of impurities, as discussed below; this is one of the main reasons for errors. Data in the figure were taken from the following references: silica, Hetherington et al. (1964) and Hofmeier and Urbain (1968); germania, Kurkjian and Douglas (1960); P_2O_5, Cormia et al. (1963); B_2O_3, Tweer et al. (1971).

Most oxide glasses show a decreasing activation energy of viscous flow with increasing temperature, as shown in Fig. 9.3. At low temperatures where the viscosity is above 10^{12} P, the activation energy E of alkali silicate and soda-

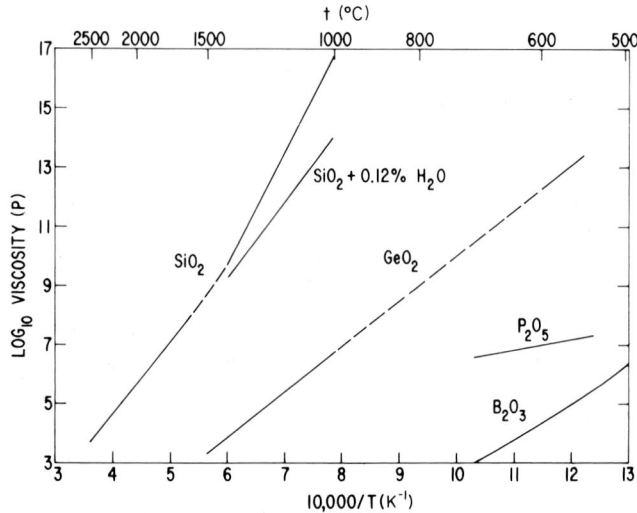

Fig. 9.2. Viscosities of various glass-forming oxides as a function of reciprocal temperature. Dashed lines show interpolations where data were not available. For references, see text.

TABLE 9.1

Activation Energies for Viscous Flow of Various Glasses

Glass	Activation energy (kcal mole^{-1})	Temperature range (°C)
Vitreous silica	170	1100–1400
	123	1600–2500
Vitreous germania	75	540–1500
Vitreous P$_2$O$_5$	41.5	545–655
Vitreous B$_2$O$_3$	83–12	26–1300

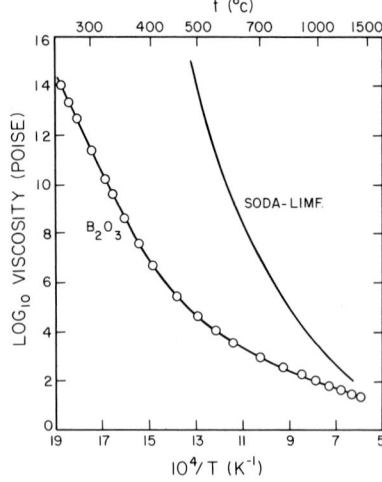

Fig. 9.3. Viscosities of boron trioxide and a soda-lime silicate glass of composition (wt. %) 21Na$_2$O–9CaO–70SiO$_2$ as a function of reciprocal temperature.

lime glasses is about 100 kcal mole^{-1}, and it decreases to 50 kcal mole^{-1} or less at high temperatures where the viscosity is about 100 P. The equation

$$\eta = \eta_0 \exp[E/(T - T_0)] \tag{4}$$

is often used instead of Eq. 3; T_0 is a constant temperature below the glass transition temperature. Equation (4) fits viscosity data for silicate glasses over wider temperature ranges than Eq. (3), but it still does not fit viscosities over the whole range from 10 to 10^{16} P with a single set of constants η_0, E, and T_0. Many theories have been suggested to explain viscosities of glasses, and especially T_0 values, but none is completely satisfactory and none can predict values of the constants.

The addition of other oxides to silica lowers its viscosity. The lowering is greatest with addition of alkali oxides and least with alkaline oxides and alumina; addition of latter oxides to an alkali silicate glass as a replacement for the alkali increases the viscosity of the glass. Zinc and boric oxides, among the multivalent oxides, are particularly effective in lowering the viscosities of silicates. Additions of alkali oxides to B_2O_3 and P_2O_5 also lower their viscosities.

TABLES AND FIGURES II

The viscosity values for binary glasses are given in Tables 9.2–9.29 and Figs. 9.4–9.7. Tables 9.30–9.115 and Figs. 9.8–9.20 show data for ternary glasses.

BIBLIOGRAPHY

Abou-El-Azm, A., and El-Batal, H. A. (1969). *Phys. Chem. Glasses* **10**, 159.
Aslanova, M. S., Chernov, V. A., and Kulakov, L. F. (1974). *Steklo Keram.* **6**, 19.
Bacon, J. F., Hasapis, A. A., and Wholley, J. W. (1960). *Phys. Chem. Glasses* **1**, 90.
Bockris, J. O'M., and Lowe, D. C. (1954). *Proc. R. Soc. London, Ser. A* **226**, 423.
Bockris, J. O'M., Mackenzie, J. D., and Kitchener, J. A. (1955). *Trans. Faraday Soc.* **51**, 1734.
Bollin, P. L. (1972). *J. Am. Ceram. Soc.* **55**, 483.
Bruckner, R., and Navarro, J. F. (1966). *Glastech. Ber.* **39**, 283.
Bulavin, I. A. (1960). *Dokl. Akad. Nauk SSSR* **130**, 133.
Cormia, R. L., Mackenzie, J. D., and Turnbull, D. (1963). *J. Appl. Phys.* **34**, 2245.
Din, A. (1968). Untersuchungen zum strukturellen Aufbau der Gläser $Na_2O \cdot xMe_nO \cdot (6 - x)$-$SiO_2$ aus der Temperaturabhangigkeit der Dichte, der Viskositat und davon abgeleiteter Groben. Dissertation, Clausthal.
Dubrovo, S. K., and Shnypikov, A. D. (1968). *Proc. Conf. Silic. Ind., 9th, Budapest* p. 459.
Eipeltauer, E., and More, A. (1960). *Radex-Rundsch.* **4**, 230.
Evstrop'ev, K. S., Gutorova, L. L., and Komleva, G. M. (1968). *Glass Ceram.* **25**, 470.
Fokin, V. M., Kalinina, A. M., and Filipovich, V. N. (1981). *J. Cryst. Growth* **52**, 115.
Frohberg, M. G., and Weber, R. (1965). *Arch. Eisenhuettenwes.* **36**, 477.
Gulityai, I. I. (1962). *Izv. Akad. Nauk SSSR, Otd. Metall. Topl.* **5**, 52.

Hagy, H. E. (1963). *J. Am. Ceram. Soc.* **46,** 93.

Heidtkamp, G., and Endell, K. (1936). *Glastech. Ber.* **14,** 89.

Hetherington, G., Jack, K. H., and Kennedy, J. C. (1964). *Phys. Chem. Glasses* **5,** 130.

Hoffman, L. C., Kupinski, T. A., Thakur, R. L., and Weil, W. A. (1952). *J. Soc. Glass Technol.* **36,** 196.

Hofmaier, G. (1968). *BHM, Berg-Huettenmaenn. Monatsh.* **113,** 269.

Hofmaier, G., and Urbain, G. (1968). *In* "Science of Ceramics," Vol. 4, p. 25. Br. Ceram. Soc., Stoke-on-Trent, England.

Hunold, K., and Bruckner, R. (1980). *Glastech. Ber.* **53,** 149.

Imoto, F., and Hirao, K. (1959). *Yogyo Kyokaishi* **67,** 381.

Ivanov, O. G., Tret'yakava, N. I., and Mazurin, O. V. (1969). *Glass Ceram.* **26,** 437.

Kadogawa, Y., Nakamura, K., Kawai, K., and Yamate, T. (1973). *Technol. Rep. Kansai Univ.* No. 14, p. 51.

Kani, K. (1935). *Proc. Imp. Acad. (Tokyo)* **11,** 334.

Karlsson, K. (1970). *Suom. Kemistil.* **43,** 479.

Kasymova, S. S. (1964). Physico-chemical properties of sodium–strontium–silicate glasses. Candidate's Dissertation, Tashkent.

Kawahara, M., Morinaga, K., and Yanagase, T. (1977). *Nippon Kinzoku Gakkaishi* **41,** 1047.

Kawahara, M., Morinaga, K., and Yanagase, T. (1979). *Nippon Kinzoku Gakkaishi* **43,** 55.

Kawahara, M., Mizoguchi, K. and Suginohara, Y. (1981). *Kyushu Kogyo Daigaku Kyushu Kogyo Daigaku Kenkyu Kogaku* No. 43, p. 53.

Konijnendijk, W. L. (1975). The structure of borosilicate glasses. Thesis, Eindhoven.

Kou, T., Mizoguchi, K., and Suginohara, Y. (1978). *Nippon Kinzoku Gakkaishi* **42,** 775.

Kurkjian, C. R., and Douglas, R. W. (1960). *Phys. Chem. Glasses* **1,** 19.

Kuroda, K., Kawahara, M., Morinaga, K., and Yanagase, T. (1982). *Nippon Kinzoku Gakkaishi* **46,** 275.

Leko, V. K., Lebedeva, R. B., Gusakova, N. K., and Pavlova, G. A. (1975). *Fiz. Khim. Stekla* **1,** 174.

Leont'eva, A. A. (1939). *Zh. Fiz. Khim.* **13,** 1020.

Li, J. H., and Uhlmann, D. R. (1970). *J. Non-Cryst. Solids* **3,** 127.

Lillie, H. R. (1933). *J. Am. Ceram. Soc.* **16,** 619.

Lillie, H. R. (1939). *J. Am. Ceram. Soc.* **22,** 367.

Machin, J. S., and Yee, T. B. (1948). *J. Am. Ceram. Soc.* **31,** 200.

Machin, J. S., and Yee, T. B. (1954). *J. Am. Ceram. Soc.* **37,** 177.

Machin, J. S., Yee, T. B., and Hanna, D. L. (1952). *J. Am. Ceram. Soc.* **35,** 322.

McMillan, P. W., Partridge, G., and Darrant, J. G. (1969). *Phys. Chem. Glasses* **10,** 153.

Matusita, K., and Tashiro, M. (1973). *Phys. Chem. Glasses* **14,** 77.

Matusita, K., Maki, T., and Tashiro, M. (1974). *Phys. Chem. Glasses* **15,** 106.

Matusita, K., Sakka, S., and Miyanishi, K. (1975). *J. Non-Cryst. Solids* **17,** 436.

Mazurin, O. V. (1969). *Likvatsionnye Yavleniya Steklakh, Tr. Vses. Simp., 1st, Leningrad, 1968* pp. 30–35.

Mazurin, O. V., and Strel'tsina, M. V. (1971). "Method of Determination of the Tie-line Directions in the Metastable Phase-Separation Regions of Ternary Systems," No. 3462–71 Dep. VINITI.

Mazurin, O. V., and Strel'tsina, M. V. (1972). *J. Non-Cryst. Solids* **11,** 199.

Mazurin, O. V., and Tret'yakova, N. I. (1970). *Inorg. Mater. (Engl. Transl.)* **6,** 1773.

Mazurin, O. V., Roskova, G. P., Strel'tsina, M. V., and Totesh, A. S. (1971). *In* "Stekloobraznaye Sistemy i Novyae Stekla na ikh osn.," p. 51. Moscow.

Mikiashvili, S. M., Tsylev, L. M., and Samarin, A. M. (1957). *Fiz.-Khim. Osn. Proizvod. Stali, Tr. Konf., 3rd, Moscow, 1955* p. 423.

Nakamura, T., Morinaga, K., and Yanagase, T. (1977). *Nippon Kinzoku Gakkaishi* **41,** 1300.

Neiman, T. S., Yinnon, H., and Uhlmann, D. R. (1982). *J. Non-Cryst. Solids* **48,** 393.

Nemilov, S. V. (1968). *Inorg. Mater. (Engl. Transl.)* **4,** 835.

Nemilov, S. V. (1969). *J. Appl. Chem. USSR (Engl. Transl.)* **42,** 46.

Nemilov, S. V., and Kasymova, S. S. (1975). *J. Appl. Chem. USSR (Engl. Transl.)* **48,** 1984.

Ouchi, Y., and Kato, E. (1979). *Nippon Kinzoku Gakkaishi* **43,** 625.

Pavlova, G. A., and Amatuni, A. N. (1975). *Inorg. Mater.* (*Engl. Transl.*) **11,** 1443.

Pavlova, G. A., Leko, V. K., Gusakova, N. K., and Komarova, L. A. (1974). *In* "Fiz.-Khim. Issled. Struk. i Svoistv Kvarts. Stekla," p. 116. Moscow.

Petrovskii, G. T. (1956). *Tr. L.I.I., Sb. Stud. Rabot* p. 37.

Poole, J. P. (1949). *J. Am. Ceram. Soc.* **32,** 230.

Poole, J. P., and Gensamer, M. (1949). *J. Am. Ceram. Soc.* **32,** 220.

Pospelov, B. A. (1954). *Zh. Fiz. Khim.* **28,** 2178.

Pospelov, B. A. (1955). *Zh. Fiz. Khim.* **29,** 70.

Pospelov, B. A. (1959). *Zh. Fiz. Khim.* **33,** 547.

Pospelov, B. A., and Evstrop'ev, K. S. (1941). *Zh. Fiz. Khim.* **15,** 125.

Pospelov, B. A., and Evstrop'ev, K. S. (1949). *In* "Physicochemical Properties of the Ternary System of Sodium Oxide, Lead Oxide and Silica," pp. 70–82. Izd. Akad. Nauk, Moscow–Leningrad.

Preston, E. (1938). *J. Soc. Glass Technol.* **22,** 45.

Rait, J. R., McMillan, Q. C., and Hay, R. (1939). *J. R. Tech. Coll. Glasgow* **4,** 449.

Rao, B. V. J. (1962). *J. Sci. Ind. Res., Sect. B* **21,** 108.

Rao, B. V. J. (1963a). *Phys. Chem. Glasses* **4,** 22.

Rao, B. V. J. (1963b). *J. Am. Ceram. Soc.* **46,** 107.

Richet, P. (1984). *Geochim. Cosmochim. Acta* **48,** 471.

Riebling, E. F. (1964a). *Can. J. Chem.* **42,** 2811.

Riebling, E. F. (1964b). *J. Chem. Phys.* **41,** 451.

Riebling, E. F. (1964c). *J. Am. Ceram. Soc.* **47,** 478.

Riebling, E. F. (1966). *J. Chem. Phys.* **44,** 2857.

Riebling, E. F. (1968a). *J. Am. Ceram. Soc.* **51,** 143.

Riebling, E. F. (1968b). *J. Am. Ceram. Soc.* **51,** 406.

Rossin, R., Bersan, J., and Urbain, G. (1964). *Rev. Int. Hautes Temp. Refract* **1,** 159.

Saringyulyan, R. S., and Kostanyan, K. A. (1969). "Viscosity and Electrical Conductivity of Glasses in Wide Intervals of Temperature." No. 902–69 Dep. VINITI.

Sasek, L. (1975). *Sb. Vys. Sk. Chem.-Technol. Praze, Chem. Technol. Silik.* **L6,** 61.

Sasek, L., Meissnerova, H., and Hoskova, V. (1975a). *Sb. Vys. Sk. Chem.-Technol. Praze, Chem. Technol. Silik.* **L6,** 153.

Sasek, L., Meissnerova, H., and Prochazka, J. (1975b). *Sb. Vys. Sk. Chem.-Technol. Praze, Chem. Technol. Silik.* **L6,** 95.

Schultz, P. C. (1976). *J. Am. Ceram. Soc.* **59,** 214.

Segers, L., Fontana, A., and Winand, R. (1979). *Electrochim. Acta* **24,** 213.

Shartsis, L., and Spinner, S. (1951). *J. Res. Natl. Bur. Stand. (U.S.)* **46,** 176.

Shartsis, L., Spinner, S., and Capps, W. (1952). *J. Am. Ceram. Soc.* **35,** 155.

Shelby, J. E. (1978). *J. Appl. Phys.* **49,** 5885.

Sheludyakov, L. N., Sarancha, E. T., and Vakhitov, A. A. (1967). *Tr. Inst. Khim. Nauk, Akad. Nauk Kaz. SSR* **15,** 158.

Shiraishi, Y., Ikeda, K., Tamura, A., and Saito, T. (1978). *Trans. Jpn. Inst. Met.* **19,** 264.

Shvaiko-Shvaikovskaya, T. P. (1968). *Steklo* No. 3, p. 93.

Shvaiko-Shvaikovskaya, T. P., Gusakova, N. K., and Mazurin, O. V. (1971a). *Inorg. Mater.* (*Engl. Transl.*) **7,** 620.

Shvaiko-Shvaikovskaya, T. P., Gusakova, N. K., and Mazurin, O. V. (1971b). Viscosity of the glass system Na_2O–CaO–SiO_2 in narrow intervals of temperature. No. 483–71 Dep. VINITI.

Shvaiko-Shvaikovskaya, T. P., Mazurin, O. V., and Bashun, Z. S. (1971c). *Inorg. Mater.* (*Engl. Transl.*) **7,** 128.

Simpson, H. E. (1959). *Glass Ind.* **40,** 454.

Simpson, H. E. (1961). *Glass Ind.* **42,** 222.

Skornyakov, M. M. (1949). *In* "Physico-chemical Properties of the System Na_2O–PbO–SiO_2," p. 39. Moscow.

Skornyakov, M. M., Kuznetsov, A. Y., and Evstrop'ev, K. S. (1941). *Zh. Fiz. Khim.* **15**, 116.

Strel'tsina, M. V. (1970). Candidate's Dissertation, Leningrad.

Sugai, M., and Somiya, S. (1982). *Yogyo Kyokaishi* **90**, 56.

Sumita, S., Takano, H., Morinaga, K., and Yanagase, T. (1982). *Nippon Kinzoku Gakkaishi* **46**, 280.

Sviridov, S. I., Roskova, G. P., Moiseev, V. V., and Zhabrev, V. A. (1976). *Sov. J. Glass Phys. Chem.* *(Engl. Transl.)* **2**, 514.

Tananaev, I. V., Skorikov, V. M., Kargin, Y. F., and Zhereb, V. P. (1978). *Inorg. Mater.* *(Engl. Transl.)* **14**, 1576.

Taylor, T. D., and Rindone, G. E. (1970). *J. Am. Ceram. Soc.* **53**, 692.

Toropov, N. A., and Bryantsev, B. A. (1965). *In* "Structural Transformations at High Temperatures," p. 205. Moscow.

Tret'yakova, N. I. (1969). Investigation of viscosity of the glass system $Na_2O-RO-SiO_2$ over a narrow temperature interval. Atoref., Candidate's Dissertation, Leningrad.

Tret'yakova, N. I., and Mazurin, O. V. (1969). "Influence of Bivalent Metal Oxides on the Viscosity of Potassium Silicate Glass," No. 995–69 Dep. VINITI.

Tweer, H., Laberge, N., and Macedo, P. B. (1971). *J. Am. Ceram. Soc.* **54**, 121.

Urbain, G. (1974). *Rev. Int. Hautes Temp. Refract.* **11**, 133.

Urbain, G., Millon, F., and Cariset, S. (1980). *C. R. Hebd. Seances Acad. Sci., Ser. C* **290**, 137.

Urbain, G., Bottinga, Y., and Richet, P. (1982). *Geochim. Cosmochim. Acta* **46**, 1061.

VanBernst, A., and Delaunois, C. (1966). *Verres Refract.* **20**, 435.

Varshal, B. G., Goikhman, V. Y., Mirskikh, L. L., Shitts, L. K., and Shitts, Y. A. (1980). *Fiz. Khim. Stekla* **6**, 734.

Yagi, S., Mizoguchi, K., and Suginohara, Y. (1980). *Bull. Kyushu Inst. Technol. Sci. Technol.* No. 40, p. 33.

Yanishevskii, V. M. (1961). *Steklo* No. 4, p. 34.

Yanishevskii, V. M. (1964). *In* "Stekloobraznaye Sostoyaniye," p. 76. Minsk.

Žagar, L., and Lüneberg, H. (1971). *Glastech. Ber.* **37**, 235.

TABLE 9.2

Viscosity of Lithium Silicate Glasses

Composition (mole % Li_2O)	Temperature (°C)	$\log \eta$ (P)	Reference[a]	Composition (mole % Li_2O)	Temperature (°C)	$\log \eta$ (P)	Reference[a]
20	426	13	1	30	493	10.5	1
	437	12.5			506	10	
	449	12			518	9.5	
	461	11.5			533	9	
	473	11			553	8.5	
	487	10.5			575	8	
	500	10		33.33	448	13.42	2
	513	9.5			460	12.64	
	531	9			465	12.35	
	547	8.5			469	12.15	
	565	8			471	12.05	
30	438	13	1		475	11.78	
	448	12.5			478	11.64	
	459	12			485	11.27	
	471	11.5			492.5	10.88	
	482	11			502	10.46	

[a] Data from (1) Nemilov (1969) and (2) Fokin *et al.* (1981).

TABLE 9.3

Viscosities of Li$_2$O–SiO$_2$ Glass Melts[a]

Composition (mole % Li$_2$O)	η (P) at given temperatures (°C)											
	1150	1200	1250	1300	1350	1400	1450	1500	1550	1600	1650	1700
20						191	132	93.3	72.4	49.0	36.3	27.5
25				141	118	81.3	56.2	39.8	28.8	21.4		
30		115	77.6	55.0	38.9	28.2	20.9	15.9				
33	123	81.3	55.0	38.9	27.5	20.0	15.1					
35	72.4	49.0	33.9	24.6	18.2	13.5	10.2	7.87				
40	35.5	26.9	20.0	15.5	12.3	9.55	7.76	6.46				
45		8.91	6.61	5.13	3.98	3.09	2.51	2.00				
50			2.88	2.24	1.78	1.41	1.15	0.96				
55			0.92	0.75	0.63	0.53	0.45					

[a] Data from Bockris *et al.* (1955).

TABLE 9.4

Parameters of the Equation $\log \eta$ (P) $= \log \eta_0 + B/T$ for the Viscosities of Sodium Silicate Glasses[a]

Composition (mole % Na$_2$O)	Temperature range (°C)	$-\log \eta_0$ (P)	$10^{-3}B$ (K)
14.73	495–645	9.919	17.806
20.13	490–590	15.619	21.432
24.82	455–635	17.848	22.759
34.53	425–575	20.604	23.980
39.48	415–545	21.780	23.675

[a] Data from Poole (1949) and Poole and Gensamer (1949).

TABLE 9.5

Viscosities of Na$_2$O–SiO$_2$ Glasses[a]

Composition (mole % Na$_2$O)	Temperatures (°C) corresponding to given $\log \eta$ (P)										
	8	8.5	9	9.5	10	10.5	11	11.5	12	12.5	13
5				838	809	781	757	732	708	683	653
13			712	693	675	658	641	624	607	592	578
20	652	629	608	588	571	551	532	517	502	488	473

[a] Data from Nemilov (1969).

TABLE 9.6

Viscosities of Sodium Silicate Glass Melts[a]

Composition (mole % Na$_2$O)	Values of $\log \eta$ (P) at given temperature (°C)						
	900	1000	1100	1200	1300	1400	1500
15			4.15	3.55	3.05	2.70	2.26
20	4.60	3.95	3.35	2.90	2.50	2.20	1.86
25	4.25	3.65	3.05	2.60	2.25	1.90	1.59
30	3.90	3.30	2.75	2.35	2.00	1.65	1.40
35	3.60	3.00	2.45	2.05	1.70	1.40	1.24
40	3.25	2.65	2.10	1.75	1.45	1.15	
45		2.10	1.55	1.25	0.90	0.65	
50		1.55	1.00	0.70	0.35	0.10	

[a] Average values based on the data sets of 10 different research laboratories. See Shvaiko-Shvaikovskaya et al. (1971c).

TABLE 9.7

Parameters of the Equation $\log \eta$ (P) $= \log \eta_0 + B/T$ for the Viscosities of Potassium Silicate Glasses[a]

Composition (mole % K_2O)	Temperature range (°C)	$-\log \eta_0$ (P)	$10^{-3}B$ (K)
8.03	570–697	13.617	22.039
10.64	505–612	15.460	22.562
18.01	502–612	16.163	22.427
18.52	497–583	16.303	22.602
24.50	485–612	16.393	22.052
29.20	473–582	18.679	23.061
34.05	457–502	29.042	30.303

[a] Data from Poole (1949) and Poole and Gensamer (1949).

TABLE 9.8

Viscosities of K_2O–SiO_2 Glasses[a]

Composition (mole % K_2O)	Temperature (°C) corresponding to given $\log \eta$ (P)										
	8	8.5	9	9.5	10	10.5	11	11.5	12	12.5	13
5	838	807	777	748	722	698	675.5	653	632	612	592
13	702	680	655	632	612	595	578	563	547	532	517
20	662	641	621	601	583	566	550	533	518	506	491
30	658	643	631	619	607	597	587	577	567	557	548

[a] Data from Nemilov (1969).

TABLE 9.9

Viscosities of Potassium Silicate Glass Melts[a]

Composition (mole % K_2O)	η (P) at given temperature (°C)													
	1100	1150	1200	1250	1300	1350	1400	1450	1500	1550	1600	1650	1700	1750
2.5											4680	2690	1780	1260
6.3									3470	1700	1070	724	537	
10.8						1860	1180	759	525	347	240			
16.9	7080	4790	2240	1380	832	708	355							
22.3	4370	2460	1410	851	537	339	229							
25.6	2190	1230	708	324	269	170	115							
33.4	1260	676	390	224	145	83.2	60.3							

[a] Data from Bockris et al. (1955).

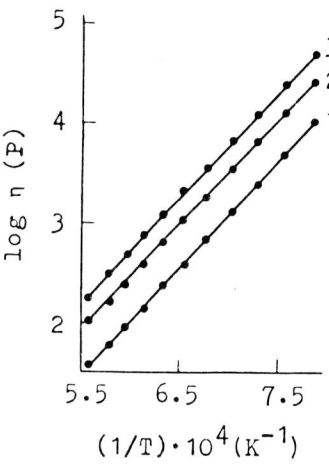

Fig. 9.4. Temperature dependence of viscosity of (1) $Rb_2O-2SiO_2$, (2) $Rb_2O-3SiO_2$, and (3) $Rb_2O-4SiO_2$ systems. Data from Sasek (1975) and Sasek *et al.* (1975a).

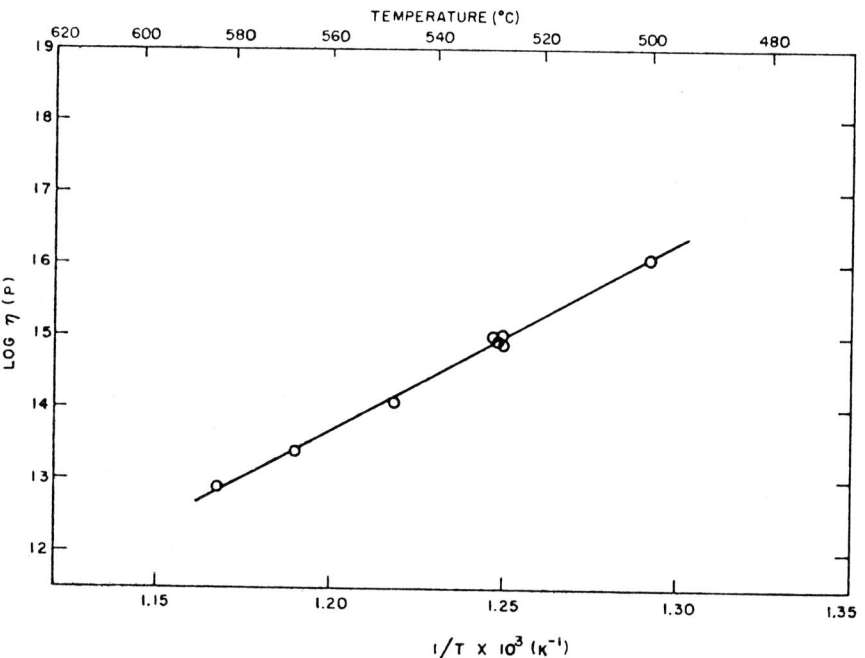

Fig. 9.5. Temperature dependence of the viscosity of glass composition (mole %) $8Rb_2O-92SiO_2$. Viscosities were determined from the stress–strain curves at low stress levels. From Li and Uhlmann (1970).

TABLE 9.10

**Softening Temperatures of
Rubidium Silicate Glasses[a]**

Composition (mole % Rb₂O)	Littleton softening temperature (°C); $\log \eta$ (P) = 7.6
12.1	908
17.6	874
24.3	828

[a] Data from Simpson (1961).

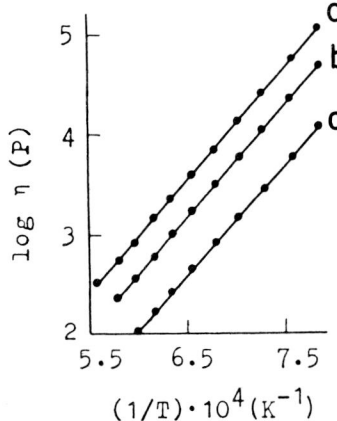

Fig. 9.6. Temperature dependence of viscosity of (a) $Cs_2O-2SiO_2$, (b) $Cs_2O-3SiO_2$, and (c) $Cs_2O-4SiO_2$ systems. Data from Sasek (1975) and Sasek *et al.* (1975b).

TABLE 9.11

Viscosities of Magnesium Silicate Glass Melts[a]

Composition (mole % MgO)	Temperature (°C)	$\ln \eta$ (P)	Composition (mole % MgO)	Temperature (°C)	$\ln \eta$ (P)
41.4	1774	1.287	50	1936	−0.221
	1789	1.173		1992	−0.439
	1827	0.9746		1995	−0.462
	1848	0.8879	65.1	2020	−1.26
	1891	0.6780		2022	−1.27
	1891	0.6729		2062	−1.36
	1947	0.5481		2066	−1.39
50	1714	0.7372		2111	−1.29
	1758	0.5324		2136	−1.52
	1781	0.4370		2138	−1.52
	1783	0.3954		2183	−1.53
	1831	0.1467		2186	−1.61
	1862	0.0564		2188	−1.63
	1884	−0.055			
	1889	−0.046			

[a] Data from Urbain *et al.* (1982).

TABLE 9.12

Viscosities of Calcium Silicate Glass Melts[a]

Composition (mole % CaO)	Temperature (°C)	ln η (P)	Composition (mole % CaO)	Temperature (°C)	ln η (P)
32	1717	2.762	50	1765	−0.064
	1763	2.510		1829	−0.324
	1811	2.230		1878	−0.504
	1858	1.966		1938	−0.720
	1899	1.739		1989	−0.856
	1972	1.381		2039	−1.04
37.6	1552	2.833		2096	−1.26
	1602	2.477	60	1811	0.837
	1653	2.189		1845	−1.00
	1703	1.901		1878	−1.08
	1753	1.635		1904	−1.22
	1803	1.362		1941	−1.24
	1853	1.105		1967	−1.35
	1903	0.8755		1993	−1.43
50	1585	0.8242		2056	−1.59
	1646	0.4886		2120	−1.81
	1691	0.2639			

[a] Data from Urbain *et al.* (1982).

TABLE 9.13

Viscosities of CaO–SiO₂ Glass Melts[a]

Composition (mole % CaO)	η (P) at given temperature (°C)							
	1450	1500	1550	1600	1650	1700	1750	1800
30.5						13.6	10.4	8.5
34.6					10.0	7.8	6.05	4.5
38.8	21.1	14.0	9.8	7.10	5.25	3.92	3.10	2.5
41.6		9.35	6.48	4.68	3.57	2.75	2.16	1.8
43.6		7.65	5.60	4.05	2.95	2.35	1.95	1.5
48.7		4.35	3.17	2.41	1.90	1.50	1.20	0.99
49.7			2.88	2.18	1.68	1.33	1.06	0.88
50.7			2.70	2.00	1.52	1.19	0.97	0.80
51.7			2.43	1.81	1.39	1.11	0.90	0.75
52.7		3.03	2.20	1.66	1.28	1.01	0.83	0.72
53.7		2.88	2.11	1.57	1.20	0.96	0.79	0.66
54.7		2.57	1.39	1.40	1.10	0.90	0.75	0.66
56.1	3.51	2.39	1.80	1.39	1.05	0.81	0.68	0.60
57.6				1.13	0.90	0.74	0.62	0.54

[a] Data from Bockris and Lowe (1954).

TABLE 9.14

Viscosities of Calcium Silicate Glass[a]

Composition (mole % CaO)	Temperature (°C)	$\log \eta$ (P)
46	879	8
	848	9
	827	10
	807	11
	787	12
	769	13

[a] Data from Varshal et al. (1980).

TABLE 9.15

Viscosities of Strontium Silicate Glass Melts[a]

Composition (mole % SrO)	η (P) at given temperature (°C)					
	1550	1600	1650	1700	1750	1800
20.1				80.4	58.2	43.1
25.4		62.0	42.2	28.0	22.4	16.6
29.7	26.9	18.6	13.8	10.2	7.76	5.78
40.5	11.5	8.05	6.19	4.79	3.73	2.95
44.4	6.30	4.88	3.58	2.81	2.14	1.82
50.3		3.18	2.47	1.95	1.57	1.30

[a] Data from Bockris et al. (1955).

TABLE 9.16

Viscosities of SrO–SiO$_2$ System[a]

Composition (mole % SrO)	Temperature (°C)	$\ln \eta$ (P)	Composition (mole % SrO)	Temperature (°C)	$\ln \eta$ (P)
50	1464	1.947	50	1789	0.1553
	1499	1.698		1842	−0.039
	1552	1.386		1898	−0.235
	1590	1.112		1913	−0.304
	1649	0.8307		2014	−0.655
	1710	0.5247		2101	−0.952

[a] Data from Urbain et al. (1982).

TABLE 9.17

Viscosities of Barium Silicate Glass Melts[a]

Composition (mole % BaO)	Temperature (°C)	ln η (P)	Composition (mole % BaO)	Temperature (°C)	ln η (P)
3.7	1632	5.131	50	1602	1.056
	1690	4.683		1646	0.8109
	1750	4.265		1700	0.5550
	1815	3.796		1737	0.3528
	1891	3.359		1797	0.0980
	1930	3.125		1876	−0.252
	1983	2.811		1914	−0.389
	2082	2.326		1991	−0.687
				2063	−0.952

[a] Data from Urbain *et al.* (1982).

TABLE 9.18

Viscosities of BaO–SiO$_2$ Glass Melts[a]

Composition (mole % BaO)	η (P) at given temperature (°C)						
	1500	1550	1600	1650	1700	1750	1800
15.4				299	195	135	100
25.1	110	66.1	43.7	30.2	21.4	16.0	12.5
30.7	32.4	22.4	15.9	11.8	8.51	6.68	5.31
33.5	26.3	18.2	12.9	9.66	7.20	5.80	4.50
40.2	17.0	11.8	8.85	6.68	5.37	4.57	3.98
42.0		7.94	5.62	4.25	3.31	2.62	2.15
49.8			2.80	2.22	1.86	1.64	1.50

[a] Data from Bockris *et al.* (1955).

TABLE 9.19

Viscosities of Lead Silicate Glasses[a]

Composition (mole % PbO)	Temperature (°C) corresponding to $\log \eta$ (P)										
	4	5	6	7	8	9	10	11	12	13	16
20					733	673	633	501	565	535	443
30			737	678	647	605	572	543	517.5	492	425
33.3	856	770	703	656	616	581	551	524	499	476	412
35	870	773	702	647	608	575	547	522	497	473	413
40.1	794	707	653	616	581	551	523	499.5	477	455	400
42.1	754	668	614	571	544	518	493	471	449	429.5	377
45				593	542	508	486.5	465	444	424	372
48				524	497	477	457	437	419	402	355
49.4				524	497	477	459	440	423	407	362
49.6				519	484.5	465	448	431	417	402	362
50			553	515	485	463	447	432	417	400	362
55					457	439	423	405	389	373	330
60					444	427	412	397	382	367	327
63					427	412	398	385	372	359	324
66					410	393	378	367	356.5	345	313
66.6					410	392	378	367	357	346	317
68					419	402	390	377	366	354	322
70					415	401	388	377	365	354	322
73					392	378.5	367	356	345	333	303

[a] Data from Nemilov (1968).

TABLE 9.20

Viscosity of PbO–SiO$_2$ System[a]

Composition (mole % PbO)	Temperature range (°C)	$\log \eta$ (P)
50	362–1290	$= 11.0 - (2.82 \times 10^4 T^{-1})$ $+ (1.99 \times 10^7 T^{-2})$

[a] Data from Neiman et al. (1982).

[b] Obtained by fitting the combined viscosity data of Nemilov (1968) in the temperature range 362–553°C and of Ouchi and Kato (1979) in the interval 800–1290°C; T is temperature (K).

TABLE 9.21

Viscosities of Manganese Silicate Glass Melts[a]

Composition (mole % MnO)	Temperature (°C)	$\ln \eta$ (P)	Composition (mole % MnO)	Temperature (°C)	$\ln \eta$ (P)
50	1591	0.9270	68.8	1432	−1.90
	1645	0.4324		1440	−1.51
	1649	0.4253		1446	−1.51
	1652	0.4246		1469	−1.56
	1655	0.3542		1480	−1.61
	1694	0.2127		1486	−1.66
	1696	0.1890		1493	−1.71
	1698	0.1740		1526	−1.83
	1700	0.1579		1539	−1.83
	1702	0.1570		1564	−1.90
	1704	0.1536		1594	−1.90
	1710	0.1484		1612	−1.97
	1713	0.1151		1620	−1.90
	1740	−0.040		1634	−1.97
	1747	−0.029		1636	−2.04

[a] Data from Urbain et al. (1982).

TABLE 9.22

Viscosities of Iron Silicate Glass Melts[a]

Composition (mole %)			Temperatures (°C) and values of $\ln \eta$ (P)							
FeO[b]	SiO₂									
60	40	Temp. (°C)	1350	1367	1372	1385	1402	1425	1432	1435
		$\ln \eta$ (P)	−0.329	−0.401	−0.416	−0.446	−0.511	−0.580	−0.616	−0.616
60	40	Temp. (°C)	1438	1440						
		$\ln \eta$ (P)	−0.635	−0.635						
65	35	Temp. (°C)	1170	1185	1192	1245	1305	1377	1402	1430
		$\ln \eta$ (P)	−0.329	−0.386	−0.416	−0.580	−0.755	−0.942	−0.994	−1.08
66.7	33.3	Temp. (°C)	1166	1203	1232	1254	1293	1295	1298	1323
		$\ln \eta$ (P)	0.0953	−0.073	−0.163	−0.249	−0.371	−0.371	−0.386	−0.462
66.7	33.3	Temp. (°C)	1328	1374	1375	1396	1406	1408	1412	
		$\ln \eta$ (P)	−0.478	−0.616	−0.598	−0.673	−0.693	−0.713	−0.693	
70	30	Temp. (°C)	1205	1208	1235	1268	1272	1302	1305	1379
		$\ln \eta$ (P)	−0.821	−0.844	−0.916	−0.994	−0.994	−1.08	−1.11	−1.24
70	30	Temp. (°C)	1382	1411	1413	1415				
		$\ln \eta$ (P)	−1.27	−1.31	−1.35	−1.31				

[a] Data from Urbain et al. (1982).
[b] Melt in equilibrium with metallic iron; all iron is given as FeO.

TABLE 9.23

Viscosities of Cobalt Silicate Glass Melt[a]

Composition (mole % CoO)	Temperature (°C)	log η (P)
0.9	2090	4.720
	2230	4.305
	2350	4.086

[a] Data from Bacon et al. (1960).

TABLE 9.24

High-Temperature Viscosities of B_2O_3–SiO_2 System at Various Temperatures

Composition (mole % B_2O_3)	Temperatures and viscosity values						
6.2	Temp. (°C)	1763	1783	1815	1840		
	$\eta \times 10^{-4}$ (P)	33.0	26.6	16.9	13.1		
10.1	Temp. (°C)	1727	1730	1736	1738	1740	1757	1768
	$\eta \times 10^{-4}$ (P)	13.3	11.2	10.9	11.4	11.0	9.07	8.57
10.1	Temp. (°C)	1775	1778	1792			
	$\eta \times 10^{-4}$ (P)	7.78	6.54	5.83			
14.6	Temp. (°C)	1691	1693	1720	1725	1752	1757	1778
	$\eta \times 10^{-4}$ (P)	3.51	3.37	2.63	2.45	1.92	1.85	1.47
14.6	Temp. (°C)	1783	1797	1800	1802	1812	1816
	$\eta \times 10^{-4}$ (P)	1.45	1.17	1.14	1.12	1.00	0.97
25.2	Temp. (°C)	1303	1329	1355	1376	1418	1444
	$\eta \times 10^{-4}$ (P)	127.0	89.8	67.4	44.5	32.0	21.9

[a] Data from Urbain et al. (1980).

TABLE 9.25

Parameters of the Equation log η (P) = log η_0 + B/T for the Viscosities of B_2O_3–SiO_2 System[a]

Composition (mole % B_2O_3)	Temperature range (°C)	$-\log \eta_0$ (P)	B (K)
42.4	1100–1460	2.3684	9823
53.1	1380–1530	1.9639	8239
62.4	1280–1460	1.9917	7687
71.9	1130–1410	1.2436	5740
82.5	1050–1360	0.8997	4576
90.0	1030–1360	0.4225	3434
93.91	1070–1350	0.6841	3655

[a] Data from Riebling (1964c).

TABLE 9.26

Viscosities of B_2O_3–SiO_2 System at Different Temperatures[a]

Composition (mole %) H₂O	B₂O₃	SiO₂		Temperatures (°C) and log η (P)											
0.139	97.69	2.17	Temp. (°C)	257	351	466	563	672	801	952	1099				
			log η (P)	13	8.0	5.202	3.944	3.198	2.648	2.193	1.876				
0.097	95.31	4.59	Temp. (°C)	258	360										
			log η (P)	13	8.0										
0.082	93.97	5.95	Temp. (°C)	264	364	469	525	581	645	710	778	858	935	1029	1105
			log η (P)	13	8.0	5.522	4.736	4.107	3.60	3.213	2.904	2.613	2.367	2.137	1.963
0.074	90.38	9.55	Temp. (°C)	266	377										
			log η (P)	13	8.0										
0.089	89.19	10.72	Temp. (°C)	270	381	473	547	613	694	795	886	996	1087		
			log η (P)	13	8.0	5.765	4.689	4.062	3.505	3.026	2.696	2.389	2.170		
0.091	84.07	15.84	Temp. (°C)	272	396										
			log η (P)	13	8.0										
0.082	77.65	22.27	Temp. (°C)	284	420	543	614	701	797	892	996	1098			
			log η (P)	13	8.0	5.662	4.864	4.182	3.644	3.226	2.891	2.611			
0.049	70.56	29.39	Temp. (°C)	301	455										
			log η (P)	13	8.0										
0.041	64.61	35.35	Temp. (°C)	310	487	587	640	696	751	844	939	1036	1128		
			log η (P)	13	8.0	6.446	5.843	5.347	4.954	4.424	3.986	3.593	3.263		
0.053	56.38	43.57	Temp. (°C)	322	525										
			log η (P)	13	8.0										
0.059	51.58	48.36	Temp. (°C)	327	566	611	664	715	792	866	961	1055	1155		
			log η (P)	13	8.0	7.479	6.909	6.443	5.816	5.357	4.835	4.381	3.955		
0.048	44.55	55.40	Temp. (°C)	343	646										
			log η (P)	13	8.0										
0.044	39.26	60.69	Temp. (°C)	355	713	696	775	864	954	1031	1112	1211			
			log η (P)	13	8.0	7.435	6.738	6.036	5.490	5.056	4.635	4.219			

[a] Data from Bruckner and Navarro (1966).

TABLE 9.27

Viscosities of Al_2O_3–SiO_2 System at Various Temperatures[a]

Composition (mole %)			Temperatures (°C) and $\ln \eta$ (P)							
Al_2O_3	SiO_2									
6.2	93.8	Temp. (°C)	1653	1703	1753	1803	1853	1903	1953	2003
		$\ln \eta$ (P)	11.29	10.57	9.878	9.250	8.594	8.013	7.438	6.897
20.2	79.8	Temp. (°C)	1653	1703	1753	1803	1853	1903	1953	2002
		$\ln \eta$ (P)	6.252	5.759	5.298	4.860	4.433	4.031	3.651	3.296
50	50	Temp. (°C)	1853	1872	1880	1882	1904	1955	2005	2050
		$\ln \eta$ (P)	1.278	1.176	1.135	1.115	1.015	0.7608	0.5188	0.3134
70	30	Temp. (°C)	1853	1903	1953	2003	2053	2103	2154	2204
		$\ln \eta$ (P)	0.4187	0.2231	0.0296	−0.160	−0.333	−0.506	−0.673	−0.816

[a] Data from Urbain et al. (1982).

TABLE 9.28

Viscosities of Germanium Silicate Glasses[a]

Composition (mole % GeO$_2$)	Temperature (°C)	log η (P)
42.2	626	14.5
	719	13
	781	12
50.6	642	14.5
	737	13
	801	12

[a] Data from Riebling (1968b).

TABLE 9.29

Arrhenius Parameters for the Viscosity of Titanium Silicate Glasses[a]

Composition (mole % TiO$_2$)	Temperature (°C)	log η (P)	Temperature range (°C)	$-$log η_0 (P)	E (kcal mole^{-1})	Reference[a]
5.7			850–1250	7.4 $\pm$ 0.9	126 $\pm$ 9	1
6.5			1100–1400	7.4 $\pm$ 0.9	126 $\pm$ 9	2
			1600–1900	7.0 $\pm$ 0.8	124 $\pm$ 6	
6.1	1200	11.21				3
	1700	6.71				
	1800	6.10				

[a] Data from (1) Pavlova et al. (1974), (2) Leko et al. (1975), and (3) Pavlova and Amatuni (1975).

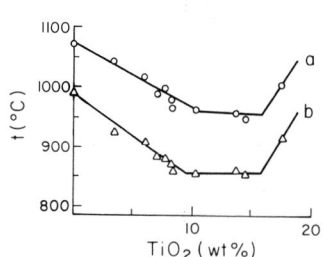

Fig. 9.7. Composition dependence of (a) annealing point, log η = 13.2 P, and (b) strain point, log η = 14.6 P, of titanium silicate glasses melted by flame hydrolysis method. From Schultz (1976).

TABLE 9.30

Parameters of the Equation $\eta = \eta_0 \exp(B/T)$ for Viscosities of $Li_2O-Na_2O-SiO_2$ Glasses[a]

Composition (mole %)			Temperature (°C) for $\log \eta = 10$	η_0 (P)	$10^{-4}B$ (K)
Li_2O	Na_2O	SiO_2			
24.27	2.91	72.82	500	1.5×10^{-16}	4.58
32.33	2.91	64.76	463	1.0×10^{-23}	5.59

[a] Data from Matusita and Tashiro (1973) and Matusita et al. (1974).

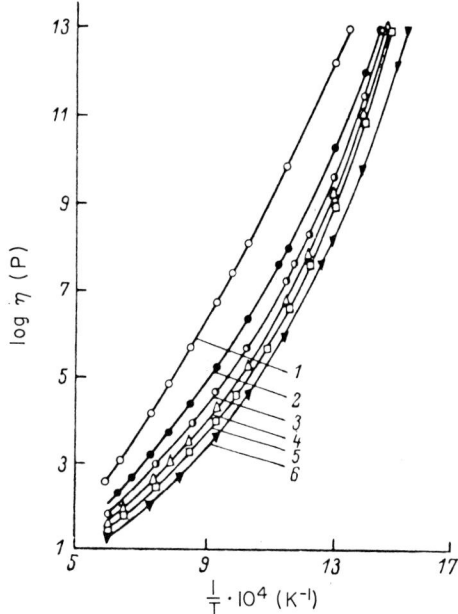

Fig. 9.8. Temperature dependence of viscosity of $x Li_2O-14.29 Na_2O-(85.71-x)SiO_2$ system with values of x: (1) 0, (2) 7.14, (3) 10.71, (4) 14.28, (5) 17.86, and (6) 21.43. From Din (1968).

TABLE 9.31

Viscosities of Lithium-Potassium Silicate Glasses[a]

| Composition (mole %) | | | Temperature (°C) for given $\log \eta$ (P) | | | | | | | | | | | |
Li₂O	K₂O	SiO₂	8	8.5	9	9.5	10	10.5	11	11.5	12	12.5	13	16
5	15	80	608	586	566	548	531	517	503	490	476	462	451	384
10	10	80	624	601	580	562	545	530	513	499	483	471	457	387
15	5	80	638	612	588	571	552	537	521	506	491	477	462	390
20	—	80	565	547	531	513	500	487	473	461	449	437	426	366
—	30	70	658	643	631	619	607	597	587	577	567	457	448	397
5	25	70	567	550	533	517	502	487	472	459	446	433	420	354
15	15	70	580	561	543	527	512	499.5	485	471	457	446	433	367
20	10	70	562	545	531	514.5	501	488	473	461	449	437	427	364
25	5	70	570	552	537	522	509	493	481	468	455	443	431	367
30	—	70	575	553	533	518	506	493	482	471	459	448	438	381

[a] Data from Nemilov (1969).

TABLE 9.32

Parameters of the Equation $\eta = \eta_0 \exp(B/T)$ for the Viscosities of Li₂O–K₂O–SiO₂ Glasses[a]

Composition (mole %)			Temperature (°C) for $\log \eta = 10$	η_0 (P)	$10^{-4}B$ (K)
Li₂O	K₂O	SiO₂			
24.27	2.91	72.82	510	2.6×10^{-17}	4.78
32.33	2.91	64.76	475	4.1×10^{-21}	5.22

[a] Data from Matusita and Tashiro (1973) and Matusita et al. (1974).

TABLE 9.33

Parameters of the Equation $\eta = \eta_0 \exp(B/T)$ for the Viscosities of Li₂O–Cs₂O–SiO₂ Glass Systems[a]

Composition (mole %)			Temperature (°C) for $\log \eta = 10$	η_0 (P)	$10^{-4}B$ (K)
Li₂O	Cs₂O	SiO₂			
24.27	2.91	72.82	430	6.0×10^{-17}	4.83
32.33	2.91	64.76	488	5.5×10^{-23}	5.63

[a] Data from Matusita and Tashiro (1973) and Matusita et al. (1974).

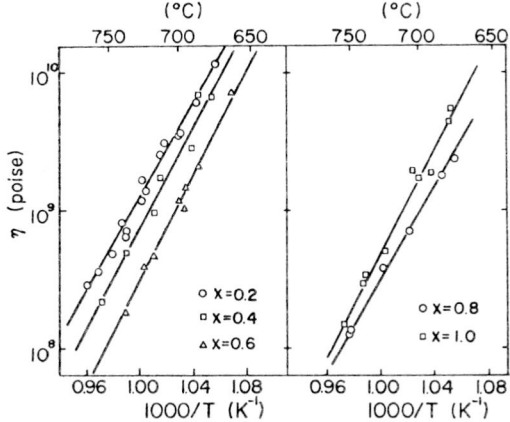

Fig. 9.9. Temperature dependence of viscosity of $(1 - x)\text{Li}_2\text{O}-x\text{Cs}_2\text{O}-9\text{SiO}_2$ glasses. From Matusita et al. (1975).

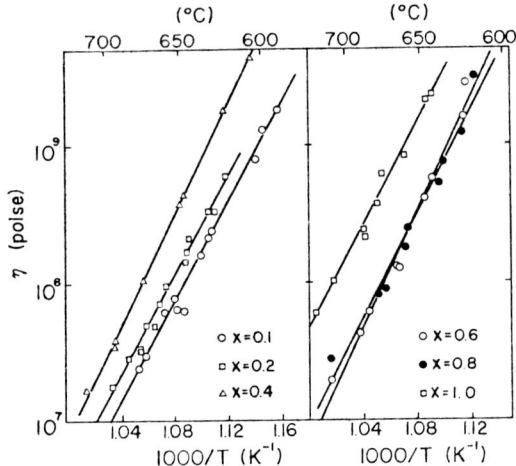

Fig. 9.10. Temperature dependence of viscosity of $(1 - x)Li_2O-xCs_2O-4SiO_2$ glasses. From Matusita *et al.* (1975).

TABLE 9.34

Parameters of the Equation $\log \eta$ (P) $= \log \eta_0 + B/T$
for the Viscosities of $Na_2O-K_2O-SiO_2$ Glasses[a]

Composition (mole %)			$-\log \eta_0$	B	Temperature range
Na_2O	K_2O	SiO_2	(P)	(K)	(°C)
4.76	12.03	83.21	13.66	19702	490–600
6.58	16.59	76.83	16.19	21033	440–550
8.38	28.07	63.55	15.98	19510	420–550
10.15	7.02	82.83	13.23	19115	490–580
12.70	9.27	78.03	16.22	20571	460–510
12.91	11.34	75.75	15.65	20202	440–550
14.26	3.20	82.54	12.47	18628	490–600
17.80	15.44	66.76	14.95	18578	420–550
18.74	4.14	77.12	17.06	21344	440–550
26.31	6.48	67.21	14.14	18017	420–550
31.14	2.78	66.08	16.63	20147	425–550

[a] Data from Poole (1949) and Poole and Gensamer (1949).

TABLE 9.35

Viscosities of Sodium–Potassium Silicate Glasses[a]

Composition (mole %)			Temperature (°C) for $\log \eta$ (P)											
Na$_2$O	K$_2$O	SiO$_2$	8	8.5	9	9.5	10	10.5	11	11.5	12	12.5	13	16
—	5	95	838	807	777	748	722	698	675.5	653	632	612	592	492
1	4	95		826	797	769	744	718	694	672	651	632	612	510
2	3	95		822	796	768	746	722	702	679.5	658	641	621.5	524
3	2	95		837	808	780	758	733	712	688	667	647.5	628	529
4	1	95			824	797	775	754	729	709	689	668	650	554
5	—	95				838	809	781	757	732	708	683	653	547
—	13	87	702	680	655	632	612	595	578	563	547	532	517	438
3	10	87	683	658	633	610	589	568	550	530	512	495	479	393
5	8	87	693	668	643	620	597	578	557	537	517	499	482	393
7	6	87	688	662	637	612	591	568	547	530	511	493	476	387
9	4	87	692	666	645	622	601	581	562	543	526	510	493	410
11	2	87		684	662	642	624	606	588	572	557	540	524	446
13	—	87			712	693	675	657.5	641	624	607	592.5	578	499.5
—	20	80	662	641	621	601	583	566	550	533	518	506	491	417
5	15	80	632	607	582	561	543	526	512	493	479	466	451	378
10	10	80	605	582	563	543	526	510	493	479	465.5	451	437	367
12	8	80	609	586	563	542	524	506	489	473	458	444	429	357
14	6	80	608	582	557	533	517	501	485	471	456	442	428.5	358
16	4	80	600	580	561	543	527	511	493	479	466	448.5	437	366
18	2	80	612	592.5	571	550	532	517	502	486	472	458	444	374
20	—	80	652	629	607.5	588	571	551	532	517	502	488	473	397

[a] Data from Nemilov (1969).

TABLE 9.36

Parameters of the Equation $\eta = \eta_0 \exp(B/T)$ for the Viscosities of Some Li$_2$O–MO–SiO$_2$ Glass Systems[a]

Composition (mole %)			Temperature (°C) for log $\eta = 10$	η_0 (P)	$10^{-4}B$ (K)
Li$_2$O	MO	SiO$_2$			
	M = Mg				
24.27	2.91	72.82	497	8.7×10^{-19}	4.97
32.33	2.91	64.76	474	4×10^{-25}	5.90
	M = Ca				
24.27	2.91	72.82	495	6.2×10^{-19}	4.97
32.33	2.91	64.76	471	6.2×10^{-24}	5.68
	M = Sr				
24.27	2.91	72.82	497	5.2×10^{-20}	5.17
32.33	2.91	64.76	471	7.7×10^{-25}	5.83
	M = Ba				
24.27	2.91	72.82	500	5.7×10^{-19}	5.01
32.33	2.91	64.76	478	3.2×10^{-24}	5.78

[a] Data from Matusita and Tashiro (1973) and Matusita et al. (1974).

TABLE 9.37

Softening Points of Lithium–Lead–Silicate Glasses[a]

Composition (mole %)			Softening temperature (°C)
Li$_2$O	PbO	SiO$_2$	
0.06	74.68	25.26	532
0.14	74.40	25.46	525
0.34	73.40	26.26	517
0.75	71.40	27.85	511

[a] Data from Abou-El-Azm and El-Batal (1969).

TABLE 9.38
Viscosities of Sodium–Copper–Silicate Glasses[a]

Composition (mole %)			Temperature (°C)	log η (P)	Reference[a]	Composition (mole %)			Temperature (°C)	log η (P)	Reference[a]
Na$_2$O	CuO	SiO$_2$				Na$_2$O	CuO	SiO$_2$			
14.3	14.3	71.4	439	12.5	1	20	10	70	501	9.62	2
			485	10.9					502	9.45	
			493	10.7					510	9.41	
			537	9.3					526	9.10	
			550	9.0					531	8.95	
20	10	70	402	13.29	2				545	9.16	
			442	11.57					552	8.79	
			453	11.07					564	8.56	
			455	10.98					576	8.09	
			462	10.99		20	20	60	429	11.7	1
			465	10.86					445	11.4	
			474	10.67					458	10.7	
			475	10.31					475	9.9	
			485	10.19					500	9.2	
			491	9.79							
			492	9.84							
			495	9.69							

[a] Data from (1) Hoffman et al. (1952) and (2) Petrovskii (1956).

TABLE 9.39

Viscosities of $Na_2O-CuO-SiO_2$ Glasses and Melts[a]

Composition (mole %)			Temperature (°C) for given $\log \eta$ (P)									
Na_2O	CuO	SiO_2	2	3	4	6	7	8	9	11	12	13
20	10	70	1303	1043	864	650	598	555	519	465	447	432
20	20	60	1140	895	736	—	—	502	469	426	410	398
30	10	60	1096	884	737	566	527	494	465	419	404	393

[a] Data from Ivanov *et al.* (1969) and Tret'yakova and Mazurin (1969).

TABLE 9.40

Viscosities of $Na_2O-BeO-SiO_2$ Glasses

Composition (mole %)			Temperature (°C)	$\log \eta$ (P)	Reference[a]
Na_2O	BeO	SiO_2			
20	10	70	524	12.67	1
			543	11.84	
			546	11.76	
			570	10.95	
			589	10.26	
			614	9.47	
			624	9.25	
			645	8.78	
			650	8.73	
29.61	7.49	62.90	437	15.33	2
			454	14.52	
			475	13.23	
			501.5	12.31	
			534	10.24	
			558	9.43	
30.35	5.14	64.51	434	15.51	2
			450	14.01	
			474	12.62	
			501	10.94	
			536	9.62	
			562	8.82	
31.16	2.63	66.21	423.5	14.81	2
			440	13.53	
			475	11.37	
			524	9.28	
			562	8.32	
			566	8.23	

[a] Data from (1) Petrovskii (1956) and (2) Pospelov (1954, 1955, 1959).

TABLE 9.41

Viscosity of Sodium–Magnesium–Silicate Glasses

Composition (mole %)			Temperature (°C)	log η (P)	Reference[a]	Composition (mole %)			Temperature (°C)	log η (P)	Reference[a]
Na₂O	MgO	SiO₂				Na₂O	MgO	SiO₂			
14.3	14.3	71.4	543	13.4	1	30.36	5.15	64.49	416	15.04	1
			571	12.3					454.5	12.38	
			598	11.3					456	12.36	
			630	10.3					456.5	12.36	
20	10	70	514	11.68	2				467	11.14	
			528	11.31					470	11.37	
			550	10.73					511	9.55	
			572	10.34					529	8.87	
			600	9.66					535	8.22	
			626	9.11		31.16	2.64	66.20	414	15.69	3
20	20	60	532	11.4	1				430	14.25	
			573	9.7					456	12.36	
			602	8.7					478	11.51	
29,60	7.52	62.88	420	15.18	3				511.5	9.78	
			433	14.08					535	8.89	
			456	12.56							
			475	11.43							
			511	9.62							
			536	8.96							
			570	8.12							

[a] Data from (1) Hoffman et al. (1952), (2) Petrovskii (1956), and (3) Pospelov (1954, 1959).

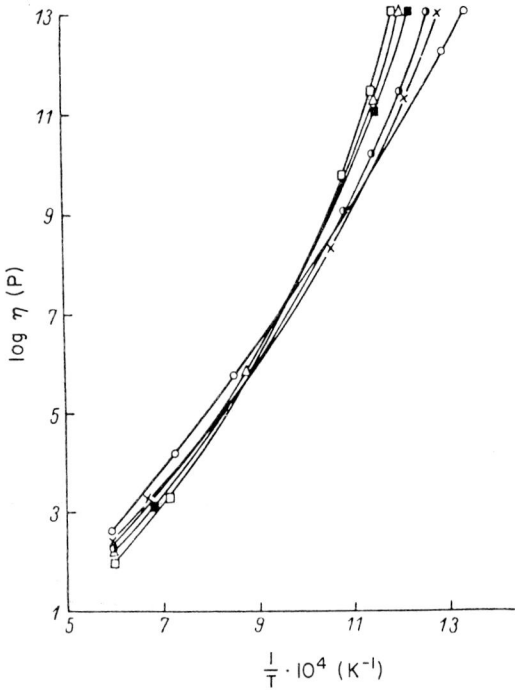

Fig. 9.11. Temperature dependence of viscosities of the glass compositions (mole %) $14.29Na_2O-xMgO-(85.71 - x)SiO_2$. Value of x: $\bigcirc$, 0.0; x, 7.11; $\bullet$, 10.71; $\triangle$, 14.29; $\blacksquare$, 17.86; and $\square$, 21.43. From Din (1968).

TABLE 9.42

Parameters of the Equation $\log \eta = \log \eta_0 + B/T$ for Viscosities of Soda-Lime Silicate Glasses[a]

Composition (mole %)			$-\log \eta_0$	$10^{-3}B$	Temperature range
Na$_2$O	CaO	SiO$_2$	(P)	(K)	(°C)
9.29	15.03	75.68	24.434	32.665	580–670
14.46	9.60	75.94	26.462	32.525	550–620
19.05	19.69	61.26	31.139	36.080	550–620
19.89	14.96	65.14	29.384	34.262	550–620
19.93	4.05	76.01	19.738	25.304	510–575
20.73	9.48	69.79	26.337	31.055	510–575
23.43	15.22	61.35	30.500	34.115	510–575
23.44	10.28	66.29	26.269	30.523	510–575
24.07	5.19	70.74	19.664	24.781	510–575

[a] Data from Poole and Gensamer (1949).

TABLE 9.43

Viscosities of Some Na_2O–CaO–SiO_2 Glasses and Melts[a]

Composition (mole %)			Temperature (°C) for given $\log \eta$ (P)											
Na_2O	CaO	SiO_2	2	3	4	5	6	7	8	9	10	11	12	13
4.6	18.4	77.0	—	—	—	—	—	—	802	764	736	710	684	657
5	20	75	—	—	—	—	—	—	840	795	756	718	688	658
10	10	80	1540	1302	—	—	—	—	—	—	—	—	—	—
10	15	75	1440	1206	1050	—	—	—	—	700	666	636	611	585
10	20	70	1350	1142	—	—	—	—	—	707	675	648	624	606
15	5	80	1480	1217	1040	—	—	—	—	—	—	—	—	—
15	10	75	1420	1167	1004	888	808	744	692	652	620	593	567	543
15	15	70	1325	1106	960	860	785	728	682	650	624	600	576	556
15	20	65	1325	1070	—	—	—	—	—	666	636	609	588	570
20	5	75	1395	1134	967	850	765	698	648	608	578	548	520	498
20	10	70	1310	1081	924	826	759	692	650	615	581	555	534	515
20	15	65	1300	1060	920	822	—	—	—	618	592	570	548	525
20	20	60	1198	1015	—	—	—	687	656	624	—	574	557	543
25	5	70	1290	1053	902	795	717	658	615	576	545	522	500	479
25	10	65	1275	1038	888	786	725	670	627	588	556	532	512	492
25	25	50	1190	980	—	—	—	—	—	590	563	536	512	488
30	5	65	1250	1016	868	762	692	636	597	560	528	506	484	464
30	10	60	1160	960	845	—	—	—	—	560	533	506	483	465
35	5	60	1140	940	823	730	—	—	—	528	507	482	460	437

[a] Data from Shvaiko-Shvaikovskaya et al. (1971a,b), Shivaiko-Shvaikovskaya (1968), and Tret'yakova and Mazurin (1969).

TABLE 9.44

Viscosities of Sodium–Strontium–Silicate System

Na$_2$O	SrO	SiO$_2$	Temperature (°C)	log η (P)	Reference[a]
20	10	70	474	13.07	1
			496	12.36	
			515	11.50	
			521	11.30	
			524	11.20	
			559	10.19	
			582	9.42	
			610	8.82	
10	20	70	1300	2.19	2
10	30	60	1300	2.76	
15	15	70	1300	2.13	
15	25	60	1300	1.88	
20	10	70	1300	2.07	
20	20	60	1300	1.57	
25	15	60	1300	1.38	

[a] Data from (1) Petrovskii (1956) and (2) Kasymova (1964).

TABLE 9.45

Viscosities of Na$_2$O–SrO–SiO$_2$ System[a]

Na$_2$O	SrO	SiO$_2$	2	3	4	6	7	8	9	11	12	13
20	10	70	1303	1068	905	718	667	622	586	536	517	501
20	20	60	1193	1012	—	—	649	615	583	535	518	505
30	10	60	1155	945	—	637	593	557	528	481	463	449

Temperature (°C) for log η (P)

[a] Data from Ivanov et al. (1969) and Tret'yakova and Mazurin (1969).

TABLE 9.46
Viscosities of Sodium–Barium Silicate Glasses[a]

Na₂O	BaO	SiO₂	Temperature (°C)	log η (P)	Reference[a]
20	10	70	440	13.38	1
			457	12.64	
			464	12.08	
			467	11.88	
			469	11.84	
			476	11.59	
			490	11.37	
			492	11.00	
			496	10.74	
			499	10.37	
			508	10.37	
			520	9.85	
			545	9.03	
			548	8.98	
			576	8.09	
20	20	60	500	10.5	2
			524	9.6	
			532	9.4	
			580	8.1	

Na₂O	BaO	SiO₂	Temperature (°C)	log η (P)	Reference[a]
29.58	7.52	62.90	410.5	15.16	3
			412.5	14.47	
			420	14.10	
			436	12.36	
			453	11.49	
			472	10.76	
			520	8.40	
			548	7.51	
			562	7.15	
30.36	5.13	64.51	410	15.13	3
			426.5	13.30	
			455.5	11.67	
			477	10.51	
			510	9.18	
			534	8.21	
31.16	2.64	66.20	410.5	15.15	3
			434	13.28	
			439.5	13.06	
			456	11.96	
			478.5	10.98	
			517	8.88	
			548	8.11	

[a] Data from (1) Petrovskii (1956), (2) Hoffman et al. (1952), (3) Pospelov (1955, 1959).

TABLE 9.47

Viscosities of Na$_2$O–BaO–SiO$_2$ System^a

Composition (mole %)			Temperature (°C) for given log η (P)									
Na$_2$O	BaO	SiO$_2$	2	3	4	6	7	8	9	11	12	13
20	10	70	1304	1065	905	676	632	595	560	505	486	470
20	20	60	1142	938	821	—	611	572	540	488	470	458
30	10	60	1143	936	793	605	560	523	493	450	435	422

^{a} Data from Ivanov *et al.* (1969) and Tret'yakova and Mazurin (1969).

TABLE 9.48

Viscosities of Sodium–Zinc–Silicate Glasses

Composition (mole %)			Temperature (°C)	log η (P)	Reference^a
Na$_2$O	ZnO	SiO$_2$			
14.3	14.3	71.4	548	12.8	1
			557	12.5	
			570	11.9	
			623	10.2	
			661	9.0	
20	10	70	495	12.26	2
			510	11.72	
			525	11.08	
			526	11.03	
			530	10.87	
			546	10.54	
			548	10.36	
			560	9.93	
			563	9.81	
			570	9.46	
			572	9.45	

^{a} Data from (1) Hoffman *et al.* (1952) and (2) Petrovskii (1956).

TABLE 9.49

Viscosities of Na$_2$O–ZnO–SiO$_2$ Glass System^a

Composition (mole %)			Temperature (°C) for given log η (P)									
Na$_2$O	ZnO	SiO$_2$	2	3	4	6	7	8	9	11	12	13
20	10	70	1375	1131	591	—	684	641	601	542	521	504
20	20	60	1270	1052	909	—	680	647	614	562	544	530
30	10	60	1192	986	843	645	598	559	531	487	468	453

^{a} Data from Ivanov *et al.* (1969) and Tret'yakova and Mazurin (1969).

TABLE 9.50

Viscosities of Sodium–Cadmium Silicate Glass[a]

Composition (mole %)			Temperature (°C)	log η (P)
Na$_2$O	CdO	SiO$_2$		
20	10	70	472	12.40
			474	12.34
			509	11.04
			531	10.51
			536	10.27
			546	9.82
			548	9.96
			590	8.72

[a] Data from Petrovskii (1956).

TABLE 9.51

Viscosities of Na$_2$O–CdO–SiO$_2$ System[a]

Composition (mole %)			Temperature (°C) for given log η (P)									
Na$_2$O	CdO	SiO$_2$	2	3	4	6	7	8	9	11	12	13
20	10	70	1334	1075	910	—	632	595	560	505	486	470
20	20	60	1175	955	820	—	599	568	540	498	482	470
30	10	60	1134	936	793	602	565	533	503	459	442	428

[a] Data from Ivanov *et al.* (1969) and Tret'yakova and Mazurin (1969).

TABLE 9.52

Viscosities of Soda–Lead–Silicate Glasses[a]

Composition (mole %)			Temperature (°C)	log η (P)	Reference[a]
Na$_2$O	PbO	SiO$_2$			
6.3	44.1	49.6	376	13.18	1
			400	11.66	
			422	10.38	
			439.5	9.65	
			465	8.47	
8.7	38.8	52.5	366	13.71	
			384.5	12.31	
			397.5	11.33	
			400	11.28	
			422	10.19	
			439.5	9.49	
			457	8.77	
9.1	35.5	55.4	351.5	14.39	
			361	13.81	
			383	12.43	
			400	11.47	
			414	10.46	
			423	10.15	
			431	9.84	
			445	9.33	

Composition (mole %)			Temperature (°C)	log η (P)	Reference[a]
Na$_2$O	PbO	SiO$_2$			
12.2	33.2	54.6	376	12.80	1
			400	11.35	
			422	10.20	
			445	9.30	
			467	8.32	
			468	8.28	
13.7	29.6	56.7	359	13.95	
			376	12.45	
			400	11.33	
			420	10.16	
			424	9.96	
			431	9.84	
			445	9.40	
			466	8.47	
			471.5	8.32	
			489	7.72	
15.3	26.9	57.8	376	13.00	
			398	11.66	
			422	10.32	
			445	9.40	

18.4	24.1	57.5					20	10	70			
			461.5	8.73						458	8.97	
			464	8.47						461	9.04	
			470	8.38						475	8.26	2
			376	12.77	1					484	10.59	
			400	11.33						490	10.40	
			422	10.15			20	10	70	496	10.25	
			428	9.96						514	9.70	
			438	9.56						516	9.64	
			460	8.76						518	9.47	
			471	8.25						522	9.60	
19.6	19.6	60.8	364	13.80						542	9.08	
			371	13.41						556	8.27	
			400	11.94			26.6	10.2	63.2	376	13.61	1
			415	10.93						400.5	12.33	
			423	10.49						414	11.45	
			440	9.90						422	11.00	
			467	8.80						438	10.45	
			489	8.20						467	9.20	
20	10	70	448	11.83	2					510	7.80	
			456	11.57						529	7.32	
			477	10.76								

[a] Data from (1) Pospelov and Evstrop'ev (1941, 1949) and Pospelov (1955) and (2) Petrovskii (1956).

TABLE 9.53

Temperatures for Given Viscosities of $Na_2O-PbO-SiO_2$ System[a]

Composition (mole %)			Temperature (°C) for given $\log \eta$ (P)									
Na_2O	PbO	SiO_2	2	3	4	6	7	8	9	11	12	13
20	10	70	1244	994	840	610	568	533	497	453	436	421
20	20	60	1052	836	712	—	512	481	456	414	398	387
30	10	60	1075	862	724	544	509	478	451	410	395	381

[a] Data from Tret'yakova (1969) and Tret'yakova and Mazurin (1969).

TABLE 9.54

Viscosities of $Na_2O-PbO-SiO_2$ Glass Systems[a]

Composition (mole %)			$\log \eta$ (P) at given temperature (°C)									
Na_2O	PbO	SiO_2	500	600	700	800	900	1000	1100	1200	1300	1400
9.0	38.7	52.3	—	4.40	3.00	2.24	1.74	1.32	1.04	0.77	0.56	0.36
9.1	35.5	55.4	—	4.64	3.24	2.41	1.89	1.50	1.15	0.91	0.68	—
12.2	33.2	54.6	—	4.68	3.35	2.53	1.98	1.59	1.29	1.02	0.75	—
13.7	29.6	56.7	—	—	3.50	2.68	2.13	1.74	1.40	1.11	0.89	0.77
15.3	26.9	57.8	—	—	3.60	2.78	2.20	1.80	1.45	1.16	0.91	—
18.4	24.1	57.5	7.28	5.08	3.76	2.85	2.29	1.87	1.53	1.24	1.00	—
19.6	19.6	60.8	7.62	—	4.09	3.20	2.61	2.17	1.82	1.50	1.22	—
26.6	10.2	63.2	8.12	—	4.50	3.61	2.95	2.48	2.09	1.77	1.46	—

[a] Data from Pospelov (1955).

TABLE 9.55

Viscosities of Sodium–Manganese Silicate Glasses[a]

Composition (mole %)			Temperature (°C)	$\log \eta$ (P)	Reference[a]	Composition (mole %)			Temperature (°C)	$\log \eta$ (P)	Reference[a]
Na_2O	MnO	SiO_2				Na_2O	MnO	SiO_2			
14.3	14.3	71.4	525	11.8	1	20	10	70	482	11.98	2
			548	10.9					484	11.68	
			557	10.6					494	11.42	
			578	9.9					516	10.59	
			603	9.1					517	10.63	
20	10	70	467	12.33	2				522	10.29	
			478	11.89					535	9.91	
									557	9.29	

[a] Data from (1) Hoffman et al. (1952) and (2) Petrovskii (1956).

TABLE 9.56

Viscosities of Na$_2$O–MnO–SiO$_2$ System[a]

Composition (mole %)			Temperature (°C) for log η (P)									
Na$_2$O	ZnO	SiO$_2$	2	3	4	6	7	8	9	11	12	13
20	10	70	1240	1008	854	687	642	602	564	508	486	468
20	20	60	1205	994	852	—	606	568	537	496	478	466
30	10	60	1120	992	843	611	566	530	499	458	437	424

[a] Data from Ivanov *et al.* (1969) and Tret'yakova and Mazurin (1969).

TABLE 9.57

Viscosities of Na$_2$O–FeO–SiO$_2$ Glasses

Composition (mole %)			Temperature (°C)	log η (P)	Reference[a]
Na$_2$O	FeO	SiO$_2$			
14.3	14.3	71.4	512	12.6	1
			544	11.1	
			558	10.6	
			590	9.6	
			606	9.2	
20	10	70	495	11.32	2
			496	11.31	
			504	11.05	
			516	10.65	
			517	10.58	
			520	10.43	
			531	10.15	
			550	9.44	
			555	9.31	

[a] Data from (1) Hoffman *et al.* (1952) and (2) Petrovskii (1956).

TABLE 9.58

Viscosities of Sodium–Cobalt–Silicate Glasses

Composition (mole %)			Temperature (°C)	$\log \eta$ (P)	Reference[a]	Composition (mole %)			Temperature (°C)	$\log \eta$ (P)	Reference[a]
Na_2O	CoO	SiO_2				Na_2O	CoO	SiO_2			
14.3	14.3	71.4	536	12.3	1	20	10	70	514	10.68	2
			569	10.9					531	10.09	
			591	10.2					540	9.77	
			596	10.1					552	9.45	
			618	9.5					560	9.41	
20	10	70	480	12.02	2				562	9.43	
			488	11.87					578	8.85	
			495	11.40					587	8.57	
			506	10.97					592	8.44	

[a] Data from (1) Hoffman *et al.* (1952) and (2) Petrovskii (1956).

TABLE 9.59

Viscosities of Na$_2$O–CoO–SiO$_2$ System[a]

Composition (mole %)			Temperature (°C) for given log η (P)										
Na$_2$O	CoO	SiO$_2$	2	3	4	6	7	8	9	11	12	13
20	10	70	1368	1103	935	717	661	616	575	516	494	477
20	20	60	1205	994	—	—	628	585	565	515	498	484
30	10	60	1173	960	813	—	—	536	505	458	440	—

[a] Data from Ivanov *et al.* (1969) and Tret'yakova and Mazurin (1969).

TABLE 9.60

Viscosities of Sodium–Nickel–Silicate Glass[a]

Composition (mole %)			Temperature (°C)	log η (P)
Na$_2$O	NiO	SiO$_2$		
20	10	70	476	12.44
			494	11.65
			538	10.20
			576	9.10
			592	8.52

[a] Data from Petrovskii (1956).

TABLE 9.61

Viscosities of Na$_2$O–NiO–SiO$_2$ Glass Melts at 1400°C[a]

Composition			Composition		
Na$_2$O/SiO$_2$	mole % NiO	log η (P)	Na$_2$O/SiO$_2$	mole % NiO	log η (P)
1/4	0	2.16	2/3	0	1.01
	5	2.04		5	0.82
	10	1.9		10	0.63
	15	1.71		16.5	0.5
	20	1.51	1/1	0	0.12
	25	1.28		5	0.0
3/7	0	1.65		10	−0.07
	5	1.49		15	−0.14
	10	1.30	3/2	0	−0.61
	14	1.14		5	−0.7
	20	0.97		10	−0.78

[a] Data from graph in Kawahara *et al.* (1979).

TABLE 9.62

Viscosities of Some $K_2O-MO-SiO_2$ Glass Systems[a]

Composition (mole %)			Temperature (°C) for log η (P)							
K_2O	MO	SiO_2	6	7	8	9	10	11	12	13
	M = Cu									
20	10	70	677	628	582	540	507	480	458	440
20	20	60			516	483	458	436	418	401
	M = Mg									
20	10	70	875	812	760	709	666	629	599	576
20	20	60			830	785	747	711	681	654
	M = Ca									
20	10	70	830	790	750	718	691	666	647	632
20	20	60			733	702	675	652	632	615
	M = Sr									
20	10	70	807	754	711	673	639	612	590	573
20	20	60			704	666	638	612	592	574
	M = Ba									
20	10	70	771	717	673	631	598	572	550	532
20	20	60			654	618	589	566	544	524
	M = Zn									
20	10	70	830	790	750	718	691	666	647	632
20	20	60			733	702	675	652	632	615
	M = Cd									
20	10	70	768	710	663	621	588	560	536	517
20	20	60			638	603	575	551	529	511
	M = Pb									
20	10	70	683	628	579	538	509	484	464	449
20	20	40			523	490	464	443	424	408
	M = Mn									
20	10	70	807	744	693	647	608	577	552	532
20	20	60			694	651	618	590	564	543
	M = Co									
20	10	70	832	777	722	674	634	603	579	560
20	20	60			740	701	667	637	610	586

[a] Data from Tret'yakova (1969), Tret'yakova and Mazurin (1969), and Ivanov *et al.* (1969).

TABLE 9.63

**Viscosity of 17.4K$_2$O–11.7CaO–70.9SiO$_2$ (mole %)
Glass System**[a]

Temperature (°C)	log η (P)	Temperature (°C)	log η (P)
550	14.88	1000	4.70
600	12.30	1050	4.25
650	10.36	1100	3.82
700	8.35	1150	3.50
750	7.56	1200	3.18
800	6.86	1250	2.88
850	6.44	1300	2.58
900	5.97	1350	2.30
950	5.25	1400	2.10

[a] Data from Saringyulyan and Kostanyan (1969).

TABLE 9.64

Viscosities of K$_2$O–SrO–SiO$_2$ Glasses[a]

Composition (mole %)			Temperature (°C) for given log η (P)				
K$_2$O	SrO	SiO$_2$	9	10	11	12	13
7.6	14.2	78.2	753	715	681	650	620
8.7	6.1	85.2	719	681	646	613	583
9.9	18.7	71.4	743	704	674	646	620
10.8	7.6	81.6	692	658	628	600	572
10.9	27.1	62.0	719	692	666	641	618
11.6	22.1	66.3	701	674	649	625	602
12.6	15.4	72.0	712	679	650	621	594
13.5	9.6	76.9	676	646	617	591	566
14	26.4	59.6	662	636	613	592	572
14.2	4.5	81.3	662	622	588	556	527
15.5	18.7	65.8	665	637	610	586	562
16.8	11.8	71.4	667	633	601	572	545
16.8	30	53.2	646	620	594	570	547
18	5.6	76.4	652	613	579	547	517
18.6	22.4	59.0	663	634	607	582	558
19.6	18.3	62.1	663	633	604	578	553
20.5	14.4	65.1	658	628	599	572	549
21.3	26.4	52.3	620	595	571	549	527
21.4	10.6	68.0	650	620	592	566	541
24	17.2	58.8	613	588	564	541	519
26.5	8.5	65.0	615	586	560	534	511

[a] Data from Nemilov and Kasymova (1975).

TABLE 9.65

Viscosities of Potassium–Lead–Silicate Glasses[a]

Composition (wt. %)[b]			Temperature	log η	Composition (wt. %)[b]			Temperature	log η	Composition (wt. %)[b]			Temperature	log η
K_2O	PbO	SiO_2	(°C)	(P)	K_2O	PbO	SiO_2	(°C)	(P)	K_2O	PbO	SiO_2	(°C)	(P)
0.4	79.0	20.6	702	2.65	3.2	62.4	34.4	1099	2.29	4.6	58.1	37.3	1204	2.15
			800	1.75				1104	2.28				1288	1.82
2.3	66.2	31.5	690	5.18				1198	1.92				1305	1.75
			697	5.13				1204	1.90					
			800	3.83				1296	1.61	6.0	54.4	39.6	740	5.55
			900	3.04				1311	1.43				750	5.40
			1000	2.43									801	4.79
			1100	1.98	4.6	58.1	37.3	734	5.44				803	4.81
			1103	1.98				760	5.06				897	3.93
			1198	1.60				798	4.58				907	3.79
			1298	1.28				802	4.56				999	3.20
			1305	1.25				883	3.74				1099	2.66
								900	3.71				1103	2.67
3.2	62.4	34.4	698	5.60				922	3.53				1199	2.25
			704	5.51				985	3.15				1210	2.23
			800	4.28				1000	3.06				1307	1.88
			804	4.24				1088	2.61					
			895	3.46				1103	2.53					
			905	3.37				1150	2.36					
			1000	2.77				1190	2.17					

[a] Data from Shartsis and Spinner (1951).

[b] Also contains 0.3 wt. % As_2O_3.

TABLE 9.66

Softening Points of $K_2O-PbO-SiO_2$ Glasses[a]

Composition (wt. %)			
K_2O	PbO	SiO_2	Softening temperature (°C)
25	5	70	694
25	10	65	669
25	20	55	620
25	40	35	491

[a] Data from Rao (1963b).

TABLE 9.67

Softening Temperatures of $Rb_2O-CaO-SiO_2$ Glasses[a]

Composition (mole %)			Softening temperature[b] (°C)	Composition (mole %)			Softening temperature[b] (°C)
Rb_2O	CaO	SiO_2		Rb_2O	CaO	SiO_2	
3.3	44.6	52.1	945	11.7	39.3	49.0	838
3.3	55.3	41.4	962	11.8	26.4	61.8	827
3.4	22.6	74.0	889	12	13.4	74.6	837
3.4	33.6	63.0	903	17.2	42.8	40.0	805
7.3	24.4	68.3	863	17.3	28.8	53.9	803
7.3	36.2	56.5	878	17.4	14.6	68.0	808
7.3	48	44.7	912	23.8	31.8	44.4	772
7.4	12.3	80.3	874	24	16	60	778
11.6	52	36.4	854	32.1	17.9	50	723

[a] Data from Simpson (1959, 1961).
[b] $\log \eta$ (P) = 7.6.

TABLE 9.68

Softening Temperatures of $Rb_2O-BaO-SiO_2$ Glasses[a]

Composition (mole %)			Softening temperature[b] (°C)
Rb_2O	BaO	SiO_2	
4.2	15.7	80.1	765
4.6	22.8	72.6	832
8.6	10.6	80.8	830
9.4	17.3	73.3	795
13.1	5.3	81.6	710
14.2	11.7	74.1	765
15.9	19.2	64.9	795
19.3	5.8	74.9	740
21.2	12.9	65.9	760
26.7	6.5	66.8	805

[a] Data from Simpson (1959, 1961).
[b] $\log \eta$ (P) = 7.6.

TABLE 9.69

Softening Temperatures of Rb$_2$O–PbO–SiO$_2$ Glasses[a]

Composition (mole %)			Softening temperature[b] (°C)
Rb$_2$O	PbO	SiO$_2$	
12.8	27	60.2	576
14.8	8.2	77.0	705
16.6	14	69.4	615
22	9.3	68.7	580

[a] Data from Simpson (1959, 1961).
[b] $\log \eta$ (P) = 7.6.

TABLE 9.70

Parameters of the Equation $\eta = \eta_0 \exp(B/T)$ for the Viscosities of Some Li$_2$O–M$_2$O$_3$–SiO$_2$ Glasses[a]

Composition (mole %)			Temperature (°C) for $\log \eta$ (P) = 10	η_0 (P)	$10^{-4}B$ (K)
Li$_2$O	M$_2$O$_3$	SiO$_2$			
M = B					
24.27	2.91	72.82	502	7.1×10^{-22}	5.54
32.33	2.91	64.76	484	5.1×10^{-22}	5.45
M = Al					
24.27	2.91	72.82	526	1.5×10^{-16}	4.74
32.33	2.91	64.76	487	8.7×10^{-22}	5.42
M = In					
24.27	2.91	72.82	531	7.6×10^{-20}	5.38
32.33	2.91	64.76	503	4.0×10^{-21}	5.43

[a] Data from Matusita and Tashiro (1973) and Matusita et al. (1974).

TABLE 9.71

Viscosities of Lithium Borosilicate Glasses

Composition (mole %)			Temperature (°C) for $\log \eta$ (P)											Reference[a]
Li$_2$O	B$_2$O$_3$	SiO$_2$	1.5	2	2.5	3	6	7	7.6	8	9	10	12	
5	80	15							484				367	1
7	78	15							499				384	
10	75	15							510				418	
20	65	15							575				503	
30	55	15							574				515	
40	45	15							534				485	
30	35	35							576				502	
30	18	52	1040	940	870	815	645	616	—	590	568	550	—	2

[a] Data from (1) Konijnendijk (1975) and (2) Kadogawa et al. 1(973).

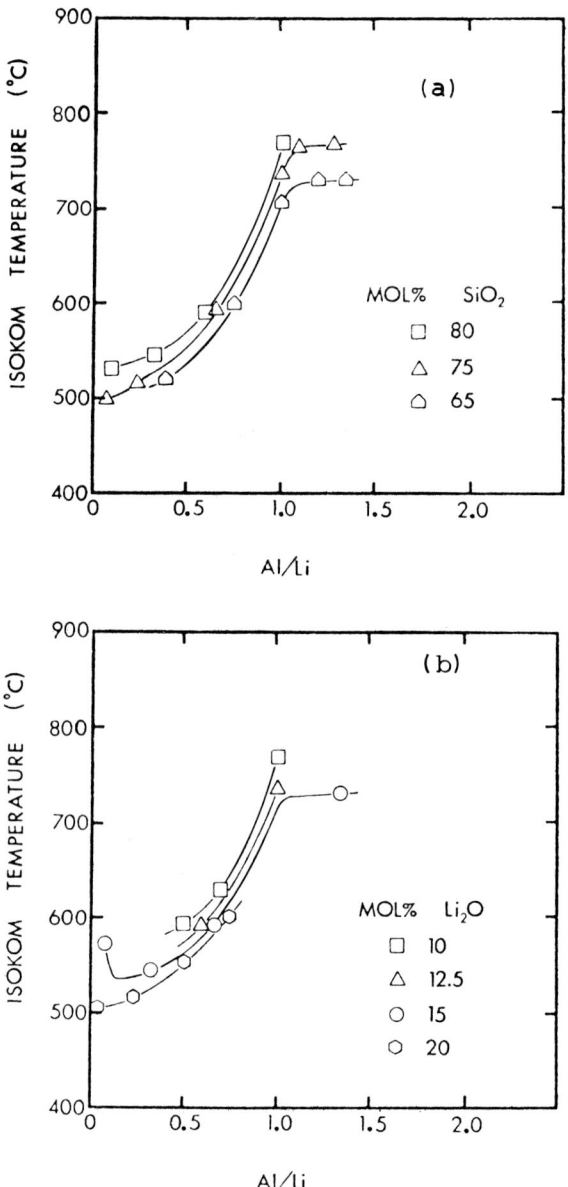

Fig. 9.12. Isokom (10^{13} P) temperature versus composition for lithium aluminosilicate glasses for (a) constant SiO_2 and (b) constant Li_2O content. From Shelby (1978).

TABLE 9.72

Viscosities of Sodium Borosilicate Glasses[a]

Composition (mole %)			Temperature (°C) for given log η (P)				
Na₂O	B₂O₃	SiO₂	7.6	10	11	12	13
1	34	65	714			490	
1	39	60		511	467	422	378
1	64	35	510			355	
2	28	70		634	593	551	510
2	38	60		550	515	480	445
3	62	35	540			375	
3	82	15	442			348	
4	21	75		765	721	674	635
4	26	70		692	652	612	574
4	31	65		661	610	558	506
4	36	60		607	563	520	494
4	41	55		574	532	490	450
5	30	65	725			543	
5	60	35	596			401	
5	70	25	498			382	
5	75	20		367	357	349	339

Composition (mole %)			Temperature (°C) for given log η (P)				
Na₂O	B₂O₃	SiO₂	7.6	10	11	12	13
5	80	15	464			366	
6	24	70		716	687	658	628
6	34	60		673	631	591	550
7	23	70		732	695	659	625
7	33	60		693	653	610	568
7	78	15	475			382	
8	17	75		717	685	653	620
8	22	70		711	684	657	630
8	27	65		741	708	674	641
8	32	60		699	660	630	592
8	37	55		680	658	636	614
9	21	70		589	566	544	521
9	31	60		665	614	563	500
10	14.5	75.5	747			622	
10	20	70		642	608	574	540
10	25	65	689			562	

(continues)

TABLE 9.72 (Continued)

Composition (mole %)			Temperature (°C) for given log η (P)					Composition (mole %)			Temperature (°C) for given log η (P)				
Na$_2$O	B$_2$O$_3$	SiO$_2$	7.6	10	11	12	13	Na$_2$O	B$_2$O$_3$	SiO$_2$	7.6	10	11	12	13
10	30	60	722	627	581	535	490	20	30	50	662			579	
10	55	35	584			445		20	45	35	615			527	
10	60	30	506			409		20	65	15	584			494	
10	70	20		424	408	393	388	20	75	5	555			478	
10	75	15	507			415		22	18	60		609	586	563	540
10	85	5	472			386		25	15	60		591	570	550	530
12	28	60		590	563	536	509	30	5	65	612			501	
13	27	60		596	573	551	529	30	11.5	58.5	621			522	
15	15	70		648	626	607	589	30	23.5	46.5	610			527	
15	25	60		615	590	567	542	30	35	35	593			522	
15	65	20		469	452	435	418	30	46.5	23.5	575			515	
17	23	60		630	605	581	558	30	55	15	579			517	
20	5	75		571	549	528	506	30	65	5	561			495	
20	10	70		618	596	574	554	40	25	35	528			457	
20	15	65	683			580									
20	20	60	689	623	603	585	567								

[a] Data from Konijnendijk (1975), Mazurin and Strel'tsina (1971, 1972), and Mazurin et al. (1971).

TABLE 9.73

Parameters of the Equation $\log \eta = \log \eta_0 + B/T$ for the Viscosities of Sodium Aluminosilicate Glasses[a]

Composition (mole %)			Temperature range (°C)	$-\log \eta_0$ (P)	$10^{-3}B$ (K)
Na_2O	Al_2O_3	SiO_2			
8.75	16.25	75	800–870	14.479	30.292
10	15	75	800–870	14.011	29.734
11.25	13.75	75	800–870	13.981	29.730
12.5	12.5	75	800–870	13.708	29.107
13.75	11.25	75	650–830	12.415	24.018
15.6	9.4	75	550–690	15.837	24.487
18.75	6.25	75	500–600	17.500	24.185
21.9	3.1	75	470–520	16.061	24.470
12.96	24.08	62.96	810–930	19.805	35.423
14.29	21.43	64.28	810–940	18.218	33.793
15.52	18.96	65.52	810–930	17.250	32.986
16.1	17.8	66.1	810–940	15.485	31.057
16.67	16.67	66.66	800–910	14.054	28.480
17.74	14.52	67.74	670–780	12.700	24.084
19.7	10.6	69.7	570–660	17.869	26.048
21.43	7.14	71.43	520–610	17.503	24.389
23.33	3.33	73.34	470–560	21.181	25.819

[a] Data from Taylor and Rindone (1970).

TABLE 9.74

Parameters of the Equation $\log \eta = \log \eta_0 + B/T$ for the High-Temperature Viscosities of $Na_2O-Al_2O_3-SiO_2$ System[a]

Composition (mole %)			Temperature range (°C)	$-\log \eta_0$ (P)	$10^{-3}B$ (K)
Na_2O	Al_2O_3	SiO_2			
9.5	15.4	75.1	1570–1680	6.50	18.870
11.0	21.9	67.1	1580–1670	6.60	17.611
11.1	14.0	74.9	1580–1690	6.05	18.434
12.3	12.5	74.2	1570–1680	6.09	18.641
13.8	19.4	66.8	1540–1690	6.33	17.689
15.0	17.9	67.1	1560–1680	6.50	18.334
16.1	9.6	74.3	1540–1680	3.09	10.836
16.3	17.2	66.5	1500–1660	6.83	18.925
17.0	22.9	60.1	1550–1660	5.98	16.365
18.5	21.3	60.2	1530–1650	6.18	16.927
18.9	13.8	67.3	1460–1640	3.97	12.576
19.9	5.4	74.7	1390–1590	2.57	8.655
20.1	19.9	60.0	1520–1640	6.49	17.638
22.1	22.1	55.8	1480–1630	6.42	17.040
22.9	10.3	66.8	1250–1500	3.28	9.909
23.0	27.6	49.4	1540–1650	6.65	16.876
24.1	16.3	59.6	1400–1600	3.93	11.788
24.7	25.2	50.1	1560–1670	6.32	16.217
27.9	4.9	67.2	1300–1460	3.07	8.527
28.1	11.9	60.0	1300–1510	3.70	10.158
29.8	19.8	50.4	1500–1620	4.66	12.610
33.9	6.0	60.1	1210–1470	3.68	9.043
35.1	14.9	50.0	1310–1500	4.19	10.646

[a] Data from Riebling (1966, 1968a).

TABLE 9.75

Viscosities of Some Na_2O–M_2O_3–SiO_2 Glass Systems

Composition (mole %)			Temperature (°C)	$\log \eta$ (P)	Reference[a]
Na_2O	M_2O_3	SiO_2			
M = Ga					
25	3.75	71.25	461	12.7	1
			488	11.8	
			513	10.7	
			555	9.1	
			581	8.4	
25	7.5	67.5	483	12.6	
			520	11.0	
			555	9.7	
			581	8.8	
			598	8.5	
25	11.25	63.75	506	12.8	
			521	12.0	
			542	11.1	
			581	9.6	
			626	8.3	
M = In					
25	3.75	71.25	545	13.0	1
			570	11.5	

Composition (mole %)			Temperature (°C)	$\log \eta$ (P)	Reference[a]
Na_2O	M_2O_3	SiO_2			
M = In					
25	3.75	71.25	606	9.7	1
			634	8.7	
			653	7.5	
25	7.5	67.5	610	13.2	
			649	11.1	
			660	10.4	
			690	9.3	
			708	8.4	
M = La					
15	5	80	1174	3.45	2
			1257	3.02	
			1338	2.68	
			1422	2.35	
16	11	73	1422	1.80	

[a] Data from (1) Hoffman *et al.* (1952) and (2) Dubrovo and Shnypikov (1968).

TABLE 9.76

Viscosities of Potassium Borosilicate Glasses

Composition (mole %)			Temperature (°C) for given $\log \eta$ (P)					Reference[a]
K_2O	B_2O_3	SiO_2	7.6	10	11	12	13	
1	34	65	665			494		1
2	38	60		550	505	463	420	2
3	82	15	435			339		1
4	36	60		616	570	522	474	2
5	60	35	536			371		1
6	34	60		613	571	530	489	2
8	32	60		549	522	496	469	2
10	14.5	75.5	816			670		1
10	25	65	706			569		1
10	30	60	674			560		1
10	55	35	533			431		1
10	60	30	512			420		1
10	75	15	486			393		1
10	85	5	469			382		1
14	51	35	579			470		1
14.5	70.5	15	509			423		1
20	15	65	732			629		1
20	45	35	611			515		1
20	65	15	547			461		1
30	5	65	655			537		1
30	11.5	58.5	639			526		1
30	55	15	548			478		1
30	65	5	529			453		1

[a] Data from (1) Konijnendijk (1975) and (2) Mazurin and Strel'tsina (1971, 1972) and Strel'tsina (1970).

TABLE 9.77

**Natural Logarithm of
Viscosities of
$12.5K_2O-12.5Al_2O_3-75SiO_2$
(mole %) System**[a]

Temperature ($^\circ$C)	$\ln \eta$ (P)
1342	14.33
1382	13.50
1443	12.40
1488	11.64
1545	10.94
1575	10.29
1581	10.16
1661	9.127
1675	8.525
1749	7.473
1794	7.073
1822	6.937
1825	6.640

[a] Data from Urbain *et al.* (1982).

TABLE 9.78

Softening Temperatures of Rubidium Borosilicate Glasses[a]

Composition (mole %)			
Rb_2O	B_2O_3	SiO_2	Softening temperature ($^\circ$C)
8.1	42.2	49.7	659
12.4	10.8	76.8	858
12.5	22.7	64.8	797
12.7	34.1	53.2	729
13	46.3	40.7	637
17.8	11.9	70.3	776
17.9	24.8	57.3	664
18.4	37.7	43.9	688
26.2	41.8	32.0	610

[a] Data from Simpson (1961).

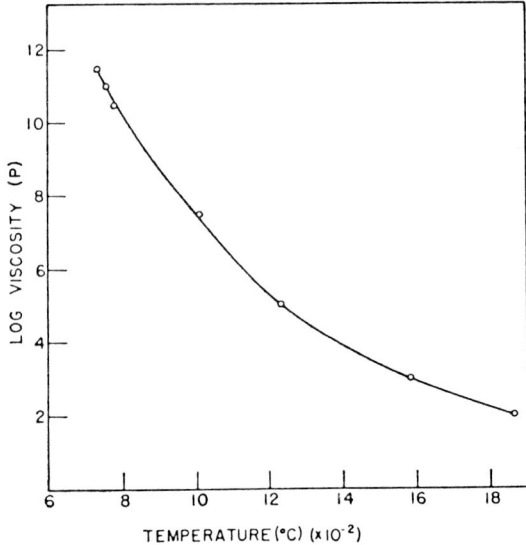

Fig. 9.13. Temperature dependence of viscosity of the glass composition (mole %) $20Cs_2O-5Al_2O_3-75SiO_2$. From Bollin (1972).

TABLE 9.79

Parameters of the Equation $\eta = \eta_0 \exp(B/T)$ for the Viscosities of $Li_2O-MO_2-SiO_2$ Glasses[a]

Composition (mole %)			Temperature (°C) for log η (P) = 10	η_0 (P)	$10^{-4}B$ (K)
Li_2O	MO_2	SiO_2			
	M = Ge				
24.27	2.91	72.82	497	3.8×10^{-20}	5.2
32.33	2.91	64.76	479	2.3×10^{-21}	5.3
	M = Ti				
24.27	2.91	72.82	519	7.9×10^{-21}	5.47
32.33	2.91	64.76	498	8.3×10^{-23}	5.68
	M = Zr				
24.27	2.91	72.82	548	2.6×10^{-19}	5.2
32.33	2.91	64.76	516	2.2×10^{-17}	4.83

[a] Data from Matusita and Tashiro (1973) and Matusita et al. (1974).

TABLE 9.80

Viscosities of $Li_2O-TiO_2-SiO_2$ System[a]

Composition			Temperature	η
Li_2O	TiO_2	SiO_2	(°C)	(P)
35	35	30	1250	0.45
			1300	0.41
			1350	0.35
			1400	0.315

[a] Data from graph in Nakamura *et al.* (1977).

TABLE 9.81

Viscosities of Sodium−Germanium−Silicate Glasses[a]

Composition (mole %)			Temperature	$\log \eta$	Composition (mole %)			Temperature	$\log \eta$
Na_2O	GeO_2	SiO_2	(°C)	(P)	Na_2O	GeO_2	SiO_2	(°C)	(P)
25	3.75	71.25	491	11.5	25	22.5	52.5	533	9.1
			504	10.8				546	8.5
			520	9.9	25	37.5	37.5	487	12.0
			543	9.3				501	11.3
			569	8.4				519	10.2
25	15	60	497	11.2				531	9.4
			511	10.3				546	8.4
			534	9.3	25	52.5	22.5	496	12.0
			543	8.8				501	11.1
			546	8.8				519	9.9
25	22.5	52.5	492	11.3				531	9.0
			495	11.1				544	8.4
			523	9.7					

[a] Data from Hoffman *et al.* (1952).

TABLE 9.82

Parameters of the Equation $\log \eta = \log \eta_0 + B/T$ for the Viscosities of $Na_2O-GeO_2-SiO_2$ Glass Melts[a]

Composition (mole %)			Temperature range (°C)	$-\log \eta_0$ (P)	B (K)
Na_2O	GeO_2	SiO_2			
23.1	31.5	45.4	1000–1400	2.44	5746
23.9	36.5	39.6		2.54	5556
24.0	45.5	30.5		2.90	5753
24.3	23.0	52.7		1.99	5390
24.4	59.5	16.1		1.77	3483
24.6	7.7	67.7		2.76	7410
24.6	15.1	60.3		2.82	7101
25.3	53.6	21.1		1.58	3271
25.4	49.9	24.7		2.60	4942

[a] Data from Riebling (1964b).

TABLE 9.83

Viscosities of Sodium–Tin–Silicate Glasses[a]

Composition (mole %)			Temperature (°C)	$\log \eta$ (P)	Composition (mole %)			Temperature (°C)	$\log \eta$ (P)
Na_2O	SnO_2	SiO_2			Na_2O	SnO_2	SiO_2		
32.04	3.24	64.72	514	12.25	30.70	6.01	63.29	567	12.44
			527	11.67				582	11.81
			540	11.11				597	11.19
			553	10.54				612	10.56
			567	9.98				628	9.92
31.96	4.75	63.29	540	12.49	31.02	8.73	60.24	620	12.47
			546	12.18				628	12.14
			560	11.61				644	11.51
			574	11.02				662	10.86
			589	10.44				679	10.22

[a] Data from Sviridov et al. (1976).

TABLE 9.84

Viscosities of Sodium–Titanium–Silicate Glasses[a]

Composition (mole %)			Temperature (°C)	log η (P)	Composition (mole %)			Temperature (°C)	log η (P)
Na_2O	TiO_2	SiO_2			Na_2O	TiO_2	SiO_2		
25	3.75	71.25	496	12.3	25	18.75	56.25	564	11.8
			518	11.1				591	10.3
			548	10.0				603	9.7
			565	9.4				614	9.1
			608	8.2				642	8.1
25	7.5	67.5	506	13.1	25	26.25	48.75	546	13.3
			520	12.2				579	10.8
			545	11.2				602	10.1
			570	10.0				608	9.7
			595	8.9				633	8.2
25	11.25	63.75	534	12.3	25	33.75	41.25	564	12.5
			548	11.4				580	11.3
			585	10.0				590	10.8
			602	9.3				606	9.5
			626	8.3				626	8.5
25	15	60	541	12.5					
			562	11.8					
			586	10.6					
			616	9.6					
			636	8.8					

[a] Data from Hoffman et al. (1952).

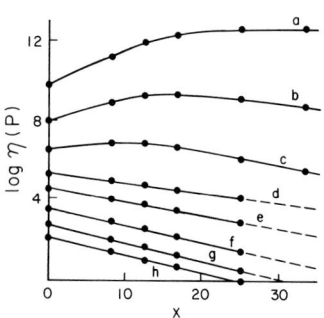

Fig. 9.14. Viscosity–composition isotherms for $Na_2O–xTiO_2–(2 − x)SiO_2$ glasses at various temperatures (°C): a, 500; b, 550; c, 600; d, 650; e, 700; f, 800; g, 900; h, 1000. From Evstrop'ev et al. (1968).

TABLE 9.85

Viscosities of Na_2O–ZrO_2–SiO_2 Glass System

Composition (mole %)			Temperature (°C)	log η (P)	Reference[a]
Na_2O	ZrO_2	SiO_2			
15	5	80	1174	4.22	1
			1257	3.73	
			1338	3.30	
			1422	2.28	
16	11	73	1215	4.64	
			1257	4.16	2
			1338	3.66	
			1422	3.08	
20	14	66	1174	4.35	
			1257	3.63	
			1338	3.04	
			1422	2.50	
25	3.75	71.25	488	12.5	2
			548	10.2	
			565	9.5	
			608	8.4	
			621	8.0	

Composition (mole %)			Temperature (°C)	log η (P)	Reference[a]
Na_2O	ZrO_2	SiO_2			
25	7.5	67.5	548	12.5	2
			577	11.4	
			594	10.9	
			625	9.8	
			668	8.7	
25	11.25	63.75	575	13.0	2
			592	12.3	
			626	11.1	
			669	9.8	
			703	8.8	
25	15	60	671	13.2	2
			708	12.0	
			742	11.0	
			783	9.5	
			814	8.9	

[a] Data from (1) Dubrovo and Shnypikov (1968) and (2) Hoffman et al. (1952).

TABLE 9.86

Viscosities of Na$_2$O–ThO$_2$–SiO$_2$ Glasses[a]

Na$_2$O	ThO$_2$	SiO$_2$	Temperature (°C)	log η (P)
25	3.75	71.25	496	12.6
			521	11.6
			548	10.6
			577	9.4
			626	8.1
25	7.5	67.5	541	12.1
			570	11.0
			591	10.2
			625	9.0
			650	8.3

Composition (mole %) shown under Na$_2$O, ThO$_2$, SiO$_2$ columns.

[a] Data from Hoffman *et al.* (1952).

TABLE 9.87

Softening Temperatures of K$_2$O–TiO$_2$–SiO$_2$ Glasses[a]

K$_2$O	TiO$_2$	SiO$_2$	Softening temperature[b] (°C)	K$_2$O	TiO$_2$	SiO$_2$	Softening temperature[b] (°C)
15	5	80	819	25	40	35	670
15	12.5	72.5	838	25	45	30	658
15	15	70	831	30	10	60	740
15	20	65	813	30	15	55	743
15	25	60	787	30	20	50	727
15	30	55	753	30	30	40	674
15	35	50	740	30	40	30	627
15	40	45	737	35	5	60	696
20	5	75	777	35	10	55	710
20	10	70	796	35	15	50	695
20	15	65	803	35	20	45	677
20	20	60	789	35	30	35	613
20	25	55	784	35	40	25	568
20	30	50	746	35	45	20	624
20	35	45	727	35	50	15	582
20	40	40	704	40	5	55	684
20	45	35	696	40	10	50	668
25	5	70	746	40	20	40	619
25	10	65	762	40	30	30	568
25	15	60	765	40	40	20	532
25	20	55	770	45	5	50	645
25	25	50	751	45	10	45	621
25	30	45	713	45	15	40	616
25	35	40	682	45	20	35	572

Composition (wt. %) shown under K$_2$O, TiO$_2$, SiO$_2$ columns in both halves.

[a] Data from Rao (1963a,b).
[b] log η (P) = 7.65.

TABLE 9.88

Parameters of the Equation $\eta = \eta_0 \exp(E/RT)$ for the Viscosities of $K_2O-TiO_2-SiO_2$ System[a]

Composition (mole %)			Temperature range ($^\circ$C)	$\eta_0 \times 10^{-4}$ (P)	E (kcal mole^{-1})
K_2O	TiO_2	SiO_2			
25	12.5	62.5	1250–1350	1051.5	19.7
25	25	50	1100–1300	6.7	31.5
33	4	63	1360–1420	39.3	32
33	7	60	1340–1400	36	31.5
33	9	58	1300–1380	15.7	33.7
33	10	57	1240–1400	1.3	39.9
33	17	50	1100–1360	1.4	35.8
33	23	44	1000–1220	4.5	31.5
33	28	39	920–1260	1.5	31.1
33	33	34	900–1200	2.7	25.9
40	6.6	53.4	960–1360	0.4	34.7
40	20	40	800–1100	0.4	26.3
50	10	40	1050–1300	6.7	26.2
50	20	30	800–1000	1.0	28.1
50	25	25	750–1050	0.4	28.2

[a] Data from VanBernst and Delaunois (1966).

TABLE 9.89

Parameters of the Equation $\eta = \eta_0 \exp(B/T)$ for the Viscosities of Some $Li_2O-M_2O_5-SiO_2$ Glasses[a]

Composition (mole %)			Temperature ($^\circ$C) for log η (P) = 10	η_0 (P)	$10^{-4}B$ (K)
Li_2O	M_2O_5	SiO_2			
	M = P				
24.27	2.91	72.82	510	2.1×10^{-19}	5.15
32.33	2.91	64.76	483	3.4×10^{-22}	5.46
	M = V				
24.27	2.91	72.82	474	4.3×10^{-20}	5.11
32.33	2.91	64.76	469	4.8×10^{-21}	5.25

[a] Data from Matusita and Tashiro (1973) and Matusita et al. (1974).

TABLE 9.90

Viscosities of Some Na₂O–M₂O₅–SiO₂ Glass Systems[a]

Composition (mole %)			Temperature (°C)	log η (P)	Composition (mole %)			Temperature (°C)	log η (P)
Na₂O	MO₂.₅	SiO₂			Na₂O	MO₂.₅	SiO₂		
M = P					**M = V**				
25.5	1.9	72.6	462	12.1	26	3.9	70.1	420	12.0
			493	10.8				443	11.1
			506	10.3				468	9.9
			532	9.5				497	8.8
			552	8.8				508	8.6
26	3.9	70.1	465	11.9	26.5	6	67.5	392	12.3
			489	11.2				422	10.8
			502	10.6				440	9.9
			536	9.4				468	8.8
			578	8.5				448	8.3
M = Sb					**M = Ta**				
25.5	1.9	72.6	461	12.0	25.5	1.9	72.6	508	12.2
			488	11.2				534	11.0
			501	10.5				563	9.9
			531	9.4				586	9.3
			565	8.6				618	8.5
26	3.9	70.1	461	11.9	26	3.9	70.1	528	12.5
			498	10.5				546	11.8
			511	9.9				585	10.5
			542	8.9				618	9.5
			544	8.6				650	8.6
M = V									
25.5	1.9	72.6	444	12.2					
			465	11.2					
			488	10.2					
			498	9.8					
			520	9.1					

[a] Data from Hoffman *et al.* (1952).

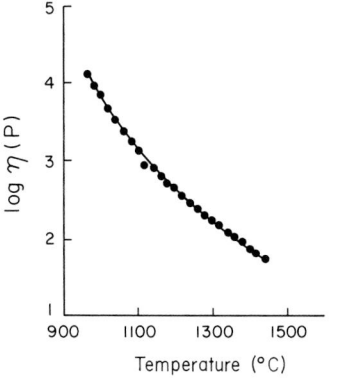

Fig. 9.15. Temperature dependence of viscosity of the glass composition (mole %) 19.16Na₂O–2.5Nb₂O₅–77.5SiO₂. From Yanishevskii (1961, 1964).

TABLE 9.91

Softening Temperatures of $K_2O-Nb_2O_5-SiO_2$ Glasses[a]

Composition (mole %)			Softening temperature[b] ($^\circ$C)	Composition (mole %)			Softening temperature[b] ($^\circ$C)
K_2O	Nb_2O_5	SiO_2		K_2O	Nb_2O_5	SiO_2	
14.3	14.3	71.4	876	20	20	60	803
14.7	11.8	73.5	890	20.8	16.7	62.5	825
15.2	9.1	75.7	915	21.7	13.1	65.2	827
15.6	6.3	78.1	893	22.7	9.1	68.2	784
16.1	3.2	80.7	833	23.8	4.8	71.4	730
16.7	16.7	66.6	842	25	25	50	754
17.2	13.8	69.0	860	26.3	21.1	52.6	766
17.9	10.7	71.4	882	27.8	16.7	55.5	758
18.5	7.4	74.1	870	29.4	11.8	58.8	738
19.2	3.8	77.0	824	31.3	6.2	62.5	710
19.2	23.1	57.7	778				

[a] Data from Rao (1962).
[b] $\log \eta$ (P) = 7.65.

TABLE 9.92

Viscosities of Magnesium–Calcium–Silicate Glass Melts[a]

Composition (wt. %)			η (P) at given temperature ($^\circ$C)					
MgO	CaO	SiO_2	1300	1350	1400	1450	1500	1550
5	35	60	76.0	52.0	35.0	23.2	14.2	10.2
5	40	55	—	—	11.5	7.5	5.5	3.5
5	45	50	—	—	6.4	4.3	3.2	2.2
5	50	45	22.0	7.0	4.5	3.4	2.5	—
10	30	60	101.0	59.0	38.5	23.2	13.6	10.2
10	35	55	28.0	15.5	9.3	6.2	4.2	3.5
10	40	50	17.0	9.5	6.0	4.2	3.0	2.2
10	45	45	—	6.5	4.5	3.2	2.3	—
10	50	40	—	—	—	80.0	21.0	13.5
15	30	55	—	16.5	9.5	6.5	4.3	3.3
15	35	50	—	9.0	6.0	4.0	3.0	2.2
15	40	45	—	6.0	4.5	3.5	2.5	—
15	45	40	—	—	—	80.0	19.0	2.4
20	25	55	26.5	16.5	10.0	6.2	4.3	3.3
20	30	50	17.0	9.5	6.5	4.4	3.5	2.5
20	35	45	—	—	4.5	3.5	2.5	2.0
20	40	40	—	—	—	—	8.0	—
25	30	45	—	—	12.5	3.6	2.5	2.0
25	35	40	—	—	—	—	3.5	2.0

[a] Data from Gulityai (1962).

TABLE 9.93

Viscosity of MgO–CaO–SiO$_2$ Glass Melt[a]

Composition (mole %)										
MgO	CaO	SiO$_2$	Temperature (°C) and natural logarithm of viscosity η (P)							
25	25	50	Temp. (°C)	1402	1540	1672	1781	1857	1931	2039
			ln η (P)	2.451	1.384	0.6831	0.2070	−0.094	−0.362	−0.673

[a] Data from Urbain *et al.* (1982). See Richet (1984) for the composition.

TABLE 9.94

Softening Temperatures of Some MO–PbO–SiO$_2$ Glass Systems[a]

Composition (mole %)			Softening temperature (°C)	Composition (mole %)			Softening temperature (°C)
MO	PbO	SiO$_2$		MO	PbO	SiO$_2$	
M = Mg				M = Sr			
0.18	74.63	25.19	528	2.36	71.93	25.71	547
0.39	74.22	25.39	536	4.87	68.71	26.42	576
0.94	72.99	26.07	558	M = Ba			
1.97	70.76	27.27	589	0.75	74.18	25.07	527
M = Ca				1.38	73.45	25.17	533
0.26	74.54	25.20	528	3.50	71.12	25.38	546
0.51	74.10	25.39	536	7.09	67.07	25.84	570
1.30	72.72	25.98	550	M = Zn			
2.75	70.26	26.99	586	0.21	74.57	25.22	528
M = Sr				0.42	74.18	25.40	531
0.46	74.42	25.12	528	1.07	72.89	26.04	536
0.93	73.80	25.27	534	2.26	70.57	27.17	566

[a] Data from Abou-El-Azm and El-Batal (1969).

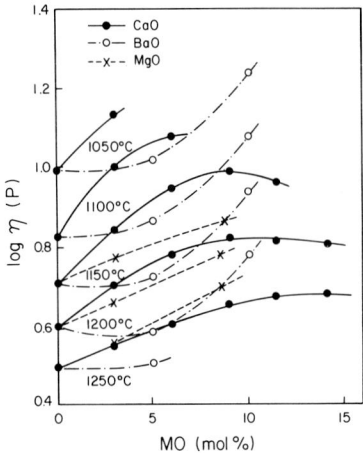

Fig. 9.16. Viscosity–composition curves of MO–PbO–SiO₂ (M = Mg, Ca, and Ba) glass melts containing 50 mole % SiO₂ at different temperatures. From Ouchi and Kato (1979).

TABLE 9.95

Viscosities of MgO–FeO–SiO₂ Glass Melts[a]

Composition (wt. %)			η (P) at given temperature (°C)					
MgO	FeO	SiO₂	1250	1300	1350	1400	1450	1500
5	40	55	—	41	21	10	4.5	2.5
10	35	55	—	50	26	12	6	3.5
10	40	50	26	13	6	3.5	2	—
15	30	55	—	62	36	19	9	5
15	35	50	36	16	8	5	3.5	—
15	40	45	21	10	4	3	2	—
20	25	55	—	—	44	24	13.5	7
20	30	50	50	20	9	6	4	—
20	35	45	28	13	6	3.5	2	—
20	40	40	50	22	8	4	2	—
25	20	55	—	—	46	28	16.5	10
25	25	50	64	24	11.5	7	5	—
25	30	45	40	20	10	6	4	—
25	35	40	—	42	13	5	3	3.5
25	40	35	—	62	18	8	4	—

[a] Data from Toropov and Bryantsev (1965).

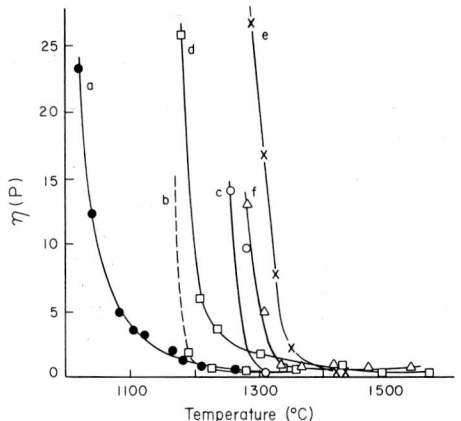

Fig. 9.17. Temperature dependence of viscosity of the compositions (wt. %): a, $10CaO-30BaO-60SiO_2$; b, $20CaO-40BaO-40SiO_2$; c, $30CaO-30BaO-40SiO_2$; d, $40CaO-20BaO-40SiO_2$; e, $50CaO-10BaO-40SiO_2$; f, $10CaO-60BaO-30SiO_2$. From Bulavin (1960).

TABLE 9.96

Viscosities of $CaO-ZnO-SiO_2$ Glass Melts at 1500°C[a]

Composition (mole % ZnO)	$\log \eta$ (P)
CaO/SiO$_2$ ratio = 2/3	
0	0.942
5	0.827
10	0.628
16.5	0.424
33.5	−0.037
CaO/SiO$_2$ ratio = 1	
0	0.471
10	0.215
20	0.021
30	−0.047

[a] Data from graph in Sumita *et al.* (1982).

TABLE 9.97

Parameters of the Equation $\ln \eta = \ln \eta_0 + E/RT$ for the Viscosities of $CaO-MnO-SiO_2$ System[a]

Composition (mole %)			$\eta_{1500°C}$ (P)	$-\ln \eta_0$ (P)	E (kcal mole^{-1})
CaO	MnO	SiO$_2$			
9.1	40.9	50.0	2.7	8.083	31.9
14.0	36.0	50.0	2.9	6.510	26.6
33.0	17.0	50.0	3.6	9.056	36.3
—	55.0	45.0	1.0	9.226	32.4
19.4	35.6	45.0	1.5	8.161	30.1
28.8	26.2	45.0	1.5	9.830	36.0
37.9	17.1	45.0	2.5	9.204	35.5
46.6	8.4	45.0	1.9	11.429	42.4
—	60.0	40.0	0.86	7.128	24.5
14.5	45.5	40.0	0.78	8.960	30.6
24.5	35.5	40.0	0.83	9.610	33.1
30.0	30.0	40.0	0.80	9.330	31.9
34.0	26.0	40.0	1.1	9.223	32.9
38.4	21.6	40.0	1.8	7.300	27.7
47.3	12.7	40.0	1.3	10.800	38.8
9.4	55.6	35.0	0.44	8.920	28.4
19.6	45.4	35.0	0.58	7.770	25.4
34.4	30.6	35.0	0.90	5.788	19.9

[a] Data from Segers et al. (1979).

TABLE 9.98

Viscosities of $CaO-CoO-SiO_2$ Glass Melts at 1550°C[a]

Composition (mole % CoO)	log η (P)
CaO/SiO$_2$ ratio = 2/3	
0	0.858
5	0.678
10	0.484
17	0.273
20	0.224
30	0.092
CaO/SiO$_2$ ratio = 1	
0	0.387
10	0.141
20	0.018

[a] Data from graph in Kuroda et al. (1982).

TABLE 9.99

Viscosities of CaO–NiO–SiO$_2$ Glass Melts[a]

Composition (mole %)			Temperature (°C)	$\log \eta$ (P)
CaO	NiO	SiO$_2$		
30	10	60	1550	0.892
35	5	60		0.914
40	0	60		0.925
25	25	50		0.258
30	20	50		0.269
33.5	16.5	50		0.285
35	15	50		0.296
40	10	50		0.339
45	5	50		0.366
50	0	50		0.435
35	25	40		−0.0044
40	20	40		0
42	18	40		0.016

Composition (mole %)			Temperature (°C)	$\log \eta$ (P)
CaO	NiO	SiO$_2$		
15	25	60	1400	1.257
20	20	60		1.198
25.5	14.5	60		1.139
30	10	60		1.099
35	5	60		1.069
40	0	60		1.0
30	20	50		0.574
33.5	16.5	50		0.515
40	10	50		0.356
45	5	50		0.247
50	0	50		0.129

[a] Data from graph in Kawahara *et al.* (1979).

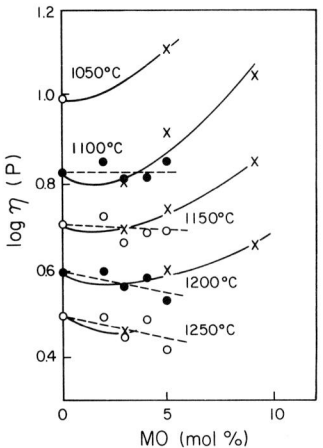

Fig. 9.18. Viscosity–composition curves of PbO–MO–SiO$_2$ (M = Co and Ni) glass melts containing 50 mole % SiO$_2$ at different temperatures; ● and ○, NiO; X, CaO. From Ouchi and Kato (1979).

TABLE 9.100

Parameters of the Equation log η = log η_0 + B/T for the Viscosities of Magnesium Aluminosilicate System

Composition (mole %)			Temperature range (°C)	$-\log \eta_0$ (P)	B (K)	Reference[a]
MgO	Al$_2$O$_3$	SiO$_2$				
13.49	11.92	74.59	1550–1700	6.10	16,901	1
15.16	13.90	70.94	1520–1690	5.96	15,930	
17.07	15.10	67.83	1540–1700	5.78	15,064	
18.67	17.45	63.88	1550–1700	5.64	14,337	
20.96	19.49	59.55	1570–1710	5.82	14,153	
23.37	21.63	55.00	1580–1710	5.50	13,059	
24.53	23.49	51.98	1570–1700	4.69	11,205	
26.80	25.62	47.58	1580–1720	5.87	13,131	
30.21	19.98	49.81	1550–1630	3.94	9,442	
37.09	12.95	49.96	1500–1700	4.50	10,266	
43.67	6.09	50.24	1570–1680	4.14	9,085	
12.5	12.5	75.0	1400–2200	5.80	16,240	2
25	25	50	1400–2200	4.97	11,454	
37.5	37.5	25	1400–2200	3.53	7,344	

[a] Data from (1) Riebling (1964a) and (2) Hofmaier (1968).

TABLE 9.101

Parameters of the Equation $\eta = \eta_0 \exp(E/RT)$ for the Viscosities of Calcium Aluminosilicate Glass Melts[a]

Composition (mole %)			Temperature range (°C)	$-\log \eta_0$ (P)	E (kcal mole⁻¹)	Composition (mole %)			Temperature range (°C)	$-\log \eta_0$ (P)	E (kcal mole⁻¹)
CaO	Al₂O₃	SiO₂				CaO	Al₂O₃	SiO₂			
4	4	92	1650–1950	6.44	98.8	28	42	30	1550–2100	4.57	45.16
4.28	10.72	85	1850–2100	5.89	86.02	30	20	50	1550–2000	5.25	47.59
7.13	17.87	75	1750–2050	4.99	67.9	30	40	30	1700–2050	4.41	43.88
8	8	84	1650–1950	6.48	88.93	30	50	20	1750–2100	4.32	41.27
10	50	40	1900–2200	4.21	44	32	32	36	1550–1950	4.85	48.54
10	60	30	1900–2100	4.08	41.5	35	15	50	1700–2000	4.35	45.75
10.97	4.03	85	1700–2100	5.70	83.08	37.61	37.61	24.78	1600–1850	5.05	48.88
11.3	58.8	29.9	1850–2100	3.96	40.61	40	20	40	1550–2000	4.31	43
12.27	12.27	75.46	1650–1850	6.17	79.32	42	28	30	1550–2100	4.57	45.16
15	35	50	1800–2000	4.75	54	43.62	43.62	12.76	1650–1950	4.68	44.41
16.8	16.8	66.4	1650–1950	5.81	69.68	50	10	40	1500–1950	3.78	37.5
18.3	6.7	75.0	1700–2100	4.32	60.29	51	34	15	1550–2150	4.07	39.16
19.2	42.3	38.5	1700–2100	4.69	46.5	51.22	18.78	30.0	1550–2000	4.12	38.15
20	50	30	1750–2050	4	41.6	58.16	11.84	30.0	1750–2000	3.27	30.75
20	60	20	1850–2200	3.71	35.68	62.19	22.81	15.0	1550–2000	3.94	37.07
21.3	6.5	72.2	1550–1850	6.21	66.4	70.63	14.37	15.0	1700–2050	2.80	23.83
21.62	21.62	56.76	1550–1950	6.04	66.54						
24.25	60.75	15.0	1800–2050	3.41	34						
25	25	50	1550–1950	5.32	57.97						
26.7	26.7	46.6	1550–1950	4.90	52.19						

[a] Data from Rossin et al. (1964).

TABLE 9.102

Parameters of the Equation $\eta = \eta_0 \exp(E/RT)$ for the Viscosities of $MO-Al_2O_3-SiO_2$ Systems[a]

Composition (mole %)			Temperature range (°C)	$-\log \eta_0$ (P)	E (kcal mole^{-1})
MO	Al_2O_3	SiO_2			
M = Sr					
12.5	12.5	75.0	1400–2200	6.15	81.6
25	25	50		5.29	59.6
37.5	37.5	25.0		4.93	51.3
M = Ba					
12.5	12.5	75.0	1400–2200	6.72	89.2
25	25	50		5.84	67.5
37.5	37.5	25.0		5.33	56.2

[a] Data from Hofmaier (1968).

TABLE 9.103

Softening Temperatures of Some $MO-Al_2O_3-SiO_2$ Glasses

Composition (mole %)			Softening temperature (°C)	Reference[a]
MO	Al_2O_3	SiO_2		
M = Sr				1
45	5	50	1000	
45	10	45	914	
50	5	45	994	
M = Pb				2
75.08	0.85	24.07	536	
75.21	1.71	23.08	539	
75.50	4.36	20.14	548	
76.12	8.65	15.23	563	

[a] Data from (1) Žagar and Lüneberg (1971) and (2) Abou-El-Azm and El-Batal (1969).

TABLE 9.104

Viscosities of the Zinc Aluminosilicate System[a]

Composition (wt. %)			$\log \eta$ (P) at given temperature (°C)		
ZnO	Al_2O_3	SiO_2	700	750	800
36.0	16.0	48.0	12.22	9.97	7.91
44.7	14.0	41.3	11.41	9.39	7.73
50.0	10.0	40.0	11.23	9.15	7.31

[a] Data from McMillan *et al.* (1969).

TABLE 9.105

Viscosities of Lead Borosilicate System[a]

Composition (wt. %)			Temperature (°C)	$\log \eta$ (P)	Composition (wt. %)			Temperature (°C)	$\log \eta$ (P)
PbO	B_2O_3	SiO_2			PbO	B_2O_3	SiO_2		
70.4	9.6	20.0	605	4.12	79.8	19.6	0.6	474	3.72
			664	3.00				511	2.44
			770	1.84				563	1.74
			870	1.29				668	0.57
70.5	10.0	19.5	572	3.97	80.0	9.6	10.4	498	4.13
			614	2.93				518	3.48
			707	1.43				568	2.30
			824	0.61				665	1.10
71.7	28.6	0.7	552	3.93				771	0.49
			614	2.50	89.5	9.7	0.8	448	2.30
			717	1.04				568	0.45
			824	0.35				591	0.29

[a] Data from Imoto and Hirao (1959).

TABLE 9.106

Softening Temperature of Lead Borosilicate Glasses[a]

Composition (mole %)			
PbO	B_2O_3	SiO_2	Softening temperature (°C)
75.31	0.58	24.11	523
75.62	1.17	23.21	492
76.60	2.96	20.44	456
78.30	6.05	15.65	422

[a] Data from Abou-El-Azm and El-Batal (1969).

TABLE 9.107

Viscosities of Some $MO-Al_2O_3-SiO_2$ Glass Melts[a]

Composition (mole %)			Temperature (°C)	η (P)
MO	Al_2O_3	SiO_2		
M = Ba				
47.4	5.5	47.1	1320	960
			1340	740
			1360	139
			1380	26
			1400	20
M = Zn				
47.4	5.5	47.1	1360	21
			1380	18
			1400	15
M = Pb				
47.4	5.5	47.1	1140	34
			1200	25
			1260	20
			1320	16
			1340	14.5
M = Fe				
47.4	5.5	47.1	1380	260
			1400	12

[a] Data from Sheludyakov et al. (1967).

TABLE 9.108

Viscosities of Manganese Aluminosilicate Glass Melts[a]

Temperatures (°C) and values of natural logarithm of viscosity η (P)

Composition (mole %)																	
MnO	Al$_2$O$_3$	SiO$_2$															
12.2	11.7	76.1	Temp. (°C)	1607	1611	1614	1630	1684	1719	1726	1760	1767	1825	1833	1879	1890	1975	1993
			ln η (P)	6.678	6.653	6.632	6.584	5.842	5.533	5.531	5.152	5.129	4.573	4.554	4.179	4.124	3.391	3.262
24	14	62	Temp. (°C)	1423	1459	1466	1488	1576	1594	1608	1642	1647	1666	1673	1698	1721	1790	1795
			ln η (P)	5.661	5.541	5.498	5.117	4.222	4.026	3.930	3.615	3.557	3.424	3.374	3.174	2.986	2.514	2.487
12	29	59	Temp. (°C)	1472	1477	1529	1530	1547	1551	1553	1576	1579	1581	1583	1585	1588	1593	1597
			ln η (P)	4.542	4.514	3.891	3.886	3.717	3.681	3.651	3.565	3.512	3.462	3.452	3.454	3.453	3.340	3.422
22.3	22.8	54.9	Temp. (°C)	1466	1468	1510	1513	1546	1584	1614	1622	1659	1666	1698	1727	1759	1779	1798
			ln η (P)	3.666	3.694	3.311	3.300	2.944	2.610	2.313	2.305	2.041	2.028	1.758	1.548	1.361	1.144	1.099
28.4	29	42.6	Temp. (°C)	1563	1591	1598	1609	1613	1620	1629	1637	1660	1685	1691	1749	1758	1819	1831
			ln η (P)	1.914	1.858	1.813	1.728	1.707	1.597	1.587	1.528	1.338	1.206	1.194	0.8713	0.8329	0.4700	0.4253
48	14	38	Temp. (°C)	1200	1220	1252	1277	1296	1317	1345	1352	1367	1410	1446	1479	1542	1597	1603
			ln η (P)	4.619	4.549	4.202	3.845	3.618	3.399	3.096	2.991	2.932	2.487	2.228	1.926	1.491	1.182	1.160

[a] Data from Urbain et al. (1982).

TABLE 9.109

Viscosities of Iron Silicate Glass Melts[a]

Analyzed composition (wt. %)				η (P) at given temperature (°C)			
FeO	Metallic Fe	Fe_2O_3	SiO_2	1250	1300	1350	1400
82.8	0.82	6.58	9.8			0.24	0.22
73.4	0.97	4.33	21.3	0.31	0.30	0.29	0.27
70.1	0.94	1.36	27.6		0.43	0.40	0.37
68.7	0.95	2.05	28.3		0.52	0.46	0.41
68.2	0.90	2.00	28.9	0.76	0.64	0.54	0.47
67.7	0.93	2.27	29.1		0.62	0.52	0.46
67.0	0.87	2.33	29.8	0.60	0.53	0.48	0.43
66.0	0.84	1.56	31.6	0.66	0.59	0.52	0.47
65.5	0.80	1.80	31.9	0.77	0.66	0.57	0.50
65.0	0.69	1.91	32.4	0.98	0.84	0.74	0.60
61.2	0.65	0.75	37.4	1.81	1.43	1.16	0.92

[a] Data from Shiraishi et al. (1978).

TABLE 9.110

Viscosities of Some
$35MO-35TiO_2-30SiO_2$ (mole %)
Glass Melt Compositions[a]

MO	Temperature (K)	$-\log \eta$ (P)
MgO	1773	0.195
	1800	0.240
	1820	0.273
CaO	1673	0.061
	1723	0.156
	1773	0.256
	1823	0.334
SrO	1773	0.217
	1800	0.262
	1818	0.29
BaO	1673	0.017
	1723	0.123

[a] Data from graph in Nakamura et al. (1977).

TABLE 9.111

Viscosities of PbO–TiO$_2$–SiO$_2$ Glass Melts[a]

PbO	TiO$_2$	SiO$_2$	Temperature (K)	$\log \eta$ (P)	PbO	TiO$_2$	SiO$_2$	Temperature (K)	$\log \eta$ (P)
35	30	35	1423	0.187	50	17	33	1323	−0.118
			1473	0.10				1373	−0.183
			1523	0.0				1423	−0.249
			1573	−0.087				1473	−0.301
40	20	40	1323	0.46	53	10	37	1223	0.150
			1373	0.34				1273	0.033
			1423	0.24				1323	−0.072
			1473	0.14				1373	−0.17
45	10	45	1273	0.80	50	30	20	1423	−0.505
			1323	0.667				1523	−0.574
			1373	0.533	57	20	23	1373	−0.505
			1423	0.433				1423	−0.554
45	25	30	1423	−0.262				1473	−0.588
			1473	−0.327	63	10	27	1223	−0.381
			1523	−0.38				1273	−0.45
			1573	−0.438				1323	−0.484

[a] Data from graph in Nakamura *et al.* (1977).

TABLE 9.112

Softening Temperatures of Some PbO–MO$_2$–SiO$_2$ Glasses[a]

PbO	MO$_2$	SiO$_2$	Softening temperature (°C)
	M = Ti		
73.77	6.55	19.68	554
74.53	2.64	22.83	532
74.73	1.37	23.90	525
	M = Zr		
72.75	5.92	21.33	589
73.41	4.07	22.52	563
74.25	1.98	23.77	546

[a] Data from Abou-El-Azm and El-Batal (1969).

TABLE 9.113

Compositions of the Glass Melts of MnO–TiO$_2$–SiO$_2$ System[a]

Glass number	Composition (mole %)			Glass number	Composition (mole %)		
	MnO	TiO$_2$	SiO$_2$		MnO	TiO$_2$	SiO$_2$
1	50	—	50	12	70	30	—
2	60	—	40	13	40	40	20
3	70	—	30	14	50	40	10
4	50	10	40	15	60	40	—
5	60	10	30	16	30	50	20
7	50	20	30	17	40	50	10
8	60	20	20	18	50	50	—
10	50	30	20				
11	60	30	10				

[a] Data from Yagi *et al.* (1980).

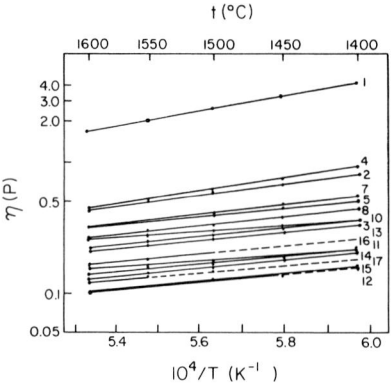

Fig. 9.19. Temperature dependence of viscosities of MnO–TiO$_2$–SiO$_2$ glass melts. For compositions see Table 9.113. From Yagi *et al.* (1980).

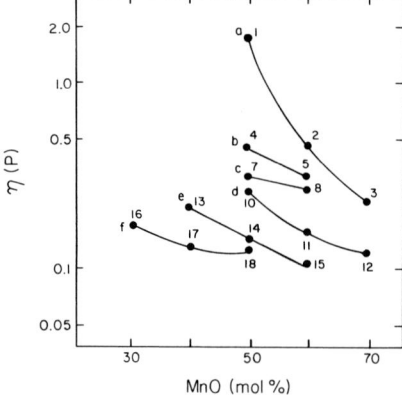

Fig. 9.20. Variations in viscosity with MnO content of MnO–TiO$_2$–SiO$_2$ glass melts at 1600°C for various TiO$_2$ contents (mole %): a, 0; b, 10; c, 20; d, 30; e, 40; f, 50. For melt compositions, see Table 9.113. From Yagi *et al.* (1980).

TABLE 9.114

Viscosities of Some Al_2O_3–M_2O_3–SiO_2 Glass Systems[a]

Composition (wt. %)			Temperature (°C) for given log η (P)		
Al_2O_3	M_2O_3	SiO_2	4	6	10
	M = La				
15	40	45	1240	1115	985
22.5	51	26.5	—	1160	840
28.5	38.5	33.0	1300	1150	910
	M = Nd				
28.5	38.5	33.0	1310	1170	930

[a] Data from Karlsson (1970).

TABLE 9.115

Viscosity of Al_2O_3–TiO_2–SiO_2 Melts[a,b]

Composition (wt. % Al_2O_3)	Temperature (°C)	log η (P)
0	1600	6
5		5.69
10		4.17
12.5		4.32
15		5.83
20		6.83

[a] Data from Sugai and Somiya (1982).

[b] The ratio of TiO_2 and SiO_2 was maintained at the eutectic composition (wt. %) $10.5TiO_2$–$89.5SiO_2$.

Chapter **10**

Elastic Properties

An ideal elastic material deforms instantaneously (more exactly, at the speed of sound) when it is subjected to a stress, and the deformation recovers instantaneously when the stress is removed. Vitreous silica is close to an ideal elastic solid at temperatures below about 1000°C under most loading conditions, because it deforms instantaneously, it does not creep (deform slowly) under continuous load, and the deformation recovers instantaneously. Most silicate glasses containing modifying oxides, such as alkalis and alkaline earths, are close to ideal elastic solids below the glass transition temperature or strain point, which is between 400 and 600°C for most commercial and laboratory glasses other than vitreous silica. However, these glasses show small delayed elastic effects that apparently are related to the motions of alkali ions. Glasses deformed by sharp indenters show delayed elastic and shear faulting effects. These types of deformation are not treated in this chapter, which focuses on the instantaneous elastic properties.

The specification of the strains resulting from applied stress requires 21 elastic constants in the most general case, but for an isotropic material like glass only two constants are required. Four constants are in common use, so they are interrelated. Hooke's law states that strain is proportional to stress, and this proportionality is found in silicate glasses over wide ranges of stress.

If a rod or bar is stretched by a uniform stress S, the proportionality constant between stress and strain ε is Young's modulus E:

$$S = E\varepsilon. \tag{1}$$

The units of E are the same as for stress, since ε is dimensionless. The ratio of lateral to longitudinal stress in this geometry is Poisson's ratio σ. If a uniform compressive stress is applied to a body, the proportionality constant between stress and strain is the bulk modulus K. The compressibility $\beta = 1/K$. For a shear stress, the proportionality constant is the shear modulus G. Some relations between E and the other elastic constants are

$$E = \frac{9KG}{3K + G} = 3K(1 - 2\sigma) = 2G(1 + \sigma). \tag{2}$$

The units of stress, pressure, and modulus used in this work are newtons (N) per square meter, or pascals (Pa). This unit is small, so megaPascals (MPa $= 10^6$ Pa) and gigaPascals (GPa $= 10^9$ Pa) are used. Conversion factors are: 1 Pa $= 10$ dyn cm^{-2}, 10^{-7} kg mm^{-2}, 10^{-11} kg cm^{-2}, 1.45×10^{-4} lb in.$^{-2}$. One bar is equal to 10^5 Pa, and is approximately 1 atm.

Elastic constants can be calculated from the deformation in bending, tension, or torsion and the measured forces. The velocity of acoustic waves in glass is related to the elastic properties by the relations

$$v_1 = (E\rho)^{1/2}, \tag{3}$$

$$v_t = (G\rho)^{1/2}, \tag{4}$$

where v_1 and v_t are the longitudinal and transverse sound velocities and ρ is the density of the glass. The measurement of the acoustic velocities in glasses by a pulse echo technique has been developed by Manghnani (1972). High accuracy is possible with this technique.

For most materials, the modulus increases with increasing pressure and decreases with increasing temperature. However, the elastic moduli for vitreous silica, Pyrex borosilicate glass, and some other glasses with low modifier concentrations increase with increasing temperature and decrease with increasing pressure. These anomalies have been explained in terms of the open structure of vitreous silica (Ernsberger, 1980). The Young's modulus of vitreous silica also increases at very high strains, whereas for most other glasses it decreases at these high strains.

Glass that is not annealed and contains residual stress has a higher modulus than does carefully annealed glass that contains no residual stress.

As modifying alkali oxides are added to vitreous silica, the modulus decreases. If di- or trivalent oxides are added to an alkali silicate glass, the modulus increases. These variations can be understood in terms of the number density (moles per unit volume) of oxygen atoms in the glass (Ray, 1971). The Young's modulus of a wide variety of oxide glasses was found to be roughly proportional to the density of oxygen atoms. If the oxygen density was greater than that of the lattice oxide (SiO_2, B_2O_3, P_2O_5), then the modulus was greater than that of lattice oxide.

There was considerable scatter in this correlation, so it is not possible to use it to predict an accurate modulus value for a glass of unknown modulus. However, it can be used to make a rough estimate of the modulus. The oxygen atom density d in moles per cubic centimeter can be calculated from the expression

$$d = \sum_i n_i w_i / M_i \tag{5}$$

where w_i is the weight fraction of oxide i in the glass, M_i is its molecular weight, and n_i is the number of atoms of oxygen per molecule (e.g., 1 for Na_2O and CaO, 2 for SiO_2). Then the unknown Young's modulus E in GPa can be roughly estimated from the equation

$$E = 72[1 + 40(d - 0.0735)]   \qquad (6)$$

where 0.0735 moles cm^{-3} is the density of oxygen in vitrous silica and 72 GPa its modulus.

II   TABLES, FIGURES, AND EQUATIONS

Tables 10.1–10.11, Fig. 10.1, and Eq. 10.1 give data on the elastic properties of binary glasses; Tables 10.12–10.45 and Figs. 10.2–10.6 give data for ternary glasses.

BIBLIOGRAPHY

Aleinkov, F. K., Paulavichus, P. B., and Slizhis, V. A. (1962a). *Tr. Akad. Nauk SSR Ser. B* **2**, 69.
Aleinkov, F. K., Slizhis, V. A., Paulavichus, P. B., and Dundzis, P. V. (1962b). *Opt.-Mekh. Prom.* No. 9, p. 38.
Aleinkov, F. K., Slizhis, V. A., Paulavichus, P. B., and Dundzis, P. V. (1963). *Stekloobraznoe Sostoyanie* **3**(2), 30.
Andreatch, P., and McSkimin, H. J. (1976). *J. Appl. Phys.* **47**, 1299.
Appen, A. A., Kozlovskaya, E. I., and Gan'Fu-Si (1961). *J. Appl. Chem. USSR (Engl. Transl.)* **34**, 942.
Baidov, V. V., Kunin, L. L., and Urman, N. S. (1968). *In* "Physical Chemistry and Electrochemistry of Molten Salts and Slags," p. 148. Leningrad.
Balashov, Y. S., and Chernyshov, A. V. (1972). *Inorg. Mater. (Engl. Transl.)* **8**, 1193.
Balashov, Y. S., Chernyshov, A. V., and Sanin, V. N. (1979). *Sov. J. Glass Phys. Chem. (Engl. Transl.)* **5**, 445.
Bloom, H., and Bockris, J. O'M. (1957). *J. Phys. Chem.* **61**, 515.
Brekhovskikh, S. M. (1959). *Glastech. Ber.*, **32**, 437.
Butta, E., and Paoletti, G. (1961). *Vetro Silic.* **5**, 21.
Deeg, E. (1958a). *Glastech. Ber.* **31**, 124.
Deeg, E. (1958b). *Glastech. Ber.* **31**, 229.
Eagan, R. J., and Swearengen, J. C. (1978). *J. Am. Ceram. Soc.* **61**, 27.
Ellern, G. A., and Pavlushkin, N. M. (1969). *Tr. Mosk. Khim.-Tekhnol. Inst., Silic.* Publ. No. 59, p. 30.
Ernsberger, F. M. (1980). *In* "Glass: Science and Technology" (D. R. Uhlmann and N. J. Kreidl, eds.), Vol. 5, p. 1, Academic Press, New York.
Gladkov, A. V. (1964). *Stekloobraznoe Sostoyanie* p. 124.
Gladkov, A. V., Tarasov, V. V., and Yunitskii, G. A. (1964). *In* "Applications of Ultraacoustics to the Investigation of Materials," Publ. No. 20, p. 181. Moscow.
Goral'nik, A. S., Kulbitskaya, M. N., Mikhailov, I. G., Fershtat, L. N., and Shutilov, V. A. (1972). *Akust. Zh.* **18**, 391.
Gutop, V. G. (1940). *Stekol'naya Prom.* No. 11/12, 24.
Hamilton, E. H., Waxler, R. M., and Nivert, J. M. (1959). *J. Res. Natl. Bur. Stand. (U.S.)* **62**, 59.

Higgins, T. J., Boesh, L. P., Volterra, V., Moynihan, C. T., and Macedo, P. B. (1973). *J. Am. Ceram. Soc.* **56**, 334.

Hirao, K., and Soga, N. (1982). *Yogyo Kyokaishi* **90**, 476.

Imaoka, M., Hasegawa, H., Hamaguchi, Y., and Kurotaki, Y. (1971). *Yogyo Kyokaishi* **79**, 164.

Karapetyan, G. O., Livshits, V. Y., and Tennison, D. G. (1979). *Sov. J. Glass Phys. Chem. (Engl. Transl.)* **5**, 278.

Karapetyan, G. O., Livshits, V. Y., Tennison, D. G., and Zhukovskaya, O. V. (1980). *Sov. J. Glass Phys. Chem. (Engl. Transl.)* **6**, 310.

Karapetyan, G. O., Livshits, V. Y., and Tennison, D. G. (1981). *Sov. J. Glass Phys. Chem. (Engl. Transl.)* **7**, 131.

Kasymova, S. S., and Plutalova, N. Y. (1975). *Inorg. Mater. (Engl. Transl.)* **11**, 616.

Kunugi, M., Soga, N., and Miyashita, A. (1979). *Rep. Asahi Glass Found. Ind. Technol.* **35**, 79.

Laberge, N. L., Gupta, P. K., and Macedo, P. B. (1975). *J. Non-Cryst. Solids* **17**, 61.

Livshits, V. Y., Tennison, D. G., and Karapetyan, G. O. (1982a). *Sov. J. Glass Phys. Chem. (Engl. Transl.)* **8**, 285.

Livshits, V. Y., Tennison, D. G., and Karapetyan, G. O. (1982b). *Sov. J. Glass Phys. Chem. (Engl. Transl.)* **8**, 422.

Livshits, V. Y., Tennison, D. G., Gukasyan, S. B., and Kostanyan, A. K. (1982c). *Sov. J. Glass Phys. Chem. (Engl. Transl.)* **8**, 463.

Makishima, A., and Mackenzie, J. D. (1975). *J. Non-Cryst. Solids* **17**, 147.

Manghnani, M. H. (1972). *J. Am. Ceram. Soc.* **55**, 360.

Manghnani, M. H., and Singh, B. K. (1974). *Proc. Int. Congr. Glass, 10th, Kyoto*, **N11**, 104.

Matusita, K., Sakka, S., Osaka, A., Soga, N., and Kunugi, M. (1974). *J. Non-Cryst. Solids* **16**, 308.

Matveev, M. A., Mazo, E. E., and Valkodatov, A. F. (1964). *Glass Ceram. (Engl. Transl.)* **21**, 312.

Matveev, M. A., Mazo, E. E., Volkodatov, A. F., and Volcek, L. K. (1965a). "Investigations in the Field of the Chemistry of Silicates and Oxide," p. 63. Moscow.

Matveev, M. A., Mazo, E. E., Volkodatov, A. F., and Volchek, L. K. (1965b). *Stekloobraznoe Sostoyanie* p. 147.

Maynell, C. A., Saunders, G. A., and Scholes, S. (1973). *J. Non-Cryst. Solids* **12**, 271.

Nassau, K., Shiever, J. W., and Krause, J. T. (1975). *J. Am. Ceram. Soc.* **58**, 461.

Nemilov, S. V. (1973). *Russ. J. Phys. Chem. (Engl. Transl.)* **47**, 831.

Nemilov, S. V., and Kasymova, S. S. (1975). *J. Appl. Chem. USSR (Engl. Transl.)* **48**, 1299.

Pavlova, G. A., and Amatuni, A. N. (1975). *Inorg. Mater. (Engl. Transl.)* **11**, 1443.

Pelous, J., and Vacher, R. (1977). *Phys. Chem. Glasses* **18**, 36.

Petrenko, Y. M., Ushakov, D. F., and Gilev, I. S. (1973). *Inorg. Mater. (Engl. Transl.)* **9**, 273.

Ray, N. H. (1971). *Int. Congr. Glass, 9th, Versailles* **1**, 655.

Shaw, R. R., and Uhlmann, D. R. (1971). *J. Non-Cryst. Solids* **5**, 237.

Shelby, J. E., and Day, D. E. (1969). *J. Am. Ceram. Soc.* **52**, 169.

Soga, N., Ota, R., and Kunugi, M. (1972). *Mech. Behav. Mater., Proc. Int. Conf., Kyoto, 1st, 1971* **4**, 366.

Soga, N., Yamanaka, H., Hisamoto, C., and Kunugi, M. (1976). *J. Non-Cryst. Solids* **22**, 67.

Sokolov, L. N., Baidov, V. V., and Kunin, L. L. (1969). In "Physical Chemistry and Electrochemistry of Fused Salts and Slags, Vol. IV, All Union Conference, Part 1, Thermodynamics and the Structure of Fused Salts," p. 299. Kiev.

Sokolov, L. N., Baidov, V. V., and Kunin, L. L. (1970). In "Properties and Structure of Molten Slags," p. 94. Moscow.

Spinner, S. (1954). *J. Am. Ceram. Soc.* **37**, 229.

Takahashi, K., and Osaka, A. (1983). *Yogyo Kyokaishi* **91**, 116.

Takahashi, K., Osaka, A., and Furuno, R. (1983). *Yogyo Kyokaishi* **91**, 199.

Thiele, A. (1955). *Glastech. Ber.* **28**, 384.

Tille, U., Frischat, G. H., and Leers, K.-J. (1978). *Glastech. Ber.* **51**, 8.

Totes, A. S., Grigor'eval, L. F., Strelcina, M. V., and Roskova, G. P. (1965). *Sklar Keram.* **15**, 370.

Yamane, M., and Okuyama, M. (1982). *J. Non-Cryst. Solids* **52**, 217.

TABLE 10.1

Elastic Properties of Lithium Silicate Glasses at Ambient Conditions and Their Pressure and Temperature Derivatives[a]

Composition (mole % Li_2O)	E (GPa)	G (GPa)	K (GPa)	σ	β (GPa^{-1})	$\left(\dfrac{dK}{dP}\right)$[b]	$\left(\dfrac{dG}{dP}\right)$[b]	$\left(\dfrac{dK}{dT}\right)$[c] (GPa deg^{-1})	$\left(\dfrac{dG}{dT}\right)$[c] (GPa deg^{-1})	$10^{-5} v_{l}$ (cm sec^{-1})	$10^{-5} v_{t}$ (cm sec^{-1})
20	75.45	31.64	40.88	0.192	0.0245	0.61	−1.20	0.0018 (0.0015)	0.0034 (−0.0038)	6.033	3.724
25	77.06	32.03	43.22	0.203	0.0231	1.39	−0.640	−0.0051 (−0.0055)	−0.0084 (−0.0091)	6.105	3.727
28	77.89	32.14	44.96	0.211	0.0222	3.32	−0.170	−0.0043 (−0.0058)	−0.0084 (−0.0083)	6.149	3.720
30	78.29	32.18	46.02	0.216	0.0217	3.22	0.183	−0.0067 (−0.0094)	−0.0089 (−0.0097)	6.176	3.715
32	78.80	32.15	47.81	0.225	0.0209	3.96	0.384	−0.011 (−0.012)	−0.0102 (−0.0106)	6.224	3.706
35	80.44	32.71	49.55	0.230	0.0202	5.33	0.537	−0.0156 (−0.0149)	−0.0112 (−0.0117)	6.298	3.722

[a] Data from M. H. Manghnani (1974 personal communication).

[b] Values determined at 25°C.

[c] Values in 25–100°C range; values shown in parentheses are in 25–200°C interval.

TABLE 10.2

Elastic Constants of Lithium Silicate Glasses[a]

Composition (mole % Li$_2$O)	E (GPa)	G (GPa)	K (GPa)	σ	$10^{-5}v_1$ (cm sec^{-1})	$10^{-5}v_t$ (cm sec^{-1})
10	74.26	30.48	43.92	0.218	6.156	3.696
15	76.42	32.15	40.89	0.158	6.090	3.774
20	76.99	31.40	46.81	0.226	6.240	3.714
25	78.44	31.74	49.43	0.236	6.313	3.714
30	78.81	31.73	50.92	0.242	6.320	3.687
35	81.28	33.10	49.79	0.228	6.316	3.750

[a] Data from Shaw and Uhlmann (1971).

TABLE 10.3

Elastic Constants of Sodium Silicate Glasses at Ambient Conditions and Their Pressure and Temperature Derivatives[a]

Composition (mole % Na_2O)	E (GPa)	G (GPa)	K (GPa)	σ	β (GPa^{-1})	$\left(\dfrac{dK}{dP}\right)^b$	$\left(\dfrac{dG}{dP}\right)^b$	$\left(\dfrac{dK}{dT}\right)^c$ (GPa deg^{-1})	$\left(\dfrac{dG}{dT}\right)^c$ (GPa deg^{-1})	$10^{-5} v_l$ (cm sec^{-1})	$10^{-5} v_t$ (cm sec^{-1})
10	65.29	27.68	33.95	0.180	0.0294	−2.16	−1.93	0.0063 (0.0063)	0.0004 (0.0004)	5.565	3.478
15	62.90	26.29	34.51	0.196	0.0290	−0.22	−1.15	0.0030 (0.0030)	−0.0021 (−0.0021)	5.457	3.355
20	61.08	25.18	35.46	0.213	0.0282	1.55	−0.46	−0.0017 (−0.0017)	−0.0041 (−0.0041)	5.382	3.251
25	59.77	24.26	37.16	0.232	0.0269	3.18	0.06	−0.0060 (−0.0060)	−0.0058 (−0.0058)	5.347	3.159
30	59.33	23.80	39.03	0.247	0.0256	4.11	0.39	−0.0097 (−0.0098)	−0.0077 (−0.0076)	5.348	3.101
35	58.18	23.02	41.02	0.264	0.0243	4.61	0.67	−0.0122 (−0.0123)	−0.0107 (−0.0106)	5.361	3.038
40	57.73	22.58	43.41	0.278	0.0230	5.83	0.84	−0.0182 (−0.0182)	−0.0108 (−0.0108)	5.401	2.993

[a] Data from M. H. Manghnani (1974, personal communication) and Manghnani and Singh (1974).
[b] Values at 25°C.
[c] Values in 25–100°C range; values in parentheses are in the 25–200°C interval.

TABLE 10.4

Elastic Constants of Potassium Silicate Glasses at Ambient Conditions and Their Pressure and Temperature Derivatives[a]

Composition (mole % K_2O)	E (GPa)	G (GPa)	K (GPa)	σ	β (GPa^{-1})	$\left(\dfrac{dK}{dP}\right)$[b]	$\left(\dfrac{dG}{dP}\right)$[b]	$\left(\dfrac{dK}{dT}\right)$[c] (GPa deg^{-1})	$\left(\dfrac{dG}{dT}\right)$[c] (GPa deg^{-1})	$10^{-5}\,v_1$ (cm sec^{-1})	$10^{-5}\,v_{t_1}$ (cm sec^{-1})
15	52.67	21.62	31.13	0.218	0.0322	1.41	−0.882	0.0041 (0.0039)	−0.0005 (−0.0004)	5.065	3.042
18	—	—	—	—	—	—	—	—	—	4.958	2.889
20	49.04	19.75	31.62	0.242	0.0316	3.35	−0.247	0.0006 (0.0006)	−0.0019 (−0.0019)	4.935	2.881
22.5	—	—	—	—	—	—	—	—	—	4.810	2.758
25	46.45	18.43	32.31	0.260	0.0310	4.13	−0.195	−0.0024 (−0.0026)	−0.0034 (−0.0034)	4.846	2.759

[a] Data from M. H. Manghnani (1974, personal communication) and Takahashi and Osaka (1983).
[b] Values at 25°C.
[c] Values in 25–100°C range; values in parentheses are in 25–200°C interval.

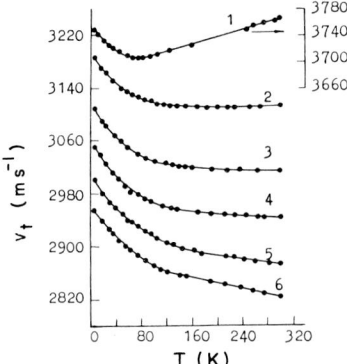

Fig. 10.1. Temperature dependence of transverse wave velocity in potassium silicate glasses. K_2O (mole %): 1, 0; 2, 13; 3, 16; 4, 19; 5, 22; 6, 25. From Balashov and Chernyshov (1972).

TABLE 10.5

Shear Modulus and Transverse Sound Velocity for Rubidium Silicate Glasses at 20°C[a]

Composition (mole % Rb_2O)	G (GPa)	$10^{-5}v_t$ (cm sec^{-1})
24	16.89	2.406
30	14.66	2.165
35	13.84	2.080

[a] Data from Nemilov (1973).

TABLE 10.6

Elastic Constants of Cesium Silicate Glasses[a]

Composition (mole % Cs_2O)	E (GPa)	G (GPa)	K (GPa)	$10^{-5}v_l$ (cm sec^{-1})	$10^{-5}v_t$ (cm sec^{-1})
16.67	39.99	15.74	29.03	4.01	2.25
20	37.67	14.86	27.02	3.78	2.13
24	—	13.73	—	—	2.06
25	32.50	12.55	26.38	3.56	1.92
30	—	11.47	—	—	—
33.33	—	—	25.31	3.18	—
35	—	10.93	—	—	1.74

[a] Data from Takahashi and Osaka (1983), Nemilov (1973), and Kunugi *et al.* (1979).

TABLE 10.7

Elastic Constants of Thallium Silicate Glasses at 20°C[a]

Composition (mole % Tl_2O)	E (GPa)	G (GPa)	σ	$10^{-5}v_t$ (cm sec^{-1})
44.62	24.36	9.49	0.286	1.187
47.00	22.70	8.66	0.312	1.150
51.62	22.06	8.63	0.280	1.124
53.35	22.50	8.56	0.314	1.075

[a] Data from Nemilov (1973).

TABLE 10.8

Elastic Constants of Some MO–SiO₂ Glasses

Composition (mole % MO)	E (GPa)	G (GPa)	K (GPa)	σ	$10^{-5}v_l$ (cm sec^{-1})	$10^{-5}v_t$ (cm sec^{-1})	Reference[a]
M = Ca							
50	—	—	—	0.235	5.825	3.425	1
M = Sr							
33.33	71.43	28.22	50.77	0.266	5.242	2.962	2
M = Ba							
30	—	25.1	—	—	—	2.678	1
33.33	64.09	25.15	47.27	0.274	4.673	2.607	2

[a] Data from (1) Nemilov (1973) and (2) Soga *et al.* (1976).

TABLE 10.9

Elastic Constants of Lead Silicate Glasses[a]

Composition (mole % PbO)	E (GPa)	G (GPa)	K (GPa)	σ	$10^{-5}v_l$ (cm sec^{-1})	$10^{-5}v_t$ (cm sec^{-1})
24.6	47.14	20.44	33.86	0.249	3.960	2.290
30.0	50.13	21.35	25.63	0.174	3.522	2.213
35.7	46.34	18.50	31.14	0.252	3.440	1.980
38.4	52.84	22.98	25.14	0.150	3.295	2.115
45.0	51.70	21.21	30.62	0.219	3.250	1.950
50.0	44.08	17.51	30.46	0.259	2.980	1.700
55.0	49.31	20.18	29.51	0.222	2.980	1.782
60.0	43.55	17.0	33.09	0.281	2.890	1.596
65.0	41.21	16.06	31.59	0.283	2.721	1.498

[a] Data from Shaw and Uhlmann (1971).

TABLE 10.10

Young's Modulus of B₂O₃–SiO₂ Glasses[a]

Composition (mole % B₂O₃)	E (GPa)	Composition (mole % B₂O₃)	E (GPa)
60	23.8	80	23.3
65	23.0	85	21.6
70	24.0	90	21.3
75	24.6	95	21.6

[a] Data from Imaoka *et al.* (1971).

EQ. 10.1

Acoustic Velocities in Al₂O₃–SiO₂ Glasses[a]

$$10^{-5}v_l \text{ (cm sec}^{-1}) = 5.962 + 0.0321M$$

$$10^{-5}v_t \text{ (cm sec}^{-1}) = 3.762 + 0.00943M$$
where M is the mole % of Al₂O₃ (0–7) in the glass

[a] Data from Nassau *et al.* (1975).

TABLE 10.11

Elastic Constants of TiO_2–SiO_2 and V_2O_5–SiO_2 Glasses[a]

Composition (mole % metal oxide)	E (GPa)	G (GPa)	K (GPa)	σ	$10^{-5} v_l$ (cm sec^{-1})	$10^{-5} v_t$ (cm sec^{-1})	dv_l/dT^b (cm sec^{-1} K^{-1})	dv_t/dT^b (cm sec^{-1} K^{-1})
TiO_2								
3.81	66	32.7	66	—	5.46	3.85	120	30
5.75	67.9	—	—	—	5.74	3.625	—	25
6.14	66	28.0	72	0.19	5.73	3.56	50	—
V_2O_5								
1	66	28.2	71	—	5.69	3.58	110	50

[a] Data from Goral'nik et al. (1972), Pavlova and Amatuni (1975), and Andreatch and McSkimin (1976).
[b] Temperature range 60–100°C.

TABLE 10.12

Elastic Constants of Li$_2$O–Na$_2$O–SiO$_2$ Glasses[a]

Composition (mole %)			E	G	K		$10^{-5}v_l$	$10^{-5}v_t$
Li$_2$O	Na$_2$O	SiO$_2$	(GPa)	(GPa)	(GPa)	σ	(cm sec^{-1})	(cm sec^{-1})
0.5	24.5	75		25.4				
1.25	23.75	75		25.7				
3.75	21.25	75		28.0				
8.75	16.25	75		28.9				
12.5	12.5	75		28.9				
16.25	8.75	75		30.3				
21.25	3.75	75		34.9				
23.75	1.25	75		32.1				
24.5	0.5	75		33.5				
5	20	75	68.0	27.9	40.3	0.219	5.663	3.398
10	15	75	72.2	29.8	41.9	0.213	5.835	3.525
15	10	75	75.3	31.2	43.0	0.208	5.972	3.627
20	5	75	77.5	32.2	43.6	0.204	6.081	3.708

[a] Data from Shelby and Day (1969) and Karapetyan *et al.* (1981).

TABLE 10.13

Elastic Constants of Lithium–Potassium–Silicate Glasses at 25°C

Composition (mole %)			G	β	
Li$_2$O	K$_2$O	SiO$_2$	(GPa)	(GPa^{-1})	Reference[a]
24.5	0.5	75	22.0		1
23.75	1.25	75	32.1		
21.25	3.75	75	28.4		
16.25	8.75	75	29.9		
12.5	12.5	75	28.5		
8.75	16.25	75	25.3		
3.75	21.25	75	21.8		
1.25	23.75	75	19.5		
0.5	24.5	75	20.5		
0.125	24.875	75	19.9		
—	25	75	18.4		
10	30	60		0.068[b]	2

[a] Data from (1) Shelby and Day (1969) and (2) Laberge *et al.* (1975).
[b] At T_g (634 K).

TABLE 10.14

Shear Moduli of Lithium–Cesium Silicate Glasses at 25°C[a]

Composition (mole %)			G
Li_2O	Cs_2O	SiO_2	(GPa)
24.5	0.5	75	29.5
23.75	1.25	75	31.6
21.25	3.75	75	29.9
16.25	8.75	75	27.1
12.5	12.5	75	26.7

[a] Data from Shelby and Day (1969).

TABLE 10.15

Elastic Constants of Sodium–Potassium Silicate Glasses

Composition (mole %)			G	E	$10^{-5}v_1$	
Na_2O	K_2O	SiO_2	(GPa)	(GPa)	(cm sec^{-1})	Reference[a]
24.875	0.125	75	24.7			1
24.5	0.5	75	26.5			
22.5	2.5	75	25.2			
20	5	75	26.8			
15	10	75	23.9			
10	15	75	24.7			
5	20	75	22.4			
2.5	22.5	75	21.7			
1.25	23.75	75	20.4			
0.5	24.5	75	20.0			
5	15	80		56.6	4.74	2
10	10	80		54.2	4.89	
15	5	80		59.2	5.01	
5	25	70		52.1	4.60	
10	20	70		53.9	4.72	
15	15	70		56.1	4.83	
20	10	70		55.2	4.87	
25	5	70		59.2	4.94	
10	30	60		47.8	4.48	
20	20	60		55.6	4.78	
30	10	60		58.2	4.88	

[a] Data from (1) Shelby and Day (1969) and (2) Deeg (1958a,b).

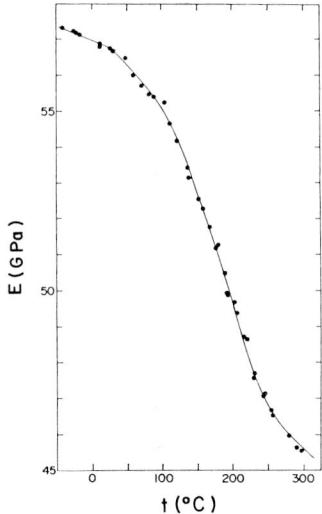

Fig. 10.2. Temperature dependence of Young's modulus measured at 240 Hz for $0.5Na_2O-0.5K_2O-3SiO_2$ glass. From Higgins *et al.* (1973).

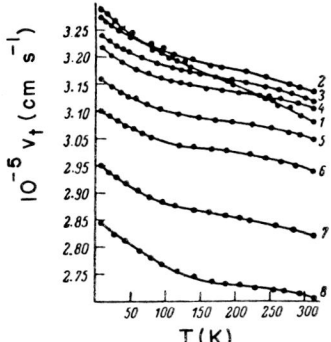

Fig. 10.3. Temperature dependence of velocity of transverse ultrasonic waves in $xNa_2O-(30-x)$ $K_2O-70SiO_2$ (mole %) glasses. Values of x: 1, 30; 2, 27; 3, 23; 4, 19; 5, 15; 6, 11; 7, 3; 8, 0. From Balashov *et al.* (1979).

TABLE 10.16

Shear Moduli of Sodium – Rubidium –
Silicate Glasses at 25°C[a]

Composition (mole %)			G
Na_2O	Rb_2O	SiO_2	(GPa)
24.5	0.5	75	25.6
12.5	12.5	75	24.5
—	25.0	75	17.2

[a] Data from Shelby and Day (1969).

TABLE 10.17

Elastic Constants of $xNa_2O-(1-x)Cs_2O-5SiO_2$ Glasses[a]

x	G (GPa)	K (GPa)	$10^{-5}v_l$ (cm sec^{-1})	$10^{-5}v_t$ (cm sec^{-1})
0	16	28.9	4.10	2.25
0.1	17.9	30	4.18	2.41
0.2	18.8	31.9	4.36	2.60
0.4	22	32.6	4.66	2.77
0.6	23.4	34.1	4.93	2.95
0.75	23.1	33.8	5.04	3.02
1.0	23.8	35.2	5.31	3.16

[a] Data from graphs in Matusita et al. (1974).

TABLE 10.18

Shear Moduli of Sodium–Cesium Silicate Glasses at 25°C[a]

Na_2O	Cs_2O	SiO_2	G (GPa)
24.5	0.5	75	23.1
23.75	1.25	75	25.8
22.5	2.5	75	23.5
21.25	3.75	75	24.7
16.25	8.75	75	23.3
12.5	12.5	75	22.4

Composition (mole %)

[a] Data from Shelby and Day (1969).

TABLE 10.19

Shear Moduli of Potassium–Rubidium Silicate Glasses at 25°C[a]

K_2O	Rb_2O	SiO_2	G (GPa)
24.5	0.5	75	19.1
23.75	1.25	75	19.6
21.25	3.75	75	20.4
12.5	12.5	75	20.6

Composition (mole %)

[a] Data from Shelby and Day (1969).

TABLE 10.20

Young's Modulus of $Li_2O-MO-SiO_2$ Glasses[a]

Li_2O	MO	SiO_2	E (GPa)	Li_2O	MO	SiO_2	E (GPa)
	M = Be				M = Ba		
16.7	16.7	66.6	35.0	15.0	15.6	69.4[d]	77.0
	M = Mg				M = Zn		
17.9	16.3	65.8[b]	86.0	17.3	17.3	65.4[e]	87.0
	M = Ca				M = Cd		
17.0	16.2	66.8[c]	84.0	16.9	17.0	66.1[b]	81.0
	M = Sr				M = Pb		
19.1	18.3	62.6[b]	83.0	17.4	17.2	65.4	65.0

Composition mole %) Composition (mole %)

[a] Data from Totes et al. (1965).
[b] Includes 0.2% $(Al_2O_3 + Fe_2O_3)$.
[c] Includes 0.3% $(Al_2O_3 + Fe_2O_3)$.
[d] Includes 0.7% $(Al_2O_3 + Fe_2O_3)$.
[e] Includes 0.1% $(Al_2O_3 + Fe_2O_3)$.

TABLE 10.21

Elastic Parameters of Some Na$_2$O–MO–SiO$_2$ Glasses[a]

Composition (mole %)			E (GPa)	G (GPa)	σ	$10^{-5}v_l$ (cm sec^{-1})	$10^{-5}v_t$ (cm sec^{-1})
Na$_2$O	MO	SiO$_2$					
	M = Be						
12	18	70	78.9	33.2	0.224	5.87	3.68
16.4	16.7	66.9	79.0				
	M = Mg						
12	18	70	68.1	28.4	0.202	5.58	3.41
19.0	15.9	65.1[b]	72.0				
	M = Sr						
12	18	70	76.5	31.0	0.232	5.58	3.30
16.4	16.2	67.4[c]	74.0				
	M = Ba						
12	18	70	75.1	30.6	0.252	5.40	3.11
	M = Zn						
12	18	70	67.0	27.6	0.222	5.25	3.14
17.0	16.7	66.3[c]	64.0				
	M = Cd						
12	18	70	68.1	27.4	0.247	5.16	2.99
17.0	14.8	68.2	68.0[b]				
	M = Pb						
12	18	70	54.5	22.6	0.217	4.15	2.50
16.7	15.9	67.4	65.0				

[a] Data from Appen *et al.* (1961) and Totes *et al.* (1965).
[b] Includes 0.3% (Al$_2$O$_3$ + Fe$_2$O$_3$).
[c] Includes 0.1% (Al$_2$O$_3$ + Fe$_2$O$_3$).

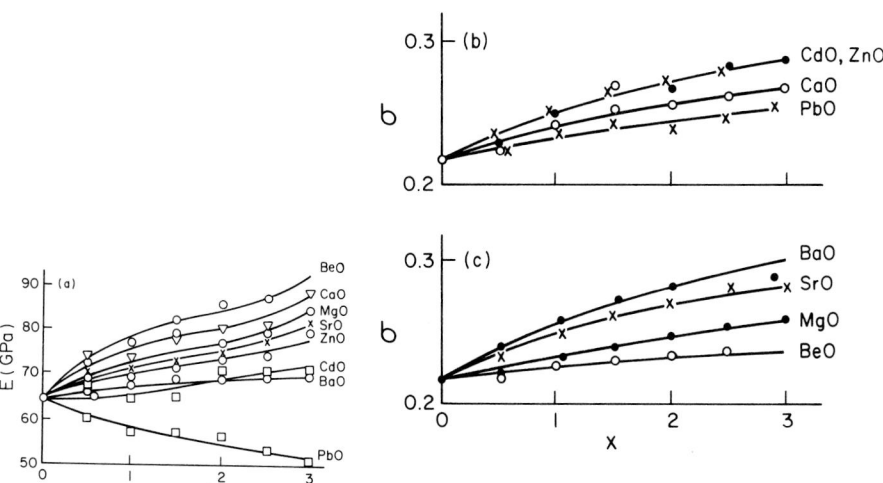

Fig. 10.4. Composition dependence of Young's modulus (E) and Poisson's ratio (σ) of Na$_2$O–xMO–5SiO$_2$ glasses. (M = Be, Mg, Ca, Sr, Ba, Zn, Cd, Pb). From Aleinkov *et al.* (1962a,b, 1963).

TABLE 10.22

Elastic Parameters of Soda Lime Silicate Glasses[a]

Composition (mole %)			$E \times 10^{-10}$ (Pa)	$G \times 10^{-10}$ (Pa)	$K \times 10^{-10}$ (Pa)	σ
Na$_2$O	CaO	SiO$_2$				
4.5	25.0	70.5	7.91	3.23	4.76	0.223
7.5	22.0	70.5	7.78	3.18	4.70	0.224
9.5	15.8	74.7	7.53	3.11	4.31	0.209
10.0	8.0	82.0	7.05	2.96	3.83	0.193
10.0	10.0	80.0	7.17	2.98	4.01	0.202
10.0	11.5	78.5	7.22	3.01	3.98	0.198
10.0	13.5	76.5	7.39	3.06	4.20	0.201
10.0	15.0	75.0	7.43	3.08	4.23	0.207
10.0	17.5	72.5	7.51	3.10	4.35	0.212
10.0	19.5	70.5	7.62	3.12	4.55	0.221
10.0	25.0	65.0	7.97	3.24	4.92	0.230
10.0	27.5	62.5	8.06	3.24	5.26	0.245
11.0	18.5	70.5	7.52	3.10	4.41	0.216
11.2	18.6	70.2	7.58	3.11	4.48	0.218
13.5	8.0	78.5	6.91	2.89	3.81	0.198
16.0	8.0	76.0	6.80	2.83	3.82	0.203
17.5	12.0	70.5	7.03	2.87	4.26	0.225
19.0	8.0	73.0	6.71	2.76	3.92	0.215
21.5	8.0	70.5	6.64	2.72	3.97	0.222

[a] Data from Tille *et al.* (1978).

TABLE 10.23

Acoustic Wave Velocities in Soda-Lime Silicate Glasses

Composition (mole %)			Temperature (°C)	$v_l \times 10^{-5}$ (cm sec^{-1})	$v_t \times 10^{-5}$ (cm sec^{-1})	Reference[a]
Na$_2$O	CaO	SiO$_2$				
12	18	70	RT[b]	5.95	3.47	1
10	10	80	RT	5.41		2
14.2	26.6	59.2	RT	5.66		
14.4	16.0	69.6	RT	5.48		
19.3	16.0	64.7	RT	5.47		
19.4	5.3	75.3	RT	5.25		
14.29	10.71	75.0	20	5.540		3
			45	5.535		
			75	5.525		
			100	5.520		
			150	5.510		
			175	5.500		
			200	5.495		
			230	5.493		
			260	5.480		
			320	5.472		
			350	5.460		
			390	5.445		
			425	5.422		
			450	5.410		
			470	5.400		
			485	5.393		
			505	5.381		
			515	5.331		
			520	5.295		
			525	5.252		
			535	5.204		
			540	5.165		
			550	5.100		

[a] Data from (1) Appen *et al.* (1961), (2) Deeg (1958b), and (3) Butta and Paoletti (1961).

[b] Room temperature.

TABLE 10.24

Acoustic Wave Velocities in Soda–Lead–Silicate Glasses

Composition (mole %)			Temperature (°C)	$v_l \times 10^{-5}$ (cm sec^{-1})	$v_t \times 10^{-5}$ (cm sec^{-1})	Reference[a]
Na$_2$O	PbO	SiO$_2$				
12	18	70	RT[b]	4.15	2.50	1
10	10	80	RT	4.65	—	2
20.7	15.3	64	RT	4.23	—	2
23.1	21.3	55.6	RT	3.99	—	2
14.29	10.71	75	20	4.360	—	3
			45	4.355		
			70	4.350		
			110	4.345		
			145	4.340		
			200	4.335		
			220	4.330		
			250	4.325		
			300	4.300		
			330	4.290		
			375	4.275		
			415	4.260		
			425	4.225		
			450	4.170		
			475	4.105		
			490	4.070		
			515	4.020		

[a] Data from (1) Appen *et al.* (1961), (2) Deeg (1958b), and (3) Butta and Paoletti (1961).
[b] Room temperature.

TABLE 10.25

Elastic Parameters of Some K_2O–MO–SiO_2 Glasses[a]

Composition (mole %)			E (GPa)	G (GPa)	σ	$10^{-5}v_l$ (cm sec^{-1})	$10^{-5}v_t$ (cm sec^{-1})
K_2O	MO	SiO_2					
	M = Be						
12	18	70	71.2	28.9	0.232	5.63	3.34
15.9	16.8	67.3[b]	68.0				
	M = Mg						
12	18	70	63.5	25.9	0.222	5.44	3.24
	M = Ca						
12	18	70	72.1	29.9	0.235	5.71	3.48
16.6	17.3	66.1	45.0				
	M = Ba						
12	18	70	61.5	24.7	0.239	4.82	2.82
16.1	15.1	68.8[c]	54.0				
	M = Zn						
12	18	70	64.6	26.2	0.235	5.23	3.08
17.0	17.3	65.7[d]	61.0				
	M = Cd						
12	18	70	61.2	24.5	0.250	4.92	2.84
17.6	16.5	65.9[d]	57.0				

[a] Data from Appen *et al.* (1961) and Totes *et al.* (1965).
[b] Includes 0.3% $(Al_2O_3 + Fe_2O_3)$.
[c] Includes 0.7% $(Al_2O_3 + Fe_2O_3)$.
[d] Includes 0.2% $(Al_2O_3 + Fe_2O_3)$.

TABLE 10.26

Elastic Parameters of K_2O–SrO–SiO_2 Glasses[a]

Composition (mole %)			E	G		$10^{-5}v_l$	$10^{-5}v_t$
K_2O	SrO	SiO_2	(GPa)	(GPa)	σ	(cm sec^{-1})	(cm sec^{-1})
5.4	10.1	84.5	66.0	27.4	0.204	5.34	3.26
7.6	14.2	78.2	67.0	27.3	0.225	5.32	3.17
8.7	6.1	85.2	63.71	26.4	0.209	5.32	3.22
9.9	18.7	71.4	67.4	27.0	0.249	5.29	3.06
10.8	7.4	81.8	62.3	25.5	0.221	5.25	3.15
10.9	27	62.1	69.04	27.2	0.270	5.28	2.96
12.6	15.3	72.1	64.0	25.6	0.247	5.27	3.05
13.4	30	56.6	70.0	27.4	0.279	5.28	2.92
13.5	9.5	77.0	61.6	25.1	0.228	5.22	3.10
14	26.4	59.6	68.6	26.8	0.277	5.28	2.94
14.2	4.5	81.3	58.7	23.9	0.228	5.20	3.08
14.7	4.6	80.7	58.5	23.7	0.235	5.19	3.06
15.5	18.7	65.8	63.8	25.3	0.262	5.22	2.96
16.8	11.8	71.4	60.0	23.9	0.254	5.18	2.97
16.8	30	53.2	68.3	26.6	0.281	5.24	2.89
17.6	26.6	55.8	66.8	26.2	0.277	5.22	2.90
18	5.6	76.4	56.1	22.4	0.254	5.13	2.95
18.6	22.4	59.0	62.1	24.2	0.281	5.14	2.84
19.6	18.3	62.1	60.4	23.6	0.277	5.16	2.86
20.5	14.4	65.1	60.1	23.7	0.265	5.15	2.91
21.3	26.4	52.3	65.9	25.6	0.285	5.23	2.86
21.4	10.6	68.0	56.6	22.4	0.262	5.08	2.88
22.3	7	70.7	54.3	21.5	0.265	5.06	2.86
24	17.2	58.8	60.3	23.5	0.281	5.16	2.86
26.5	8.5	65.0	54.4	21.3	0.277	5.07	2.82

[a] Data from Nemilov and Kasymova (1975) and Kasymova and Plutalova (1975).

TABLE 10.27

Elastic Parameters of Potassium–Lead–Silicate Glasses[a]

Composition (mole %)			E	G	K		$10^{-5}v_l$
K_2O	PbO	SiO_2	(GPa)	(GPa)	(GPa)	σ	(cm sec^{-1})
2.9	35.4	61.7[b]	54.11	22.0	33.4	0.230	3.369
3.8	31.6	64.6[b]	54.26	22.1	33.2	0.228	3.478
5.3	27.9	66.8[b]	54.09	22.3	32.6	0.213	3.582
6.6	25.2	68.2[b]	53.54	22.3	29.8	0.200	3.653
16.6	17.3	66.1	44.1				

[a] Data from Spinner (1954), Makishima and Mackenzie (1975), and Totes et al. (1965).
[b] Also contains 0.3% As_2O_3.

TABLE 10.28

Elastic Constants of Lithium Aluminosilicate Glasses[a]

Composition (mole %)			E	G	K		$10^{-5}v_l$	$10^{-5}v_t$
Li$_2$O	Al$_2$O$_3$	SiO$_2$	(GPa)	(GPa)	(GPa)	σ	(cm sec^{-1})	(cm sec^{-1})
25	30	45	90.5	36.6	57.8	0.239	6.596	3.865
25	25	50	82.5	33.9	48.6	0.217	6.298	3.786
29	14	57	87.6	35.6	54.7	0.233	6.511	3.843
20	20	60	85.7	34.9	52.3	0.227	6.405	3.805
19	20	61	86.3	35.1	52.7	0.227	6.419	3.815
30	9	61	86.2	35.0	53.6	0.232	6.472	3.824
17	21	62	86.7	35.3	53.1	0.228	6.442	3.823
15	22	63	87.3	35.5	53.7	0.229	6.463	3.831
17	20	63	85.8	35.1	52.0	0.225	6.402	3.815
32	5	63	84.2	34.2	51.8	0.229	6.401	3.793
13	22	65	88.8	36.1	54.8	0.230	6.508	3.854
15	20	65	87.5	35.6	53.9	0.230	6.469	3.833
17	18	65	84.9	34.6	51.0	0.223	6.361	3.800
17.5	17.5	65	84.5	34.6	50.8	0.223	6.354	3.795
19	16	65	84.3	34.6	50.7	0.223	6.368	3.802
21	14	65	85.7	35.0	52.0	0.225	6.409	3.817
23	12	65	85.2	34.8	51.5	0.224	6.395	3.813
25	10	65	85.3	34.7	52.1	0.227	6.422	3.816
27	8	65	84.2	34.4	50.8	0.224	6.368	3.798
29	6	65	83.8	34.1	51.4	0.228	6.393	3.794
31	4	65	83.0	33.8	50.6	0.227	6.357	3.780
33	2	65	82.4	33.5	50.6	0.229	6.355	3.769
13	21	66	81.2	33.2	48.5	0.221	6.284	3.759
25	9	66	84.5	34.5	51.0	0.224	6.381	3.805
13	20	67	88.6	36.1	54.9	0.231	6.508	3.851
15	18	67	84.6	34.7	50.5	0.221	6.354	3.803
11	20	69	87.0	35.5	52.7	0.225	6.442	3.837
15	16	69	82.8	34.0	48.7	0.217	6.291	3.784
15	15	70	82.8	34.0	48.5	0.216	6.293	3.789
17	13	70	83.0	34.1	48.2	0.213	6.289	3.796
23	7	70	82.4	33.8	48.5	0.217	6.293	3.784
25	5	70	81.8	33.6	48.2	0.217	6.284	3.777
15	12	73	82.6	34.1	47.8	0.212	6.277	3.795
15	10	75	81.8	33.8	46.8	0.209	6.250	3.791
15	8	77	80.1	33.4	44.4	0.200	6.173	3.782
15	6	79	78.8	33.0	43.0	0.195	6.127	3.773
15	4	81	77.6	32.6	41.6	0.189	6.079	3.763

[a] Data from Livshits *et. al.* (1982b) and Karapetyan *et. al.* (1979, 1980).

TABLE 10.29

Elastic Constants of Sodium Borosilicate Glasses[a]

xNa$_2$O$-$B$_2$O$_3$$-ySiO_2$		E	G	K	$10^{-5}v_l$	$10^{-5}v_t$
x	y	(GPa)	(GPa)	(GPa)	(cm sec^{-1})	(cm sec^{-1})
0.07	1	34.28	13.59	23.92	4.534	2.578
0.2	1	45.33	17.94	31.63	5.072	2.884
0.3	1	56.14	22.85	34.48	5.358	3.178
0.35	1	58.95	23.69	38.37	5.502	3.202
0.4	1	62.93	25.28	41.07	5.641	3.280
0.45	1	67.22	27.46	40.63	5.686	3.390
0.5	1	71.07	28.87	44.19	5.833	3.446
0.55	1	72.27	28.98	47.59	5.934	3.440
0.6	1	75.24	30.19	49.42	6.013	3.489
0.7	1	76.11	30.61	49.36	6.007	3.500
0.9	1	71.57	28.50	48.86	5.873	3.364
1.0	1	70.49	28.04	48.37	5.815	3.325
1.2	1	63.89	25.05	47.38	5.616	3.127
1.3	1	63.10	24.67	47.54	5.605	3.104
0.1	2	41.90	17.25	24.50	4.758	2.867
0.2	2	47.57	19.22	30.05	5.092	2.991
0.3	2	54.08	21.47	37.48	5.346	3.098
0.4	2	62.54	25.23	40.04	5.650	3.306
0.45	2	66.61	26.70	41.69	5.730	3.377
0.5	2	71.10	29.15	42.26	5.819	3.488
0.55	2	75.31	30.89	44.64	5.943	3.566
0.6	2	73.70	29.78	46.73	5.940	3.486
0.65	2	76.59	31.25	46.49	5.973	3.556
0.7	2	76.50	31.11	47.13	5.969	3.536
0.9	2	75.70	30.66	47.55	5.927	3.489
1.0	2	76.16	30.86	47.70	5.926	3.492
1.1	2	73.25	29.57	46.57	5.850	3.426
1.2	2	72.55	29.14	47.43	5.850	3.400
1.3	2	70.49	28.18	47.17	5.791	3.350

[a] Data from Takahashi et al. (1983).

TABLE 10.30

Young's Moduli of Sodium Borosilicate Glasses[a]

Composition (mole %)			E	Composition (mole %)			E
Na_2O	B_2O_3	SiO_2	(GPa)	Na_2O	B_2O_3	SiO_2	(GPa)
10	10	80	81.5	20	60	20	61.1
10	15	75	78.6	20	65	15	56.2
10	20	70	76.4	20	70	10	50.9
10	25	65	73.9	20	75	5	51.3
10	30	60	63.1	25	50	25	71.8
10	35	55	60.3	25	55	20	73.2
10	40	50	54.1	25	60	15	68.8
10	45	45	50.3	25	65	10	65.1
10	50	40	47.1	25	70	5	67.4
10	55	35	47.4	30	5	65	73.5
10	60	30	46.3	30	10	60	75.1
10	65	25	44.9	30	15	55	75.8
10	70	20	38.6	30	20	50	77.2
10	75	15	37.8	30	25	45	82.4
10	80	10	36.5	30	30	40	84.5
10	85	5	32.8	30	35	35	78.9
20	5	75	76.2	30	40	30	73.5
20	10	70	78.0	30	45	25	79.9
20	15	65	82.8	30	50	20	75.1
20	20	60	83.8	30	55	15	72.4
20	25	55	83.5	30	60	10	70.9
20	30	50	84.2	30	65	5	70.5
20	35	45	76.1				
20	40	40	75.5				
20	45	35	73.3				
20	50	30	70.0				
20	55	25	67.2				

[a] Data from Imaoka *et al.* (1971).

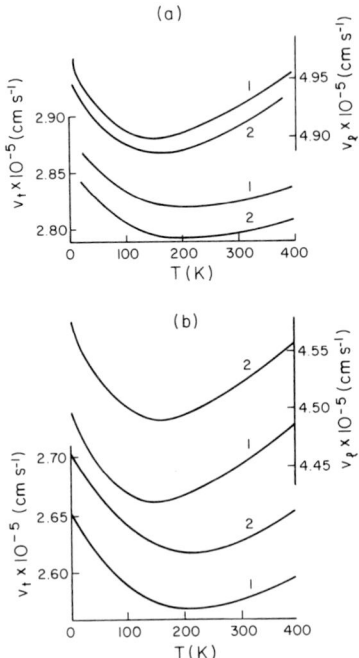

Fig. 10.5. The temperature dependence of the velocity of 12-MHz longitudinal and transverse ultrasound waves in (a) $10.0Na_2O-50.5B_2O_3-39.7SiO_2$ (wt. %) and (b) $4.4-Na_2O-48.9B_2O_3-46.6SiO_2$ (wt. %) glasses. Time of heat treatment at 550°C: 1, 0; 2, 48 hr. From Maynell *et al.* (1973).

TABLE 10.31

Elastic Constants of Sodium Aluminosilicate Glasses[a]

Composition (mole %)			E (GPa)	G (GPa)	K (GPa)	σ	$10^{-5}v_l$ (cm sec^{-1})	$10^{-5}v_t$ (cm sec^{-1})
Na_2O	Al_2O_3	SiO_2						
15	5	80	65.3	27.1	37.0	0.206	5.570	3.390
15	10	75	70.9	29.7	38.6	0.194	5.697	3.510
15	15	70	71.9	29.7	41.6	0.212	5.737	3.469
15	20	65	73.7	30.4	42.5	0.211	5.850	3.540
15	25	60	77.3	31.5	47.0	0.226	6.000	3.568
25	5	70	63.5	25.8	39.4	0.231	5.480	3.240
25	10	65	67.0	27.3	41.1	0.228	5.610	3.330
25	15	60	68.4	27.8	41.8	0.227	5.640	3.350
25	20	55	70.9	29.4	40.5	0.208	5.680	3.450
25	25	50	73.9	30.4	43.2	0.215	5.790	3.490
25	30	45	78.5	31.9	49.0	0.233	6.026	3.556
17.5	17.5	65	71.2	29.3	41.8	0.216	5.746	3.458
20	15	65	71.3	29.8	39.0	0.195	5.670	3.490
30	5	65	63.4	25.5	41.3	0.244	5.500	3.200
32	5	63	63.9	25.8	41.3	0.242	5.516	3.220

[a] Data from Livshits *et al.* (1982a,c).

TABLE 10.32

Elastic Constants of Sodium Aluminosilicate Glasses

Al/Na ratio	E (GPa)	G (GPa)
$20Na_2O-xAl_2O_3-(80-x)SiO_2$		
0	60.6	24.4
0.25	65	26.7
0.5	69.8	28.1
1.0	73.2	30.6
1.2	77.5	—
1.3	—	32
$(40-x)Na_2O-xAl_2O_3-60SiO_2$		
0	62.6	24.8
0.25	69.2	27.6
0.67	73	29.5
1.0	73.9	30.2
1.3	78	31.4

[a] Data from graph in Eagan and Swearengen (1978).

TABLE 10.33

Elastic Properties of Potassium Borosilicate Glasses[a]

$xK_2O-B_2O_3-ySiO_2$		E (GPa)	G (GPa)	K (GPa)	$v_1 \times 10^{-5}$ (cm sec^{-1})	$v_t \times 10^{-5}$ (cm sec^{-1})
x	y					
0.1	1	33.91	13.46	22.53	4.508	2.568
0.3	1	44.96	17.56	34.16	5.074	2.802
0.4	1	51.89	20.68	35.28	5.225	2.997
0.5	1	55.92	22.51	36.16	5.312	3.098
0.6	1	58.66	23.42	39.47	5.440	3.131
0.7	1	60.07	24.06	39.76	5.436	3.146
0.8	1	57.37	23.02	37.67	5.310	3.082
1.0	1	52.40	21.12	33.66	5.040	2.946
1.1	1	49.08	19.90	30.67	4.876	2.876
0.3	2	48.67	19.57	31.81	5.138	2.986
0.4	2	56.09	23.07	32.86	5.271	3.174
0.5	2	58.99	24.09	35.64	5.400	3.220
0.55	2	62.15	25.17	39.07	5.545	3.264
0.6	2	64.55	26.31	39.37	5.551	3.300
0.7	2	65.44	26.32	42.45	5.632	3.281
0.8	2	65.21	26.10	43.37	5.637	3.257
0.9	2	63.84	25.53	42.60	5.567	3.213
1.0	2	60.98	24.53	39.57	5.395	3.143
1.1	2	61.06	24.36	41.26	5.456	3.136
1.3	2	51.80	20.46	36.86	5.101	2.881

[a] Data from Takahashi et al. (1983).

TABLE 10.34

Velocities of Acoustic Waves in Potassium Borosilicate Glasses[a]

Composition (mole %)			$10^{-5}v_l$	$10^{-5}v_t$
K_2O	B_2O_3	SiO_2	(cm sec^{-1})	(cm sec^{-1})
16	4	80	5.37	3.20
16	8	76	5.60	3.41
16	12	72	5.82	3.48
16	16	68	5.90	3.57
16	20	64	5.75	3.43
16	32	52	5.67	3.22

[a] Data from Appen *et al.* (1961).

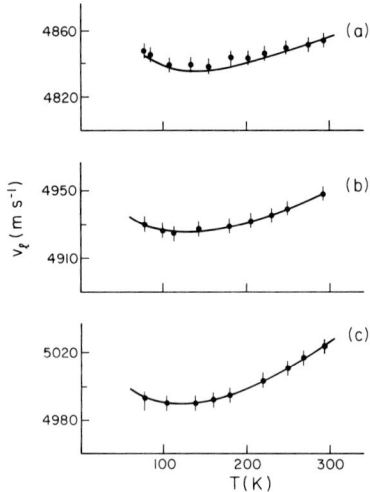

Fig. 10.6. Temperature dependence of longitudinal sound velocity in potassium borosilicate glasses of various compositions (mole %):
(a) $10K_2O-50B_2O_3-40SiO_2$;
(b) $8.33K_2O-41.66B_2O_3-50SiO_2$;
(c) $6.66K_2O-33.33B_2O_3-60SiO_2$.
Frequency 30 GHz. From Pelous and Vacher (1977).

TABLE 10.35

Compressibility of Potassium Aluminosilicate Glass[a]

Composition (mole %)			β
K_2O	Al_2O_3	SiO_2	(GPa^{-1})
10	30	60	0.076[b]

[a] Data from Laberge *et al.* (1975).
[b] At T_g (740 K).

TABLE 10.36

Young's Moduli of $Li_2O-ZrO_2-SiO_2$ Glasses[a]

Composition (mole %)			E
Li_2O	ZrO_2	SiO_2	(GPa)
20	—	80	72
20	2.5	77.5	76
20	5	75	80
20	7.5	72.5	84
20	10	70	86.5
20	12.5	67.5	88.5

[a] Data from graph in Ellern and Pavlushkin (1969).

TABLE 10.37

Elastic Properties of Na_2O–TiO_2–SiO_2 Glasses and Their Pressure and Temperature Derivatives[a]

Composition (mole %)			E^b (GPa)	G^b (GPa)	K^b (GPa)	σ^b	dE/dP^c	dG/dP^c	dK/dP^c	$d\sigma/dP^c$ (GPa^{-1})	dG/dT^d (GPa °C^{-1})	dK/dT^d (GPa °C^{-1})	$d\sigma/dT \times 10^{5\ d}$ (°C^{-1})
Na_2O	TiO_2	SiO_2											
30.0	20.0	50.0	75.44	30.21	50.00	0.249	2.29	0.53	4.85	0.0157	−0.0071	−0.0081	1.7
25.0	20.0	55.0	77.95	31.41	50.11	0.241	1.76	0.37	3.86	0.0134	−0.0051	−0.0070	0.7
17.5	20.0	62.5	81.80	33.42	49.39	0.224	0.47	−0.15	2.59	0.0124	−0.0036	−0.0025	1.3
27.5	25.0	47.5	78.98	31.59	52.64	0.250	2.03	0.51	4.00	0.0119	−0.0063	−0.0080	1.0
22.5	25.0	52.5	83.07	33.45	52.95	0.239	1.56	0.29	3.64	0.0125	−0.0048	−0.0057	0.9
17.5	25.0	57.5	84.75	34.45	52.34	0.230	0.73	−0.06	3.05	0.0129	−0.0038	−0.0035	1.0

[a] Data from Manghnani (1972).

[b] At 25°C and 0.1 MPa.

[c] At 25°C and 0–0.7 GPa pressure range.

[d] In the temperature interval 25–300°C.

TABLE 10.38

Longitudinal and Transverse Sound Velocities and Their Pressure Derivatives for Na$_2$O–TiO$_2$–SiO$_2$ Glasses[a]

Composition (mole %)			$10^{-5}v_l{}^b$ (cm sec^{-1})	$10^{-5}v_t{}^b$ (cm sec^{-1})	$(dv_1/dP) \times 10^{-5c}$ (cm sec^{-1} GPa^{-1})	$(dv_t/dP) \times 10^{-5c}$ (cm sec^{-1} GPa^{-1})
Na$_2$O	TiO$_2$	SiO$_2$				
30.0	20.0	50.0	5.731	3.315	0.120	−0.002
25.0	20.0	55.0	5.794	3.386	0.0803	−0.0124
17.5	20.0	62.5	5.895	3.516	0.0169	−0.0419
27.5	25.0	47.5	5.806	3.353	0.0969	−0.0035
22.5	25.0	52.5	5.906	3.461	0.0670	−0.0164
17.5	25.0	57.5	5.955	3.526	0.0343	−0.0355

[a] Data from Manghnani (1972).
[b] At 25°C and 0.1 MPa.
[c] At 25°C and 0–0.7 GPa pressure range.

TABLE 10.39

Elastic Parameters of Some MO–M'O–SiO$_2$ Glasses[a]

Composition (mole %)			E (GPa)	G (GPa)	K (GPa)	σ	$10^{-5}v_l$ (cm sec^{-1})	$10^{-5}v_t$ (cm sec^{-1})	$\left(\dfrac{dK}{dP}\right)_T{}^b$
MO	M'O	SiO$_2$							
MgO	CaO	SiO$_2$							
12.5	37.5	50	94.95	37.05	72.30	0.281	6.507	3.590	
25	25	50	99.14	38.81	74.11	0.277	6.649	3.692	
MgO	SrO	SiO$_2$							
12.5	37.5	50	87.22	34.39	62.70	0.268	5.546	3.122	
25	25	50	90.74	35.49	68.25	0.278	5.916	3.278	
MgO	BaO	SiO$_2$							
12.5	37.5	50	71.37	27.88	54.04	0.280	4.756	2.629	
25	25	50	79.52	30.97	61.24	0.284	5.295	2.910	
MgO	FeO	SiO$_2$							
25	25	50			76.2				4.6
CaO	BaO	SiO$_2$							
16.67	16.67	66.66	71.36	28.21	50.59	0.265	5.226	2.955	
25	25	50	79.52	31.36	57.06	0.268	5.204	2.931	
CaO	FeO	SiO$_2$							
25	25	50			77.8				4.5
SrO	BaO	SiO$_2$							
16.67	16.67	66.66	67.15	26.42	48.81	0.271	4.898	2.746	

[a] Data from Soga et al. (1972, 1976).
[b] Pressure range 0.1–100 MPa.

TABLE 10.40

**Young's Moduli of
MgO–Al$_2$O$_3$–SiO$_2$ Glasses**[a]

Composition (mole %)			E
MgO	Al$_2$O$_3$	SiO$_2$	(GPa)
35	10	55	104.4
32.5	12.5	55	106.4
30	15	55	106.8
27.5	17.5	55	103.9
25	20	55	104.1
40	10	50	107.4
37.5	12.5	50	108.9
35	15	50	109.4
32.5	17.5	50	109.2
30	20	50	108.1
40	12.5	47.5	109.9
37.5	15	47.5	111
35	17.5	47.5	109.7

[a] Data from graphs in Matveev et al. (1964, 1965a,b).

TABLE 10.41

Elastic Parameters of Calcium Aluminosilicate Glasses[a]

Composition (mole %)			E	G	K	
CaO	Al$_2$O$_3$	SiO$_2$	(GPa)	(GPa)	(GPa)	σ
32.1	2.9	65.0	87.2	34.5	61.2	0.262
32.8	6.0	61.2	90.1	35.2	67.7	0.278
34.2	12.5	53.2	93.3	36.2	73.1	0.287
35.8	19.7	44.5	98.8	38.2	78.8	0.291
37.5	27.5	35.0	103.5	40.0	83.6	0.293
39.4	36.1	24.5	108.6	41.9	88.5	0.295

[a] Data from Yamane and Okuyama (1982).

TABLE 10.42

Elastic Constants of
$(40 - x)CaO-xAl_2O_3-60SiO_2$ Glasses[a]

Al$_2$O$_3$/CaO ratio	E (GPa)	G (GPa)
0.25	93.1	35.8
0.5	91.8	35.2
1.0	93.7	36
1.3	98.2	37.8

[a] Data from graph in Eagan and Swearengen (1978).

TABLE 10.43

Young's Moduli of
$SrO-Al_2O_3-SiO_2$ Glasses[a]

Composition (mole %)			E
SrO	Al$_2$O$_3$	SiO$_2$	(GPa)
30	15	55	78.1
32.5	12.5	55	83.2
35	10	55	82.2
37.5	7.5	55	81.6
40	5	55	80.3
35	15	50	83.6
37.5	12.5	50	84
40	10	50	82.8
42.5	7.5	50	82.5
45	5	50	81

[a] Data from graphs in Matveev et al. (1964, 1965a,b).

TABLE 10.44

Elastic Parameters of Zinc Borosilicate Glasses[a]

Composition (mole %)			E	G	
ZnO	B$_2$O$_3$	SiO$_2$	(GPa)	(GPa)	σ
55	35	10	83.66	32.27	0.296
59	23	18	85.65	32.69	0.310

[a] Data from Hamilton et al. (1959).

TABLE 10.45

Elastic Parameters of Some $PbO-M_2O_3-SiO_2$ Glasses[a]

Composition (wt. %)			E	G		$10^{-5}v_l$	$10^{-5}v_t$	
PbO	M$_2$O$_3$	SiO$_2$	(GPa)	(GPa)	σ	(cm sec^{-1})	(cm sec^{-1})	Reference[a]
	B$_2$O$_3$							
70	10	20	55.53	21.59	0.2856	3.765	2.063	1
70	25	5	75.24	31.22	0.2055	4.113	2.515	
80	10	10	79.88	34.34	0.1626	3.769	2.392	
80	15	5	77.62	34.05	0.1413	3.707	2.396	
85	10	5	77.33	29.54	0.3093	2.982	2.140	
	Bi$_2$O$_3$							
75.3	19.6	5.1	39.5		0.253			2

[a] Data from (1) Thiele (1955) and (2) Brekhovskikh (1959).

Chapter 11

Microhardness

The hardness of a material is intuitively easy to grasp, but its exact measurement is difficult. The ability of a harder material to scratch a softer one is the basis for the Moh's scale of hardness from 1 (soft) to 10 (hard). This scale is nonlinear, especially at the high end. Silicate glasses fall between apatite (5) and quartz (7).

Many hardness measurements have been made by indentation methods, for example, by a hard square pyramid with a sharp point (Vickers hardness). The hardness is measured by measuring the length of the diagonals of the indentation in the glass surface for a fixed load, and are reported as load per unit area. The smaller the area of the indentation for a fixed load, the harder is the material. Indentation of a glass surface by a Vickers point forms cracks in the glass surface and a permanent deformation. The mechanism of this deformation is discussed in the section on plastic deformation in the chapter on strength. The hardness is therefore a complicated mixture of resistance to fracture and to elastic and plastic deformation. Reproducible measurements are difficult to make and depend on the surface condition of the glass, especially its state of hydration.

The most useful hardness measurements are made by comparing surfaces treated in the same way with the same measuring method and load. Absolute hardness values are harder to compare because of experimental variables.

TABLES II

Tables 11.1–11.15 provide data on microhardness of binary glasses; Tables 11.16–11.58 on ternary glasses.

BIBLIOGRAPHY

Ainsworth, L. (1954a). *J. Soc. Glass Technol.* **38**, 501.
Ainsworth, L. (1954b). *J. Soc. Glass Technol.* **38**, 536.
Aleinkov, F. K., Paulavichus, P. B., and Slizhis, V. A. (1962a). *Liet. TSR Mokslu Akad. Darbo, Ser.* B **2**, 69.
Aleinkov, F. K., Slizhis, V. A., Paulavichus, P. B., and Dundzis, P. V. (1962b). *Opt.-Mekh. Prom.* No. 9, p. 38.
Aleinkov, F. K., Slizhis, V. A., Paulavichus, P. B., and Dundzis, P. V. (1963). *Stekloobraznoe Sostoyaniye* **3**(2), 30.
Aleksandrov, V. I., Borik, M. A., Dechev, G. K., Markov, N. I., Myzina, V. A., Osiko, V. V., Tatarintsev, V. M., and Khoaakovskaya, R. Y. (1980). *Sov. J. Glass Phys. Chem. (Engl. Transl.)* **6**, 117.
Alekseeva, O. S., Bokin, P. Y., Govorova, R. A., Korelova, A. I., and Nikandrova, G. A. (1965). *Stekloobraznoe Sostoyanie* **4**, 382.
Appen, A. A. (1954). *Steklo Keram.* No. 3, p. 7.
Aslanova, M. S., Dorzhiev, D. B., Sapozhkova, L. A., Gorbachev, V. V., Bystrikov, A. S., Petrakov, V. N., and Fertikov, V. I. (1982). *Sov. J. Glass Phys. Chem. (Engl. Transl.)* **8**, 26.
Botvinkin, O. K., and Demichev, S. A. (1964). *Steklo* No. 2, p. 1.
Botvinkin, O. K., and Yaroker, K. G. (1966). *Steklo* No. 3, p. 1.
Dubrovo, S. K., and Lileev, I. S. (1960). *J. Appl. Chem. USSR (Engl. Transl.)* **33**, 1461.
El-Batal, H. A., Ghoneim, N. A., Ammar, M. M., and Halawa, M. M. (1980). *Cent. Glass Ceram. Res. Inst. Bull.* **27**, 72.
El-Batal, H. A., Ghoneim, N. A., Ammar, M. M., and Halawa, M. M. (1982). *Sprechsaal* **115**, 223.
Ellern, G. A., and Pavlushkin, N. M. (1969). *Tr. Mosk. Khim.-Tekhnol. Inst., Silic.* Publ. No. 59, p. 30.
Frischat, G. H., Ozmen, E., Richter, T., and Michels, B.-D. (1982). *Glastech. Ber.* **55**, 119.
Frohlich, F., Grau, P., and Grellmann, W. (1977). *Phys. Status Solidi A* **42**, 79.
Georoff, A. N., and Babcock, C. L. (1973). *J. Am. Ceram. Soc.* **56**, 97.
Gottardi, V., Locardi, B., Bianchini, A., and Martini, P. L. (1968). *Glass. Technol* **9**, 139.
Hanada, T., Goto, S., and Soga, N. (1981). *Nippon Kagaku Kaishi* **13**, 1577.
Imaoka, M., Hasegawa, H., Hamaguchi, Y., and Kurotaki, Y. (1971). *Yogyo Kyokaishi* **79**, 164.
Kasymova, S. S., and Milyukov, E. M. (1974). *Uzb. Khim. Zh.* **3**, 10.
Kennedy, C. R., Bradt, R. C., and Rindone, G. E. (1980). *Phys. Chem. Glasses* **21**, 99.
Komiyama, T. (1974). *Kogyo Gijutsu Shikensho Kiho* **25**, 84.
Kranich, J. F., and Scholze, H. (1976). *Glastech. Ber.* **49**, 135.
Maksimova, O. S., Korzunova, L. V., and Milberg, Z. P. (1975). *In* "Neorganik. Stekla, Pokrytiya i Materialy," Publ. No. 1, p. 103. Riga.
Matveev, M. A., Rzhevyusskaya, T. L., and Milevskaya, R. N. (1970). *Stekloobraznoe Sostoyaniye* **7**, 276.
Mochida, N., Takahashi, K., and Shibusawa, S. (1980). *Yogyo Kyokaishi* **88**, 583.
Mukherjee, S. P., Zarzycki, J., and Traverse, J. P. (1976). *J. Mater. Sci.* **11**, 341.
Myuller, R. L., and Pronkin, A. A. (1963). *J. Appl. Chem. USSR (Engl. Transl.)* **36**, 1144.
Neely, J. E. (1969). Ph.D. Thesis, Rensselaer Polytech. Inst., Troy, New York.
Petrenko, Y. M., Ushakov, D. F., and Gilev, I. S. (1973). *Inorg. Mater. (Engl. Transl.)* **9**, 273.
Petzold, A., Wihsmann, F. G., and VanKamptz, H. (1961). *Glastech. Ber.* **34**, 56.
Takahashi, K., Mochida, N., and Hatta, G. (1975). *Yogyo Kyokaishi* **83**, 103.
Takahashi, K., Mochida, N., Matsui, H., Takeuchi, S., and Gohshi, Y. (1976). *Yogyo Kyokaishi* **84**, 482.
Takahashi, K., Mochida, N., and Yoshida, Y. (1977). *Yogyo Kyokaishi* **85**, 330.
Totesh, A. S., Grigor'eva, L. F., Strel'dina, M. V., and Roskova, G. P. (1961). "Tr. VI Koordinatsion. Sovesh. po Shlifovke i Palirovke Stekla," p. 102. Saratov.
Totesh, A. S., Grigor'eva, L. F., Strel'dina, M. V., and Roskova, G. P. (1965). *Sklar Keram.* **15**, 370.
Vakhrameev, V. I., and Evstrop'ev, K. S. (1969). *Inorg. Mater. (Engl. Transl.)* **5**, 82.

TABLE 11.1

Diamond Pyramid Microhardness of Lithium Silicate Glasses Measured under a Load of 50 g[a]

Composition (mole % Li_2O)	Microhardness H (kg mm^{-2})
23.4	475
34.4	460

[a] Data from Alekseeva et al. (1965).

TABLE 11.2

Vickers Microhardness of Lithium Silicate Glasses Measured under Various Loads[a]

Composition (mole % Li_2O)	Load (g)	Hardness H (kg mm^{-2})
33.33	300	535 ± 15
	500	535 ± 10
	700	525 ± 10
	1000	525 ± 10

[a] Data from Neely (1969).

TABLE 11.3

Knoop Microhardness of Lithium Silicate Glasses Measured under a Load of 100 g[a]

Composition (mole % Li_2O)	Microhardness H (kg mm^{-2})
20	475 ± 10
24.8	445 ± 11
28.5	440 ± 20
33	435 ± 11

[a] Data from graph in Kennedy et al. (1980).

TABLE 11.4

Knoop Microhardness of Sodium Silicate Glasses

Composition (mole % Na_2O)	Microhardness H (kg mm^{-2})	Reference[a]	Composition (mole % Na_2O)	Microhardness H (kg mm^{-2})	Reference[a]
14.3	360 ± 15	1[b]	15	419 ± 10	2[c]
15	347 ± 8		20	413 ± 11	
20	325 ± 9		24.8	403 ± 5	
25	302 ± 13		28.5	396 ± 11	
30	296 ± 9		33	375 ± 9	

[a] Data from (1) Kranich and Scholze (1976) and (2) Kennedy et al. (1980).
[b] Measured under loads of 25, 50, 100, 200, and 500 g.
[c] Data read from graph; measured under a load of 100 g.

TABLE 11.5
Vickers Pyramid Microhardness of Sodium Silicate Glasses

Composition (mole % Na$_2$O)	Microhardness H (kg mm^{-2})	Reference[a]	Composition (mole % Na$_2$O)	Load (g)	Microhardness H (kg mm^{-2})	Reference[a]
25	423 ± 4	1[b]	30	200	438	2
30	413 ± 3			500	450	
35	414 ± 4		40	200	404	
40	394 ± 2			500	423	
45	378 ± 2					

[a] Data from (1) Mochida et al. (1980) and (2) Imaoka et al. (1971).
[b] Measured under loads of 50, 100, 200, and 300 g.

TABLE 11.6

Diamond Pyramid Microhardness of Sodium Silicate Glasses Measured under Loads of 30, 50, 70, and 100 g[a]

Composition (mole % Na_2O)	Microhardness H (kg mm^{-2})	Composition (mole % Na_2O)	Microhardness H (kg mm^{-2})
11.8	505	30.8	354
15.7	442	34.6	364
19.5	405	36.4	376
23.3	375	38.3	384
27.0	362	42.1	388

[a] Data from Ainsworth (1954a).

TABLE 11.7

Vickers Microhardness of Potassium Silicate Glass[a]

Composition (mole % K_2O)	Load (g)	Microhardness H (kg mm^{-2})
33.33	200	330 ± 10
	6000	336 ± 2

[a] Data from Neely (1969).

TABLE 11.8

Knoop Microhardness of Potassium Silicate Glasses Measured under a Load of 100 g[a]

Composition (mole %)	Microhardness H (kg mm^{-2})
15	325 ± 10
20	300 ± 6
24.8	280 ± 10
28.5	261 ± 12
33	250 ± 12

[a] Data from graph in Kennedy et al. (1980).

TABLE 11.9

Diamond Pyramid Microhardness of Potassium Silicate Glasses Measured under Loads of 30, 50, 70, and 100 g[a]

Composition (mole % K_2O)	Microhardness H (kg mm^{-2})	Composition (mole % K_2O)	Microhardness H (kg mm^{-2})
12.1	406	19.6	364
15.8	374	23.0	369
17.7	359	26.4	337

[a] Data from Ainsworth (1954a).

TABLE 11.10

Microhardness of Rubidium Silicate Glass[a]

Composition (mole % Rb_2O)	Microhardness L_2VH[b] (kg mm^{-2})
25	252.9

[a] Data from Frischat *et al.* (1982).

[b] The Vickers hardness index L_2VH was measured using a dynamic technique of Frohlich *et al.* (1977) with continuous recording of load and depth of penetration during increasing and decreasing of load. The value of L_2VH was independent of the load and of glass cracking. For glasses and ceramics, L_2VH values were 2–3 times lower than the conventional Vickers hardness.

TABLE 11.11

Vickers Microhardness of Calcium Silicate Glasses[a]

Composition (mole % CaO)	Microhardness H (kg mm^{-2})
50.0	620 ± 10
55.0	620 ± 15

[a] Data from Neely (1969).

TABLE 11.12

Vickers Microhardness of Lead Silicate Glasses[a]

Composition (mole % PbO)	Microhardness H (kg mm^{-2})
60	452
65	437
70	421
75	406
80	390
85	375
90	361

[a] Data from El-Batal *et al.* (1980, 1982).

TABLE 11.13

Vickers Microhardness of B_2O_3–SiO_2 Glasses[a]

Composition (mole % B_2O_3)	Microhardness H (kg mm^{-2}) under given loads (g)		Composition (mole % B_2O_3)	Microhardness H (kg mm^{-2}) under given loads (g)	
	200	500		200	500
60	345	328	80	271	239
65	297	293	85	267	239
70	279	251	90	257	231
75	269	237	95	253	227

[a] Data from Imaoka *et al.* (1971).

TABLE 11.14

Vickers Microhardness of La$_2$O$_3$–SiO$_2$ Glasses Prepared by Oxides Melting and Gel Methods[a,b]

Oxides melting method		Gel method	
Composition (wt. % La$_2$O$_3$)	Microhardness H (kg mm^{-2})	Composition (wt. % La$_2$O$_3$)	Microhardness H (kg mm^{-2})
5.75	681 ± 10	5	685 ± 10
10.75	663 ± 17	10	667 ± 17
16	672 ± 20	15	678 ± 18
20.75	693 ± 18	20	697 ± 14
25.75	701 ± 32	25	723 ± 16
31	655 ± 20	30	669 ± 15

[a] Data from graph in Mukherjee *et al.* (1976).
[b] Measured under the load of 100 g.

TABLE 11.15

Vickers Microhardness of P$_2$O$_5$–SiO$_2$ Glasses Measured under Loads of 50, 100, 200, and 300 g[a]

Composition (mole % PO$_{2.5}$)	Microhardness H (kg mm^{-2})
17	608 ± 20
28	425 ± 22
42	495 ± 17
49	623 ± 15
51	663 ± 15

[a] Data from graph in Takahashi *et al.* (1976).

TABLE 11.16

Vickers Microhardness of Li$_2$O–K$_2$O–SiO$_2$ Glasses[a]

Composition (mole %)			Microhardness H (kg mm^{-2})
Li$_2$O	K$_2$O	SiO$_2$	
11.11	22.22	66.67	460 ± 5
16.67	16.66	66.66	495 ± 10

[a] Data from Neely (1969).

TABLE 11.17

Diamond Pyramid Hardness of
$Na_2O-K_2O-SiO_2$ Glasses[a]

Composition (mole %)			Microhardness H[b]
Na_2O	K_2O	SiO_2	(kg mm^{-2})
20	—	80	405
16	4	80	420
12	8	80	429
8	12	80	413
4	16	80	394
—	20	80	364

[a] Data from Ainsworth (1954a).
[b] Measured under loads of 30, 50, 70, and 100 g.

TABLE 11.18

Microhardness of $Na_2O-Rb_2O-SiO_2$ Glasses[a]

Composition (mole %)			Microhardness L_2VH[b]
Na_2O	Rb_2O	SiO_2	(kg mm^{-2})
6.3	18.7	75	280.4
12.5	12.5	75	231.5
18.7	6.3	75	265.1

[a] Data from Frischat et al. (1982).
[b] The Vickers hardness index L_2VH was measured using a dynamic technique of Frohlich et al. (1977) with continuous recording of load and depth of penetration during increasing and decreasing of load. The value of L_2VH was independent of the load and of glass cracking. For glasses and ceramics, L_2VH values were 2–3 times lower than the conventional Vickers hardness.

TABLE 11.19

**Microhardness of $Li_2O-MO-SiO_2$ Glasses Measured
under Loads of 50, 100, and 150 g[a]**

Composition (mole %)			Microhardness H
Li_2O	MO	SiO_2	(kg mm^{-2})
	M = Be		
16.7	16.7	66.6[b]	716
	M = Mg		
17.9	16.3	65.8[c]	736
	M = Ca		
17.0	16.2	66.8[d]	622
	M = Sr		
19.1	18.3	62.6[c]	580
	M = Ba		
15.0	15.6	69.4[e]	582
	M = Zn		
17.3	17.3	65.4[f]	790
	M = Cd		
16.9	17.0	66.1[c]	609

[a] Data from Totesh *et al.* (1961, 1965).
[b] No analysis.
[c] Also includes 0.2% $(Al_2O_3 + Fe_2O_3)$.
[d] 0.3% $(Al_2O_3 + Fe_2O_3)$.
[e] 0.7% $(Al_2O_3 + Fe_2O_3)$.
[f] 0.1% $(Al_2O_3 + Fe_2O_3)$.

TABLE 11.20

Microhardness of $Li_2O-SrO-SiO_2$ Glasses[a]

Composition (mole %)			Microhardness H (kg mm^{-2}) under given load	
Li_2O	SrO	SiO_2	50 g	100 g
20	—	80	640	578
17.5	2.5	80	674	599
15	5	80	740	636
12.5	7.5	80	740	665
10	10	80	787	698
8	12	80	886	738
30	—	70	577	542
27.7	2.3	70	608	554
25	5	70	667	596
22.5	7.5	70	702	596
20	10	70	705	598
17.7	12.3	70	709	602
15	15	70	740	658
12.8	17.2	70	759	690
10	20	70	740	640
8	22	70	708	614
5.5	24.5	70	709	614

[a] Data from graph in Kasymova and Milyukov (1974).

TABLE 11.21

Vickers Microhardness of $Li_2O-PbO-SiO_2$ Glasses[a]

Composition (mole %)			Vickers microhardness H (kg mm^{-2})
Li_2O	PbO	SiO_2	
1	74	25	407
2	73	25	410
5	70	25	416
10	65	25	423

[a] Data from El-Batal et al. (1980, 1982).

TABLE 11.22

Microhardness of $Na_2O-CuO-SiO_2$ Glass[a]

Composition (mole %)			Microhardness H (kg mm^{-2})
Na_2O	CuO	SiO_2	
14.3	14.3	71.4	508[b]

[a] Data from Petzold et al. (1961).
[b] Measured under loads of 10, 30, 50, 70, and 100 g.

TABLE 11.23

**Microhardness of $Na_2O-xBeO-5SiO_2$
Glasses Measured under
a Load of 100 g[a]**

x	Microhardness H (kg mm^{-2})
0.5	525
1.0	577
1.5	643
2.0	688
2.5	715

[a] Data from graph in Aleinkov *et al.*
(1962a,b, 1963).

TABLE 11.24

Microhardness of $Na_2O-MgO-SiO_2$ Glasses

Composition (mole %)			Microhardness H	
Na$_2$O	MgO	SiO$_2$	(kg mm^{-2})	Reference[a]
12.5	12.5	75	601	1[c]
14.3	14.3	71.4	494	1
16	20	64	500	2[d]
18	10	72	498	3[d]
19.0	15.9	65.1[b]	535	4[e]
7.1	22.9	70	573	5[f]
10	20	70	562	5
10.5	19.5	70	560	5
15	15	70	544	5
18.8	11.2	70	533	5
11.3	23.7	65	539	5
11.7	23.3	65	537	5
13.2	21.8	65	508	5
15.9	19.1	65	489	5
16.6	18.4	65	482	5
21.9	13.1	65	470	5
25.8	9.2	65	467	5
10.6	32.4	57	572	5
17.2	25.8	57	535	5
17.9	25.1	57	525	5
21.5	21.5	57	472	5
22.2	20.8	57	457	5
23.2	19.8	57	445	5
27	16	57	428	5
11	11	78	558	5
11	17.2	71.8	541	5
11	24	65	542	5
11	29.6	59.4	542	5
11	34.3	54.7	560	5

(continues)

TABLE 11.24 (*Continued*)

Composition (mole %)			Microhardness H	Reference[a]
Na$_2$O	MgO	SiO$_2$	(kg mm^{-2})	
16	14.8	69.2	497	5
16	19	65	493	5
16	22.5	61.5	502	5
16	25.7	58.3	528	5
16	28	56	538	5
16	30.5	53.5	547	5
22	13.1	64.9	457	5
22	15.3	62.7	461	5
22	17.7	60.3	468	5
22	19.9	58.1	479	5
22	21.4	56.6	483	5
22	24.8	53.2	492	5

[a] Data from (1) Petzold *et al.* (1961), (2) Appen (1954), (3) Ainsworth (1954a), (4) Totesh *et al.* (1961, 1965), and (5) Botvinkin and Yaroker (1966).
[b] Also includes 0.3% (Al$_2$O$_3$ + Fe$_2$O$_3$).
[c] Measured under loads of 10, 30, 50, 70, 100 g.
[d] Measured under loads of 30, 50, 70, 100 g.
[e] Measured under loads of 50, 100, 150 g.
[f] Measured under loads of 100 g.

TABLE 11.25

Knoop Microhardness of Na$_2$O–CaO–SiO$_2$ Glasses Measured under Loads of 100, 200, 300, and 500 g[a]

Composition (mole %)			Microhardness H
Na$_2$O	CaO	SiO$_2$	(kg mm^{-2})
16.5	18.5	65	564.8
33	2	65	396.3
15	17.5	67.5	562.4
12.5	17.5	70	561.2
14.5	15.5	70	561.5
16.5	13.5	70	477.6
20	10	70	445.1
26	4	70	395.4
28	2	70	359.5
17	11	72	465.8
20	7.5	72.5	436.5
24.5	3	72.5	359.0
14	10	76	471.0
19	5	76	440.2
10	10	80	462.5
15	5	80	426.4

[a] Data from Georoff and Babcock (1973).

TABLE 11.26

**Diamond Pyramid Microhardness
of $Na_2O-CaO-SiO_2$ Glasses Measured
under Loads of 30, 50, 70, and 100 g[a]**

Composition (mole %)			Microhardness H
Na_2O	CaO	SiO_2	(kg mm^{-2})
19	5	76	501
18	10	72	562
10.1	20.0	69.9	620
15	15	70	605
17.5	12.5	70	583
19.6	10.0	70.4	554
25	5	70	487
17	15	68	603
16	20	64	594
15	25	60	588
14.5	27.5	58	575
14	30	56	570

[a] Data from Ainsworth (1954a).

TABLE 11.27

**Microhardness of $Na_2O-xSrO-5SiO_2$
Glasses Measured under a Load
of 100 g[a]**

x	Microhardness H (kg mm^{-2})
0.5	482
1.0	521
1.5	550
2.0	570
2.5	584
3.0	597

[a] Data from graphs in Aleinkov et al.
(1962a,b, 1963).

TABLE 11.28

**Microhardness of $Na_2O-xBaO-5SiO_2$
Glasses Measured under a Load
of 100 g[a]**

x	Microhardness H (kg mm^{-2})
1	492
1.5	501
2	505
2.5	514
3	515

[a] Data from graphs in Aleinkov et al.
(1962a,b, 1963).

TABLE 11.29

**Microhardness of $Na_2O-xZnO-5SiO_2$
Glasses Measured under a Load
of 100 g[a]**

x	Microhardness H (kg mm^{-2})
1	508
1.5	527
2	546
2.5	553
3	565

[a] Data from graphs in Aleinkov et al.
(1962a,b, 1963).

TABLE 11.30

**Microhardness of $Na_2O-xCdO-5SiO_2$
Glasses Measured under
a load of 100 g[a]**

x	Microhardness H (kg mm^{-2})
1	505
1.5	518
2	531
2.5	544
3	551

[a] Data from graphs in Aleinkov et al.
(1962a,b, 1963).

TABLE 11.31

Microhardness of Na$_2$O–PbO–SiO$_2$ Glasses

Composition (mole %)			Microhardness H	
Na$_2$O	PbO	SiO$_2$	(kg mm^{-2})	Reference[a]
10	50	40	270	1[b]
14	30	56	409	
16	20	64	440	
17	15	68	447	
18	10	72	445	
19	5	76	432	
12.5	12.5	75	522	2[c]
14.3	14.3	71.4	438	
16.7	15.9	67.4	412	3[d]

[a] Data from (1) Ainsworth (1954a), (2) Petzold *et al.* (1961), and (3) Totesh *et al.* (1961, 1965).
[b] Measured under loads of 30, 50, 70, 100 g.
[c] Measured under loads of 10, 30, 50, 70, 100 g.
[d] Measured under loads of 50, 100, 150 g.

TABLE 11.32

Vickers Microhardness of Na$_2$O–PbO–SiO$_2$ Glasses[a]

Composition (mole %)			Microhardness H
Na$_2$O	PbO	SiO$_2$	(kg mm^{-2})
1	74	25	408
2	73	25	413
5	70	25	421
10	65	25	431

[a] Data from El-Batal *et al.* (1980, 1982).

TABLE 11.33

Microhardness of $K_2O-MO-SiO_2$ Glasses

Composition (mole %) K₂O	MO	SiO₂	Microhardness H (kg mm^{-2})	Reference[a]
	M = Be			
15.9	16.8	67.3[d]	533[b]	1
	M = Mg			
12.5	12.5	75	547[c]	2
	M = Ca			
12.5	12.5	75	611[c]	2
16.3	16.9	66.8[e]	492[b]	1
	M = Ba			
12.5	12.5	75	491[c]	2
16.1	15.1	68.8[f]	464[b]	1
	M = Zn			
17	17.3	65.7[e]	549[b]	1
	M = Cd			
17.6	16.5	65.9[e]	460[b]	1

[a] Data from (1) Totesh *et al.* (1961, 1965) and (2) Petzold *et al.* (1961).
[b] Measured under loads of 50, 100, and 150 g.
[c] Measured under loads of 10, 30, 50, 70, and 100 g.
[d] Also includes 0.3% $(Al_2O_3 + Fe_2O_3)$.
[e] Also includes 0.2% $(Al_2O_3 + Fe_2O_3)$.
[f] Also includes 0.7% $(Al_2O_3 + Fe_2O_3)$.

TABLE 11.34

Microhardness of $K_2O-SrO-SiO_2$ Glasses[a]

K₂O	SrO	SiO₂	50 g	100 g
15.4	4.6	80	528	509
12.6	7.4	80	542	523
6.8	13.2	80	622	556
25	5	70	347	306
22.6	7.4	70	407	356
20.2	9.8	70	520	457
18.6	11.4	70	514	457
15	15	70	633	570
11	19	70	583	533

[a] Data from graph in Kasymova and Milyukov (1974).

TABLE 11.35

Vickers Microhardness of K$_2$O–PbO–SiO$_2$ Glasses[a]

Composition (mole %)			Vickers microhardness H
K$_2$O	PbO	SiO$_2$	(kg mm^{-2})
1	74	25	411
2	73	25	415
5	70	25	424
10	65	25	439

[a] Data from El-Batal et al. (1980, 1982).

TABLE 11.36

Vickers Microhardness of K$_2$O–PbO–SiO$_2$ Glasses[a]

K$_2$O:SiO$_2$ (by wt.) = 1:4		K$_2$O:SiO$_2$ (by wt.) = 1:3	
wt. % PbO	Microhardness H (kg mm^{-2})	wt. % PbO	Microhardness H (kg mm^{-2})
1.2	427	4.8	380
4.7	467	10	456
9.5	489	20	489
14	495	25	488
19.5	470	29	462
29	454	34.5	447
39	436	44	386

[a] Data from graph in Gottardi et al. (1968).

TABLE 11.37

Microhardness of Cs$_2$O–CaO–SiO$_2$ Glass[a]

Composition (mole %)			Microhardness L$_2$ VH[b]
Cs$_2$O	CaO	SiO$_2$	(kg mm^{-2})
17.3	8.6	73.9	298.8

[a] Data from Frischat et al. (1982).
[b] The Vickers hardness index L$_2$VH was measured using a dynamic technique of Frohlich et al. (1977) with continuous recording of load and depth of penetration during increasing and decreasing of load. The value of L$_2$VH was independent of the load and of glass cracking. For glasses and ceramics, L$_2$VH values were 2–3 times lower than the conventional Vickers hardness.

TABLE 11.38

Vickers Microhardness of Sodium Borosilicate Glasses[a]

Composition (mole %)			Microhardness H (kg mm^{-2}) under given loads	
Na$_2$O	B$_2$O$_3$	SiO$_2$	200 g	500 g
10	10	80	706	696
10	15	75	733	705
10	20	70	713	696
10	25	65	673	663
10	30	60	603	588
10	35	55	552	539
10	40	50	501	497
10	45	45	465	456
10	50	40	408	400
10	55	35	400	400
10	60	30	396	392
10	65	25	384	369
10	70	20	358	340
10	75	15	317	296
10	80	10	281	282
10	85	5	274	275
20	5	75	509	507
20	10	70	574	578
20	15	65	641	641
20	20	60	660	656
20	25	55	669	660
20	30	50	651	634
20	35	45	621	584
20	40	40	590	566
20	45	35	576	545
20	50	30	553	533
20	55	25	520	511
20	60	20	473	451
20	65	15	440	436
20	70	10	384	388
20	75	5	378	373
25	50	25	558	543
25	55	20	509	505
25	60	15	501	499
25	65	10	482	483
25	70	5	458	450
30	5	65	553	539
30	10	60	574	562
30	15	55	608	600
30	20	50	625	641
30	25	45	644	634
30	30	40	462	626
30	35	35	618	583
30	40	30	572	571
30	45	25	577	569
30	50	20	593	580
30	55	15	593	541
30	60	10	521	488
30	65	5	481	466

[a] Data from Imaoka *et al.* (1971).

TABLE 11.39

Microhardness of $Na_2O-B_2O_3-SiO_2$ Glasses Measured under Loads of 30, 50, 70, and 100 g[a]

Composition (mole %)			Microhardness H
Na_2O	B_2O_3	SiO_2	$(kg\ mm^{-2})$
12.8	27.0	60.2	534
13.9	22.4	63.7	578
15.0	17.8	67.2	591
16.0	13.3	70.7	588
17.1	8.8	74.1	554
17.7	6.5	75.8	513
18	10	72.0	578
18.2	4.3	77.5	478

[a] Data from Ainsworth (1954a,b).

TABLE 11.40

Microhardness of Sodium Aluminosilicate Glasses

Composition (mole %)			Microhardness H	
Na_2O	Al_2O_3	SiO_2	$(kg\ mm^{-2})$	Reference[a]
16	22	62	505.7	1[b]
17.5	20.5	62	493.8	
19	19	62	479.6	
23	15	62	489.5	
28	10	62	504.4	
20	10	70	580	2
25	10	65	560	
30	10	60	566	
30	15	55	565	
30	20	50	568	
25	25	50	747	3[c]
28.6	14.3	57.1	646	
30.3	9.1	60.6	567	

[a] Data from (1) Georoff and Babcock (1973), (2) Myuller and Pronkin (1963), and (3) Dubrovo and Lileev (1960).
[b] Measured under loads of 100, 200, 300 g.
[c] Measured under load of 70 g.

TABLE 11.41

Vickers Microhardness of $Na_2O-Ga_2O_3-SiO_2$ Glasses[a]

Composition (mole %)			Microhardness H
Na_2O	Ga_2O_3	SiO_2	(kg mm^{-2})
21.7	35	43.3	454.1
23.3	30	46.7	464.7
25	25	50	478.4
25	50	25	464.7
26.7	20	53.3	439.0
30	10	60	399.9
30	40	30	456.7
32.5	35	32.5	487.0
35	30	35	517.1
40	20	40	456.7
45	10	45	373.4

[a] Data from Hanada et al. (1981).

TABLE 11.42

Vickers Microhardness of $Na_2O-M_2O_3-SiO_2$ Glasses[a]

Composition (mole %)			Microhardness H
Na_2O	M_2O_3	SiO_2	(kg mm^{-2})
	M = Sc		
30	10	60	585
	M = Y		
30	10	60	546
	M = La		
30	10	60	509.5
	M = In		
30	10	60	535
	M = Sb		
30	10	60	451.6
	M = Bi		
30	10	60	433.7

[a] Data from Takahashi et al. (1975).

TABLE 11.43

Microhardness of $3Na_2O-xM_2O_3-7SiO_2$ Glasses[a]

x (mole %)	Microhardness H (kg mm^{-2})	x (mole %)	Microhardness H (kg mm^{-2})
M = Nd		M = Yb	
1.3	485	1.3	600
3.7	573	3.7	610
4.8	650	4.7	638
6.0	682	6.0	656
7.0	670	9.0	688
9.0	680		

[a] Data from graph in Komiyama (1974).

TABLE 11.44

Microhardness of Na_2O–Fe_2O_3–SiO_2 Glasses[a]

Fe^{3+}/Fe$_{total}$ = 0.9				Fe^{3+}/Fe$_{total}$ = 0.5			
Composition (mole %)			Microhardness H (kg mm^{-2})	Composition (mole %)			Microhardness H (kg mm^{-2})
Na$_2$O	Fe$_2$O$_3$	SiO$_2$		Na$_2$O	Fe$_2$O$_3$	SiO$_2$	
20	0.5	79.5	459 ± 9	20	0.25	79.75	431 ± 6
20	2.5	77.5	468 ± 9	20	2.5	77.5	440 ± 6
20	5	75	477 ± 8	20	5	75	446 ± 8
20	7.5	72.5	521 ± 8	20	7.5	72.5	477 ± 8
20	10	70	530 ± 6	20	10	70	508 ± 8

[a] Data from graph in Petrenko et al. (1973).

TABLE 11.45

Microhardness of $Li_2O-ZrO_2-SiO_2$ Glasses Measured under a Load of 100 g[a]

Composition (mole %)			Microhardness H
Li_2O	ZrO_2	SiO_2	(kg mm^{-2})
20	—	80.0	575
20	2.5	77.5	615
20	5.0	75.0	638
20	7.5	72.5	677
20	10.0	70.0	710

[a] Data from graph in Ellern and Pavlushkin (1969).

TABLE 11.46

Vickers Microhardness of $Na_2-GeO_2-SiO_2$ Glasses[a]

Composition (mole %)			Microhardness H
Na_2O	GeO_2	SiO_2	(kg mm^{-2})
20	—	80	428
20	10	70	445 ± 14
20	20	60	472 ± 14
20	30	50	478 ± 14
20	40	40	501 ± 14
20	50	30	489 ± 14
20	60	20	484 ± 15
20	70	10	467 ± 10
20	80	—	462 ± 15
30	10	60	445 ± 15
30	20	50	472 ± 15
30	30	40	479 ± 14
30	40	30	503 ± 13
30	50	20	489 ± 14
30	60	10	486 ± 15
30	70	—	467 ± 12

[a] Data from graph in Takahashi et al. (1977).

TABLE 11.47

Microhardness of Na_2O–SnO_2–SiO_2 Glasses

Composition (mole %)			Microhardness H	
Na_2O	SnO_2	SiO_2	(kg mm^{-2})	Reference[a]
15	—	85	450	1
15	2.5	82.5	508	
15	5.0	80	575	
15	7.5	77.5	645	
20	10	70	593	2[b]
30	5	65	493	
30	10	60	590	

[a] Data from graphs in (1) Vakhrameev and Evstrop'ev (1969) and (2) Takahashi et al. (1977).
[b] Vickers microhardness.

TABLE 11.48

Vickers Microhardness of Na_2O–TiO_2–SiO_2 Glasses[a]

Composition (mole %)			Microhardness H	Composition (mole %)			Microhardness H
Na_2O	TiO_2	SiO_2	(kg mm^{-2})	Na_2O	TiO_2	SiO_2	(kg mm^{-2})
20	5	75	438 ± 12	30	5	65	438 ± 12
20	10	70	509 ± 14	30	10	60	507 ± 14
20	15	65	534 ± 15	30	15	55	533 ± 15
20	20	60	538 ± 13	30	20	50	538 ± 13
20	25	55	559 ± 8	30	25	45	559 ± 8
20	30	50	596 ± 10	30	30	40	596 ± 10
20	40	40	606 ± 11	30	40	30	607 ± 11

[a] Data from graph in Takahashi et al. (1977).

TABLE 11.49
Microhardness of $Na_2O-ZrO_2-SiO_2$ Glasses[a]

Composition (mole %)			Microhardness H (kg mm^{-2}) at given loads		
Na_2O	ZrO_2	SiO_2	20 g	50 g	100 g
14.8	1.3	83.9	478	492	535
15.2	3.8	81.0	542	557	580
15.6	6.6	77.8	573	596	625
16.0	9.4	74.6	598	623	660
16.5	12.5	71.0	609	636	720
23.2	5.1	71.7	—	—	620
25.4	3.8	70.8	—	—	608
27.6	2.5	69.9	—	—	590
29.7	1.3	69.0	—	—	580
18.5	8.0	73.5	—	—	635
26.4	8.0	65.6	354	388	530
31.8	8.0	60.2	336	368	515
37.2	8.0	54.8	323	356	505
39.8	8.0	52.2	323	349	499

[a] Data from Botvinkin and Demichev (1964).

TABLE 11.50
Microhardness of $Na_2O-ThO_2-SiO_2$ Glasses[a]

Composition (mole %)			Vickers microhardness H (kg mm^{-2})
Na_2O	ThO_2	SiO_2	
20	5	75	505[b]
30	5	65	505

[a] Data from Takahashi et al. (1977).
[b] Read from figure.

TABLE 11.51
Microhardness of $Na_2O-TeO_2-SiO_2$ Glasses[a]

Composition (mole %)			Vickers microhardness H (kg mm^{-2})	Composition (mole %)			Vickers microhardness H (kg mm^{-2})
Na_2O	TeO_2	SiO_2		Na_2O	TeO_2	SiO_2	
25	4.8	70.2	407	30	15	55	366
25	9.8	65.2	391	30	20.2	49.8	347
25	19.6	55.4	360	30	24.9	45.1	335
25	24.9	50.1	343	35	5	60	388
30	4.8	65.2	398	35	10	55	367
30	10	60	383	40	5	55	372
				45	5	50	342

[a] Data from Mochida et al. (1980).

TABLE 11.52

Microhardness of $Li_2O-Nb_2O_5-SiO_2$ Glasses Measured under a Load of 50 g[a]

Composition (mole %)			Microhardness H
Li_2O	Nb_2O_5	SiO_2	(kg mm^{-2})
30	5	65	500
30	10	60	560
30	15	55	540
35	15	50	540
35	20	45	530
40	20	40	515

[a] Data from Maksimova et al. (1975).

TABLE 11.53

Microhardness of $Na_2O-Nb_2O_5-SiO_2$ Glasses Measured under a Load of 50 g[a]

Composition (mole %)			Microhardness H
Na_2O	Nb_2O_5	SiO_2	(kg mm^{-2})
30	5	65	430
30	10	60	510
30	15	55	510
35	15	50	480
35	20	45	480
40	20	40	450

[a] Data from Maksimova et al. (1975).

TABLE 11.54

Microhardness of $K_2O-Nb_2O_5-SiO_2$ Glasses Measured under a Load of 50 g[a]

Composition (mole %)			Microhardness H
K_2O	Nb_2O_5	SiO_2	(kg mm^{-2})
30	5	65	150
30	10	60	420
30	15	55	450
35	15	50	380
35	20	45	375
40	20	40	200

[a] Data from Maksimova et al. (1975).

TABLE 11.55

Vickers Microhardness of Some $MO-PbO-SiO_2$ Glasses[a]

Composition (mole %)			Microhardness H	Composition (mole %)			Microhardness H
MO	PbO	SiO_2	(kg mm^{-2})	MO	PbO	SiO_2	(kg mm^{-2})
M = Mg				M = Sr			
1	74	25	410	5	70	25	422
2	73	25	414	10	65	25	434
5	70	25	427	M = Ba			
10	65	25	443	1	74	25	408
M = Ca				2	73	25	411
1	74	25	410	5	70	25	420
2	73	25	414	10	65	25	430
5	70	25	423	M = Zn			
10	65	25	439	1	74	25	408
M = Sr				2	73	25	411
1	74	25	409	5	70	25	417
2	73	25	412	10	65	25	426

[a] Data from El-Batal et al. (1980, 1982).

TABLE 11.56

Microhardness of Magnesium Aluminosilicate Glasses with Constant SiO_2 Content of 69.5 mole %[a]

Mole ratio Al_2O_3/MgO	Microhardness H (kg mm^{-2})
0.48	667
0.66	650
1.01	618
1.29	638
1.70	672

[a] Data from graph in Aslanova et al. (1982).

TABLE 11.57

Microhardness of $BaO-TiO_2-SiO_2$ Glasses Measured under a Load of 100 g[a]

Composition (mole %)			Microhardness H
BaO	TiO_2	SiO_2	(kg mm^{-2})
65	5	30	489
60	10	30	540

[a] Data from Matveev et al. (1970).

TABLE 11.58

Microhardness of $Al_2O_3-La_2O_3-SiO_2$ Glasses Measured under a Load of 100 g[a]

Composition (mole %)			Microhardness H	Composition (mole %)			Microhardness H
Al_2O_3	La_2O_3	SiO_2	(kg mm^{-2})	Al_2O_3	La_2O_3	SiO_2	(kg mm^{-2})
20	25	55	454	19	12	69	482
22.5	22.5	55	474	20.5	10.5	69	509
23.5	21.5	55	495	21	10	69	527
25	20	55	520	23	8	69	545
27	18	55	533	24.5	6.5	69	573
28.5	16.5	55	564	26	5	69	600
32.5	12.5	55	612				
35	10	55	645				
37	8	55	620				

[a] Data from graph in Aleksandrov et al. (1980).

Chapter 12

Strength

INTRODUCTION I

The strength of glass depends upon its surface condition. Tiny flaws (cracks) in the surface lead to weakening and failure by brittle fracture (see definitions below). Glass also becomes weaker with time under stress (fatigue). The theoretical strength of glass is very high ($\sim 2.4 \times 10^{10}$ Pa for vitreous silica and $\sim 1.6 \times 10^{10}$ Pa for commercial soda-lime glass), but these strengths are rarely even approached in practice because of the surface flaws.

A typical strength of glass in an inert atmosphere (no water) is 70 MPa (10,000 psi). However, fatigue under stress rapidly reduces this strength. At half the inert stress, commercial soda-lime glass will fail in 10 s or less in 50% relative humidity. Thus the effective strength is reduced to one-third or less of the inert strength. The maximum design strength for glass pieces in ambient air is therefore usually 20 MPa (3000 psi) or less.

This sensitivity to surface condition and fatigue means that strength is not an intrinsic property of glass, and a table of strengths of different glass compositions is not useful, because the strength depends strongly on the history of treatment, handling of the surface, and relative humidity. Composition has only a small effect on fracture strength, although it does have considerable influence on fatigue.

In this chapter the following subjects are discussed: testing methods, strength distributions, surface treatment and strength, fracture surfaces, plastic deformation and definitions, fatigue, and life predictions. These discussions are based on a review (Doremus, 1982), which should be consulted for more detailed discussions of these matters, atomic models of fracture and fatigue, and additional references.

II   TESTING METHODS

In brittle materials, failure usually takes place at much lower tensile stresses than compressive. In principle the simplest test is to pull a rod, putting a uniform tensile stress on it. In practice there are difficulties in gripping the sample and, except for fine fibers, one must make the central length of the rod smaller in diameter to be certain that failure does not occur at the grips. Making such samples is difficult and expensive, so other tests are used in which the stress distribution is not as simple as in a tensile test. For long fibers the direct tensile test is the simplest and most frequently used, since gripping problems can be overcome in a variety of ways (Proctor *et al.*, 1967; Cameron, 1968; Maurer, 1975; Kurkjian *et al.*, 1976).

The most common tests are in bending (flexure), with either three or four points of stress application. Rod samples are much better than bars because chips at the corners of the bars can initiate fracture. In the bending of the rod with four points, the tensile stress parallel to the sample axis is maximum on a line on the sample surface opposite and between the inner points. The stress S along this line for small strains is given by the simple bending formula

$$S = 8P/\pi D^3, \tag{1}$$

where P is the force at each contact point and D is the sample diameter. The stress on the sample surface decreases with the angle θ with the midplane,

$$S_s = S(\cos\theta) \approx S(1 - 2\theta/\pi)^{1/2}, \tag{2}$$

and the stress decreases approximately linearly from the surface to the sample axis. Thus the bend tests favor failure starting at the sample surface, where the tensile stress is higher. It is always wise to examine the failed specimen, if possible, to make certain that failure did not take place at the contact points, where the stress is uncertain.

The four-point bend test is preferable to the three-point test because the four-point test assesses more of the specimen surface and is less prone to failure at the contact points.

The diametral compression test provides an attractive alternative to bend tests for some materials (Varder and Finnie, 1975; Thomas *et al.*, 1980). In this test a right circular cylinder is compressed along its diameter by two flat plates, giving a maximum tensile stress normal to the loading direction across the loading diameter. This stress is present across the diameters of both flat ends of the cylinder and also across a plane from one end of the cylinder to another, so that both surface and volume failures are possible. A pad of material softer than the compressing plates must be inserted between the sample and plates to prevent failure at the plates.

Compressive tests are sometimes used for glass, but failure usually takes place at a corner contact of the sample with the compressing plates.

Presumably, failure is caused by a tensile stress that is difficult to calculate, so these tests give scattered and unreliable results.

In principle the simplest method of studying fatigue in glass is a static test in which a constant load is imposed on a set of samples and the time of failure of each sample is recorded. Such tests can be done in bending, simple tension, or diametral loading; bending is easiest for rods and tubes and requires simple test equipment. Although static tests are the closest to practical situations and are the least complicated in principle, they have some disadvantages. They require a large number of samples for each of several stresses to define stress–failure time relationships because of the large spread in failure times at a particular stress. Often it is difficult or expensive to obtain the required number of samples. Static tests must also be carried out over a large spread of times, leading to long delays in obtaining results.

These disadvantages have led to a search for alternative ways to test fatigue in glass and other brittle oxides. Two methods have been widely used. In the first, samples are tested at different loading rates; the dependence of fracture strength on loading rate is related to the stress–failure time relation in static loading. This method requires fewer samples and less time, but a testing machine with a linear load application is needed. The main disadvantage of this technique is that it can be used for only a narrow range of failure stresses because of limits in loading rates available. Thus this method is good for surveys of materials and rapid or preliminary tests, but static fatigue tests are needed for complete definition of strength–failure time relationships and comparisons with theories.

Another method is to measure directly the velocity at which large (~ 1 cm) cracks propagate under different applied loads. This method has been widely used, but experimental and theoretical treatments suggest that results from this technique are not always directly comparable with the fatigue behavior of practical materials with much smaller flaws (a few micrometers long).

STRENGTH DISTRIBUTIONS   III

When the failure stress of a number of glass samples is measured under identical conditions of sample preparation and loading, the failure stresses vary over a wide range of a factor of 2 or more. It is desirable to express this distribution of strengths in statistical terms. The strengths are often distributed symmetrically, so they can be described by a normal (Gaussian) distribution:

$$P = [1/(d\sqrt{2\pi})] \exp[-(S - S_m)^2/2d^2], \tag{3}$$

where P is the probability of finding a sample of strength (failure stress) S, S_m is the strength of greatest probability ($P_m = 1/d\sqrt{2\pi}$), and d is a measure of the spread of the distribution (called the standard deviation) and is equal to the root mean square of deviations from the mean strength. The integral of Eq. (3) gives the fracture F of samples that breaks below the stress S,

$$F = \tfrac{1}{2}\{1 + \text{erf}[(S - S_m)/\sqrt{2}d]\},$$

where erf is the error function,

$$\text{erf } X = \frac{2}{\sqrt{\pi}} \int_0^X \exp(-\lambda^2) \, d\lambda.$$

Recently another distribution function called the Weibull distribution has become quite popular in describing strength distributions:

$$F = 1 - \exp[-(S/S_0)^m], \tag{4}$$

where S_0 is a scaling factor and m is a measure of the spread of the distribution—smaller m, broader distribution. For m larger than about 3, the Weibull distribution is nearly symmetrical and closely resembles the normal distribution. A convenient way to examine the fit of strength data to a Weibull distribution and to calculate m is to transform Eq. (4) to

$$\log[-\ln(1 - F)] = m \log S - m \log S_0. \tag{5}$$

Weibull parameters and correlation coefficients calculated from a regression analysis of Eq. (5) for different glass compositions, surface treatments, and temperatures are summarized in Table 12.1. Compositions of the glasses are given in Table 12.2.

Strengths of glass fibers carefully drawn are much higher than for ordinary glass; E-glass has a strength of from 1 to 4×10^9 N m^{-2} at room temperature (Cameron, 1968), and silica fibers have even higher strengths (Proctor et al., 1967; Maurer, 1975; Kurkjian et al., 1976). The distributions of strength of these fibers in short lengths can be quite narrow (Thomas, 1958; Mould, 1958), with a coefficient of variation (standard deviation divided by mean strength) of 1 to 2%. On the other hand, flame-polished or etched glass rods with high strengths can have very broad and nonuniform distributions (Pavelchek and Doremus, 1974).

IV SURFACE TREATMENT, TEMPERATURE, AND STRENGTH

When a glass surface is abraded or bombarded with particles, the strength is reduced. Different treatments give quite different strengths, as shown in Table 12.3.

If glass samples are carefully prepared and kept "pristine," they have very high strengths. Many results up to 10^{10} Pa have been reported (Proctor et al., 1967), and Hillig (1962) measured strengths up to 1.5×10^{10} Pa at $-196°C$ for

TABLE 12.1

Mean Strength and Weibull Modulus for Strengths of Glasses[a]

Run	Abraded	Number of samples	Temperature (°C)	Crosshead speed (cm/min)	Mean strength (MN/m^2)	Coefficient of variation (%)	m^b	R^2 for Weibull fit[c]
				Soda-lime silicate				
1	no	57	−196	0.13	245	25	4.5	0.98
2	yes	49	−196	0.13	137	16	6.6	0.87
3	yes	29	−196	0.13	236	17	6.5	0.86
4	yes	104	23	0.13	135	20	6.4	0.96
5	yes	58	23	0.025	132	18	5.9	0.96
6	yes	90	23	0.013	112	18	6.7	0.94
7	yes	123	23	0.0013	103	20	6.2	0.94
8	no	37	23	0.13	166	28	4.2	0.95
				Pyrex borosilicate				
9	no	87	−196	0.05	246	21	5.5	—
10	no	21	23	0.25	134	17	6.4	—
11	no	21	23	0.051	123	15	6.0	—
12	no	21	23	0.013	118	20	6.4	—
13	yes	85	23	0.127	117	14	7.8	—
				FN borosilicate				
14	yes	100	−196	—	102	6.9	7.3	—

[a] From Doremus (1982).
[b] Least-squares fit to Weibull distribution calculated from Eq. (5).
[c] Calculated from Eq. (5).

TABLE 12.2

Glass Compositions (wt. %)

Glass	SiO_2	Na_2O	K_2O	CaO	MgO	Al_2O_3	B_2O_3	BaO
Soda-lime Kimble R-6	68	15	—	5	4	3	1	2
Pyrex borosilicate Corning 7740	81	4	—	—	—	2	13	—
G.E. FN Corning 7052	66	4	3	—	—	3	24	—
E-Glass	54	1	—	21	1	15	8	—

TABLE 12.3

Influence of Abrasion Method on Fracture of Soda-Lime Glass[a]

Abrasion method	Average fracture stress at $-196°C$ (MN/m^2)	$\log t_{1/2}$ (s)
Severe grit blast	86	0.94
Mild grit blast	93	0.46
Perpendicular to stress, 600-grit paper	134	−2.67
320 Grit paper	95	−1.41
150 Grit paper	70	−0.25
Parallel to stress, 150-grit paper	165	−0.85

[a] From Doremus (1982). Data from Mould and Southwick (1959).

TABLE 12.4

Temperature Dependence of the Fracture Strength of Soda-Lime Glass Bubbles Inside the Glass as Determined by the Bubble Method[a]

Temperature (K)	Fracture strength (GPa)
76	0.74
195	0.65
240	0.42
295	0.33
495	0.27
595	0.19

[a] From Ernsberger (1969).

vitreous silica rods 1 mm in diameter. These samples were flame drawn from larger rods and carefully protected from contact with solids.

The strength of soda-lime glass as a function of temperature was determined by Ernsberger (1969) for failure at oblate bubbles inside the glass as shown in Table 12.4. The strengths of other high-strength glasses have a similar temperature dependence (Proctor et al., 1967; Cameron, 1968). However, abraded glasses of low strength often show little temperature dependence of fracture strength. It seems likely that these temperature effects are caused by the same mechanisms as delayed failure, as discussed below. Hillig (1962) found no difference in glass strength between −196°C and −268°C, at which temperatures delayed failure is absent.

OBSERVATIONS OF FRACTURE SURFACES AND SURFACE DAMAGE   V

Markings on fracture surfaces can give information about the origin and direction of fracture propagation, as well as about material inhomogeneities and local stress effects. The study of these markings is called fractology, and is discussed on the scale of the optical microscope by Frechette (1972). The determination of the macroscopic origin of fracture, directions of fracture propagation, and stress distributions will not be discussed here.

A fracture surface on glass shows several characteristic regions—an origin usually on the glass surface, a smooth region called the mirror, a misty region outside the mirror, and finally a region of coarse surface roughness, called hackle, as shown in Fig. 12.1. The distance r from the fracture origin to the mirror–mist boundary (or the distance to the mist–hackle boundary) is related to the fracture strength S by the empirical equation

$$S\sqrt{r} = A \tag{6}$$

(Shand, 1954, 1959; Levengood, 1958; Mecholsky *et al.*, 1974; Kirchner *et al.*, 1976), where A is a constant for a particular brittle material and temperature. Values of about 2×10^6 N m$^{-3/2}$ were found for A for glass at room temperature. Apparently, A is not significantly dependent on temperature or glass composition.

Equation (6) is of considerable practical value because it allows one to make an accurate estimate of fracture stress without having to measure it directly. Thus for pieces with complex shapes where a stress analysis is difficult, or where load measurements are not possible, Eq. (6) is invaluable.

If a hard sphere is pressed against a glass surface, a crack around the contact region first develops and then grows into the glass in a conical shape. These cracks are called Hertzian cracks because the elastic stress distribution for contact of a hard sphere on a plate was calculated by Hertz. Papers on this subject are listed by Lawn and Wilshaw (1975) and Bilby (1980).

If the indenter is sharp rather than spherical, the crack pattern is entirely different (Lawn and Wilshaw, 1975; Peter, 1964). A "median vent" crack propagates into the glass directly under the indenter during loading. When the load is released, winglike cracks ("lateral vents") propagate radially from the contact region. A region of the glass directly under the indenter is deformed in a particular way that is discussed in more detail in the next section.

The impact of a hard spherical projectile gives damage with features of both spherical and sharp indenters. At low impact velocities, Hertzian cone cracks are formed, but at higher velocities the cone cracks turn up at their ends and finally resemble the radian winglike cracks; at high velocities, median cracks also propagate down onto the glass.

When a hard object such as a diamond indenter is dragged across a glass surface, median and lateral cracks form along its track much as under an indenter (Ernsberger, 1972; Kerkhof, 1970). Abrasion of a hard surface against the glass also gives this kind of linear damage.

Fig. 12.1. Fracture surface of soda-lime silicate glass, broken in bending at ambient temperature (25°C). The marker is 100 μm. From Doremus

PLASTIC DEFORMATION IN GLASS VI

Tensile and bend tests of glasses to stresses close to the theoretical strength showed no evidence of any nonrecoverable deformation (Hillig, 1962; Krause et al., 1979). However, indentation of glass with a sharp point leads to some regions of deformation under the indenter that do not recover when the load is removed (Neely and Mackenzie, 1968). This deformation in fused silica is apparently a "densification" that recovers completely on heating. In glasses with modifiers (alkali and alkaline earth oxides), this deformation can be partly a densification (Ernsberger, 1968), but only a small part of the deformation recovers upon heating (Izumitani, 1979). More detailed examination of this deformed region under a sharp indenter showed that it is made up of flow or fault lines, between which there are only elastic strains (Peter, 1970; Dick, 1970; Ernsberger, 1977; Hagan, 1979, 1980).

Some concise definitions will perhaps help to clarify the discussion of "plasticity" in glass and distinguish between the following different kinds of deformation.

(1) *Elastic*—instantaneous and completely recoverable upon removing the load.

(2) *Anelastic*—recoverable but requiring a measurable time for deformation after application or removal of the load.

(3) *Viscous*—the uniform nonrecoverable flow of a liquid or amorphous solid subjected to a shearing stress. The rate of deformation is usually directly proportional to the load.

(4) *Plastic*—the nonrecoverable nonuniform deformation of a crystalline solid. Occurs preferentially on certain lattice planes and in certain lattice directions, and results from dislocation motion. Can occur at low stresses.

(5) *Faulting*—nonrecoverable nonuniform deformation along certain planes in a solid that are determined by the applied loads. Occurs only under a compressive stress close to the ultimate strength of the solid.

It is also useful to define *crack* as a flaw or defect in a solid, usually planar, in which solid planes are separated by a few or more lattice distances over a certain area. With these definitions the shear deformation of glass under a concentrated compressive load, as distinguished from other types of deformation, is called faulting, in analogy to geological faulting. The excellent studies of Hagan (1979, 1980) show that this deformation has the following additional characteristics:

(1) The fault lines are extremely narrow, probably no more than a few atomic distances thick.

(2) The fault lines meet at 110°, instead of the 90° required by the traditional ideal elastic–plastic behavior.

(3)   Kinks at intersections of flow lines show that the fault planes are not cracks (see definition above) with free surfaces.

(4)   The fault lines can act as nucleating sites for cracks.

In view of these characteristics, these fault lines resemble geological faults and are quite different from slip in plastic deformation of crystals. Thus it seems important to distinguish this faulting deformation from plastic deformation in ductile crystals.

Viscous flow is much too slow in glass to cause appreciable deformation at ambient temperatures. The viscosity of commercial soda-lime glass at 490°C is about 10^{15} P, with an effective activation energy of greater than 400 kJ mole^{-1}; the activation energy increases with decreasing temperature. Thus the viscosity at room temperature is greater than 10^{59} P.

Scanning electron micrographs of glass surfaces after scoring with diamond points show shapes that resemble those from metal working (Peter, 1970; Dick, 1970). To quote Ernsberger (1977, p. 306), "they were able to demonstrate the existence of coherent spiral turnings, buttery-looking grooves, and blocky extrusions. The eye is convinced; only the mind can retain reservations."

Nevertheless, these deformations are very different from plastic deformation in metals and other ductile materials, since they arise only from sharp points and do not occur with large-scale stressing in tensile and bending modes. They probably result from a combination of densification and shear faulting. Likewise, furrows and pile-up after scribing (Marsh, 1964) are probably combinations of these modes of deformation; in none of these cases is there any viscous flow or plastic deformation in the sense of the above definitions.

VII   FATIGUE

Fatigue of glass depends strongly on the ambient humidity, the glass composition, the thermal, abrasion, and atmospheric history of the glass surface, and of course on the applied stress. Some representative fatigue times for soda-lime glass are given in Tables 12.3 and 12.5 for a variety of conditions and show a variation of fatigue time $t_{1/2}$ of more than six orders of magnitude. The time $t_{1/2}$ is the time to failure at an applied stress that is one-half the mean fracture stress at $-196°C$ (liquid nitrogen), where there is no fatigue. This method of defining $t_{1/2}$ allows comparison of fatigue times for samples of widely different inert strengths.

The results in Table 12.5 show that fatigue times increase strongly as the ambient concentration of water is reduced. The type, severity, and direction of abrasion had a strong affect on both inert fracture strength and fatigue times, as shown in Table 12.3.

TABLE 12.5

Fatigue Times for Abraded Soda-Lime Silicate Glass[a]

Number	Relative humidity (%)	Average fracture stress at $-196°C$		$\log t_{1/2}$ (s)	Remarks
		MN/m^2	10^3 psi		
1	100	118	17.1	1.67	Tested right after abrasion
2	50	118	17.1	2.46	Tested right after abrasion
3	100	160	23.2	-0.66	
4	50	160	23.2	0.48	
5	Water	75	10.9	0.94	Aged 20–60 s in water
6	Water	86	12.4	0.94	
7	43	86	12.4	2.30	
8	0.5	86	12.4	3.54	
9	Water	90	13.1	2.30	Heated at 470°C for 3 hr in vacuum of 10^{-5} mm Hg, stored in dry nitrogen, then tested in water.
10	50	187	27.1	1.47	Heated to 400°C for 1 hr after abrasion, then held 24 hr in water.

[a] All samples were aged 24 hr in water after abrasion and before testing unless otherwise noted. From Doremus (1982). Data from Mould and Southwick (1959) and Pavelcheck and Doremus (1976).

Fatigue times decrease as the sample temperature is increased, but there is no simple temperature dependence (Doremus, 1982).

Because of the large influence of sample history on fatigue, it is difficult to compare the fatigue response of different glass compositions. In Table 12.6 some results for different glasses are listed; unfortunately, the histories of the different compositions were not all the same. Nevertheless, it appears that all

TABLE 12.6

Failure Times for Different Glasses at ∼50% Relative Humidity and 25°C

Glass	Treatment	$\log t_{1/2}$ (s)	Strength at $-196°C$ (MN/m^2)	Reference
Vitreous silica	Centerless ground	4.8	100	Burke et al. (1971)
Pyrex	As received	2.4	241	Friedman et al. (1982)
E-Borosilicate	As drawn	2	5.6×10^3	Schmitz and Metcalfe (1966)
FN borosilicate	Centerless ground	1.5	100	Burke et al. (1971)
Soda-lime	Various abrasions, aged in water	1.0–2.0	70–150	Mould and Southwick (1959); Pavelcheck and Doremus (1976)

the glasses except for fused silica have roughly the same fatigue sensitivity. Abraded silica is less sensitive; however, pristine or etched silica has a lower $\log t_{1/2}$ (< 2.0) (Doremus, 1982).

VIII LIFE PREDICTION

An important practical problem is to predict the life of a piece of glass under stress. Several studies have shown that there is no separate effect of cyclical stress on glass fatigue, so the fatigue results from a cumulative process depending on the time at different applied stresses. Thus the problem of predicting life can be limited to experiments at different fixed stresses, or static fatigue experiments.

The problem is therefore to examine the functional dependence of failure times on stress and to find a reliable function that can be extrapolated to longer times and lower stresses than are convenient experimentally.

Failure times show a wide spread in values for identical testing conditions because of the distribution of strengths described previously. Many studies show that the logarithms of the failure times for glass are approximately symmetrically distributed, so the log of failure time is the parameter used for statistical analysis of fatigue. The arithmetic mean of the log of failure times for samples tested under identical conditions is the "best" statistical measure of the "true" failure time (Doremus, 1982). The more tests made, the more reliable is the mean value found; $d/\sqrt{n}$, where d is the standard deviation of test results and n is the number of tests, is a measure of the reliability of the mean.

Several different functional dependencies of failure time on stress have been suggested:

$$\log t = a - bS/S_N, \tag{7}$$

$$\log t = c - n \log(S/S_N), \tag{8}$$

$$\log t = d + gS_N/S, \tag{9}$$

$$\log t = h + p(S_N/S)^2. \tag{10}$$

Over the stress ranges studied experimentally it is often difficult to choose from among these equations, yet if they are extrapolated to times of a few years, the sample lives predicted from each of the equations are very different. Thus it is important to choose the best equation if one wishes to extrapolate or to examine theories.

The simplest way to examine the equations is to calculate a regression equation by the least-squares method in the form of Eqs. (7)–(10) and compare the correlation coefficients from the regression. The mean $\log t$ values for Pyrex glass in Table 12.7 were examined in this way—see Table 12.8. There is a clear increase in goodness of fit in the order Eqs. (7), (8), (9), and (10).

Often data are not extensive or reliable enough to find any differences between correlation coefficients, and a more sensitive method is needed. Pavelchek and Doremus (1976) suggested that the variation of the spread (standard deviation) of log t values with stress is a sensitive measure of the functional dependence of failure time on stress. The applied stress S can be considered constant compared to the spread in inert strengths (S_N values) for different specimens. Thus the differential of S/S_N is

$$d(S/S_N) = -(S \, dS_N)/S_N^2.$$

If the differential of log t is considered to equal the standard deviation of log t (designated by δ log t) and $dS_N/\bar{S}_N$ is the coefficient of variation (standard

TABLE 12.7

Standard Deviation of Log Failure Time as a Function of $S/\bar{S}_N{}^a$

$S/\bar{S}_N$	Mean log t	Standard deviation	$S/\bar{S}_N$	Mean log t	Standard deviation
Fused silica, sets of 100 samples[b]			Pyrex borosilicate (*continued*)[c]		
0.70	2.22	0.76	0.531	1.102	0.902
0.65	3.06	0.74	0.504	0.924	0.863
0.60	3.57	0.63	0.489	1.182	1.239
0.55	4.04	0.97	0.471	1.731	1.065
0.50	4.73	0.93	0.466	1.829	1.337
			0.452	1.898	1.028
FN borosilicate, sets of 100 samples[b]			0.435	2.373	1.496
0.550	0.77	0.78	0.425	1.740	1.253
0.525	1.20	0.83	0.408	2.374	1.300
0.500	1.51	0.85	0.394	3.214	1.634
0.475	2.02	0.86	0.389	3.573	1.540
0.450	2.67	1.00	0.371	3.744	1.484
0.400	3.96	1.39	0.352	4.261	1.106
			0.337	4.811	1.177
Pyrex borosilicate[c]			0.329	5.317	1.179
0.728	0.530	0.629			
0.673	0.390	0.715	Soda-lime silicate[d]		
0.647	0.289	0.783	0.461	1.80	0.75
0.608	0.447	0.882	0.424	2.40	1.1
0.583	1.040	0.845	0.378	2.98	1.1
0.568	0.887	0.995	0.332	4.00	1.2
0.542	1.296	0.994	0.298	5.36	2.4

[a] From Doremus (1982).
[b] From Burke et al. (1971).
[c] From Friedman et al. (1982).
[d] From Jakus et al. (1978).

TABLE 12.8

**Comparison of Correlation Coefficients for
Different Equations for Failure Times of
Pyrex Glass at 60% Relative Humidity
and 25°C**

Equation number	Correlation coefficient R^2
(7)	0.805
(8)	0.869
(9)	0.917
(10)	0.944

deviation divided by the mean) of the inert strength $(\delta S_N)/\bar{S}_N$, then the following relations are found from Eqs. (7)–(10):

$$\delta \log t = b \frac{S}{S_N} \frac{\delta S_N}{\bar{S}_N}, \tag{11}$$

$$\delta \log t = n \frac{\delta S_N}{\bar{S}_N}, \tag{12}$$

$$\delta \log t = g \frac{\bar{S}_N}{S} \frac{\delta S_N}{\bar{S}_N} \tag{13}$$

$$\delta \log t = 2p \frac{\bar{S}_N}{S} \frac{\delta S_N}{\bar{S}_N} \tag{14}$$

These equations show that for the direct exponential of Eq. (7), the spread in log t values should decrease as the stress decreases; that for the power law of Eq. (8), the spread in log t should remain constant with changes in stress; and for the inverse exponentials of Eqs. (9) and (10), the spread in log t should increase as the stress decreases. In these equations, $\bar{S}_N$ denotes the mean value of strengths at $-196°C$.

For every set of sufficiently reliable glass static fatigue data, the spread in log t values increases as the stress decreases. Some examples are shown in Table 12.7 for several different glasses; more extensive results for soda-lime glass are given by Pavelchek and Doremus (1976).

The conclusion from a direct comparison using correlation coefficients and a regression analysis, from the variation of standard deviations of log t with stress, and from long-term tests (Doremus, 1982; Friedman *et al.*, 1982), is that Eqs. (9) and (10) are superior to Eqs. (7) and (8) for fitting static fatigue data. It would seem that any extrapolation of failure times longer than experimental for fatigue of glass should start with one of these equations. A large number of calculations involving fatigue in glass have been based on Eq. (8), a power law, probably because of mathematical simplicities; although this equation may fit fatigue data over short ranges of applied stress, it is unwise to use it for extrapolations, for which its validity should be even poorer than for measured failure times.

BIBLIOGRAPHY

Bilby, B. A. (1980). *J. Mater. Sci.* **15**, 535.

Burke, J. E., Doremus, R. H., Hillig, W. B., and Turkalo, A. M. (1971). *In* "Ceramics in Severe Environments" (W. W. Kriegel and H. Palmour, eds.), p. 435. Plenum, New York.

Cameron, N. M. (1968). *Glass Technol.* **9**, 14, 121.

Dick, E. (1970). *Glastech. Ber.* **43**, 16.

Doremus, R. H. (1982). Fracture and fatigue of glass. *In* "Glass III" (M. Tomozawa and R. H. Doremus, eds.), Treatise on Materials Science and Technology, Vol. 22, p. 169. Academic Press, New York.

Ernsberger, F. M. (1968). *J. Am. Ceram. Soc.* **51**, 545.

Ernsberger, F. M. (1969). *Phys. Chem. Glasses* **10**, 240.

Ernsberger, F. M. (1977). *J. Non-Cryst. Solids* **25**, 295.

Frechette, V. D. (1972). *In* "Introduction to Glass Science" (L. D. Pye, H. J. Stevens, and W. C. LaCourse, eds.), p. 451. Plenum, New York.

Friedman, G. S., Cushman, K., and Doremus, R. H. (1982). *J. Mater. Sci.* **17**, 994.

Hagan, J. T. (1979). *J. Mater. Sci.* **14**, 462.

Hagan, J. T. (1980). *J. Mater. Sci.* **15**, 1417.

Hillig, W. B. (1962). *In* "Modern Aspects of the Vitreous State" (J. D. Mackenzie, ed.), Vol. II, p. 152. Butterworth, London.

Izumitani, T. (1979). *In* "Glass II" (M. Tomozawa and R. H. Doremus, eds.), Treatise on Materials Science and Technology, Vol. 17, p. 115. Academic Press, New York.

Jakus, K., Coyne, D. C., and Ritter, J. E. (1978). *J. Mater. Sci.* **13**, 2071.

Kerkhof, F. (1970). "Bruchvorgange in Glasern." Verlag Dtsch. Glastech. Ges., Frankfurt.

Kirchner, H. P., Gruver, R. M., and Sotter, W. A. (1976). *Philos. Mag.* **33**, 775.

Krause, J. T., Testardi, L. R., and Thurston, R. N. (1979). *Phys. Chem. Glasses* **20**, 135.

Kurkjian, C. R., Albarino, R. V., Krause, J. T., Vazirani, H. N., DiMarcello, F. V., Torza, S., and Schonhorn, H. (1976). *Appl. Phys. Lett.* **28**, 588, 712.

Lawn, B., and Wilshaw, R. (1975). *J. Mater. Sci.* **10**, 1049.

Levengood, W. C. (1958). *J. Appl. Phys.* **29**, 820.

Marsh, D. M. (1964). *Proc. R. Soc. London, Ser. A* **282**, 33.

Maurer, R. D. (1975). *Appl. Phys. Lett.* **27**, 220.

Mecholsky, J. J., Freiman, S. W., and Rice, R. W. (1974). *J. Am. Ceram. Soc.* **57**, 440.

Mould, R. E. (1958). *J. Appl. Phys.* **29**, 1263.

Mould, R. E., and Southwick, R. D. (1959). *J. Am. Ceram. Soc.* **42**, 542, 582.

Neely, J. E., and Mackenzie, J. D. (1968). *J. Mater. Sci.* **3**, 603.

Pavelchek, E. K., and Doremus, R. H. (1974). *J. Mater. Sci.* **9**, 1803.

Pavelchek, E. K., and Doremus, R. H. (1976). *J. Non-Cryst. Solids* **20**, 305.

Peter, K. (1964). *Glastech. Ber.* **37**, 333.

Peter, K. (1970). *J. Non-Cryst. Solids* **5**, 103.

Proctor, B. A., Whitney, I., and Johnson, J. W. (1967). *Proc. R. Soc. London, Ser. A* **297**, 534.

Schmitz, G. K., and Metcalfe, A. G. (1966). *Ind. Eng. Chem., Prod. Res. Dev.* **5**, 1.

Shand, E. B. (1954). *J. Am. Ceram. Soc.* **37**, 52.

Shand, E. B. (1959). *J. Am. Ceram. Soc.* **42**, 474.

Thomas, M. B., Doremus, R. H., Jarcho, M., and Salsbury, R. L. (1980). *J. Mater. Sci.* **15**, 891.

Thomas, W. F. (1958). *Nature (London)* **181**, 1006.

Vardar, O., and Finnie, I. (1975). *Int. J. Fracture* **11**, 495.

Part **IV**
Electrical and Transport Properties

Chapter **13**

Electrical Conductivity

INTRODUCTION I

Glass is often used as an electrical insulator, so that knowledge of its electrical properties is important. In most oxide glasses, and all those considered here, the electrical conductivity results from ionic motion. In certain special compositions containing transition-metal oxides such as vanadium and iron, electrons carry the current. Many nonoxide glasses, based on sulfides, selenides, or tellurides, are also electronic conductors.

Monovalent cations are the most mobile ions in oxide glasses. The conducting ion is sodium in most commercial glasses. Lithium ions are also quite mobile in oxide glasses. Potassium and hydronium ions are also mobile, although much less than lithium and sodium in most oxide glasses. Even in glasses with no nominal addition of monovalent cations, monovalent cations almost always carry the current as impurities; in vitreous silica containing less than one part per million of sodium and lithium ions, these ions are the current carriers. [See Doremus (1973) for a review.]

The most common method for measuring electrical conductivity of glasses uses metallic electrodes deposited on the glass surface, either from the vapor or as a metallic (silver, gold, or platinum) paste. These electrodes are "blocking," because the ions in the glass do not move through the metal electrodes, and so are blocked at the surface. With direct current these blocking electrodes lead to strong polarization effects, so that the current decreases sharply with time for a constant applied voltage. Alternating current must therefore be used. Above about 200°C for samples of a few millimeters thickness a frequency of 1000 Hz is satisfactory. There is a range of frequencies over which the measured conductivity is constant, which is an indication that the measured conductivity is also the dc value.

A Wheatstone bridge is often used to measure the unknown resistance. At lower temperatures certain precautions are necessary. Surface conductivity can become important; to measure only volume or bulk conductivity, an additional (guard) electrode can be deposited on the surface around one electrode and grounded. To measure high resistivities at low temperatures, thin glass films or bulbs can be used.

A less common but more direct method is to use electrodes that supply or take up ions at the glass surfaces (nonblocking electrodes). Fused salts, for example alkali nitrates, make suitable electrolytes; closed-end tubes of glass are a convenient geometry. A direct current can be used for an accurate measurement of the conductivity with non-blocking electrodes. The greatest error comes from uncertainties in the electrode area and sample thickness.

The conductivity σ of a material is related to its dimensions and resistance R,

$$\sigma = (1/R)(L/A), \tag{1}$$

where A is the electrode area and L the sample thickness. The resistivity is $\rho = 1/\sigma$. The units of ρ are Ω cm and of σ, Ω^{-1} cm^{-1}.

Below the glass transition temperature (annealing point), the electrical conductivity of glass usually fits an Arrhenius relation,

$$\sigma = \sigma_0 \exp(-E/RT), \tag{2}$$

where E (the activation energy) and σ_0 are not functions of temperature. Figure 13.1 shows that above the glass transition temperature the conductivity no longer fits Eq. (2).

Over wide ranges of temperature Eq. (2) is not obeyed for some glasses, even below the glass transition temperature. In vitreous silica, E varies from about 150 kJ mole^{-1} below 280°C to 100 kJ mole^{-1} above 600°C (see Chapter 2).

The intersection of the lines on the log ρ versus $1/T$ plot at low and intermediate temperatures has been used to determine a transition temperature. However, this temperature of about 430°C for the glass in Fig. 13.1 is considerably lower than the temperature of about 490°C, where the viscosity is 10^{13} P, which is the more conventional transition temperature. Thus it appears that ionic transport changes character at a lower temperature than thermal expansion, the more usual property used as a criterion for transition temperature.

Stress influences the electrical conductivity of glass; a quenched glass is expanded and has a higher conductivity compared to an annealed specimen. The effects of hydrostatic compression and quenching expansion on the electrical conductivity of a soda-lime glass are shown in Fig. 13.2. The log conductivity is linear with stress, since small volume changes are proportional to the stress S. Thus,

$$\sigma = \sigma_0 \exp(V^*S/RT), \tag{3}$$

where σ_0 and V^*, the activation volume, are independent of stress at constant temperature. From Fig. 13.2, V^* is greater for quenching expansion than for hydrostatic compression ($V^* = 3.7$ cm^3 mole^{-1}). For binary alkali silicate

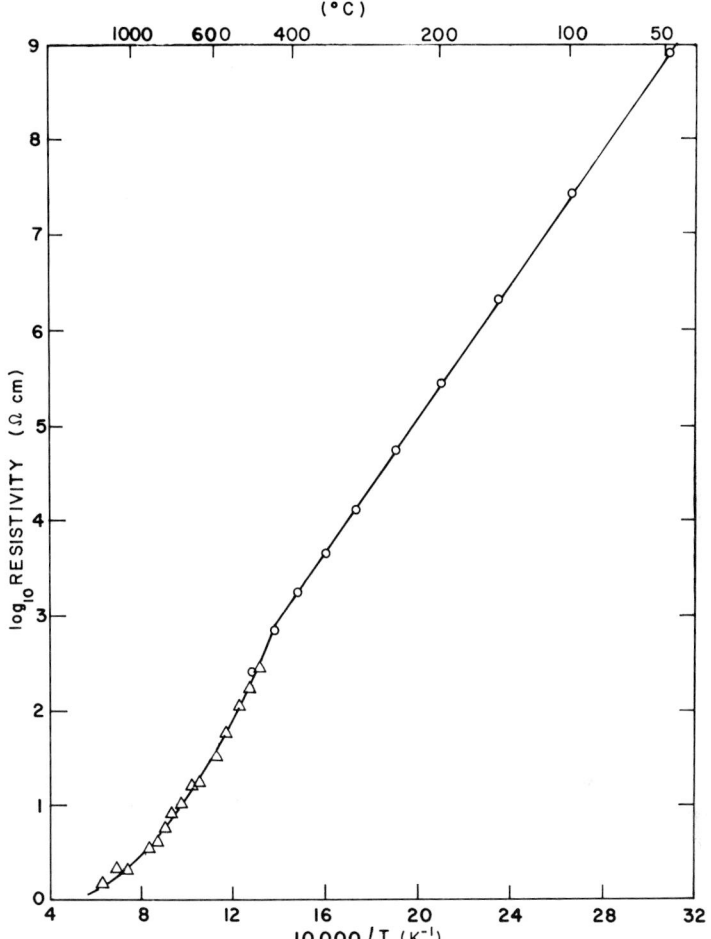

Fig. 13.1. Electrical resistivity of a $26.5Na_2O-73.5SiO_2$ (wt. %) glass as a function of reciprocal temperature. $\bigcirc$, Seddon *et al.* (1932); $\triangle$, Babcock (1934).

glasses, V^* depends strongly on the size of the alkali ion: lithium, 1.0 cm^3 mole^{-1}; sodium 3.4; potassium, 6.0 (Hamann, 1965).

In ionic conductors the electrical conductivity is related to the diffusion coefficient D of the ions by the Einstein equation,

$$\sigma = Z^2 F^2 Dc/fRT, \tag{4}$$

where Z is the valence of the conducting ions, c their concentration, and F the Faraday. A correction factor f is sometimes needed for glasses; f is unity for glasses with low alkali concentrations, and decreases to 0.3 to 0.5 for higher alkali concentrations. With Eq. (4), data on tracer diffusion coefficients of ions can be used to calculate the electrical conductivity of the glass in which

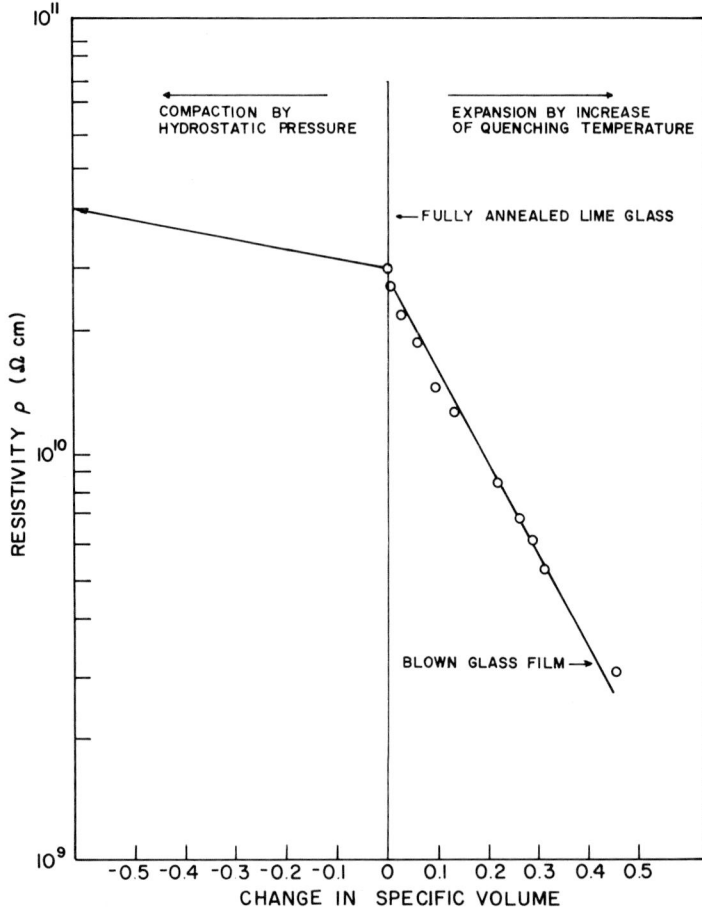

Fig. 13.2. Electrical resistivity of a commercial soda-lime silicate glass (Corning 0080; see Chapter 3, Table 3.1, for composition) as a function of changes in its specific volume caused by hydrostatic pressure or by quenching from various temperatures. From Charles (1963).

they are conducting, as long as only one ionic species carries current. If several current carrying ions are present, σ is a sum:

$$\sigma = \frac{F^2}{RT} \sum_i \frac{Z_i^2 D_i c_i}{f_i}. \tag{5}$$

Equations (4) and (5) show that as the concentration c of alkali ions in a silicate glass increases, the conductivity increases. The diffusion coefficient D of alkali ions also increases as the alkali concentration increases, giving a further increase in conductivity.

The addition of an oxide of a higher-valent metal ion to an alkali silicate glass leads to a decrease in the ionic mobility of the alkali ion. The decrease is related to the ionic size of the added ion for divalent ions; the largest change was found for barium ions, with a decreasing effect for lead, strontium, and calcium ions and a relatively small change for addition of magnesium, zinc, or beryllium oxides. Addition of aluminum oxide increases the ionic mobility and decreases its activation energy.

When a second alkali oxide is added to an alkali silicate glass, the conductivities decrease sharply. This "mixed-alkali" effect is shown in Fig. 13.3 for mixed binary alkali silicate glasses. Measurements of diffusion coefficients in such glasses show that the mobility of each ion is decreased by the addition of the other. The decrease in conductivity results from these reductions in mobility, but the reason for the mobility reductions in terms of the mechanism of diffusion is uncertain.

Reviews of electrical conductivity of glass are in Owen (1963) and Doremus (1973); ionic diffusivity is reviewed in Frischat (1975) and Terai and Hayami (1975).

TABLES AND FIGURES II

Tables 13.1–13.19 and Fig. 13.4 present data on the electrical conductivity of binary glasses, and Tables 13.20–13.123 and Figs. 13.5–13.15 of ternary glasses.

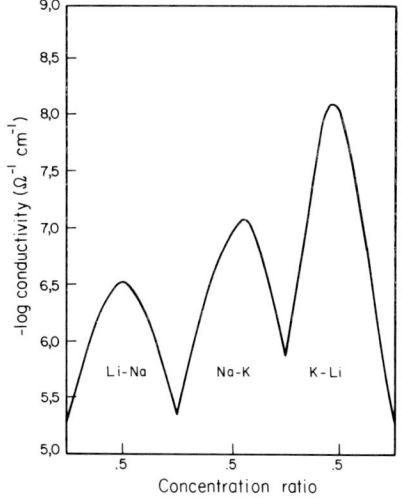

Fig. 13.3. The electrical conductivity of mixed binary alkali silicate glasses containing 26 mole % alkali and 74 mole % SiO_2 at 250°C. From Lengel and Boksay (1963).

BIBLIOGRAPHY

Aleinikov, F. K. (1970). *In* "Stekloobraznaye Sostoyaniye," p. 157. Erevan.
Aleinikov, F. K., Vaytkus, Y. P., and Zhikyavichyute, I. J. (1967). *Liet. TSR Mokslv Akad. Darb.*, *Ser. B* **2**, p. 75.
Appen, A. A., and Gan'Fu-Si (1959). *Fiz. Tverd. Tela (Leningrad)* **1**, 1529.
Appen, A. A., and Gan'Fu-Si (1960). *In* "Stekloobraznaye Sostoyaniye," p. 493. Moscow.
Bahat, D. (1969). *J. Mater. Sci.* **4**, 855.
Barton, J. L. (1966). *Verres Refract.* **20**, 328.
Belyustin, A. A., and Valkov, S. E. (1967). *Sov. Electrochem.* (*Engl. Transl.*) **3**, 781.
Bezrodn'iy, V. G., and Kudyshkina, A. S. (1974). *In* "Investigations of Processes of Improved Technology in the Production of Polymer Materials and Glass," p. 78. Ivanovo.
Blank, K. (1966). *Glastech. Ber.* **39**, 489.
Bockris, J. O'M., Kitchener, J. A., Ignatowicz, S., and Tomlinson, J. W. (1948). *Discuss. Faraday Soc.* **4**, 265.
Bockris, J. O'M., Kitchener, J. A., Ignatowicz, S., and Tomlinson, J. W. (1952). *Trans. Faraday Soc.* **48**, 75.
Boricheva, V. N. (1956). Investigation of electrical conductivity of certain silicate glasses in wide temperature range. Avtoref., Candidate's Dissertation, Leningrad.
Botvinkin, O. K., and Yaroker, K. G. (1966). *Steklo* **3**, 1.
Charles, R. J. (1963). *J. Am. Ceram. Soc.* **46**, 235.
Charles, R. J. (1966). *J. Am. Ceram. Soc.* **49**, 55.
Cheng, Chang-ying (1965). *J. Chin. Silic. Soc.* **4**, 58.
Danilova, N. P. (1970). "Electrical Conductivity of Non-Alkali Gallium-Aluminosilicate Glasses," 2332–70 Dep. VINITI.
Deshkovskaya, A. A., and Bobkova, N. M. (1974). *In* "Stekloobraznaye Sostoyaniye," p. 103. Erevan.
Doremus, R. H. (1973). "Glass Science." Wiley, New York.
El-Bayoumi, O. H., and MacCrone, R. K. (1976). *J. Am. Ceram. Soc.* **59**, 386.
Ershov, O. S., and Shul'ts, M. M. (1974). *Inorg. Mater.* (*Engl. Transl.*) **10**, 492.
Ershov, O. S., and Shul'ts, M. M. (1975). *Inorg. Mater.* (*Engl. Transl.*) **11**, 266.
Ershov, O. S., Dimakov, I. V., Markova, T. P., and Shul'ts, M. M. (1972). *Inorg. Mater.* (*Engl. Transl.*) **8**, 1606.
Erznkyan, E. A. (1971). *In* "Stekloobraznie Sistemy i Novye Stekla na ikh Osnove," p. 66. Moscow.
Erznkyan, E. A., and Kostanyan, K. A. (1968). *Arm. Khim. Zh.* **21**, 759.
Erznkyan, E. A., and Kostanyan, K. A. (1970). *In* "Stekloobraznaye Sostoyaniye," p. 199. Erevan.
Evstrop'ev, K. K. (1970). *In* "Stekloobraznaye Sostoyaniye ," p. 25. Erevan.
Evstrop'ev, K. K., and Kharyuzov, V. A. (1961). *Dokl. Akad. Nauk SSSR* **136**, 140.
Evstrop'ev, K. K., and Pavlovskii, V. K. (1967). *Inorg. Mater.* (*Engl. Transl.*) **3**, 592.
Evstrop'ev, K. K., Kuznetsov, A. Y., and Mel'nikova, I. G. (1951). *Zh. Tekh. Fiz.* **21**, 104.
Evstrop'ev, K. K., Veksler, G. N., and Kondrat'eva, B. S. (1974). *Dokl. Akad. Nauk SSSR* **215**, 902.
Foex, M. (1944). *C. R. Hebd. Seances Acad. Sci.* **218**, 196.
Frischat, G. H. (1975). "Ionic Diffusion in Oxide Glasses." Trans. Tech. Publ., Bay Village, Ohio.
Geokchyan, O. K. (1972). Investigations of electrical conductivity of non-alkaline and slightly alkaline glass in fused conditions. Avtoref., Candidate's Dissertation, Erevan.
Grechanik, L. A., Famberg, E. A., and Zertsalova, I. N. (1962). *Fiz. Tverd. Tela (Leningrad)* **4**, 454.
Grechanik, L. A., Famberg, E. A., and Zertsalova, I. N. (1963). *J. Appl. Chem. USSR* (*Engl. Transl.*) **36**, 88.
Gupta, Y. P., and Mishra, U. D. (1969). *J. Phys. Chem. Solids* **30**, 1327.
Hakim, R. M., and Uhlmann, D. R. (1967). *Phys. Chem. Glasses* **8**, 174.
Hakim, R. M., and Uhlmann, D. R. (1971). *Phys. Chem. Glasses* **12**, 132.
Hamann, S. D. (1965). *Aust. J. Chem.* **18**, 1.
Haven, Y., and Verkerk, V. (1965). *Phys. Chem Glasses* **6**, 38.
Hayward, P. J. (1976). *Phys. Chem. Glasses* **17**, 54.

Hayward, P. J. (1977). *Phys. Chem. Glasses* **18**, 1.
Hirayama, C., and Berg, D. (1961). *Phys. Chem. Glasses* **2**, 145.
Hirayama, C., and Berg, D. (1963). *J. Am. Ceram. Soc.* **46**, 85.
Hogarth, C. A., and Khan, M. N. (1978). *J. Mater. Sci.* **13**, 402.
Hunold, K., and Bruckner, R. (1980). *Glastech. Ber.* **53**, 149.
Ivanov, A. O., and Galant, E. I. (1963). *Opt.-Mekh. Prom.* No. 3, p. 43.
Ivanov, A. O., and Galant, E. I. (1964). *In* "Electrical Properties and Structure of Glass," Material for the IV All-Union Conference on the Glassy State, p. 44. Moscow-Leningrad.
Jain, H., Peterson, N. L., and Downing, H. L. (1983). *J. Non-Cryst. Solids* **55**, 283.
Karapetyan, G. O., Tsekhomskii, V. A., and Yudin, D. M. (1963). *Fiz. Tverd. Tela (Leningrad)* **5**, 627.
Karpechenko, V. G. (1965). "Issledovaniya v Oblasti Khimii Silikatov i Okislov" (Studies in the Chemistry of Silicates and Oxides), p. 84. Izd. Nauka, Leningrad.
Kasymova, S. S. (1974). *J. Appl. Chem. USSR* **47**, 2033.
Keshishyan, T. N., Smirnova, O. M., Fainberg, E. A., and Varshal, B. G. (1975). *Inorg. Mater. (Engl. Transl.)* **11**, 797.
Kharyuzov, V. A. (1959). *Opt.-Mekh. Prom.* No. 7, p. 31.
Kharyuzov, V. A., Mazurin, O. V., and Zubkova, N. M. (1960). *In* "Stekloobraznaye Sostoyaniye," p. 158. Moscow.
Kheifets, V. S. (1965). *Russ. J. Phys. Chem. (Engl. Transl.)* **39**, 809.
Kirakosyan, S. S. (1974). *In* "Stekloobraznaye Sostoyaniye," p. 78. Erevan.
Kokubo, T., Yamashita, K., and Tashiro, M. (1973). *Yogyo Kyokaishi* **81**, 132.
Kolitsch, A., Richter, E., Gehrke, E., Hinz, W., and Müller, W. (1980). *Silikattechnik* **31**, 136.
Kondrat'ev, Y. N., and Chernysh, N. V. (1966). *Inorg. Mater. (Engl. Transl.)* **2**, 1398.
Kondrat'ev, Y. N., and Smirnova, L. A. (1970a). *Inorg. Mater. (Engl. Transl.)* **6**, 298.
Kondrat'ev, Y. N., and Smirnova, L. A. (1970b). *Inorg. Mater. (Engl. Transl.)* **6**, 459.
Kondrat'ev, Y. N., and Smirnova, L. A. (1970c). *Inorg. Mater. (Engl. Transl.)* **6**, 294.
Kostanyan, K. A., and Erznkyan, E. A. (1964). *Izv. Akad. Nauk Arm. SSR* **17**, 613.
Kostanyan, K. A., and Geokchyan, O. K. (1968). *Arm. Khim. Zh.* **21**, 230.
Kostanyan, K. A., and Kirakosyan, S. S. (1974). *Arm. Khim. Zh.* **27**, 547.
Kostanyan, K. A., and Saringyulyan, R. S. (1974). *In* "Elektrokhimiya Rasplavei," p. 193. Moscow.
Kumata, K., Namikawa, H., Nakajima, T., and Munakata, M. (1968). *Yogyo Kyokaishi* **76**, 363.
Kutateladze, K. S. (1974). *Int. Congr. Glass,* [*Pap.*], *10th, Kyoto* **7**, 81.
Kuznetsov, A. Y., and Tsekhomskii, V. A. (1962). *Opt.-Mekh. Prom.* No. 7, p. 27.
Kuznetsov, A. Y., and Tsekhomskii, V. A. (1964). *In* "Electrical Properties and Structure of Glass," Material for the IV All-Union Conference on the Glassy State, p. 105. Moscow-Leningrad.
Kuznetsov, A. Y., and Tsekhomskii, V. A. (1965). *In* "Electrical Properties and Structure of Glass," Vol. 4, p. 136. Consultants Bureau, New York.
Kuznetsova, M. G. (1972a). *Proizvod. Tekh. Stroit. Stekla* No. 2, p. 74.
Kuznetsova, M. G. (1972b). Investigations of the physico-chemical properties of the glass systems $RO-MnO-SiO_2$ and $RO-MnO-B_2O_3$. Candidate's Dissertion, Leningrad.
Kuznetsova, M. G., and Evstrop'ev, K. S. (1972). *Inorg. Mater. (Engl. Transl.)* **8**, 302.
Leko, V. K. (1967). *Inorg. Mater. (Engl. Transl.)* **3**, 1079.
Leko, V. K. (1970). *In* "Stekloobraznaye Sostoyaniye," p. 46. Erevan.
Lengyel, B., and Boksay, Z. (1961). *Z. Phys. Chem.* **217**, 357.
Lengyel, B., and Boksay, Z. (1963). *Z. Phys. Chem.* **223**, 49.
Lengyel, B., Boksay, Z., and Gallyas, F. (1961). *Magy. Tud. Akad. Kem. Tud. Oszt. Kozl.,* 35.
Lengyel, B., Boksay, Z., and Varga, M. (1969). *Z. Phys. Chem. (Leipzig)* **242**, 37.
Lor'yan, S. G., Saringyulyan, R. S., and Kostanyan, K. A. (1977). *Sov. J. Glass Phys. Chem. (Engl. Transl.)* **3**, 559.
Makarova, T. M., Mazurin, O. V., and Malchanov, V. S. (1960). *Izv. Vuzoz., Khim. Khim. Tekhnol.* **3**, 1072.
Mashkovich, M. D., and Varshal, B. G. (1969). *Izv. Akad. Nauk SSSR, Neorg. Mater.* **5**, 335.
Matusita, K., Takayama, S., and Sakka, S. (1980). *J. Non-Cryst. Solids* **40**, 149.

Matveev, M. A., Rzevyusskaya, T. L., and Milevskaya, R. N. (1970). *In* "Stekloobraznaye Sostoyaniye," p. 276. Erevan.

Mazurin, O. V. (1953). Study of the electrical conductivity of silicate glasses. Avtoref., Candidate's Dissertaion, Leningrad.

Mazurin, O. V. (1962). "Electrical Properties of Glass," Tr. Leningr. Technol. Inst. im. Lensoveta, No. 62, p. 139.

Mazurin, O. V., and Borisovskii, E. S. (1957). *Zh. Tekh. Fiz.* **27,** 275.

Mazurin, O. V., and Brailovskaya, R. V. (1960). *Fiz. Tverd. Tela (Leningrad)* **2,** 1477.

Mazurin, O. V., and Brailovskii, V. B. (1959). *Izv. Vuzov., Fiz.* **1,** 117.

Mazurin, O. V., and Kasymova, S. S. (1975). *Inorg. Mater. (Engl. Transl.)* **11,** 433.

Mazurin, O. V., and Petrovskii, G. T. (1956a). *Tr. L. T. I., Sb. Stud. Rabot* p. 30.

Mazurin, O. V., and Petrovskii, G. T. (1956b). *Tr. L. T. I., Sb. Stud. Rabot* p. 51.

Mazurina, E. K., and Evstrop'ev, K. S. (1967). *Izv. Vuzov., Khim. Khim. Tekhnol.* **6,** 673.

Mazurina, E. K., and Evstrop'ev, K. S. (1970). In "Stekloobraznaye Sostoyaniye," p. 195. Erevan.

Melling, P. J., and Duncan, J. F. (1979). *Phys. Chem. Glasses* **20,** 102.

Mel'nikova, I. G., Kuznetsov, A. Y., and Brinberg, V. A. (1950). *Zh. Fiz. Khim.* **24,** 1294.

Moiseev, V. V., and Zhabrev, V. A. (1965). *Silic. Ind.* **30,** 495.

Moiseev, V. V., and Zhabrev, V. A̧. (1969). *Inorg. Mater. (Engl. Transl.)* **5,** 793.

Murawski, L. (1982). *J. Mater. Sci.* **17,** 2155.

Myuller, R. L., and Leko, V. K. (1965). *In* "Khimiya Tverdogo Tela," p. 151. Leningrad.

Myuller, R. L., and Pronkin, A. A. (1965a). *In* "Khimiya Tverdogo Tela," p. 134. Leningrad.

Myuller, R. L., and Pronkin, A. A. (1965b). *In* "Khimiya Tverdogo Tela," p. 173. Leningrad.

Namikawa, H. (1969). *Yogyo Kyokaishi* **77,** 46.

Namikawa, H. (1971). *Zairyo Kagaku* **8,** 49.

Namikawa, H. (1975). *J. Non-Cryst. Solids* **18,** 173.

Namikawa, H., and Kumata, K. (1968). *Yogyo Kyokaishi* **76,** 64.

Negodaev, G. D., and Malinin, V. R. (1974). *In* "Stekloobraznaye Sostoyani," p. 146. Erevan.

Nemilov, S. V., and Kasymova, S. S. (1975a). *J. Appl. Chem. USSR (Engl. Transl.)* **48,** 1299.

Nemkovich, I. K., Yasinskii, L. G., and Levchenya, A. A. (1974). Steklo, Sitally Silik. *Mater.* **3,** 60, 72.

Otto, K. (1966). Phys. Chem. Glasses **7,** 29.

Otto, K., and Milberg, M. E. (1967). *J. Am. Ceram. Soc.* **50,** 513.

Otto, K., and Milberg, M. E. (1968). *J. Am. Ceram. Soc.* **51,** 326.

Owen, A. E. (1963). Electric conduction and dielectric relaxation in glass. *In* "Progress in Ceramic Science" (J. E. Burke, ed.), p. 77. Macmillan, New York.

Parfenov, A. I., Klimov, A. F., and Mazurin, O. V. (1959). *Vestn. Leningr. Univ., Fiz. Khim.* No. 10, p. 129.

Pavlova, G. A. (1958a). *Tr. L. T. I., Work Field Chem. Chem. Technol.* Publ. No. 46, p. 56.

Pavlova, G. A. (1958b). *Izv. Vyssh. Uchebn. Zaved., Khim. Khim. Tekhnol.,* No. **5,** 82.

Pavlova, G. A., and Amatuni, A. N. (1975). *Inorg. Mater. (Engl. Transl.)* **11,** 1443.

Petrovskii, G. (1956). *Tr. L. T. I., Sb. Stud. Rabot* p. 37.

Petrovskii, G. (1959). *Silikaty (Prague)* **3,** 336.

Phillips, S. V., and McMillan, P. W. (1965). *Glass Technol.* **6,** 46.

Pronkin, A. A. (1965). *In* "Khimiya Tverdogo Tela" (Z. V. Borisova, ed.), p. 125. Leningrad Univ., Leningrad.

Rana, M. A., and Douglas, R. W. (1961). *Phys. Chem. Glasses* **2,** 179.

Rao, B. V. J. (1962). *J. Am. Ceram. Soc.* **45,** 555.

Ravaine, D., Diard, J. P., and Souquet, J. L. (1975). *J. C. S. Faraday II* **71,** 1935.

Riebling, E. F. (1966). *J. Electrochem. Soc.* **113,** 920.

Saringyulyan, R. S., and Kostanyan, K. A. (1969). "Viscosity and Electrical Conductivity of Glasses in Wide Intervals of Temperature," 902-69 Dep. VINITI.

Saringyulyan, R. S., and Kostanyan, K. A (1970). *Arm. Khim. Zh., Khim. Tekhnol.* **23,** 928.

Saringyulyan, R. S., and Kostanyan, K. A. (1971). *In* "Stekloobraznaye Sostoyaniye," p. 289. Moscow.

Saringyulyan, R. S., and Kostanyan, K. A. (1974). *In* "Stekloobraznaye Sostoyaniye," p. 31. Erevan.

Schwartz, M., and Mackenzie, J. D. (1966). *J. Am. Ceram. Soc.* **49**, 582.

Seddon, E., Tippett, E. J., and Turner, W. E. S. (1932). *J. Soc. Glass Technol.* **16**, 450.

Segers, L., Fontana, A., and Winand, R. (1975). *Silic. Ind.* **40**, 341.

Sheludyakov, L. N. (1967). *Vestn. Akad. Nauk Kaz. SSR* **23**, 43.

Strauss, S. W. (1956). *J. Res. Natl. Bur. Stand.* (*U.S.*) **56**, 183.

Strauss, S. W., Moore, D. G., Harrison, W. N., and Richards, L. E. (1956). *J. Res. Natl. Bur. Stand.* (*U.S.*) **56**, 135.

Sviridov, S. I., Roskova, G. P., Moiseev, V. V., and Zhabrev, V. A. (1976). *Sov. J. Glass Phys. Chem.* (*Engl. Transl.*) **2**, 514.

Tarlakov, Y. P., Pronkin, A. A., Kekeliya, D. I., Verulashvili, R. D., and Kogan, V. E. (1980). *Izv. Akad. Nauk Gruz. SSR, Ser. Khim.* **6**, 166.

Taylor, H. E. (1959). *J. Soc. Glass Technol.* **43**, 124.

Terai, R. (1968a). *Yogyo Kyokaishi* **76**, 189.

Terai, R. (1968b). *Osaka Kogyo Gijutsu Shikensho Kiho* **19**, 283.

Terai, R. (1969). *Phys. Chem. Glasses* **10**, 146.

Terai, R. (1970). *Osaka Kogyo Gijutsu Shikensho Kiho* **21**, 64.

Terai, R. (1971). *Osaka Kogyo Gijutsu Shikensho Kiho* **22**, 73.

Terai, R. (1972). *Osaka Kogyo Gijutsu Shikensho Kiho* **23**, 36.

Terai, R. (1975). *Rep. Govt. Ind. Res. Inst. Osaka* 347.

Terai, R., and Hayami, R. (1975). *J. Non-Cryst. Solids* **18**, 217.

Terai, R., and Kitaoka, T. (1968). *Yogyo Kyokaishi* **76**, 11.

Terai, R., and Kitaoka, T. (1969). *Osaka Kogyo Gijutsu Shikensho Kiho* **20**, 58.

Terai, R., Kitaoka, T., and Ueno, T. (1969a). *Yogyo Kyokaishi* **77**, 88.

Terai, R., Kitaoka, T., and Ueno, T. (1969b). *Osaka Kogyo Gijutsu Shikensho Kiho* **20**, 198.

Tickle, R. E. (1967). *Phys. Chem. Glasses* **8**, 101.

Topping, J. A., and Murthy, M. K. (1974). *J. Am. Ceram. Soc.* **57**, 281.

Toropov, N. A., and Bryantsev, B. A. (1965). *In* "Structural Transformations at High Temperatures," p. 205. Moscow.

Tsekhomskaya, T. S. (1966). Candidate's Dissertation, Leningrad.

Tsekhomskii, V. A. (1966). Polyconducting glasses based on the oxides of iron and titanium. Candidate's Dissertation, Leningrad.

Tsekhomskii, V. A., Mazurin, O. V., and Evstrop'ev, K. K. (1963). *Fiz. Tverd. Tela* (*Leningrad*) **5**, 586.

Urnes, S. (1959). *Glass Ind.* **40**, 237.

Vakhrameev, V. I. (1968). *Steklo* No. 3, p. 84.

Vakhrameev, V. I. (1970). *In* "Stekloobraznaye Sostoyaniye," p. 58. Erevan.

Vakhrameev, V. I., and Evstrop'ev, K. S. (1969). *Inorg. Mater.* (*Engl. Transl.*) **5**, 82.

Vargin, V. V., and Antonova, E. A. (1956). *Zh. Prikl. Khim.* (*Leningrad*) **29**, 1749.

Wejnarth, A. (1934). *Trans. Electrochem. Soc.* **65**, 177.

Zertsalova, I. N. (1964). *In* "Electrical Properties and Structure of Glass," Material for the IV All-Union Conference on the Glassy State, p. 112. Moscow-Leningrad.

Zertsalova, I. N. (1965). *In* "Electrical Properties and Structure of Glass," Vol. 4, p. 141. Consultants Bureau, New York.

Zhabrev, V. A. (1970). Studies of diffusion processes in glass by radioactive tracer technique. Avtoref., Candidate's Dissertation, Leningrad.

Zhabrev, V. A., and Moiseev, V. V. (1965). *In* "Stekloobraznaye Sostoyaniye," p. 288. Moscow.

Zhabrev, V. A., and Moiseev, V. V. (1970). *In* "Stekloobraznaye Sostoyaniye," Vol. 5, p. 143.

Zhunina, L. A., Kuz'menkov, M. I., and Yaglov, V. N. (1974). "Pyroxene Glass Ceramics." Minsk.

TABLE 13.1

Arrhenius Parameters for the Electrical Conductivity of Lithium Silicate Glasses[a]

Composition (mole % Li_2O)	$-\log \sigma$ at 100°C (Ω^{-1} cm^{-1})	Temperature range (°C)	$\log \sigma_0$ (Ω^{-1} cm^{-1})	E (kcal mole^{-1})
6.7	14.48	250–320	2.87	29.6
14.7	7.23	55–130	1.50	14.9
30.7	6.62	25–90	1.94	14.6
39.8	5.99	25–60	2.16	13.9

[a] Data from Charles (1963, 1966) and Namikawa (1975).

TABLE 13.2

Electrical Conductivities of Lithium Silicate Glasses

Composition (mole % Li_2O)	$-\log \sigma$ (Ω^{-1} cm^{-1}) at temperature (°C)				Reference[a]
	150	250	300	350	
10		5.38		4.28	1
20		4.36		3.36	
30		4.17		3.19	
40		3.54		2.62	
16.6	5.95		3.95		2
20	6.18		3.80		
30	5.55		3.52		
33.3	5.24		3.36		
36	4.93		3.23		

[a] Data from (1) Otto and Milberg (1968) and (2) Mazurin and Borisovskii (1957).

TABLE 13.3

Arrhenius Parameters for the Electrical Conductivity of Sodium Silicate Glasses

Composition (mole % Na_2O)	$-\log \sigma$ at 100°C (Ω^{-1} cm^{-1})	Temperature range (°C)	σ_0 (Ω^{-1} cm^{-1})	E (kcal mole^{-1})	Reference[a]
20		$100-T_g$	66.3	17.01	1
24			102	16.71	
26			119	16.37	
30			110	14.96	
33			96.7	14.18	
40			142	13.55	
7.9		$<T_g$	25	17.98	2
13.0			35	17.29	
17.4			36	17.29	
18.6			129	22.36	
18.9			145	22.36	
24.3			155	17.52	
26.2			186	16.83	
30.0			309	16.37	
40.0			141	13.6	
16.4		200–450	120	17.7	3
19.8			127	16.8	
24.9			96	15.3	
34.0			116	13.6	
7.8	9.89		3.5	17.0	4
15.1	8.15		22.4	16.2	
30.2	6.58		83.2	14.5	
44.2	5.15		131.8	12.4	
14	8.13	50–450		16.8	5
21	7.85			17.1	
22	7.80			17.1	
25	7.41			16.2	
30	6.58			15.3	
33.3	6.23			14.5	
40	6.51			13.9	

[a] Data from (1) Blank (1966), (2) Haven and Verkerk (1965), (3) Terai (1969, 1975) and Terei et al. (1969a,b), (4) Charles (1966) and Namikawa (1975), and (5) Ravaine et al. (1975).

TABLE 13.4

Electrical Conductivities of Sodium Silicate Glasses at Various Temperatures

Composition (mole % Na_2O)	$-\log \sigma$ (Ω^{-1} cm^{-1}) at temperature (°C)				Reference[a]
	150	250	300	350	
5	10.45		7.33		1
13	6.96		4.79		
20	6.45		4.36		
27	5.87		3.94		
30	5.48		3.64		
33.3	5.06		3.34		
36	4.89		3.22		
40	4.58		2.97		
45	4.33		2.69		
48	4.09		2.58		
7.5	7.59		5.30		2
10	7.35		5.18		
13	6.90		4.77		
20	6.80		4.64		
25	6.05		4.03		
30	5.75		3.78		
10		6.14		4.96	3
20		4.85		3.80	
30		4.42		3.46	
40		3.59		2.66	

[a] Data from (1) Mazurin and Barisovskii (1957), (2) Evstrop'ev and Pavlovskii (1967), and (3) Otto and Milberg (1968).

TABLE 13.5

Arrhenius Parameters for the Electrical Conductivity of Potassium Silicate Glasses[a]

Composition (mole %)	$-\log \sigma$ at 100°C (Ω^{-1} cm^{-1})	Temperature range (°C)	$\log \sigma_0$ (Ω^{-1} cm^{-1})	E (kcal mole^{-1})
7.6	11.98		1.21	22.5
14.2	9.49		1.24	18.3
30.4	6.81		1.57	14.3
41.7	5.97		1.65	13.0
20	8.98	50–450		19.0
25	7.68	50–450		16.3
33.33	6.51	50–450		14.8

[a] Data from Charles (1966), Namikawa (1975), and Ravaine et al. (1975).

TABLE 13.6

Electrical Conductivities of Potassium Silicate Glasses at Various Temperatures

Composition (mole % K_2O)	$-\log \sigma$ (Ω^{-1} cm^{-1}) at temperature (°C)						Reference[a]
	150	200	250	300	350	400	
3	13.59			9.97			1
5	11.29			8.33			
7.5	9.60			7.02			
10	8.68			6.24			
15	8.34			5.75			2
20	7.55			5.27			
23	6.98			4.75			
27	6.57			4.28			
33	5.77			3.76			
40	5.56			3.58			
10			7.47		6.10		3
20			5.47		4.39		
30			4.34		3.38		
40			4.00		3.01		
12.3		7.20		5.90		4.95	4
16.8		6.75		5.34		4.28	
19.1		5.94		4.76		3.93	
21.5		5.70		4.55		3.74	
24.7		5.44		4.30		3.45	
27.3		5.29		4.14		3.34	

[a] Data from (1) Evstrop'ev and Pavlovskii (1967), (2) Mazurin and Barisovskii (1957), (3) Otto and Milberg (1968), and (4) Kostanyan and Erznkyan (1964).

TABLE 13.7

Arrhenius Parameters for the Electrical Conductivity of Rubidium Silicate Glasses

Composition (mole % Rb_2O)	$-\log \sigma$ at 100°C (Ω^{-1} cm^{-1})	$\log \sigma_0$ (Ω^{-1} cm^{-1})	E (kcal mole^{-1})	Reference[a]
7.7	13.34	0.90	24.3	1
15.0	10.15	1.05	19.1	
31.5	6.99	1.80	15.0	
39.4	6.48	1.90	14.3	
20		1.48	19.1	2
24		1.50	17.3	

[a] Data from (1) Charles (1966) and Namikawa (1975), and (2) Blank (1966).

TABLE 13.8

Electrical Conductivities of Rubidium Silicate Glasses at Various Temperatures

Composition (mole % Rb_2O)	$-\log \sigma$ (Ω^{-1} cm^{-1}) at temperature (°C)				Reference[a]
	150	250	300	350	
5	13.72		10.12		1
10	9.68		7.10		
15	8.38		5.95		
20	7.53		5.26		
10				7.119	2
20		5.886		4.77	
30		4.469		3.523	
40		4.161		3.137	

[a] Data from (1) Evstrop'ev and Pavlovskii (1967) and (2) Otto and Milberg (1968).

TABLE 13.9

Arrhenius Parameters for the Electrical Conductivity of Cesium Silicate Glasses

Composition (mole % Cs_2O)	$-\log \sigma$ at 100°C (Ω^{-1} cm^{-1})	σ_0 (Ω^{-1} cm^{-1})	E (kcal mole^{-1})	Reference[a]
6.57	15.53	2.95	27.3	1
12.8	11.52	8.13	21.2	
27.9	8.20	15.14	16.0	
12.5		4.33	20.6	2
14.3		9.47	20.4	
16.7		10.1	19.0	
24		18.5	17.1	3

[a] Data from (1) Charles (1966) and Namikawa (1975), (2) Terai (1969, 1970), and (3) Blank (1966).

TABLE 13.10

Electrical Conductivities of Cesium Silicate Glasses[a]

Composition (mole % Cs_2O)	$-\log \sigma$ (Ω^{-1} cm^{-1}) at temperature (°C)	
	250	350
10	8.48	7.08
20	5.886	4.796
30	4.77	3.77
40	4.456	3.432

[a] Data from Otto and Milberg (1968).

TABLE 13.11

Electrical Conductivities of Thallium Silicate Glasses

Composition (mole % Tl_2O)	$-\log \sigma$ (Ω^{-1} cm^{-1}) at temperature (°C)			Reference[a]
	200	250	350	
13.0		7.658	6.167	1
20.5		5.553	4.357	
30.1		5.222	4.022	
39.2		4.745	—	
25	6.06			2
33.3	5.56			
42.9	5.50			
53.8	5.13			
66.7	5.10			

[a] Data from (1) Otto and Milberg (1968) and (2) Karpechenko (1965).

TABLE 13.12

Electrical Conductivity and Its Arrhenius Activation Energy for Magnesium Silicate and Calcium Silicate Glasses[a]

Composition (mole % MO)	$-\log \sigma$ at 450°C (Ω^{-1} cm^{-1})	Temperature range (°C)	E (kcal mole^{-1})
MgO			
50	9.48[b]	200–500	36.4
CaO			
40	8.56[d]	280–450	33.54 ± 0.18
45	8.36[d]	280–450	32.36 ± 0.18
50	8.1[d](12.2[c])	280–450	32.08 ± 0.17
55	8.1[d]	280–450	31.23 ± 0.17

[a] Data from Schwartz and Mackenzie (1966) and Myuller and Leko (1965).
[b] At 400°C.
[c] At 300°C.
[d] Values read from graph.

TABLE 13.13

Arrhenius Parameters for the Electrical Conductivity of Barium Silicate Glasses[a]

Composition (mole % BaO)	σ_0 (Ω^{-1} cm^{-1})	E (kcal mole^{-1})
33.33	1.86	33.1
35	5.75	32.7

[a] Data from Matusita et al. (1980) and Namikawa (1971, 1975).

TABLE 13.14

Electrical Conductivity of Barium Silicate
Glasses[a]

Composition (mole % BaO)	$-\log \sigma^b$ (Ω^{-1} cm^{-1}) at temperature (°C)		
	400	500	600
30	10.36	8.89	7.83
40	9.69	8.33	7.32
50	9.0	7.72	6.73

[a] Data from Evstrop'ev and Kharyuzov (1961).
[b] Values read from graph.

TABLE 13.15

Electrical Conductivities of Cadmium
Silicate Glasses[a]

Composition (mole % CdO)	$-\log \sigma$ (Ω^{-1} cm^{-1}) at 300°C
40	8.6
50	8.5
70	7.35

[a] Data from Tarlakov et al. (1980).

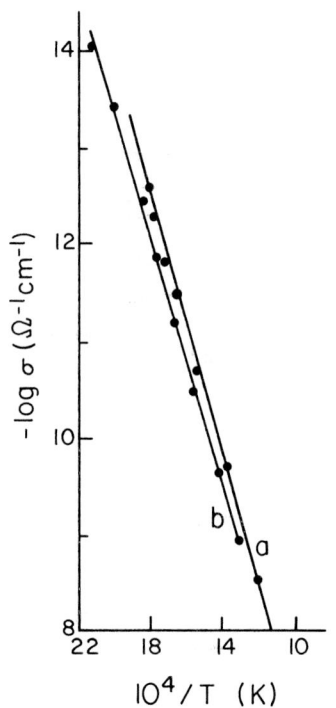

Fig. 13.4. Temperature dependence of the electrical conductivity of the glass systems; (a) BaO–2.65SiO$_2$ and (b) BaO–2SiO$_2$. [From Mashkovich and Varshall (1969).]

TABLE 13.16

Electrical Conductivities of Lead Silicate Glasses at Various Temperatures[a]

Composition (mole % PbO)	$-\log \sigma$ (Ω^{-1} cm^{-1}) at temperature given (°C)		
	150	200	300
30		12.94	10.44
35		12.10	9.89
40		11.54	9.48
45		11.05	9.10
50	12.30	10.69	8.80
55		10.33	8.43
60		10.04	8.11
65		9.76	7.81

[a] Data from Grechanik *et al.* (1963), Pavlova (1958b), and Karpechenko (1965).

TABLE 13.17

Electrical Conductivity and Its Arrhenius Parameters for 3.48PbO–2SiO$_2$ Glass[a]

Temperature (°C)	$-\log \sigma$ (Ω^{-1} cm^{-1})	Temperature range (°C)	σ_0 (Ω^{-1} cm^{-1})	E (kcal mole^{-1})
256	8.827	256–350	91.2	26.1
276	8.403			
296	8.092			
319	7.688			
350	7.162			

[a] Data from Melling and Duncan (1979).

TABLE 13.18

Electrical Conductivity and Its Arrhenius Parameters for M_2O_3–SiO_2 Systems[a]

Composition (mole % M_2O_3)	Temperature range (°C)	$\log \sigma_0$ (Ω^{-1} cm^{-1})	E (kcal mole^{-1})	$-\log \sigma$ (Ω^{-1} cm^{-1}) at temperature (°C)						
				700	900	1100	1300	1500	1700	1900
$M = B$										
2.39	1500–1900	1.49	20.4		5.30	4.72	4.40	4.02	3.76	3.56
4.80	1220–1900	2.18	17.4		5.64	5.16	4.56	4.30	4.10	3.94
9.48	990–1900	2.09	18.6		5.74	5.08	4.69	4.40	4.16	3.98
17.29	1040–1900	2.20	16.5		5.65	4.82	4.48	4.22	4.00	3.84
$M = Al$										
1.69	970–1900	1.45	17.8	5.74	4.82	4.29	3.94	3.67	3.46	3.28
3.34	1010–1900	1.59	16.0	5.34	4.65	4.15	3.76	3.56	3.36	3.20
6.73	1010–1900	1.77	14.2	5.38	4.54	4.02	3.74	3.52	3.34	3.20

[a] Data from Lor'yan et al. (1977).

TABLE 13.19

Electrical Conductivities of Titanium Silicate Glass[a]

Composition (mole % TiO_2)	$-\log \sigma \ (\Omega^{-1} \ cm^{-1})$ at temperature (°C)	
	100	300
6.1	12.3	10.2

[a] Data from Pavlova and Amatuni (1975).

TABLE 13.20

Values of Arrhenius Parameters for the Electrical Conductivity of Lithium–Sodium Silicate Glasses[a]

Composition (mole %)			Temperature range (°C)	$\log \sigma_0$ $(\Omega^{-1} \ cm^{-1})$	E (kcal mole^{-1})
Li_2O	Na_2O	SiO_2			
10.3	23	66.7	20–250	3.7	22.1
16.6	16.7	66.7		3.6	23.3
21.3	12	66.7		3.3	22.4
25.3	8	66.7		3.4	21.4
29.3	4	66.7		3.1	19.1

[a] Data from Pronkin (1965).

TABLE 13.21

Electrical Conductivities of Lithium–Sodium Silicate Glasses[a]

Composition (mole %)			$-\log \sigma \ (\Omega^{-1} \ cm^{-1})$ at temperature (°C)	
Li_2O	Na_2O	SiO_2	150	300
8.33	25	66.67	7.26	4.55
12.49	20.83	66.68	7.94	4.91
16.66	16.66	66.68	8.17	5.11
20.83	12.49	66.68	7.94	4.93
25	8.33	66.67	7.36	4.59

[a] Data from Mazurin (1953) and Mazurin and Borisovskii (1957).

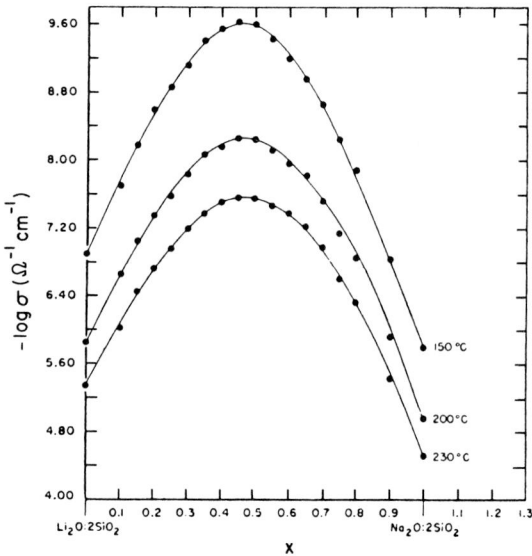

Fig. 13.5. Electrical conductivity–composition isotherms at (a) 150°C, (b) 200°C, and (c) 230°C for glasses in the system $(1 - x)$-Li$_2$O–xNa$_2$O–2SiO$_2$. [From Strauss (1956).]

TABLE 13.22

Values of Arrhenius Parameters for the Electrical Conductivity of Lithium–Potassium Silicate Glasses[a]

Composition (mole %)			Temperature range (°C)	$\log \sigma_0$ (Ω^{-1} cm^{-1})	E (kcal mole^{-1})
Li$_2$O	K$_2$O	SiO$_2$			
5	8	87	20–250	1.6	25.8
7	6	87		1.3	26.5
9	4	87		1.6	24.7

[a] Data from Pronkin (1965).

TABLE 13.23

Electrical Conductivities of Li$_2$O–K$_2$O–SiO$_2$ Glasses[a]

Composition (mole %)			$-\log \sigma$ (Ω^{-1} cm^{-1}) at temperature (°C)			
Li$_2$O	K$_2$O	SiO$_2$	150	200	250	300
3.7	13	83.3	10.68	9.35	8.28	7.34
6.7	10	83.3	11.79	10.40	9.19	8.20
9.7	7	83.3	11.09	10.44	9.22	8.15
12.7	4	83.3	9.93	8.59	7.52	6.59

[a] Data from Aleinikov et al. (1967).

TABLE 13.24

Electrical Conductivities of Lithium–Potassium Silicate Glasses

Composition (mole %)			$-\log \sigma$ (Ω^{-1} cm^{-1}) at temperature (°C)		Reference[a]	Composition (mole %)			$-\log \sigma$ (Ω^{-1} cm^{-1}) at temperature (°C)		Reference[a]
Li$_2$O	K$_2$O	SiO$_2$	150	300		Li$_2$O	K$_2$O	SiO$_2$	150	300	
5	15	80	9.70	6.75	1	10.12	16.87	73.01	9.90	6.80	2
5	25	70	8.12	5.40		12.49	20.83	66.68	9.24	6.24	
10	10	80	11.60	7.97		13.5	13.5	73.0	10.58	7.25	
10	20	70	9.80	6.66		15	25	60	8.86	5.86	
15	5	80	9.66	6.45		16.66	16.66	66.68	10.08	6.81	
15	15	70	10.75	7.20		16.87	10.12	73.01	9.75	6.60	
20	10	70	9.51	6.32		20	20	60	9.68	6.40	
25	5	70	7.35	4.85		20.25	6.75	73.0	8.18	5.46	
3.37	23.62	73.01	7.60	5.10	2	20.83	12.49	66.68	9.36	6.17	
4.17	29.16	66.67	7.09	4.73		23.62	3.37	73.01	6.79	4.46	
5	35	60	6.80	4.43		25	8.33	66.67	7.71	5.04	
6.75	20.25	73.0	8.68	5.92		25	15	60	8.65	5.64	
8.33	25	66.67	8.23	5.53		29.16	4.17	66.67	6.23	4.03	
10	30	60	7.90	5.14		30	10	60	7.24	4.60	
						35	5	60	5.72	3.61	

[a] Data from (1) Evstrop'ev (1970) and (2) Mazurin (1953) and Mazurin and Borisovskii (1957).

TABLE 13.25

Electrical Conductivities of Lithium–Cesium Silicate Glasses[a]

Composition (mole %)			$-\log \sigma$ (Ω^{-1}) at temperature (°C)		Composition (mole %)			$-\log \sigma$ (Ω^{-1} cm^{-1}) at temperature (°C)	
Li_2O	Cs_2O	SiO_2	150	300	Li_2O	Cs_2O	SiO_2	150	300
3	14	83	7.74	5.35	24	6	70	7.04	4.63
6	21	73	8.74	6.15	24	9	67	7.88	5.17
9	18	73	9.78	6.54	27	3	70	6.02	3.85
12	15	73	11.31	8.08	27	6	67	6.71	4.28
15	12	73	11.44	8.16	30	3	67	5.82	3.72
18	9	73	10.15	6.91	30	6	64	5.99	3.87
21	6	73	7.83	5.14	30	9	61	6.63	4.25
24	3	73	6.51	4.36					

[a] Data from Parfenov et al. (1959).

TABLE 13.26

Electrical Conductivities and Activation Energies of Li_2O–Cs_2O–SiO_2 Glasses[a]

Composition (mole %)			σ (Ω^{-1} cm^{-1}) at temperature (°C)		E
Li_2O	Cs_2O	SiO_2	250	350	(kcal mole^{-1})
5	10	85	4.0×10^{-11}	7.143×10^{-8}	37.3
7.5	7.5	85	5.556×10^{-11}	3.57×10^{-8}	36.5
10	5	85	1.563×10^{-10}	7.143×10^{-8}	35.0

[a] Data from Hakim and Uhlmann (1967).

TABLE 13.27

Arrhenius Parameters for the Electrical Conductivity of Sodium–Potassium Silicate Glasses

Composition (mole %)			$\log \sigma_0$	E	Temperature range	
Na_2O	K_2O	SiO_2	$(\Omega^{-1}\,cm^{-1})$	$(kcal\ mole^{-1})$	$(°C)$	Reference[a]
1	4	95	0.5	24.5	20–250	1
2	3	95	1.1	25.7		
3	2	95	1.4	24.8		
3	10	87	1.5	23.3		
4	1	95	1.2	24.7		
5	8	87	2.0	24.9		
7	6	87	2.2	24.4		
9	4	87	1.95	23.3		
11	2	87	1.85	20.7		
12	12	76	3.40	25.2	$100 - T_g$	2
16.67	16.67	66.67	4.22	25.6		
18	6	76	3.00	22.4		

[a] Data from (1) Pronkin (1965) and (2) Blank (1966).

TABLE 13.28

Electrical Conductivites of Some $Na_2O–K_2O–SiO_2$ Glasses[a]

Composition (mole %)			$-\log \sigma\ (\Omega^{-1})$ at temperature $(°C)$			
Na_2O	K_2O	SiO_2	150	200	250	300
12.5	37.5	50.0	8.35	6.84	5.64	—
25	25	50	9.12	7.66	5.81	5.30
37.5	12.5	50.0	8.31	6.90	5.80	—

[a] Data from Aleinikov et al. (1967).

TABLE 13.29

Electrical Conductivities of Sodium–Potassium Silicate Glasses

Composition (mole %)			$-\log\sigma\ (\Omega^{-1}\ cm^{-1})$ at temperature (°C)			Reference[a]
Na_2O	K_2O	SiO_2	150	300	400	
5	15	80	9.11	6.11	4.86	1
7.5	12.5	80	9.46	6.34	5.03	
10	10	80	9.37	6.30	5.00	
12.5	7.5	80	9.16	6.10	8.84	
15	5	80	8.45	5.56	4.35	
3	10	87	9.77	6.68		2
4	12	84	9.79	6.60		
5	8	87	9.99	6.87		
7	6	87	9.95	6.82		
8	8	84	9.97	6.85		
9	4	87	9.45	6.39		
10	6	84	9.65	6.50		
11	2	87	8.42	5.65		
12	4	84	8.81	5.87		
2.5	10	87.5	10.17	7.08		3
5	10	85	10.0	6.86		
5	15	80	9.31	6.30		
7.5	12.5	80	9.59	6.45		
10	2.5	87.5	8.35	5.71		
10	5	85	9.07	6.13		

Composition (mole %)			$-\log\sigma\ (\Omega^{-1}\ cm^{-1})$ at temperature (°C)			Reference[a]
Na_2O	K_2O	SiO_2	150	300	400	
10	10	80	8.85	5.82	—	3
10	15	75	8.16	5.36		
10	20	70	7.60	4.87		
12.5	7.5	80	9.31	6.16		
15	5	80	8.68	5.68		
15	10	75	8.86	5.73		
20	10	70	8.22	5.21		
6.75	20.25	73	8.29	5.43	—	4
8.33	25	66.67	7.87	4.93		
10	30	60	7.53	4.53		
12.49	20.83	66.68	8.35	5.21		
13.5	13.5	73	9.06	5.83		
16.66	16.66	66.68	8.40	5.20		
20	20	60	8.11	4.89		
20.25	6.75	73	7.93	5.14		
20.83	12.49	66.68	7.87	4.92		
25	8.33	66.67	7.05	4.36		
25	15	60	7.47	4.48		
30	10	60	6.73	4.07		
35	5	60	5.71	3.53		

[a] Data from (1) Pavlova (1958a,b), (2) Makarova et al. (1960), (3) Evstrop'ev (1970), and (4) Mazurin (1953) and Mazurin and Borisovskii (1957).

TABLE 13.30

Electrical Conductivities of Sodium–Rubidium Silicate Glasses[a]

Composition (mole %)			$-\log \sigma\ (\Omega^{-1}\ cm^{-1})$ at temperature (°C)	
Na₂O	Rb₂O	SiO₂	150	300
Na$_2$O	Rb$_2$O	SiO$_2$	150	300
5	15	80	10.16	6.80
10	2.5	87.5	9.21	6.28
10	5	85	9.83	6.70
10	15	75	10.0	6.72
10	10	80	9.40	6.22
15	5	80	8.34	5.43

[a] Data from Evstrop'ev (1970).

TABLE 13.31

Values of Arrhenius Parameters for the Electrical Conductivity of Na$_2$O–Rb$_2$O–SiO$_2$ Glasses[a]

Composition (mole %)			Temperature range (°C)	σ_0 ($\Omega^{-1}\ cm^{-1}$)	E (kcal mole^{-1})
Na$_2$O	Rb$_2$O	SiO$_2$			
12	12	76	$100-T_g$	828	26.5
18	6	76		159	24.2
16.67	16.67	66.66		8680	27.1

[a] Data from Blank (1966).

TABLE 13.32

Parameters of the Equation $\sigma = (B/T)\exp(-E/RT)$ for the Electrical Conductivity of Sodium–Cesium Silicate Glasses[a]

Composition (mole %)			$B \times 10^{-5}$ ($\Omega^{-1}\ cm^{-1}\ K$)	E^b (kcal mole^{-1})
Na$_2$O	Cs$_2$O	SiO$_2$		
25	—	75	1.36 ± 0.07	16.9
22.325	2.675	75	2.49 ± 0.10	19.2
19.55	5.45	75	3.66 ± 0.22	21.4
18.95	6.05	75	3.26 ± 0.16	21.9
16.3	8.7	75	3.64 ± 0.25	24.4
14.875	10.125	75	4.34 ± 0.41	26.0
6.425	18.575	75	0.70 ± 0.035	23.05
—	25	75	0.149 ± 0.007	16.7

[a] Data from Jain et al. (1983).
[b] Temperature range 100–425°C.

TABLE 13.33

Values of Electrical Conductivity and Its Arrhenius Parameters of Sodium–Cesium Silicate Glasses

Composition (mole %)			$-\log \sigma \ (\Omega^{-1} \ cm^{-1})$ at temperature (°C)		Temperature range (°C)	$\log \sigma_0$ $(\Omega^{-1} \ cm^{-1})$	E (kcal mole^{-1})	Reference[a]
Na$_2$O	Cs$_2$O	SiO$_2$	250	350				
5	10	85	8.799	7.0			32.0	1
7.5	7.5	85	9.519	7.255			33.7	
10	5	85	8.708	6.748			31.0	
12.5	12.5	75			170–400	7.3 ± 0.3	30 ± 3	2

[a] Data from (1) Hakim and Uhlmann (1967) and (2) Negodaev and Malinin (1974).

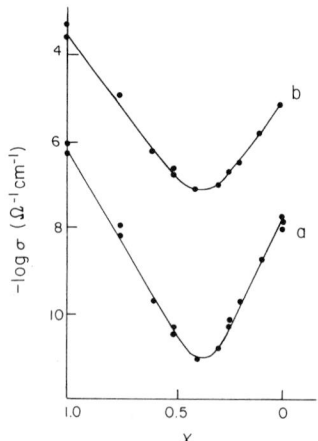

Fig. 13.6. Electrical conductivity–composition isotherms at (a) 200°C and (b) 400°C for glasses in the system xNa$_2$O–$(1 - x)$-Cs$_2$O–5SiO$_2$. [From Terai (1972).]

TABLE 13.34

Values of Electrical Conductivity and Activation Energy of Potassium–Cesium Silicate Glasses[a]

Composition (mole %)			$\sigma \ (\Omega^{-1} \ cm^{-1})$ at temperature (°C)		E (kcal mole^{-1})
K$_2$O	Cs$_2$O	SiO$_2$	250	350	
5	10	85	1.818×10^{-9}	1.5873×10^{-7}	31.1
7.5	7.5	85	5.5555×10^{-10}	1.0×10^{-7}	32.5
10	5	85	3.2258×10^{-9}	2.5×10^{-7}	30.5

[a] Data from Hakim and Uhlmann (1967).

TABLE 13.35

**Values of Electrical Conductivity and Its Arrhenius Activation Energy
of Potassium–Thallium Silicate Glasses[a]**

Composition (mole %)			$-\log \sigma \ (\Omega^{-1} \ cm^{-1})$ at 200°C	E (kcal mole^{-1})
$KO_{0.5}$	$TlO_{0.5}$	SiO_2		
10	40	50	6.47	18.3
20	30	50	7.81	25.6
40	10	50	6.33	19.6

[a] Data from Karpechenko (1965).

TABLE 13.36

**Values of Electrical Conductivity and Activation Energy of Rubidium–Cesium
Silicate Glasses[a]**

Composition (mole %)			$\sigma \ (\Omega^{-1} \ cm^{-1})$ at temperature (°C)		E (kcal mole^{-1})
Rb_2O	Cs_2O	SiO_2	250	350	
5	10	85	5.555×10^{-9}	3.333×10^{-7}	29.0
7.5	7.5	85	3.2258×10^{-9}	2.5×10^{-7}	30.2
10	5	85	1.0×10^{-8}	5.0×10^{-7}	28.7

[a] Data from Hakim and Uhlmann (1967).

TABLE 13.37

Values of Arrhenius Parameters for the Electrical Conductivity of Li_2O–MgO–SiO_2 Glass[a]

Composition (mole %)			Temperature range (°C)	$\log \sigma_0$ ($\Omega^{-1} \ cm^{-1}$)	E (kcal mole^{-1})
Li_2O	MgO	SiO_2			
20	30	50	20–600	2.8	23.05

[a] Data from Myuller and Leko (1965).

TABLE 13.38

Electrical Conductivities of Glasses in the System
$x\text{Li}_2\text{O}-y\text{MgO}-(100 - x - y/2)\text{SiO}_2$ **(mole %)**[a]

x	y	$-\log \sigma\ (\Omega^{-1}\ \text{cm}^{-1})$ at 150°C	x	y	$-\log \sigma\ (\Omega^{-1}\ \text{cm}^{-1})$ at 150°C
15	0	6.24	15	15	8.40
15	1	6.76	25	2	7.10
15	2	7.21	25	10	7.57
15	3	7.60	33	2	6.00
15	4	7.85	33	5	6.33
15	5	7.65	33	10	6.60
15	10	8.15	33	15	7.00

[a] Data from Leko (1970).

TABLE 13.39

Values of Arrhenius Parameters for the Electrical Conductivity of
$\text{Li}_2\text{O}-\text{CaO}-\text{SiO}_2$ **Glasses**[a]

Composition (mole %)			$\log \sigma_0$ $(\Omega^{-1}\ \text{cm}^{-1})$	E^b (kcal mole^{-1})
Li_2O	CaO	SiO_2		
1	49	50	0.8	32.3
2	48	50	2.4	35.7
3	47	50	2.4	35.7
4	46	50	2.3	33.4
6	44	50	2.0	32.3
7	43	50	2.3	31.1
8	42	50	2.4	30.0
10	40	50	2.3	28.8
20	30	50	2.5	23.1

[a] Data from Myuller and Leko (1965).
[b] 20–600°C.

TABLE 13.40

Electrical Conductivities of Glasses in the System
$x\text{Li}_2\text{O}-y\text{CaO}-(100 - x - y/2)\text{SiO}_2$ **(mole %)**[a]

x	y	$-\log \sigma\ (\Omega^{-1}\ \text{cm}^{-1})$ at 150°C	x	y	$-\log \sigma\ (\Omega^{-1}\ \text{cm}^{-1})$ at 150°C
15	2	7.15	33	2	5.90
15	5	7.63	33	5	6.41
15	10	8.20	33	10	6.57
15	15	8.95	33	15	7.03

[a] Data from Leko (1970).

TABLE 13.41

Values of Arrhenius Parameters for the Electrical Conductivity of Li$_2$O–BaO–SiO$_2$ Glasses[a]

Composition (mole %)			σ_0	E
Li$_2$O	BaO	SiO$_2$	(Ω^{-1} cm^{-1})	(kcal mole^{-1})
26.67	6.67	66.66	203	16.9
20.00	13.34	66.66	204	20.3
13.34	20.00	66.66	106	29.8
6.67	26.67	66.66	245	32.9

[a] Data from Matusita *et al.* (1980).

TABLE 13.42

Electrical Conductivities of Glasses in the System xLi$_2$O–yBaO–$(100 - x - y/2)$SiO$_2$ (mole %)[a]

x	y	$-\log \sigma$ (Ω^{-1} cm^{-1}) at 150°C	x	y	$-\log \sigma$ (Ω^{-1} cm^{-1}) at 150°C
15	2	7.15	25	2	7.00
15	5	7.92	25	10	8.50
15	10	9.10	33	2	5.87
15	15	9.96	33	10	7.31

[a] Data from Leko (1970).

TABLE 13.43

Electrical Conductivities of Glasses in the System xLi$_2$O–yMO–$(100 - x - y/2)$SiO$_2$ (mole %)[a]

x	y	$-\log \sigma$ (Ω^{-1} cm^{-1}) at 150°C	x	y	$-\log \sigma$ (Ω^{-1} cm^{-1}) at 150°C
	M-Zn			M-Cd	
25	5	6.90	25	5	6.77
25	10	7.15	25	10	7.21
25	15	7.55	25	15	7.64

[a] Data from Leko (1970).

TABLE 13.44

Values of Arrhenius Parameters for the Electrical Conductivity of Lithium–Lead Silicate Glass[a]

Composition (mole %)			Temperature range (°C)	$\log \sigma_0$ (Ω^{-1} cm^{-1})	E (kcal mole^{-1})
Li$_2$O	PbO	SiO$_2$			
13	13	74	20–350	2.6	24.5

[a] Data from Myuller and Pronkin (1965b).

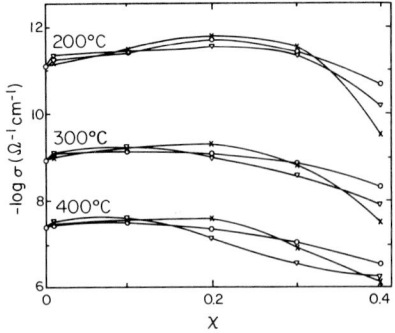

Fig. 13.7. Electrical conductivity–composition isotherms of the glass systems $xM_2O-(1-x)PbO-SiO_2$ at three temperatures: O, Li_2O; ∇, Na_2O; X, K_2O. [From Strauss *et al.* (1956).]

TABLE 13.45

Values of Electrical Conductivity and Its Arrhenius Activation Energy of $Na_2O-BeO-SiO_2$ Glasses[a]

Composition (mole %)			$-\log \sigma \ (\Omega^{-1} \ cm^{-1})$ at 150°C	E (kcal mole^{-1})
Na_2O	BeO	SiO_2		
10	20	70	7.6	17.2
20	10	70	6.5	15.3
20	20	60	6.7	15.7
30	10	60	5.6	14.4
30	20	50	5.7	13.1

[a] Data from Mazurin and Brailovskaya (1960).

TABLE 13.46

Electrical Conductivities of $Na_2O-BeO-SiO_2$ Glasses

Composition (mole %)			$-\log \sigma \ (\Omega^{-1} \ cm^{-1})$ at temperature (°C)									Reference[a]
Na_2O	BeO	SiO_2	300	400	500	600	700	800	1000	1200	1400	
15	10	75	5.14	4.20	3.48	2.72	2.09	1.59	0.98	0.63	0.42	1
15	15	70	5.14	4.14	3.46	2.80	2.16	1.63	0.99	0.63	0.42	
15	20	65	5.22	4.24	3.53	2.87	2.22	1.72	1.06	0.70	0.48	
15	25	60	5.30	4.29	3.56	2.91	2.25	1.68	0.97	0.61	0.37	
15	30	55	5.43	4.40	3.65	2.98	2.34	1.73	0.98	0.60	0.37	
16	16	68	4.88									2
20	12	68	4.46									
24	8	68	4.13									
28	4	68	3.77									

[a] Data from (1) Mazurina and Evstrop'ev (1967, 1970) and (2) Lengyel *et al.* (1969).

TABLE 13.47

Values of Electrical Conductivity and Its Arrhenius Activation Energy of $Na_2O-MgO-SiO_2$ Glasses[a]

Composition (mole %)			$-\log \sigma$ $(\Omega^{-1} cm^{-1})$ at 150°C	E (kcal mole^{-1})
Na_2O	MgO	SiO_2		
10	10	80	7.6	17.7
10	20	70	7.9	18.2
20	10	70	6.9	16.9
20	20	60	7.0	16.9
30	10	60	6.0	14.9
30	20	50	6.3	15.6

[a] Data from Mazurin and Brailovskaya (1960).

TABLE 13.48

Electrical Conductivities of $Na_2O-MgO-SiO_2$ Glasses[a]

Composition (mole %)			$-\log \sigma$ $(\Omega^{-1} cm^{-1})$ at 150°C	Composition (mole %)			$-\log \sigma$ $(\Omega^{-1} cm^{-1})$ at 150°C
Na_2O	MgO	SiO_2		Na_2O	MgO	SiO_2	
7.3	23.1	69.6	10.892	22.2	20.5	57.3	7.214
10.6	32.5	56.9	11.168	22.3	20.3	57.4	9.128
11.2	23.3	65.5	10.64	22.4	20.8	56.8	7.168
11.6	15.0	73.4	8.692	25.1	14.7	60.2	6.51
14.7	14.7	70.6	7.334	25.2	5.6	69.2	6.294
16.7	21.5	61.8	7.311	25.2	9.7	65.1	6.48
17.5	19.6	62.9	7.752	25.3	14.3	60.4	6.551
19.5	7.2	73.3	6.313	25.5	33.5	41.0	9.755
21.8	13.2	65.0	7.156	26.5	9.6	63.9	6.405
22.1	4.9	73.0	6.497	39.4	10.3	50.3	5.703
				39.7	5.2	55.1	5.119

[a] Data from Botvinkin and Yaroker (1966).

TABLE 13.49

Temperature Dependence of Electrical Conductivity of Na_2O–MgO–SiO_2 Glasses

Composition (mole %)			$-\log \sigma$ (Ω^{-1} cm^{-1}) at temperature (°C)												Reference[a]
Na_2O	MgO	SiO_2	150	200	250	300	400	500	600	700	800	1000	1200	1400	
15	5	80				5.02	4.00	3.20	2.34	1.82	1.43	1.05	0.77	0.56	1
15	10	75				4.91	3.91	3.17	2.37	1.83	1.41	0.93	0.62	0.41	
15	15	70				5.03	4.07	3.33	2.50	1.87	1.40	0.87	0.55	0.32	
15	20	65				5.45	4.38	3.61	2.80	2.02	1.52	0.83	0.51	0.29	
15	25	60				5.71	4.63	3.84				0.87	0.48	0.23	
15	30	55				6.10	4.96								
15	35	50				6.76									
10	20	70	8.75	7.92	7.30	6.13									2
13.3	16.7	70	8.02	7.23	6.68	5.38									
16.7	13.3	70	7.23	6.42	5.82	5.20									
20	10	70	7.05	6.30	5.71										
16	16	68				4.91									3
20	12	68				4.66									
24	8	68				4.25									
28	4	68				3.83									

[a] Data from (1) Mazurina and Evstrop'ev (1967, 1970), (2) Cheng (1965), and (3) Lengyel and Boksay (1963).

TABLE 13.50

Values of Electrical Conductivity and Its Arrhenius Parameters of Soda-Lime–Silicate Glasses

Composition (mole %)			$-\log \sigma \, (\Omega^{-1} \, cm^{-1})$ at 150°C	$\log \sigma_0$ $(\Omega^{-1} \, cm^{-1})$	E (kcal mole^{-1})	Reference[a]
Na$_2$O	CaO	SiO$_2$				
2	48	50		0.5	32.3[b]	1
5	45	50		1.2	33.4[b]	
8	42	50		2.0	33.4[b]	
12	38	50		1.9	28.8[b]	
10	10	80	9.0		18.6	2
10	20	70	10.0		21.2	
20	10	70	7.4		18.0	
20	20	60	8.3		19.5	
30	10	60	6.2		15.6	
30	20	50	6.9		16.8	

[a] Data from (1) Myuller and Leko (1965) and (2) Mazurin and Brailovskaya (1960).
[b] 20–600°C.

TABLE 13.51

Electrical Conductivities of Soda-Lime Silicate Glasses at Various Temperatures

Composition (mole %)			$-\log \sigma \, (\Omega^{-1} \, cm^{-1})$ at temperature (°C)							Reference[a]
Na$_2$O	CaO	SiO$_2$	150	200	250	300	400	500	600	
15	10	75				5.63	4.51	3.68	2.73	1
15	15	70				5.85	4.69	3.80	2.91	
15	20	65				6.25	5.08	4.17	3.22	
15	25	60				6.55	5.32	4.41	3.52	
15	30	55				6.76	5.49	4.55		
15	35	50				7.13	5.80	4.82		
10	20	70	10.05	9.10	8.40	7.74				2
13.3	16.7	70	8.86	7.97	7.33	6.74				
16.7	13.3	70	7.95	7.09	6.42	5.90				
20	10	70	7.44	6.58	6.00	5.48				
16	16	68				6.00				3
20	12	68				5.10				
24	8	68				4.50				
28	4	68				3.93				
17	28	55	9.65			6.68				4

[a] Data from (1) Mazurina and Evstrop'ev (1967, 1970), (2) Cheng (1965), (3) Lengyel and Boksay (1963) and Lengyel et al. (1969), and (4) Mazurin (1962).

TABLE 13.52

Values of Electrical Conductivity and Its Arrhenius Activation Energy of Na$_2$O–SrO–SiO$_2$ Glasses[a]

Composition (mole %)			$-\log \sigma \, (\Omega^{-1} \, cm^{-1})$ at 150°C	E (kcal mole^{-1})
Na$_2$O	SrO	SiO$_2$		
10	10	80	9.5	20.4
10	20	70	10.9	23.6
20	10	70	7.7	18.4
20	20	60	9.0	21.3
30	10	60	6.6	17.4
30	20	50	7.6	19.4

[a] Data from Mazurin and Brailovskaya (1960).

TABLE 13.53

Electrical Conductivities of Na$_2$O–SrO–SiO$_2$ Glasses at Various Temperatures

Composition (mole %)			$-\log \sigma$ (Ω^{-1} cm^{-1}) at temperature (°C)											Reference[a]
Na$_2$O	SrO	SiO$_2$	150	200	250	300	400	500	600	700	800	1000	1200	1400	
15	10	75				6.04	4.90	3.97	3.02	2.38	1.88	1.22	0.85	0.59	1
15	15	70				6.37	5.16	4.17	3.17	2.44	1.86	1.19	0.78	0.48	
15	20	65				6.86	5.55	4.59	3.35	2.60	2.08	1.32	0.76	0.42	
15	25	60				7.28	5.90	4.87							
15	30	55				7.76	6.26	5.17							
15	35	50				8.11	6.60	5.48							
10	20	70	10.75	9.78	9.00	8.29									2
13.3	16.7	70	9.56	8.60	7.88	7.23									
16.7	13.3	70	8.37	7.50	6.76	6.21									
20	10	70	7.80	6.90	6.27	5.68									
16	16	68				6.29									3
20	12	68				5.34									
24	8	68				4.63									
28	4	68				3.96									

[a] Data from (1) Mazurina and Evstrop'ev (1967, 1970), (2) Cheng (1965), and (3) Lengyel et al. (1969).

TABLE 13.54

Values of Arrhenius Parameters for the Electrical Conductivity of $Na_2O-BaO-SiO_2$ Glasses

Composition (mole %)			σ_0 $(\Omega^{-1} cm^{-1})$	E (kcal mole^{-1})	$-\log \sigma$ $(\Omega^{-1} cm^{-1})$ at 150°C	Reference[a]
Na_2O	BaO	SiO_2				
26.67	6.67	66.66	205	17.4		1
20.00	13.34	66.66	340	20.9		
13.34	20.00	66.66	418	26.4		
6.67	26.67	66.66	24.9	30.6		
10	10	80		22.1	9.9	2
10	20	70		24.8	11.6	
20	10	70		19.6	8.0	
20	20	60		22.2	9.3	
30	10	60		17.3	6.7	
30	20	50		19.7	7.9	

[a] Data (1) Matusita et al. (1980) and (2) Mazurin and Brailovskaya (1960).

TABLE 13.55

Electrical Conductivities of $Na_2O-BaO-SiO_2$ Glasses at Various Temperatures

Composition (mole %)			$-\log \sigma$ $(\Omega^{-1} cm^{-1})$ at temperature (°C)				Reference[a]
Na_2O	BaO	SiO_2	150	200	250	300	
10	20	70	11.50	10.43	9.60	8.86	1
13.3	16.7	70	10.15	9.10	8.31	7.67	
16.7	13.3	70	8.85	7.87	7.10	6.52	
20	10	70	8.06	7.16	6.52	5.96	
16	16	68				6.69	2
20	12	68				5.47	
24	8	68				4.81	
28	4	68				4.06	
17	28	55	10.91			7.54	3
17	13	70	8.89			6.09	4

[a] Data from (1) Cheng (1965), (2) Lengyel et al. (1961), (3) Mazurin (1962), and (4) Makarova et al. (1960).

TABLE 13.56

Values of Electrical Conductivity and Its Arrhenius Activation Energy of $Na_2O-ZnO-SiO_2$ Glasses[a]

Composition (mole %)			$-\log \sigma$ $(\Omega^{-1} cm^{-1})$ at 150°C	E (kcal mole^{-1})
Na_2O	ZnO	SiO_2		
10	10	80	7.8	17.2
10	20	70	8.0	17.6
20	10	70	6.8	16.4
20	20	60	6.9	16.6
30	10	60	5.8	15.1
30	20	50	6.2	15.7

[a] Data from Mazurin and Brailovskaya (1960).

TABLE 13.57

Electrical Conductivities of Na$_2$O–ZnO–SiO$_2$ Glasses at Various Temperatures

Composition (mole %)			$-\log \sigma$ (Ω^{-1} cm^{-1}) at given temperature (°C)												Reference[a]
Na$_2$O	ZnO	SiO$_2$	150	200	250	300	400	500	600	700	800	1000	1200	1400	
15	10	75				4.95	3.95	3.20	2.33	1.75	1.35	0.77	0.44	0.24	1
15	15	70				4.88	3.91	3.20	2.40	1.80	1.37	0.77	0.44	0.24	
15	20	65				5.22	4.20	3.46	2.59	1.93	1.44	0.83	0.50	0.28	
15	25	60				5.53	4.42	3.61	2.79	2.03	1.49	0.82	0.40	0.18	
15	30	55				5.69	4.60	3.79							
15	35	50				5.87	4.84	3.90							
10	20	70	8.50	7.75	7.18	6.61									2
13.3	16.7	70	7.80	7.01	6.43	5.95									
16.7	13.3	70	7.08	6.31	5.76	5.30									
20	10	70	6.90	6.17	5.63	5.14									
17	28	55	7.49			5.15									3
16	16	68				4.88									4
20	12	68				4.57									
24	8	68				4.25									
28	4	68				3.82									
17	13	70	7.03			4.87									5

[a] Data from (1) Mazurina and Evstrop'ev (1967, 1970), (2) Cheng (1965), (3) Mazurin (1962), (4) Lengyel et al. (1969), and (5) Makarova et al. (1960).

TABLE 13.58

Values of Arrhenius Parameters for the Electrical Conductivity of Na_2O–CdO–SiO_2 Glass[a]

Composition (mole %)			Temperature range (°C)	σ_0 (Ω^{-1} cm^{-1})	E (kcal mole^{-1})
Na_2O	CdO	SiO_2			
20	10	70	200–450	645	20.7

[a] Data from Terai and Kitaoka (1968, 1969) and Terai (1971).

TABLE 13.59

Values of Electrical Conductivity and Its Arrhenius Parameters for Sodium–Lead–Silicate Glasses[a]

Composition (mole %)			$-\log \sigma$ (Ω^{-1} cm^{-1}) at 150°C	$\log \sigma_0$ (Ω^{-1} cm^{-1})	E (kcal mole^{-1})	Reference
Na_2O	PbO	SiO_2				
10	10	80	9.6	—	21.9	Mazurin and Brailovskaya (1960)
10	20	70	10.9	—	24.0	
20	10	70	7.9	—	19.9	
20	20	60	9.0	—	22.1	
30	10	60	6.5	—	17.3	
30	20	50	7.3	—	19.1	
5	13	72	—	2.2	25.0	Myuller and Pronkin (1965b)

[a] See also Fig. 13.7.

TABLE 13.60

Electrical Conductivities of Sodium–Lead–Silicate Glasses[a]

Composition (mole %)			$-\log \sigma$ (Ω^{-1} cm^{-1}) at temperature (°C)		Composition (mole %)			$-\log \sigma$ (Ω^{-1} cm^{-1}) at temperature (°C)	
Na_2O	PbO	SiO_2	200	300	Na_2O	PbO	SiO_2	200	300
2	28	70	12.49	10.26	10	40	50	10.92	8.94
2	33	65	12.04	9.98	10	50	40	10.52	8.42
2	38	60	11.70	9.48	10	55	35	10.18	8.23
2	58	40	10.05	8.16	12	28	60	9.91	7.95
4	36	60	11.70	9.48	14	26	60	9.43	7.54
5	25	70	11.28	9.20	15	15	70	7.93	6.20
5	30	65	11.66	9.52	15	20	65	8.62	6.74
5	45	50	11.00	9.01	15	25	60	8.94	7.14
5	60	35	9.88	7.98	15	35	50	9.74	7.84
6	34	60	11.52	9.35	15	45	40	9.86	7.66
6	54	40	10.30	8.22	15	50	35	9.92	7.88
8	32	60	11.22	9.07	20	10	70	6.71	5.13
10	20	70	9.66	7.78	20	15	65	7.28	5.58
10	25	65	10.26	8.10	20	20	60	8.06	6.26
10	30	60	10.80	8.70	20	30	50	8.26	6.46
					20	40	40	8.12	6.27

[a] Data from Grechanik et al. (1962, 1963).

TABLE 13.61

Electrical Conductivities of Some $Na_2O-MO-SiO_2$ Glasses at Various Temperatures[a]

Composition (mole %)			$-\log \sigma$ (Ω^{-1} cm^{-1}) at given temperature (°C)						
Na_2O	MO	SiO_2	150	200	300	400	450	500	600
	M = Cd								
20	10	70		6.90	5.24	4.06	3.58	2.85	2.04
28	4	68			4.00				
24	8	68			4.68				
20	12	68			5.09				
16	16	68			6.47				
	M = Mn								
18	2	80	7.35		4.83				
18	4	78	7.57		4.93				
18	6	76	7.75		5.06				
18	8	74	7.90		5.14				
18	10	72	8.11		5.28				
20	10	70		6.60	5.15	4.14	3.62	2.95	2.00
25	10	65	6.82		4.40				
	M = Fe								
20	10	70		6.40	4.98	3.98	3.60	3.15	2.15
	M = Co								
20	10	70		6.10	4.72	3.70	3.15	2.62	1.73
	M = Ni								
20	10	70		6.35	4.92	3.90	3.48	2.78	2.00
28	4	68			3.62				
24	8	68			4.00				
	M = Cu								
20	10	70		7.28	5.75	4.36	3.68	3.06	2.20
28	4	68			4.12				
24	8	68			4.85				
20	12	68			5.54				
16	16	68			6.11				

[a] Data from Petrovskii (1956), Mazurin and Petrovskii (1956a,b), Lengyel *et al.* (1969), and Lengyel and Boksay (1961).

TABLE 13.62

Electrical Conductivities of Glasses in the System
xK$_2$O–yMgO–$(100 - x - y/2)$SiO$_2$ (mole %)

		$-\log \sigma$ (Ω^{-1} cm^{-1}) at temperature (°C)		
x	y	150	300	Reference[a]
15	2	9.80		1
15	5	9.55		
15	10	9.17		
15	15	8.95		
15	20	9.16		
15	30	9.80		
*[b]	*[b]	8.09	5.72	2

[a] Data from (1) Leko (1970) and (2) Makarova *et al.* (1960).

[b] Composition (mole %): 17 K$_2$O–13 MgO–70 SiO$_2$.

TABLE 13.63

Parameters of the Equation $\rho/T = (\rho/T)_0 \exp(E/RT)$ for the Electricity Resistivity of K$_2$O–CaO–SiO$_2$ Glasses[a]

Composition (mole %)				
K$_2$O	CaO	SiO$_2$	$-\log(\rho/T)_0$	E (kcal mole^{-1})
8.2	19.7	72.1	4.1	28.6
13.1	14.8	72.1	3.8	22.6
18.3	9.9	71.8	4.2	19.7

[a] Data from Barton (1966).

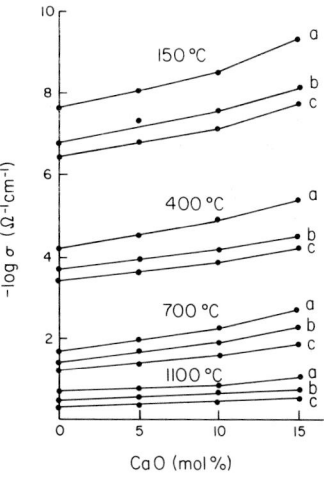

Fig. 13.8. Electrical conductivity–composition isotherms at four temperatures for the system K$_2$O–CaO–SiO$_2$. Mole % K$_2$O: (a) 20, (b) 25, (c) 30. [From Boricheva (1956).]

TABLE 13.64

Values of Electrical Conductivity and Its Arrhenius Parameters of $K_2O-SrO-SiO_2$ Glasses[a]

Composition (mole %)			$-\log \sigma$ at temperature		$\log \sigma_0$	E
K_2O	SrO	SiO_2	150°C	300°C		(kcal mole^{-1})
5.4	10.1	84.5	14.10	10.10	1.14	29.43
7.6	14.2	78.2	13.08	9.10	2.08	29.29
8.7	6.1	85.2	11.38	8.14	0.96	23.84
9.9	18.7	71.4	12.80	9.00	1.67	27.96
10.8	7.6	81.6	10.60	7.46	1.36	23.11
10.9	27.1	62.0	13.60	9.52	1.94	30.02
11.6	22.1	66.3	12.50	8.80	1.60	27.23
12.6	15.4	72.0	11.12	7.70	1.91	25.16
13.4	30.0	56.6	11.52	7.86	2.42	26.93
13.5	9.6	76.9	10.13	7.05	1.60	22.66
14.0	26.4	59.6	11.04	7.52	2.37	25.90
14.2	4.5	81.3	9.40	6.52	1.57	21.19
15.5	18.7	65.8	10.16	7.02	1.80	23.11
16.8	11.8	71.4	9.15	6.40	1.33	20.24
16.8	30.0	53.2	10.46	7.26	1.73	23.55
17.6	26.6	55.8	9.94	6.84	1.87	22.81
18.0	5.6	76.4	8.08	5.50	1.75	18.99
18.6	22.4	59.0	9.42	6.50	1.70	21.48
19.6	18.3	62.1	9.00	6.20	1.66	20.60
20.5	14.4	65.1	8.64	5.95	1.61	19.80
21.3	26.4	52.3	9.18	6.36	1.56	20.75
21.4	10.6	68.0	8.40	5.88	1.20	18.54
22.3	6.9	70.8	7.90	5.48	1.32	17.80
23.1	3.4	76.5	7.24	4.96	1.44	16.18
24.0	17.2	58.8	7.88	5.42	1.49	18.10
26.5	8.5	65.0	6.93	4.60	1.95	17.15

[a] Data from Kasymova (1974).

TABLE 13.65

Electrical Conductivity of $K_2O-BaO-SiO_2$ Glasses[a]

Composition (mole %)			$-\log \sigma \ (\Omega^{-1} \ cm^{-1})$ at 300°C	Composition (mole %)			$-\log \sigma \ (\Omega^{-1} \ cm^{-1})$ at 300°C
K_2O	BaO	SiO_2		K_2O	BaO	SiO_2	
2.5	30	67.5	11.71	7.5	30	62.5	10.94
2.5	40	57.5	10.94	7.5	40	52.5	10.31
5	20	75	11.71	10	20	70	9.89
5	30	65	11.29	10	30	60	10.2
5	40	55	10.54	10	40	50	9.8
7.5	20	72.5	11.06				

[a] Data from graph in Mazurin (1962).

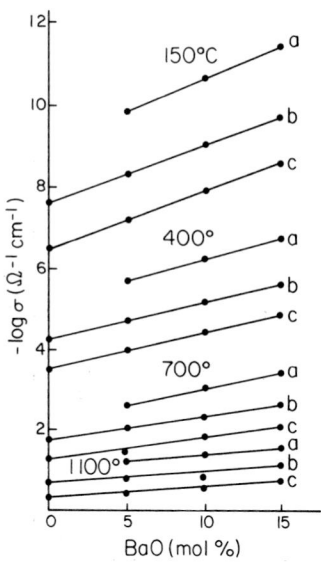

Fig. 13.9. Electrical conductivity–composition isotherms at various temperatures for the system $K_2O–BaO–SiO_2$. Mole % K_2O: (a) 15, (b) 20, and (c) 30. [From Boricheva (1956).]

TABLE 13.66

Electrical Conductivity of $K_2O–ZnO–SiO_2$ Glasses

Composition (mole %)			$-\log \sigma$ $(\Omega^{-1}$ cm$^{-1})$ at temperature (°C)		Reference[a]
K_2O	ZnO	SiO_2	150	300	
17	13	70	8.15	5.75	1
20	10	70	7.48[b]		2
20	20	60	7.5[b]		
20	30	50	7.94[b]		
20	40	40	8.43[b]		

[a] Data from (1) Makarova *et al.* (1960) and (2) Mazurin (1962).
[b] Values read from graph.

TABLE 13.67
Electrical Conductivities of Potassium–Lead–Silicate Glasses

Composition (mole %)			$-\log \sigma$ (Ω^{-1} cm^{-1}) at temperature (°C)									Reference[a]
K$_2$O	PbO	SiO$_2$	150	300	400	500	600	800	1000	1200	1400	
17	13	70	10.67	7.45								1
1.7	45.8	52.5			5.90	4.10	2.74	1.37	0.68	0.23		2
2.8	43	54.2			5.71	4.02	2.74	1.34	0.69	0.23		
5	37.5	57.5			8.10	6.40	4.83	2.63	1.54	0.95	0.55	
6.7	33.3	60				6.50	5.00	2.91	1.75	1.17	0.73	
7.7	30.8	61.5			8.07	6.40	4.89	2.87	1.72	1.13	0.71	
9	27.5	63.5				7.47	5.05	3.08	1.95	1.33	0.90	
10	10	80			6.73	5.44	4.34	2.92	2.17	1.63	1.22	
10	20	70				5.47	4.26	2.74	1.94	1.38	0.95	
10	25	65				6.03	4.75	2.95	1.94	1.33	0.90	
10	30	60				5.53	4.19	2.42	1.51	0.97	0.58	
10	40	50			6.82	5.00	3.40	1.70	0.96	0.49		
10	50	40				4.30	2.32	1.14	0.50			
11	22.5	66.5				5.85	4.60	2.90	1.91	1.35	0.93	
13.3	16.7	70			6.91	5.32	4.10	2.59	1.76	1.24	0.87	
18	5	77		5.72	4.58	3.47	2.67	1.64	1.02	0.67	0.42	
20	10	70		6.06	4.73	3.38	2.43	1.43	0.90	0.49	0.20	
20	20	60				3.38	2.39	1.32	0.77	0.36	0.06	
20	30	50			4.97	3.36	2.31	1.16	0.59	0.19		
20	40	40		6.81	5.40	4.36	3.54	1.08	0.54	0.14		
30	10	60		5.32	3.43	2.29	1.60	0.77	0.34	0.02		
30	20	50			3.44	2.26	1.44	0.66	0.21			

[a] Data from (1) Makarova et al. (1960), (2) Saringyulyan and Kostanyan (1970, 1971, 1974), and Kostanyan and Saringyulyan (1974).

TABLE 13.68

Arrhenius Parameters for the Electrical Conductivity of $K_2O-PbO-SiO_2$ Glass[a,b]

Composition (mole %)			Temperature range (°C)	$\log\sigma_0$ (Ω^{-1} cm^{-1})	E (kcal mole^{-1})
K_2O	PbO	SiO_2			
5	13	82	20–350	2.7	30.9

[a] Data from Myuller and Pronkin (1965b).
[b] See also Fig. 13.7.

TABLE 13.69

Electrical Conductivities of $K_2O-MnO-SiO_2$ Glasses[a]

Composition (mole %)			$-\log\sigma$ (Ω^{-1} cm^{-1}) at temperature (°C)	
K_2O	MnO	SiO_2	150	300
25	—	75	6.88	4.63
25	10	65	7.35	4.88

[a] Data from Mazurin and Petrovskii (1956a,b).

TABLE 13.70

Values of Electrical Conductivity and Its Arrhenius Activation Energy of $Tl_2O-PbO-SiO_2$ Glasses[a]

Composition (mole %)			$-\log\sigma$ (Ω^{-1} cm^{-1}) at 200°C	E (kcal mole^{-1})
$TlO_{0.5}$	PbO	SiO_2		
10	30	60	12	29.0
10	40	50	10.76	25.4
20	30	50	10.5	27.7
30	20	50	8.8	25.4
30	30	40	9.13	27.4
40	10	50	6.6	17.7
40	30	30	8.6	21.6

[a] Data from Karpechenko (1965).

TABLE 13.71

Electrical Conductivities of Lithium Borosilicate Glasses[a]

Composition (mole %)			$-\log \sigma \ (\Omega^{-1} \ cm^{-1})$ at 350°C	Composition (mole %)			$-\log \sigma \ (\Omega^{-1} \ cm^{-1})$ at 350°C
Li_2O	B_2O_3	SiO_2		Li_2O	B_2O_3	SiO_2	
5	45	50	6.686	15	35	50	4.353
5	70	25	7.243	15	75	10	6.421
5	85	10	<8.595	20	30	50	3.886
10	40	50	5.250	20	55	25	4.703
10	65	25	6.742	20	70	10	5.298
10	80	10	<7.818	25	25	50	3.033
14.5	60.5	25	5.883	25	50	25	3.124
				25	65	10	4.248

[a] Data from Otto (1966).

TABLE 13.72

Values of Electrical Conductivity and Its Arrhenius Activation Energy of Lithium Aluminosilicate Glasses[a]

Composition (mole %)			Temperature range (°C)	$-\log \sigma \ (\Omega^{-1} \ cm^{-1})$ at 150°C	E (kcal mole^{-1})
Li_2O	Al_2O_3	SiO_2			
10	10	80	150–450	5.8	15.5
13	13	74		5.7	14.9
17	2	81		6.95	17.9
17	3	80		6.95	18.2
17	8	75		6.9	17.3
17	11	72		6.6	16.4
17	13	70		5.9	14.9
17	17	66		5.6	15.1
17	21	62		5.9	16.7
17	24	59		6.2	17.9
21	21	58		5.6	15.3
25	5	70		6.4	17.2
25	9	66		6.2	18.3
25	15	60		6.4	17.8
25	18	57		6.0	16.0
25	25	50		5.1	16.0
25	29	46		5.7	18.2

[a] Data from Kondrat'ev and Smirnova (1970a,b) and Kondrat'ev and Chernysh (1966).

TABLE 13.73

Electrical Conductivities of Lithium Gallosilicate Glasses[a]

Composition (mole %)			$-\log \sigma$ (Ω^{-1} cm^{-1}) at 150°C
Li_2O	Ga_2O_3	SiO_2	
20	2.5	77.5	6.4
20	5	75	6.4
20	10	70	7.0
20	12.5	67.5	6.9
20	15	65	6.7
20	20	60	6.5
20	22.5	57.5	6.7
20	25	55	6.7
20	27.5	52.5	7.0

[a] Data from Tsekhomskaya (1966).

TABLE 13.74

Values of Arrhenius Parameters for the Electrical Conductivity of $Li_2O-Fe_2O_3-SiO_2$ Glasses[a]

Composition (mole %)			Temperature range (°C)	$-\log \sigma$ at 150°C (Ω^{-1} cm^{-1})	E (kcal mole^{-1})	$\log \sigma_0$ (Ω^{-1} cm^{-1})
Li_2O	Fe_2O_3[b]	SiO_2				
24	1	75	50–300	6.25	6.44	1.5
24	2	74		6.6	6.94	1.9
24	3	73		6.8	6.94	1.9
24	7	69		7.2	7.93	2.0
24	10	66		7.35	7.93	2.0
24	14	62		7.3	7.43	1.6
24	15	61		7.25	7.43	1.5
24	16	60		7.2	7.43	1.6
24	17	59		7.3	6.94	1.0
24	18	58		7.2	6.44	0.5
24	19	57		5.2	1.98	−2.9
24	20	56		4.4	1.98	−2.0
24	21	55		3.3	1.49	−1.6

[a] Data from Belyustin and Valkov (1967).
[b] Calculated as Fe_2O_3; contains $FeO + Fe_2O_3$.

TABLE 13.75

Values of Arrhenius Parameters for the Electrical Conductivity of Sodium Borosilicate Glasses

Composition (mole %)			Temperature range (°C)	$\log \sigma_0$ (Ω^{-1} cm^{-1})	E (kcal mole^{-1})	Reference[a]
Na$_2$O	B$_2$O$_3$	SiO$_2$				
7	83	10	120–180	1.52	32.4	1
13	77	10	120–160	2.59	30.1	
16	74	10	80–150	2.00	26.1	
18	72	10	60–130	1.89	24.1	
25	65	10	40–120	2.29	20.1	
10	25	65	60–170	1.56	23.4	
7	13	80	20–150	1.40	22.4	2
13	13	74		0.90	18.9	
17	13	70		1.05	18.0	
25	13	62		2.10	17.3	

[a] Data from (1) Namikawa (1975), Kumata *et al.* (1968), and (2) Myuller and Pronkin (1965a).

TABLE 13.76

Electrical Conductivities of Sodium Borosilicate Glasses

Composition (mole %)			Temperature (°C)	$-\log \sigma$ (Ω^{-1} cm^{-1})	Reference[a]
Na$_2$O	B$_2$O$_3$	SiO$_2$			
4	28	68	150	12.23	1
12	20	68		9.07	
16	4	80		7.8	
16	12	72		7.97	
16	16	68		8.26	
16	20	64		8.42	
16	24	60		8.50	
16	32	52		9.0	
20	12	68		7.4	
17.91	5.97	76.12	300	4.79	2
21.74	4.35	73.91		4.28	
25.35	2.67	71.98		3.87	
28.77	1.37	69.86		3.55	
5	45	50	350	7.35	3
10	40	50		6.78	
10	65	25		7.63	
15	35	50		5.50	
15	60	25		6.64	
20	30	50		4.70	
20	55	25		5.51	
25	25	50		3.92	
25	50	25		4.62	

[a] Data from (1) Appen and Gan'Fu-Si (1959, 1960), (2) Lengyel *et al.* (1969), and (3) Otto (1966).

TABLE 13.77

Arrhenius Parameters for the Electrical Conductivity of Sodium Aluminosilicate Glasses

Composition (mole %)			Temperature range (°C)	$\log \sigma_0$ (Ω^{-1} cm^{-1})	E (kcal mole^{-1})	Reference[a]
Na$_2$O	Al$_2$O$_3$	SiO$_2$				
21.7	2.72	75.58	100–400	1.71	8.2	1
22	5.5	72.5		2.09	8.2	
22.64	11.32	66.04		2.05	8.1	
24.03	24.03	51.94		1.29	6.7	
29.85	10.45	59.70	355–490	2.615	15.9	2

[a] Data from (1) Keshishyan *et al.* (1975) and (2) Kolitsch *et al.* (1980).

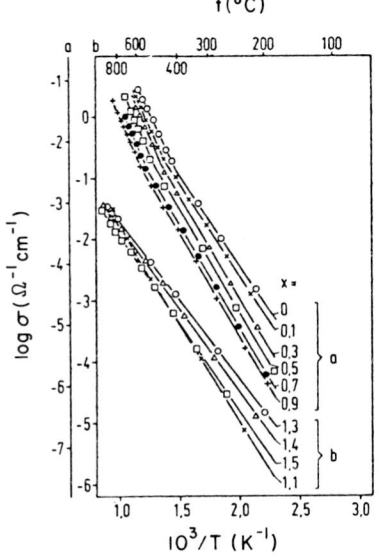

Fig. 13.10. Temperature dependence of the electrical conductivity of $(2.5 - x)\text{Na}_2\text{O}-x\text{Al}_2\text{O}_3-5\text{SiO}_2$ glasses. From Hunold and Bruckner (1980).

TABLE 13.78

**Electrical Conductivities of Sodium
Aluminosilicate Glasses[a]**

Composition (mole %)			$-\log \sigma$ $(\Omega^{-1}$ $cm^{-1})$ at temperature (°C)		
Na_2O	Al_2O_3	SiO_2	400	500	600
6.70	6.68	86.62	3.36	2.83	2.44
8.37	8.33	83.30	3.23	2.70	2.31
10.04	10.00	79.96	3.11	2.58	2.22
11.13	11.10	77.77	3.20	2.62	2.21
12.00	4.00	84.00	4.30	3.51	2.73
12.24	8.16	79.60	3.87	3.19	2.61
12.50	12.50	75.00	3.08	2.54	2.13

[a] Data from Hayward (1976, 1977).

TABLE 13.79

Electrical Conductivities of Sodium Aluminosilicate Glasses

Composition (mole %)			$-\log \sigma$ $(\Omega^{-1}$ $cm^{-1})$ at temperature (°C)								Reference[a]
Na_2O	Al_2O_3	SiO_2	200	300	400	600	800	1000	1200	1400	
20	6	74	5.77		3.48	1.97	1.29	0.95	0.66	0.32	1
20	10	70	5.61		3.40	1.95	1.30	0.98	0.53	0.32	
20	16	64	5.00		3.12	1.73	1.12	0.79	0.56	0.32	
20	20	60	4.57		2.84	1.56	1.03	0.74	0.52	0.32	
20	22	58	4.78		2.97	1.66	1.08	0.77	0.56	0.32	
20	24	56	4.90		3.09	1.76	1.16	0.84	0.58	0.32	
17.91	5.97	76.12		4.45							2
21.74	4.35	73.91		4.23							
25.35	2.67	71.98		3.97							
28.77	1.37	69.86		3.72							

[a] Data from (1) Kirakosyan (1974) and Kostanyan and Kirakosyan (1974), and (2) Lengyel et al. (1969).

TABLE 13.80

Electrical Conductivities of Sodium Gallosilicate Glasses[a]

Composition (mole %)			$-\log \sigma$ (Ω^{-1} cm^{-1}) at temperature (°C)	
Na$_2$O	Ga$_2$O$_3$	SiO$_2$	150	300
13	1.5	85.5	7.37	4.97
13	3	84	7.62	5.18
13	4.5	82.5	7.58	5.15
13	6	81	7.50	5.06
13	9	78	6.94	4.74
13	13	74	6.30	4.30
13	17	70	6.93	4.74

[a] Data from Ivanov and Galant (1963, 1964).

TABLE 13.81

Values of Electrical Conductivity and Its Activation Energy of Sodium Gallosilicate Glasses[a]

Composition (wt. %)			$-\log \sigma$ at 150°C (Ω^{-1} cm^{-1})	Temperature range (°C)	E (kcal mole^{-1})
Na$_2$O	Ga$_2$O$_3$	SiO$_2$			
30.48	9.90	57.78	5.37	100–250	13.8
29.30	12.63	57.08	5.74		14.3
28.83	16.50	53.92	5.68		14.4
25.70	22.88	51.36	5.69		14.6

[a] Data from Zhabrev and Moiseev (1965, 1970) and Zhabrev (1970).

TABLE 13.82

Arrhenius Parameters for the Electrical Conductivity of Na$_2$O–La$_2$O$_3$–SiO$_2$ Glass[a]

Composition (mole %)			Temperature range (°C)	σ_0 (Ω^{-1} cm^{-1})	E (kcal mole^{-1})
Na$_2$O	La$_2$O$_3$	SiO$_2$			
21	5.3	73.7	200–450	559	17.1

[a] Data from Terai (1971) and Terai et al. (1969a,b).

TABLE 13.83

Electrical Conductivities of Some
$Na_2O-M_2O_3-SiO_2$ Glasses[a]

Composition (mole %)			$-\log \sigma$ at 300°C
Na_2O	M_2O_3	SiO_2	$(\Omega^{-1}\ cm^{-1})$
	M = La		
17.91	5.97	76.12	5.06
21.74	4.35	73.91	4.55
25.35	2.67	71.98	4.13
28.77	1.37	69.86	3.73
	M = Bi		
17.91	5.97	76.12	5.73
21.74	4.35	73.91	4.73
25.35	2.67	71.98	4.15
28.77	1.37	69.86	3.74
	M = Cr		
25.35	2.67	71.98	3.81
28.77	1.37	69.86	3.57

[a] Data from Lengyel *et al.* (1969).

TABLE 13.84

Electrical Conductivities of $Na_2O-Fe_2O_3-SiO_2$ Glasses

Composition (mole %)			$-\log \sigma$ $(\Omega^{-1}\ cm^{-1})$ at temperature (°C)		Reference[a]
Na_2O	Fe_2O_3	SiO_2	150	300	
16	12	72	7.708	—	1
13	0.5	86.5	7.25	—	2
13	1	86	7.50	5.11	
13	2	85	8.00	5.43	
13	3	84	8.10	5.54	
13	4	83	8.15	5.49	
13	5	82	8.25	5.58	
13	7	80	8.02	5.43	
13	9	78	7.90	5.34	
13	11	76	7.73	5.24	
13	14.5	72.5	7.70	5.15	
13	17	70	7.67	5.21	

[a] Data from (1) Kutateladze (1974) and (2) Ivanov and Galant (1963, 1964).

TABLE 13.85

Values of Electrical Conductivity and Its Arrhenius Parameters of Potassium Borosilicate Glasses

Composition (mole %) K$_2$O	B$_2$O$_3$	SiO$_2$	$-\log \sigma$ (Ω^{-1} cm^{-1}) at 300°C	Temperature range (°C)	$\log \sigma_0$ (Ω^{-1} cm^{-1})	E (kcal mole^{-1})	Reference[a]
7	13	80		20–150	1.00	27.7	1
13	13	74			1.15	24.2	
17	13	70			0.50	21.3	
25	13	62			1.80	18.0	
19	19	62	7.644	50–450		23.5	2
19	37	44	7.889			26.7	
19	41	40	7.509			24.3	

[a] Data from (1) Myuller and Pronkin (1965a) and Ravaine et al. (1975).

TABLE 13.86

**Values of Arrhenius Parameters for the Electrical Conductivity
of Potassium Aluminosilicate Glasses**

Composition (mole %)			Temperature range (°C)	$\log \sigma_0$ (Ω^{-1} cm^{-1})	E (kcal mole^{-1})	Reference[a]
K_2O	Al_2O_3	SiO_2				
21.7	2.72	75.58	100–400	2.57	8.4	1
22	5.5	72.5		1.96	8.6	
22.64	11.32	66.04		0.90	7.8	
29.85	10.45	59.70	355–490	2.086	15.8	2

[a] Data from (1) Keshishyan *et al.* (1975) and (2) Kolitsch *et al.* (1980).

TABLE 13.87

Electrical Conductivities of Potassium Aluminosilicate Glasses[a]

Composition (mole %)			$-\log \sigma$ (Ω^{-1} cm^{-1}) at temperature (°C)		
K_2O	Al_2O_3	SiO_2	400	500	600
6.69	6.67	86.64	5.89	5.11	4.52
8.34	8.34	83.32	5.63	4.86	4.28
10.00	10.00	80.00	5.30	4.54	3.98
11.11	11.11	77.78	4.91	4.16	3.63
12.00	4.00	84.00	5.68	4.76	3.80
12.24	8.16	79.60	5.39	4.54	3.87
12.50	12.50	75.00	4.46	3.69	3.20

[a] Data from Hayward (1976, 1977).

TABLE 13.88

Electrical Conductivity of K_2O–Fe_2O_3–SiO_2 Glass[a]

Composition (mole %)			$-\log \sigma$ (Ω^{-1} cm^{-1}) at 150°C
K_2O	Fe_2O_3	SiO_2	
16	12	72	8.652

[a] Data from Kutateladze (1974).

TABLE 13.89

Electrical Conductivity of $Li_2O-SnO_2-SiO_2$ Glasses[a]

Composition (mole %)			
Li_2O	SnO_2	SiO_2	$-\log\sigma\ (\Omega^{-1}\ cm^{-1})$ at 150°C
33.3	1.0	65.7	5.23
33.3	2.0	64.7	5.15
33.3	3.0	63.7	5.16
33.3	4.0	62.7	5.26

[a] Data from graph in Vakhrameev (1968, 1970).

TABLE 13.90

Values of Parameters of the Equation $\sigma = (\sigma_0/T)\exp(-E/RT)$ for the Electrical Conductivity of $Na_2O-SnO_2-SiO_2$ Glasses[a]

Composition (mole %)			$-\log\sigma\ (\Omega^{-1}\ cm^{-1})$ at temperature (°C)			E
Na_2O	SnO_2	SiO_2	150	250	350	(kcal mole^{-1})
32.04	3.24	64.72	5.23	3.97	3.12	13.8
31.96	4.75	63.29	5.23	3.96	3.10	14.2
30.70	6.01	63.29	5.14	3.83	2.94	14.4
31.02	8.73	60.24	5.10	3.84	2.97	14.0

[a] Data from Sviridov et al. (1976).

TABLE 13.91

Electrical Conductivity of $Na_2O-SnO_2-SiO_2$ Glasses[a]

Composition (mole %)			$-\log\sigma\ (\Omega^{-1}\ cm^{-1})$ at 150°C	Composition (mole %)			$-\log\rho\ (\Omega^{-1}\ cm^{-1})$ at 150°C
Na_2O	SnO_2	SiO_2		Na_2O	SnO_2	SiO_2	
15	2.5	82.5	6.96	25	2.5	72.5	6.11
15	5.0	80.0	7.28	25	5.0	70.0	6.24
15	7.5	77.5	7.17	25	7.5	68.5	6.09
20	2.5	77.5	6.42	25	10.0	65.0	6.11
20	5.0	75.0	6.58	30	2.5	67.5	5.67
20	7.5	72.5	6.42	30	5.0	65.0	5.85
20	10.0	70.0	6.45	30	7.5	62.5	5.52
				30	10.0	60.0	5.52

[a] Data from graph in Vakhrameev and Evstrop'ev (1969).

TABLE 13.92

Electrical Conductivities of Sodium Titanium Silicate Glasses[a]

Composition (mole %)			$-\log \sigma$ (Ω^{-1} cm^{-1}) at 29°C	Composition (mole %)			$-\log \sigma$ (Ω^{-1} cm^{-1}) at 29°C
Na_2O	TiO_2	SiO_2		Na_2O	TiO_2	SiO_2	
21.7	24.9	53.4	10.839	26.2	12.9	61.0	9.968
22.6	37.4	40.3	10.845	27.6	25.8	46.6	9.623
23	30.2	46.8	10.732	27.7	15.8	56.5	9.279
23.3	5.4	71.5	10.415	28	35.8	36.3	9.591
23.6	21.8	54.8	10.230	29.1	31.4	39.7	9.556

[a] Data from Hirayama and Berg (1961).

TABLE 13.93

Values of Parameters of the Equation $\sigma = (\sigma_0/T) \exp(-E/RT)$ for the Electrical Conductivity of $Na_2O-TiO_2-SiO_2$ Glasses[a]

Composition (mole %)			$-\log \sigma$ (Ω^{-1} cm^{-1}) at temperature (°C)			E (kcal mole^{-1})
Na_2O	TiO_2	SiO_2	150	250	350	
32.04	3.24	64.72	5.38	4.04	3.12	15.0
32.50	5.00	62.50	5.74	4.27	3.29	16.0
31.68	6.21	62.11	5.55	4.12	3.16	15.6
30.30	9.09	60.61	5.72	4.29	3.34	15.5

[a] Data from Sviridov et al. (1976).

TABLE 13.94

Electrical Conductivities of Some $Na_2O-MO_2-SiO_2$ Glasses[a]

Composition (mole %)			$-\log \sigma$ (Ω^{-1} cm^{-1}) at temperature (°C)				
Na_2O	MO_2	SiO_2	150	300	400	500	600
	M = Ti						
20	10	70	6.62	4.60	3.80	3.18	2.46
	M = Zr						
20	10	70	6.32	4.28	3.48	2.86	2.30
	M = Ce						
20	10	70	6.90	4.64	3.70		

[a] Data from Petrovskii (1956) and Mazurin and Petrovskii (1965b).

TABLE 13.95

Electrical Conductivity of $Na_2O-ZrO_2-SiO_2$ Glasses[a]

Composition (mole %)			
Na_2O	ZrO_2	SiO_2	$-\log \sigma$ (Ω^{-1} cm^{-1}) at 150°C
25	5	70	6.05
25	10	65	5.82
25	15	60	5.57
30	5	65	5.57
30	10	60	5.46
30	15	55	5.50
30	20	50	5.70
40	5	55	5.11
40	10	50	4.75
40	15	45	5.39

[a] Data from graph in Kheifets (1965).

TABLE 13.96

Electrical Conductivity of $K_2O-SnO_2-SiO_2$ Glasses[a]

Composition (mole %)			$-\log \sigma$ (Ω^{-1} cm^{-1}) at 150°C
K_2O	SnO_2	SiO_2	
20	2.5	77.5	7.92
20	5.0	75.0	8.10
20	7.5	72.5	7.95

[a] Data from graph in Vakhrameev (1968, 1970).

TABLE 13.97

Electrical Conductivities of $Na_2O-Nb_2O_5-SiO_2$ Glasses

Composition (mole %)			$-\log \sigma$ at 26°C (Ω^{-1} cm^{-1})	Composition (mole %)			$-\log \sigma$ at 26°C (Ω^{-1} cm^{-1})
Na_2O	Nb_2O_5	SiO_2		Na_2O	Nb_2O_5	SiO_2	
12.6	12.5	74.9	11	25	5	70	9.57
15	7.5	77.5	10.77	25	25	50	10.32
15	15	70	10.79	26	20	54	9.89
19.1	19.1	61.8	10.70	29	9	62	9.04
19.8	26	54.2	10.97	29.9	5	65.1	8.62
19.9	5	75.1	10.30	29.9	16	54.1	9.20
20	10	70	10.20	30	20	50	9.43
21.9	7	71.1	9.74	33.9	19.1	47.0	8.78
22.1	16	61.9	10.15	35.1	11	53.9	8.26
22.9	23	54.1	10.45	38.1	12	49.9	9.0
24.9	13	62.1	9.71	40.9	13.6	45.5	8.34

[a] Data from Hirayama and Berg (1963).

TABLE 13.98

Electrical Conductivity of CuO–PbO–SiO$_2$ Glasses[a]

Composition (mole %)			$-\log \sigma\ (\Omega^{-1}\ cm^{-1})$ at 200°C	Composition (mole %)			$-\log \sigma\ (\Omega^{-1}\ cm^{-1})$ at 200°C
CuO	PbO	SiO$_2$		CuO	PbO	SiO$_2$	
1	54	45	11.14	1	44	55	11.14
2	53	45	11.31	2	43	55	11.49
3	52	45	11.37	3	42	55	11.54
5	50	45	11.31	5	40	55	11.48
8	47	45	11.37	8	37	55	11.2
10	45	45	11.26	10	35	55	11.03
				12	33	55	10.8
15	40	45	10.34	15	30	55	10.51
19.5	35.5	45	9.77	19.5	25.5	55	9.8

[a] Data from graph in Ershov *et al.* (1972).

TABLE 13.99

Electrical Conductivities of MgO–CaO–SiO$_2$ Glasses

Composition (mole %)			$\log \sigma\ (\Omega^{-1}\ cm^{-1})$ at temperature (°C)	
MgO	CaO	SiO$_2$	150	350
13.56	25.4	61.04	14.5	11.1
13.63	20.45	65.92	14.6	11.2
20.44	19.9	59.66	14.53	11.03
20.69	14.52	64.79	14.6	11.15
26.37	14.29	59.34	14.43	11.00
26.66	10.00	63.34	14.6	11.1
33.0	9.57	57.43	10.9	14.2
33	14	53	14.05	10.06

[a] Data from Zhunina *et al.* (1974).

TABLE 13.100

Electrical Conductivities of Some xMO–$(1 - x)$PbO–2SiO$_2$ Glass Systems[a]

x	$-\log \sigma\ (\Omega^{-1}\ cm^{-1})$ at 300°C
M = Sr	
0.25	10.3
0.50	11.33
0.75	9.4
M = Zn	
0.124	10.36
0.24	10.6
M = Cd	
0.124	10.65
0.25	10.75
0.37	10.79

[a] Data from graph in Aleinikov (1970).

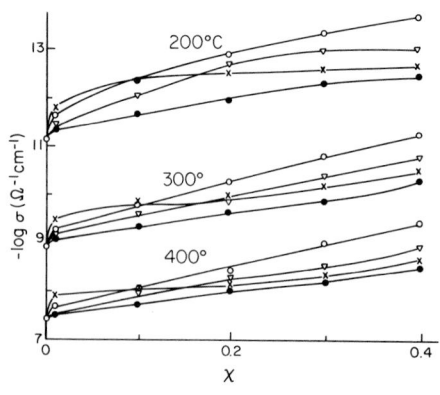

Fig. 13.11. Electrical conductivities of glasses of compositions xMO–$(1 - x)$PbO–SiO$_2$ at three temperatures: O, MgO; X, SrO; ▽, CaO; ●, BaO. From Strauss *et al.* (1956).

TABLE 13.101

Values of Electrical Conductivity and the Arrhenius Parameters for Its Temperature Dependence for Some MO–M'O–SiO$_2$ Glass Systems

Composition (mole %)			$-\log \sigma$ at 300°C (Ω^{-1} cm^{-1})	Temperature range (°C)	$\log \sigma_0$ (Ω^{-1} cm^{-1})	E (kcal mole^{-1})	Reference[a]
MO	M'O	SiO$_2$					
M = Ca	M' = Ba						
30	20	50	11.6	20–600		31.6	1
M = Mg	M' = Pb						
30	15	55		20–350	1.1	31.1	2
M = Ca	M' = Pb						
20	30	50	11.1	20–600		32.3	1
30	15	55		20–350	1.4	32.8	2
M = Ba	M' = Pb						
15	30	55	10.5	20–350	1.3	31.9	2
20	30	50	7.56[b]	20–600		30.2	1
17.5	17.5	65	7.37[b]				
20	20	60	7.02[b]				
22.5	22.5	55	7.20[b]				
23.3	20	56.7	6.46[b]				
25	25	50					

[a] Data from (1) Myuller and Leko (1965) and (2) Myuller and Pronkin (1965b).
[b] At 500°C.

TABLE 13.102

Electrical Conductivities of CaO–MnO–SiO$_2$ Glasses[a]

Composition (mole %)			$-\log \sigma$ (Ω^{-1} cm^{-1}) at temperature (°C)	
MnO	CaO	SiO$_2$	200	300
5	40	55	11.00	8.90
10	30	60	11.92	9.75
10	40	50	10.56	8.50
15	30	55	10.80	8.90
20	20	60	12.80	10.50
20	30	50	10.70	8.72
20	40	40	10.05	8.00
25	20	55	11.40	9.50
30	10	60	13.45	11.15
30	20	50	11.32	9.35
30	30	40	10.30	8.30
35	10	55	13.15	10.87
40	10	50	12.85	10.65
40	20	40	11.15	9.22
45	10	45	12.60	10.40
45	20	35	10.80	9.02

[a] Data from Bezrodn'iy and Kudyshkina (1974).

TABLE 13.103

Electrical Conductivities of SrO–MnO–SiO$_2$ Glasses[a]

Composition (mole %)			$-\log \sigma$ (Ω^{-1} cm^{-1}) at 300°C	Composition (mole %)			$-\log \sigma$ (Ω^{-1} cm^{-1}) at 300°C
SrO	MnO	SiO$_2$		SrO	MnO	SiO$_2$	
5	40	55	8.03	25	30	45	7.69
10	30	60	8.19	30	10	60	11.21
10	40	50	8.25	30	20	50	8.21
15	30	55	7.99	30	30	40	7.10
15	40	45	7.64	35	10	55	10.65
20	20	60	9.25	35	20	45	7.74
20	30	50	8.51	35	30	35	6.48
20	40	40	7.64	40	10	50	10.41
25	10	65	11.60	40	20	40	6.73
25	20	55	8.36	45	10	45	10.15

[a] Data from Kuznetsova (1972a,b).

TABLE 13.104

Electrical Conductivities of BaO–MnO–SiO$_2$ Glasses[a]

Composition (mole %)			$-\log \sigma$ (Ω^{-1} cm^{-1}) at 300°C	Composition (mole %)			$-\log \sigma$ (Ω^{-1} cm^{-1}) at 300°C
BaO	MnO	SiO$_2$		BaO	MnO	SiO$_2$	
5	40	55	9.19	25	20	55	10.01
10	30	60	9.52	25	30	45	8.03
10	40	50	8.67	30	10	60	10.95
15	30	55	9.01	30	20	50	9.01
15	40	45	8.27	35	10	55	10.56
20	20	60	10.18	35	20	45	9.73
20	30	50	8.67	40	10	50	10.23
20	40	40	7.31	40	20	40	9.51
25	10	65	11.12	45	10	45	10.01

[a] Data from Kuznetsova (1972a,b).

TABLE 13.105

Electrical Conductivities of ZnO–MnO–SiO$_2$ Glasses[a]

Composition (mole %)			$-\log \sigma$ (Ω^{-1} cm^{-1}) at 300°C
ZnO	MnO	SiO$_2$	
5	45	50	8.21
10	40	50	8.56
20	30	50	9.51
30	20	50	10.84
40	10	50	12.03
45	10	45	12.21

[a] Data from Kuznetsova (1972a,b).

TABLE 13.106

Electrical Conductivities of CdO–MnO–SiO$_2$ Glasses[a]

Composition (mole %)			$-\log \sigma$ (Ω^{-1} cm^{-1}) at 300°C
CdO	MnO	SiO$_2$	
10	40	50	8.31
20	30	50	8.73
30	20	50	9.46
35	10	55	11.32
40	10	50	11.12
45	10	45	10.81
50	10	40	9.93
55	10	35	8.65
60	10	30	8.41

[a] Data from Kuznetsova (1972a,b).

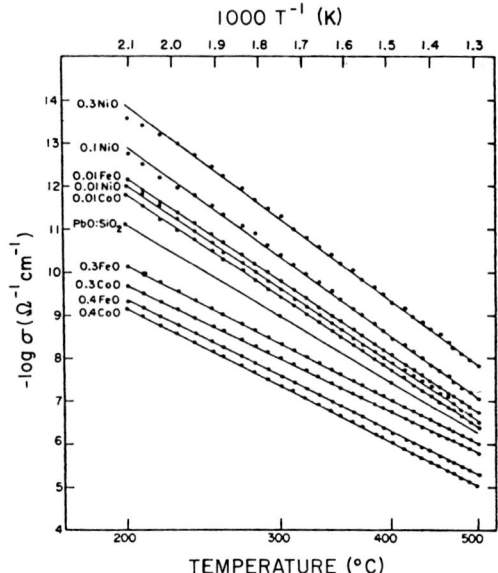

Fig. 13.12. Temperature dependence of electrical conductivity of the glass compositions: $(1 - x)PbO-xFeO-SiO_2$; $(1 - x)PbO-xCoO-SiO_2$; and $(1 - x)PbO-xNiO-SiO_2$ where x varies from 0 to 0.4. From Strauss *et al.* (1956).

TABLE 13.107

Electrical Conductivity and Its Arrhenius Activation Energy of Magnesium Aluminosilicate Glass[a]

Composition (mole %)			$-\log \sigma$ (Ω^{-1} cm^{-1}) at 300°C	Temperature range (°C)	E (kcal mole^{-1})
MgO	Al$_2$O$_3$	SiO$_2$			
54	8.1	37.9	13.8	20–600	33.4

[a] Data from Myuller and Leko (1965).

TABLE 13.108

Electrical Conductivities of Calcium Aluminosilicate Glasses[a]

Composition (mole %)			$-\log \sigma$ (Ω^{-1} cm^{-1}) at temperature (°C)	
CaO	Al$_2$O$_3$	SiO$_2$	300	450
20	10	70	12.55	9.90
20	20	60	12.50	9.80
20	25	55	13.00	10.25
30	10	60	13.60	10.90
40	10	50	12.82	10.25
40	20	40	13.75	10.90
40	30	30	13.50	11.00
50	10	40	11.90	9.50

[a] Data from Danilova (1970).

TABLE 13.109

Electrical Conductivities of Calcium Gallosilicate Glasses[a]

Composition (mole %)			$-\log \sigma$ (Ω^{-1} cm^{-1}) at temperature (°C)	
CaO	Ga$_2$O$_3$	SiO$_2$	300	450
20	10	70	12.95	10.78
20	15	65	13.15	11.10
20	20	60	12.88	11.24
20	25	55	13.22	11.14
20	30	50	13.80	10.98
30	10	60	12.84	10.55
40	10	50	13.55	10.60
40	20	40	13.68	10.70
40	30	30	13.72	10.68
40	40	20	13.55	10.60
40	45	15	13.27	10.35
50	10	40	12.65	9.88

[a] Data from Danilova (1970).

TABLE 13.110

Values of Electrical Conductivity and Its Arrhenius Activation Energy of CaO–Fe$_2$O$_3$–SiO$_2$ Glasses[a]

Composition (mole %)			$-\log \sigma$ (Ω^{-1} cm^{-1}) at 300°C	Temperature range (°C)	E (kcal mole^{-1})
CaO	Fe$_2$O$_3$	SiO$_2$			
29.5	17	53.5	4.29	100–400	13.6
33.3	11.2	55.5	6.107		15.1
34.3	14.3	51.4	4.609		15.2

[a] Data from Murawski (1982) and Zertsalova (1965).

TABLE 13.111

Electrical Conductivities of Some MO–Al$_2$O$_3$–SiO$_2$ Glasses[a]

Composition (wt %)			Temperature (°C)	$-\log \sigma$ (Ω^{-1} cm^{-1})
MO	Al$_2$O$_3$	SiO$_2$		
M = Sr				
31.1	31.6	36.6	354.7	14.17
			405.1	13.15
			454.0	12.23
			509.0	11.49
M = Ba				
40.6	26.6	31.8	347.0	14.21
			394.7	13.06
			444.7	12.06
			498.7	11.21

[a] Data from Bahat (1969).

TABLE 13.112

Electrical Conductivities of Barium Borosilicate Glasses[a]

Composition (mole %)			$-\log\sigma$ (Ω^{-1} cm^{-1}) at temperature (°C)						
BaO	B$_2$O$_3$	SiO$_2$	500	550	900	1000	1100	1200	1300
30	5	65	9.45	8.82	—	—	3.11	2.35	1.77
30	12	58	—	—	—	—	2.64	2.09	1.65
30	15	55	9.85	9.16	—	3.45	2.48	1.97	1.55
30	18	52	—	—	—	3.30	2.26	1.79	1.40
30	20	50	10.0	9.31	—	3.10	2.18	1.74	1.37
30	25	45	10.0	9.35	—	2.80	2.06	1.60	1.26
30	30	40	10.21	9.51	3.26	2.56	1.98	1.55	1.21
30	40	30	10.32	9.66	3.0	2.26	1.69	1.28	—
30	50	20	10.29	9.56	2.76	2.0	1.51	—	—
30	60	10	10.24	9.50	2.60	1.89	1.40	—	—
35	15	50	9.06	8.42	—	—	—	1.67	1.30
35	20	45	9.32	8.64	—	2.70	1.98	1.52	1.17
40	5	55	8.40	7.82	—	—	—	—	—
40	10	50	8.42	7.82	—	—	—	1.62	1.25
40	55	5	9.48	8.8	—	—	—	—	—

[a] Data from Kharyuzov (1959) and Kharyuzov et al. (1960).

TABLE 13.113

Electrical Conductivities of Barium Gallosilicate Glasses[a]

Composition (mole %)			$-\log\sigma$ (Ω^{-1} cm^{-1}) at temperature (°C)	
BaO	Ga$_2$O$_3$	SiO$_2$	300	450
20	10	70	13.68	11.68
20	15	65	14.16	12.06
20	20	60	13.92	11.58
20	25	55	13.80	11.40
20	30	50	13.45	11.24
30	10	60	13.80	10.88
30	30	40	13.42	11.24
40	10	50	12.43	9.70
40	20	40	13.00	10.33
40	30	30	13.50	10.83
40	40	20	13.36	10.60
40	45	15	12.40	9.70

[a] Data from Danilova (1970).

TABLE 13.114

**Values of Electrical Conductivity and Its Arrhenius Activation Energy
of BaO–Fe$_2$O$_3$–SiO$_2$ Glasses[a]**

Composition (mole %)			$-\log \sigma$ (Ω^{-1} cm^{-1}) at temperature (°C)		E (kcal mole^{-1})
BaO	Fe$_2$O$_3$	SiO$_2$	150	300	
21	10	69	8.4	6.1	
30	20	50	7.4	5.3	
39.5	1	59.5	14.4	10.6	
45	10	45	9.2	6.8	
32.5	15	52.5	7.8	5.7	15.4
35	10	55	8.9	6.6	17.1
37.5	5	57.5	10.7	8.1	19.1
39	2	59		10.2	22.8

[a] Data from Murawski (1982), Kuznetsov and Tsekhomskii (1965), and Tsekhomskii (1966).

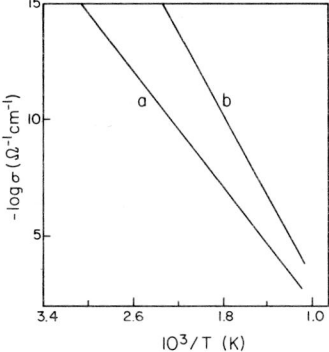

Fig. 13.13. Temperature dependence of the electrical conductivity of (a) 19.4CdO–77.7Bi$_2$O$_3$–2.9SiO$_2$ and (b) 37.3CdO–36.5Bi$_2$O$_3$–26.4SiO$_2$ glasses (wt. %). From Rao (1962).

TABLE 13.115

Electrical Conductivities of Lead Aluminosilicate Glasses[a]

Composition (mole %)			$-\log \sigma$ (Ω^{-1} cm^{-1}) at temperature (°C)	
PbO	Al$_2$O$_3$	SiO$_2$	250	300
30	20	50	—	11.90
30	30	40	—	12.37
40	10	50	11.93	10.86
40	20	40	12.54	11.39
40	30	30	—	11.62
50	10	40	11.00	9.96
50	20	30	11.52	10.45
50	30	20	11.98	10.86
50	40	10	—	11.70
60	10	30	10.50	9.49
60	20	20	10.84	9.81

[a] Data from Mazurin and Brailovskii (1959).

TABLE 13.116

Electrical Conductivities of PbO–Fe$_2$O$_3$–SiO$_2$ Glasses[a]

Composition (mole %)			$-\log \sigma$ (Ω^{-1} cm^{-1}) at temperature (°C)	
PbO	Fe$_2$O$_3$	SiO$_2$	30	220
46.97	8.75	44.28	4.11	1.25
51.87	—	48.13	4.66	3.81
51.94	0.09	47.97	4.61	3.56
52.66	0.91	46.43	4.76	3.78
53.46	1.87	44.67	4.72	3.57

[a] Data from El-Bayoumi and MacCrone (1976).

TABLE 13.117

Electrical Conductivities of Lead Ferrisilicate Glasses

Composition (mole %)			$-\log \sigma$ (Ω^{-1} cm^{-1}) at temperature (°C)			
PbO	Fe$_2$O$_3$	SiO$_2$	150	200	300	Reference[a]
30	10	60		6.70	5.51	1
32	8	60		7.33	6.08	
34	6	60		8.08	6.74	
36	4	60		8.78	7.36	
38	2	60		10.55	8.88	
39	1	60		11.73	9.81	
30	10	60	7.6		5.6	2
40	10	50	7.9		5.8	
40	20	40	6.1		4.3	
42.5	15	42.5	6.5		4.8	
45	10	45	8.1		6.0	
47.5	5	47.5	9.7		7.4	
49	2	49	11.7		8.6	
49.5	1	49.5	12.52		8.94	

[a] Data from (1) Grechanik *et al.* (1962) and (2) Karapetyan *et al.* (1963).

TABLE 13.118

Values of Electrical Conductivity and Its Arrhenius Activation Energy of PbO–Fe$_2$O$_3$–SiO$_2$ Glasses[a]

Composition (mole %)			$-\log \sigma$ (Ω^{-1} cm^{-1}) at 200°C	Temperature range (°C)	E (kcal mole^{-1})
PbO	Fe$_2$O$_3$	SiO$_2$			
22	8	70	6.87	100–400	16.0
62	8	30	7.55		17.3

[a] Data from Zertsalova (1964).

TABLE 13.119

Electrical Conductivities of MnO–Fe$_2$O$_3$–SiO$_2$ Glasses[a]

Composition (mole %)			$-\log \sigma$ (Ω^{-1} cm^{-1}) at temperature (°C)	
MnO	Fe$_2$O$_3$	SiO$_2$	200	300
45	5	50	8.9	7.5
45	8	47	9.3	7.4
50	5	45	9.3	7.6
54	5	41	8.8	7.3
60	5	35	8.5	6.9

[a] Data from Ershov and Shul'ts (1975).

TABLE 13.120

**Values of Electrical Conductivity and Its Arrhenius Activation Energy of
BaO–TiO$_2$–SiO$_2$ Glasses**

Composition (mole %)			$-\log \sigma$ (Ω^{-1} cm^{-1}) at temperature (°C)				E^b (kcal mole^{-1})	Reference[a]
BaO	TiO$_2$	SiO$_2$	150	350	400	500		
30	28	42	7.6	—	—	—	7.1	1
30	33	37	6.9	—	—	—	6.6	
31	19	50	9.8	—	—	—	8.1	
65	5	30	—	9.56	8.92	—	—	2
25	35	40	—	—	10.16	8.79	—	3

[a] Data from (1) Tsekhomskii (1966), (2) Matveev *et al.* (1970), and (3) Mazurin (1962).
[b] 150–300°C.

TABLE 13.121

Electrical Conductivity and Its Activation Energy of PbO–GeO$_2$–SiO$_2$ Glasses[a]

Composition (mole %)			$-\log \sigma$ at 350°C (Ω^{-1} cm^{-1})	Temperature range (°C)	Activation energy E (kcal mole^{-1})
PbO	GeO$_2$	SiO$_2$			
50	—	50	8.55	250–450	28.2
50	10	40	7.85	250–450	27.9
50	20	30	7.83	250–450	27.9
50	30	20	7.63	250–450	27.3
50	40	10	7.90	250–450	25.6

[a] Data from graph in Topping and Murthy (1974).

TABLE 13.122

Electrical Conductivity of PbO–TiO$_2$–SiO$_2$ Glasses[a]

Composition (mole %)			Temperature (°C)	$-\log \sigma$ (Ω^{-1} cm^{-1})
PbO	TiO$_2$	SiO$_2$		
40	3	57	74	14.41
			99	13.54
			125	12.74
			147	12.17
			172	11.48
40	6	54	72	14.92
			87	14.25
			104	13.72
			126	13.05
			149	12.13
40	9	51	76	15.18
			102	14.12
			125	13.40
			150	12.55
40	12	48	75	15.44
			102	14.26
			120	13.64

[a] Data from Mel'nikova et al. (1950).

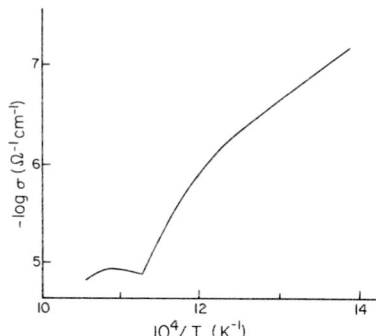

Fig. 13.14. Effect of temperature on the electrical conductivity of 40PbO–25TiO$_2$–35SiO$_2$ (mole %) glass. From Kokubo et al. (1973).

TABLE 13.123

Electrical Conductivities and Its Arrhenius Parameters for B_2O_3–Al_2O_3–SiO_2 System[a]

Composition (wt. %)			Temperature range (°C)	$\log \sigma_0$ (Ω^{-1} cm^{-1})	E (kcal mole^{-1})	$-\log \sigma$ (Ω^{-1} cm^{-1}) at temperature (°C)						
B_2O_3	Al_2O_3	SiO_2				700	900	1100	1300	1500	1700	1900
2.75	2.78	94.04	1550–1900	2.57	8.8	5.53	4.69	4.29	4.00	3.68	3.54	3.45
5.44	5.45	88.75	1290–1900	2.28	10.2	5.10	4.52	4.10	3.68	3.54	3.46	3.32
5.46	11.43	82.93	1240–1900	2.33	10.8	5.20	4.54	4.20	3.84	3.67	3.54	3.42
5.36	14.22	80.04	890–1900	1.58	14.4	5.18	4.48	4.12	3.80	3.54	3.34	3.18
9.03	10.81	79.82	1200–1900	1.94	13.5	5.16	4.46	4.09	3.82	3.60	3.43	3.28
10.80	8.87	79.98	1060–1900	1.94	13.7	5.26	4.52	4.12	3.84	3.62	3.45	3.30
11.82	5.51	82.28	1080–1900	2.60	8.2	5.14	4.50	3.90	3.74	3.61	3.52	3.42
14.21	5.47	79.91	1160–1900	2.55	9.0	5.36	4.60	4.02	3.80	3.64	3.52	3.42

[a] Data from Lor'yan et al. (1977).

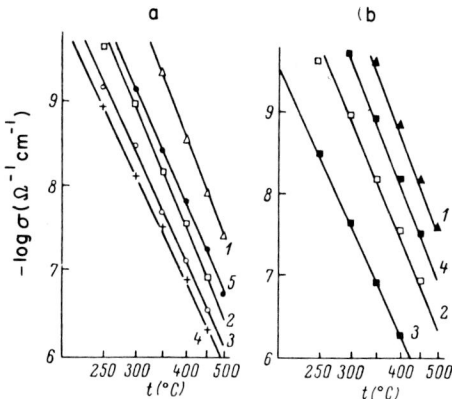

Fig. 13.15. Temperature dependence of electrical conductivity of $Al_2O_3-CeO_2-SiO_2$ glasses of various compositions: (a) 60 mole % SiO_2, (b) 20 mole % Al_2O_3. Mole % of CeO_2: 1, 15; 2, 20; 3, 25; 4, 30; 5, 35. [From Nemkovich *et al.* (1974).]

Chapter 14

Dielectric Properties

I INTRODUCTION

The charge Q on a conductor is proportional to the potential difference V between this conductor and other conductors within the electric field of the charge, so that

$$Q = CV, \tag{1}$$

where C is called the capacitance. The capacitance depends on the geometrical arrangement of the conductors and the material in between them called the dielectric and is measured in farads, which are coulombs per volt. The property of the material that determines the capacitance is called the dielectric constant ε. For a vacuum ε is called ε_0, the dielectric constant of free space, and equals 8.85×10^{-12} F m^{-1}. If the conductors are flat plates of area A, a distance d apart, then the capacitance of this "condenser" or "capacitor" is given by

$$C = A(\varepsilon/d). \tag{2}$$

If an alternating voltage is impressed across the plates of a condenser with a vacuum dielectric, the current in the circuit lags behind the voltage by 90°. If the dielectric is not a vacuum, the current often is not exactly 90° out of phase with the voltage, giving rise to a power loss. This property can be described by considering the dielectric constant to be a complex number with a real and imaginary part:

$$\varepsilon^* = \varepsilon' - i\varepsilon''. \tag{3}$$

In this case the capacitance given by Eq. (2) can also be considered to be a complex quantity. The loss angle θ by which the current deviates from a 90° lag with the voltage is given by

$$\tan \theta = \varepsilon''/\varepsilon'. \tag{4}$$

EXPERIMENTAL MEASUREMENT   II

The complex dielectric constant is usually measured with an ac bridge. A schematic diagram of a simple bridge is given in Fig. 14.1. The variable capacitors C_N and C_B are alternatively adjusted until the detector shows a minimum current, first without the sample X and then with it. At low loss the real part of the capacitance C' of the sample is just the difference between the two readings of C_N, and tan θ is proportional to the difference between the readings of C_B and C_N. For tan θ greater than about 0.1, these readings must be corrected by adding squared terms. If the sample is the dielectric of a capacitor with plates of area A a distance d apart, then ε' is given by $C'd/A$ from Eq. (2), and ε'' can be calculated from tan θ and Eq. (4). Many other bridge designs are available commercially, with more complicated circuits to give direct and even automatic readings of ε' and ε'' or the real and imaginary parts of the impedance Z of the sample.

The type of electrodes used in measuring the dielectric loss of glass can substantially influence the results. Electrodes usually have been films of metal deposited on the glass. With this type of electrode it is necessary to make a correction for the contribution of the normal dc resistivity of the sample to the apparent value of ε''. This correction is inversely proportional to the frequency, so that at low frequency it is much larger than the other losses in the glass, and makes results at these frequencies quite unreliable. Another disadvantage of the film electrodes is that surface conductivity can contribute to the loss. This contribution can be avoided by a third "ring" electrode, but this electrode leads to additional complications in the measuring circuit. Silver electrodes can also react with the glass.

At high frequencies (above about 10^6 Hz at room temperature) the real part of the dielectric constant changes little with frequency, and the imaginary part is small. However, at lower frequencies there is a loss peak, and ε' decreases in the frequency range of the loss.

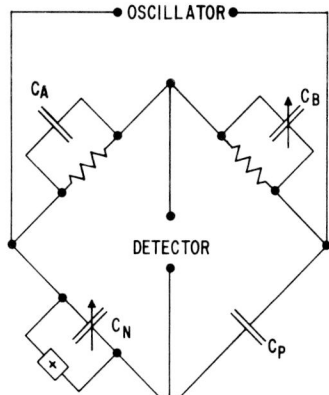

Fig. 14.1. Schematic diagram of a bridge circuit used to measure dielectric properties.

III   EFFECT OF TEMPERATURE

The loss peak and dispersion curve of ε' shift to higher frequencies as the temperature is changed; the shift of the peak maximum gives a measure of the temperature dependence of the loss. For glasses, the frequency at which the loss is a maximum increases exponentially with reciprocal temperature following the relation

$$f_m = A \exp(-E/RT), \tag{5}$$

where E is the activation energy, R is the gas constant, T is the temperature, and A is independent of temperature. The activation energies of dielectric relaxation and electrical conduction for silicate glasses are almost the same.

IV   TABLES AND FIGURES

Tables 14.1–14.12 and Figs. 14.2–14.15 show the dielectric properties of binary glasses. The dielectric properties of ternary glasses are given in Tables 14.13–14.65 and Figs. 14.16–14.27.

BIBLIOGRAPHY

Amrhein, E. M. (1963). *Glastech. Ber.* **36**, 425.
Amrhein, E. M. (1965). *Phys. Non-Cryst. Solids, Proc. Int. Conf., Delft, 1964* p. 283. (J. A. Prins, ed.). North-Holland, Amsterdam.
Amrhein, E. M. (1969). *Glastech. Ber.* **42**, 52.
Appen, A. A., and Bresker, R. I. (1952). *Zh. Tekh. Fiz.* **22**, 946.
Appen, A. A., and Gan'Fu-Si (1959). *Fiz. Tverd. Tela* (Leningrad) **1**, 1529.
Appen, A. A., and Gan'Fu-Si (1960). *In* "Stekloobraznoe Sostoyanie," p. 493. Moscow-Leningrad.
Bahat, D. (1969). *J. Mater. Sci.* **4**, 855.
Bezrodn'iy, B. G., and Kudyshkina, A. S. (1974). *In* "Investigations of Processes of Improved Technology in the Production of Polymer Materials and Glass," p. 78. Ivanov.
Bischoff, F. (1955). *Glastech. Ber.* **28**, 98.
Botvinkin, O. K., and Yaroker, K. T. (1966). *Steklo* No. 3, p. 1.
Brailovskaya, R. V. (1965). *In* "Elektronie Priborostroenie," p. 93. Moscow.
Cohen, B. M., Uhlmann, D. R., and Shaw, R. R. (1973). *J. Non-Cryst. Solids* **12**, 177.
Dgebuadze, T. P. (1967a). *Steklo* No. 7, p. 126.
Dgebuadze, T. P. (1967b). Investigation of liquation phenomena in alkali-borosilicate glasses studied from their electrical properties. Candidate's Dissertation, Leningrad.
El-Bayoumi, O. H., and MacCrone, R. K. (1976). *J. Am. Ceram. Soc.* **59**, 386.
Graham, P. W. L., Cross, L. E., and Rindone, G. E. (1969). *Phys. Chem. Glasses* **10**, 217.
Hagiwara, H., and Oyamada, R. (1976). *Yogyo Kyokaishi* **84**, 264.
Hakim, R. M., and Uhlmann, D. R. (1973). *Phys. Chem. Glasses* **14**, 81.
Higgins, T. J., Boesh, L. P., Volterra, V., Moynihan, C. T., and Macedo, P. B. (1973). *J. Am. Ceram. Soc.* **56**, 334.
Hirayama, C., and Berg, D. (1961). *Phys. Chem. Glasses* **2**, 145.
Hirayama, C., and Berg, D. (1963). *J. Am Ceram. Soc.* **46**, 85.

Hogarth, C. A., and Khan, M. N. (1978). *J. Mater. Sci.* **13**, 402.

Ioffe, V. A. (1952). *Dokl Akad, Nauk SSSR* **87**, 405.

Ioffe, V. A. (1954). *Zh. Tekh. Fiz.* **24**, 611.

Ioffe, V. A. (1955). *In* "Stroenie Stekla," p. 258. Moscow.

Kuznetsova, M. G. (1972a). *Proizvod. Tekh. Stroit. Stekla* No. 2, p. 74.

Kuznetsova, M. G. (1972b). Investigations of the physico-chemical properties of the glass systems RO–MnO–SiO$_2$ and RO–MnO–B$_2$O$_3$. Candidate's Dissertation, Leningrad.

Kuznetsova, M. G., and Evstrop'ev, K. S. (1972). *Inorg. Mater.* (*Engl. Transl.*) **8**, 302.

Leko, V. K. (1965). *In* "Stekloobraznoe Sostoyanie," p. 280. Moscow.

Leko, V. K. (1967). *Inorg. Mater.* (*Engl. Transl.*) **3**, 1645.

Mashkovich, M. D. (1969). "Electrical Properties of Inorganic Dielectrics in the Hyper Frequency Range." Moscow.

Mashkovich, M. D. (1970). *In* "Stekloobraznoe Sostoyanie," p. 75. Erevan.

Mashkovich, M. D., and Smelyanskaya, E. N. (1965). *Fiz. Tverd. Tela* (*Leningrad*) **7**, 1008.

Mashkovich, M. D., and Smelyanskaya, E. N. (1967). *Fiz. Tverd Tela* (*Leningrad*) **9**, 1249.

Mashkovich, M. D., and Udovenko, N. G. (1965). *Fiz. Tverd. Tela* (*Leningrad*) **7**, 524.

Mashkovich, M. D., and Varshal, B. G. (1969). *Inorg. Mater.* (*Engl. Transl.*) **5**, 277.

Matusita, L., and Sakka, S. (1974). *Int. Congr. Glass* [*Pap.*], *10th, Kyoto* No. 8, 44.

Mazurin, O. V. (1962). "Electrical Properties of Glasses." Leningrad.

Naidenov, A. P. (1963). *Steklo* No. 3, p. 75.

Namikawa, H. (1970). *J. Non-Cryst. Solids* **18**, 173.

Namikawa, H. (1974). *J. Non-Cryst. Solids* **14**, 88.

Namikawa, H. (1975a). *Denshi Gijutsu Sogo Kenkyusho Kenkyu Hokoku* No. 757.

Namikawa, H. (1975b). *J. Non-Cryst. Solids* **18**, 173.

Namikawa, H. (1975c). *Yogyo Kyokaishi* **83**, 500.

Navias, L., and Green, R. L. (1946). *J. Am. Ceram. Soc.* **29**, 267.

Pavlova, G. A., and Amatuni, A. N. (1975). *Inorg. Mater.* (*Engl. Transl.*) **11**, 1443.

Prasad, R. S., and Isard, J. O. (1967). *Phys. Chem. Glasses* **8**, 218.

Rao, B. V. J. (1962a). *J. Am. Ceram. Soc.* **45**, 555.

Rao, B. V. J. (1962b). *J. Sci. Ind. Res. Sect. B* **21**, 108.

Rao, B. V. J. (1963a). *J. Am. Ceram. Soc.* **46**, 107.

Rao, B. V. J. (1963b). *J. Am. Ceram. Soc.* **46**, 107.

Rao, B. V. J. (1963c). *Phys. Chem. Glasses* **4**, 22.

Rao, B. V. J. (1964). *Glass Technol.* **5**, 67.

Rauch, H. W., Commons, C. H., and Blau, H. H. (1959). *J. Am. Ceram. Soc.* **42**, 113.

Rinehart, D. W., and Bonino, J. J. (1959). *J. Am. Ceram. Soc.* **42**, 107.

Schultz, P. C., and Mizzoni, M. S. (1973). *J. Am. Ceram. Soc.* **56**, 65.

Shabanova, E. B. (1967). *Tr. Gor'k. Politekh. Inst.* **23**, 38.

Sheludyakov, L. N. (1967). *Vestn. Akad. Nauk Kaz, SSR* **23**, 43.

Stevels, D. (1961). "Electrical Properties of Glass," p. 75. Moscow.

Stevels, J. M. (1948). "Progress in Theory of Physical Properties of Glass." New York.

Stevels, J. M. (1952). *Verres Refract.* **6**, 3.

Stockdale, G. F. (1953). *Univ. Illinois Bull.* **50**(60), 411.

Svanson, S. E., and Johansson, R. (1970). *Acta Chem. Scand.* **24**, 775.

Taylor, H. E. (1959). *J. Soc. Glass Technol.* **43**, 124.

Terai, R. (1972). *Osaka Kogyo Gijutsu Shikensho Kiho* **23**, 36.

Topping, J. A., and Isard, J. O. (1971). *Phys. Chem. Glasses* **12**, 145.

Vakhrameev, V. I. (1968). *Steklo* No. 3, p. 84.

Vakhrameev, V. I. (1970). *In* "Stekloobraznoe Sostoyanie," p. 58. Erevan.

Vakhrameev, V. I., and Evstrop'ev, K. S. (1969). *Inorg. Mater.* (*Engl. Transl.*) **5**, 82.

Volger, J., and Stevels, J. M. (1956). *Phillips Res. Rep.* **11**, 452.

Von Hippel, A. R. (1954). "Dielectric Materials and Applications." Technol. Press, New York.

Watanabe, K., Sumiyoshi, Y., and Anbo, E. (1970). *Yogyo Kyokaishi* **78**, 165.

TABLE 14.1

**Dielectric Properties of
12.8 Li$_2$O–87.2SiO$_2$ Glass
at 25°C at Various Frequencies[a]**

Frequency (Hz)	$10^4 \tan \delta$	ε'
10^2	9700	9.94
10^3	3600	6.54
10^4	1000	5.45
10^5	310	5.1
10^6	174	4.95
10^7	124	4.92
3×10^8	79	4.9
10^{10}	102	4.8

[a] Data from Von Hippel (1954).

TABLE 14.2

**Dielectric Parameters of Lithium Silicate
Glasses at 20°C and 0.45 GHz[a]**

Composition (mole % Li$_2$O)	$10^4 \tan \delta$	ε'
16	63	5.45
20	89	5.78
26	110	6.85
36	196	8.15

[a] Data from Appen and Bresker (1952).

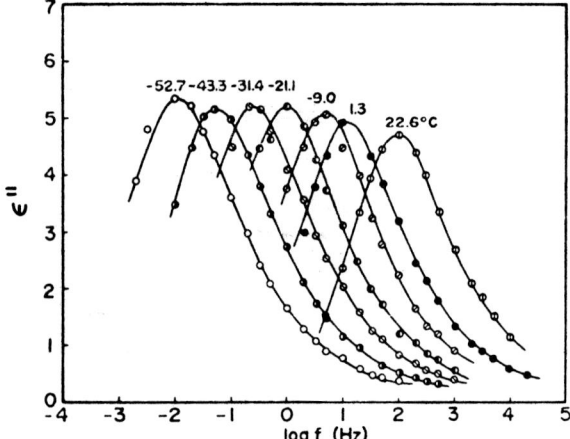

Fig. 14.2. Effect of frequency on the dielectric loss of 30Li$_2$O–70SiO$_2$ glass at various temperatures. From Namikawa (1974).

TABLE 14.3

Dielectric Parameters of Sodium Silicate Glasses at Various Frequencies and 17–19°C[a]

Composition (mole %)		Frequency (kHz)	$10^4 \tan \delta$	ε'
Na$_2$O	SiO$_2$			
11.9	88.1	9.8	260	5.78
		55	—	5.67
		140	160	5.60
		510	—	5.55
		1800	115	5.50
		8100	—	5.45
		41500	94	5.40
		90500	91	5.40
15.8[b]	84.2[b]	9.6	320	6.56
		54	—	6.45
		138	200	6.38
		500	—	6.28
		1770	148	6.22
		8000	—	6.14
		40500	122	6.08
		89000	—	6.08
19.6	80.4	9.3	360	7.32
		53	—	7.13
		133	240	7.02
		480	—	6.90

Composition (mole %)		Frequency (kHz)	$10^4 \tan \delta$	ε'
Na$_2$O	SiO$_2$			
19.6	80.4	1720	160	6.78
		7800	—	6.70
		39000	130	6.65
		86000	—	6.65
23.4	76.6	9.0	470	8.06
		51.5	—	7.88
		130	280	7.72
		465	—	7.58
		1670	175	7.46
		7550	—	7.33
		37500	135	7.26
		83000	—	7.25
27.2	72.8	8.7	620	8.95
		49.5	—	8.60
		126	350	8.38
		450	—	8.22
		1620	210	8.09
		7350	—	7.95
		36000	150	7.83
		81000	145	7.80

[a] Data from Taylor (1959).
[b] Temperature 19–21°C.

455

TABLE 14.4
Dielectric Constants of Sodium Silicate Glasses at Various Frequencies and Temperatures[a]

Composition (mole %)		Temperature (°C)	ε' at given frequencies (kHz)											
Na$_2$O	SiO$_2$		0.1	0.16	0.25	0.4	0.63	1.0	1.6	2.5	4.0	6.3	10	16
11.9	88.1	17		6.5		6.3		6.2		5.9		5.9	5.9	
		44	8.5	8.0		7.4		6.9		6.6		6.3	6.2	
		60	10.4	9.6		8.4		7.6		7.0		6.6	6.5	
		74	12.4	11.4		9.7		8.5		7.6		7.0	6.8	
		90	14.3	13.3	12.4	11.5	10.6	9.8	9.1	8.5	8.1	7.7	7.3	7.1
		108	15.6	14.9	14.2	13.4	12.5	11.7	10.8	10.0	9.3	8.7	8.2	7.8
		122	16.2	15.6	15.0	14.4	13.7	13.0	12.1	11.3	10.5	9.7	9.1	8.5
		137	16.5	16.0	15.5	15.0	14.5	13.9	13.2	12.5	11.7	10.9	10.1	9.4
		156	17.0	16.3		15.3		14.4		13.5		12.2	11.4	
		170	18.1	17.0		15.6		14.7		13.8		12.8	12.2	
		180	19.3	17.7		15.9		14.9		14.1		13.3	12.8	
		197	23.0	19.7		16.7		15.2		14.4		13.6	13.3	
		221		27.5		19.1		16.2		14.8		13.9	13.5	
		246				25.3		18.9		16.3		14.9	14.4	
15.8	84.2	18			7.6	7.4		7.0		6.8		6.6	6.5	
		55	12.1	11.1		9.7		8.7		8.0		7.5	7.4	
		75	15.3	14.2		12.0		10.3		9.1		8.3	8.0	
		89	16.9	16.0		13.9		11.8		10.2		9.0	8.6	
		101	17.8	17.2	16.4	15.5	14.5	13.4	12.4	11.5	10.6	9.9	9.3	8.8
		112	18.2	17.7	17.1	16.4	15.6	14.7	13.6	12.6	11.7	10.8	10.1	9.5
		130	18.7	18.2		17.2		16.1		14.5		12.6	11.7	

19.6	153	19.8	18.9		17.8		16.9		15.9		14.6	13.8	
	175		20.1		18.2		17.2		16.4		15.6	15.1	
	208				21.2		18.5		17.4		16.5	16.2	
80.4	17		8.7		8.2		7.9		7.7		7.5	7.5	
	41	11.7	10.9		9.8		9.0		8.4		8.0	7.9	
	56	14.3	13.1		11.2		10.0		9.1		8.5	8.3	
	73	17.6	16.2		13.6		11.6		10.3		9.3	8.9	
	84	19.4	18.2	17.0	15.6	14.0	13.2	12.1	11.3	10.6	10.1	9.5	9.2
	99	20.6	19.9	19.1	18.0	16.8	15.5	14.2	13.1	12.1	11.3	10.6	10.0
	109	20.8	20.4	19.8	19.1	18.2	17.0	15.8	14.5	13.4	12.4	11.5	10.6
	116	20.9	20.5	20.1	19.6	18.9	17.9	16.7	15.5	14.3	13.2	12.2	11.3
	136	21.0	20.6		20.0		19.1		17.8		15.6	14.4	
	156	21.8	20.9		20.0		19.5		18.8		17.5	16.5	
	179	26.5	23.0		20.4		19.5		18.9		18.4	17.9	
	205		35.7		23.7		20.5		19.4		18.9	18.6	
	234				37.4		24.3		20.7		19.6	19.2	
23.4	19	11.0	10.5		9.7		9.1		8.7		8.4	8.3	
	34	13.2	12.3		10.9		9.9		9.3		8.8	8.6	
	48	16.2	14.8		12.7		11.1		10.1		9.4	9.1	
	59	18.9	17.3		14.6		12.4		11.0		10.0	9.6	
76.6	74	21.6	20.3	18.9	17.4	16.1	14.7	13.5	12.6	11.7	11.5	10.5	10.1
	89	23.5	22.3	21.2	19.9	18.7	17.3	15.9	14.7	13.5	12.6	11.8	11.0
	101	24.3	23.3	22.4	21.4	20.3	19.1	17.8	16.5	15.2	14.0	13.0	12.1
	116	23.6	23.0	22.5	21.9	21.3	20.5	19.6	18.6	17.4	16.2	15.0	13.6
	138	25.9	24.2		22.5		21.5		20.3		18.7	17.7	
	158		27.5		23.1		21.6		20.7		19.7	19.0	
	185				28.0		22.9		21.3		20.4	20.1	
	222						33.3		23.9		21.1	20.5	

(continues)

TABLE 14.4 (Continued)

Composition (mole %)		Temperature (°C)	ε' at given frequencies (kHz)											
Na$_2$O	SiO$_2$		0.1	0.16	0.25	0.4	0.63	1.0	1.6	2.5	4.0	6.3	10	16
27.2	72.8	17	13.6	12.7		11.3		10.4		9.7		9.3	9.1	
		27	15.7	14.4		12.5		11.2		10.3		9.7	9.4	
		34	17.1	15.7		13.4		11.8		10.7		9.9	9.7	
		43	19.5	17.9		15.2		13.0		11.6		10.5	10.2	
		50	20.8	19.4		16.5		14.0		12.3		11.0	10.6	
		63	23.2	21.9	20.6	19.2	17.8	16.4	15.2	14.0	13.0	12.2	11.6	11.0
		74	24.7	23.4	22.2	21.0	19.8	18.4	17.0	15.8	14.6	13.5	12.6	11.8
		85	26.0	24.7	23.5	22.3	21.2	20.1	18.8	17.5	16.2	14.9	13.9	12.9
		96	26.6	25.6		23.5		21.5		19.4		16.9	15.6	
		116	27.8	26.1		24.2		22.6		21.0		19.1	18.0	
		133		28.4		24.4		22.7		21.6		20.3	19.3	
		156						23.4		22.0		21.0	20.5	
		173				27.6		25.7		22.7		21.6	21.2	

^{a} Data from Taylor (1959).

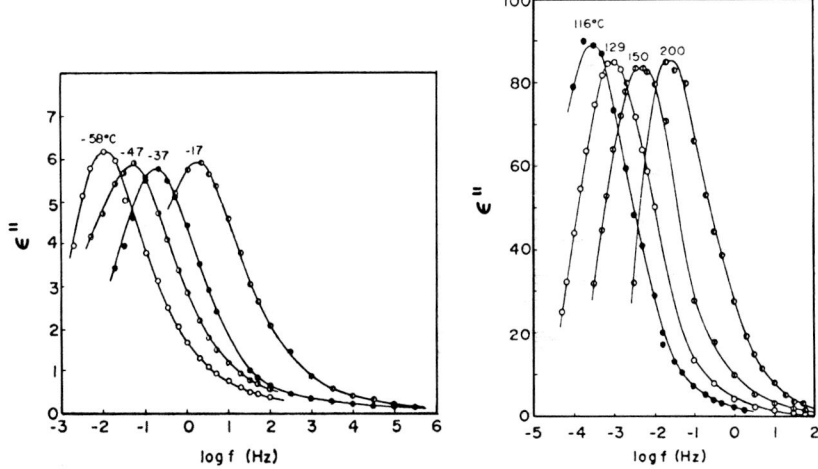

Fig. 14.3. Effect of frequency on the dielectric loss of $30Na_2O-70SiO_2$ (mole %) glass at various temperatures. From Namikawa (1975b).

TABLE 14.5

Dielectric Parameters of Potassium Silicate Glasses at Various Frequencies at Room Temperature[a]

Frequency f (kHz)	$10^4 \tan \delta$ for composition (mole % K_2O)					ε' for composition (mole % K_2O)				
	13.8	17.5	21.5	25.6	29.8	13.8	17.5	21.5	25.6	29.8
1	311	532	981	2148	4892	7.20	8.26	10.20	12.20	18.72
3	246	394	685	1394	3002	7.09	8.04	9.74	11.11	14.72
5	222	346	580	1143	2389	7.03	7.94	9.54	10.70	13.59
10	194	291	458	866	1712	6.96	7.81	9.31	10.22	12.41
30	160	227	344	609	1134	6.88	7.69	9.09	9.81	11.52
50	150	209	304	517	926	6.83	7.62	8.99	9.64	11.17
100	140	184	259	413	715	6.79	7.56	8.87	9.44	10.76
300	125	160	230	310	513	6.74	7.51	8.81	9.26	10.48

[a] Data from Stockdale (1953).

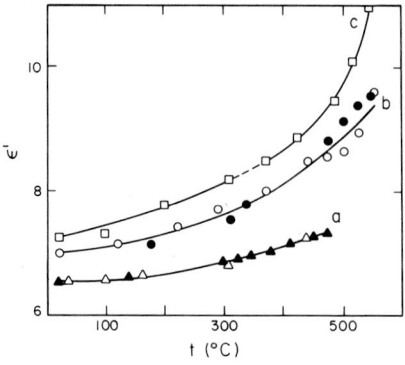

Fig. 14.4. Temperature dependence of dielectric constant of potassium silicate glasses containing (a) 20, (b) 24, (c) 26 mole % K_2O; frequency 10 GHz; $\bigcirc$, $\triangle$, $\square$, heating cycle; $\bullet$, $\blacktriangle$, cooling cycle. From Amrhein (1963).

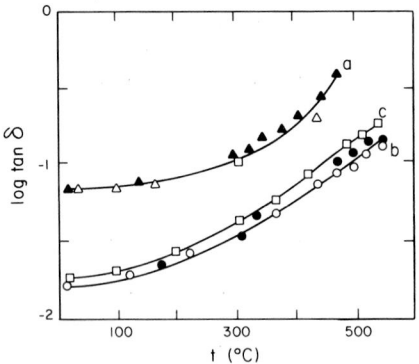

Fig. 14.5. Temperature dependence of loss tangent of potassium silicate glasses containing (a) 20, (b) 24, and (c) 26 mole % K_2O; frequency 10 GHz; $\bigcirc$, $\triangle$, $\square$, heating cycle; $\bullet$, $\blacktriangle$, cooling cycle. From Amrhein (1963).

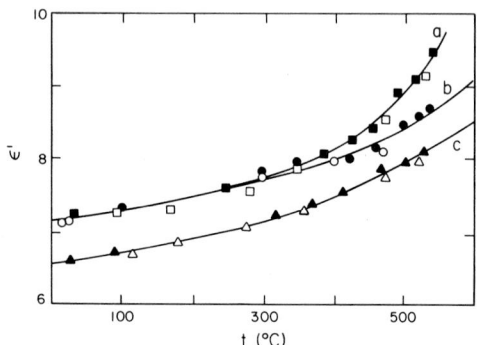

Fig. 14.6. Effect of temperature on the dielectric constant of rubidium silicate glasses containing (a) 26, (b) 24, and (c) 20 mole % Rb_2O; frequency 10 GHz; $\bigcirc$, $\triangle$, $\square$, heating cycle; $\bullet$, $\blacktriangle$, $\blacksquare$, cooling cycle. From Amrhein (1963).

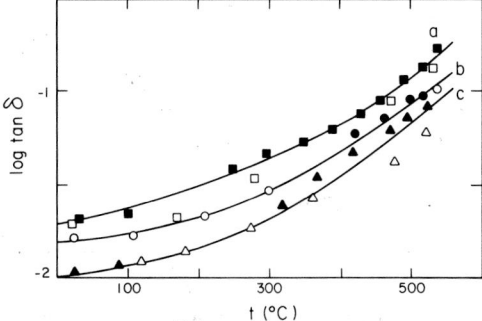

Fig. 14.7. Effect of temperature on the loss tangent of rubidium silicate glasses containing (a) 26, (b) 24, and (c) 20 mole % Rb_2O; frequency 10 GHz; ○, △, □, heating cycle; ●, ▲, ■, cooling cycle. From Amrhein (1963).

TABLE 14.6

Dielectric Properties of
12.8Rb_2O–87.2SiO_2 (mole %) Glass
at 25°C at Various Frequencies[a]

Frequency (Hz)	$10^4 \tan \delta$	ε'
10^2	98	5.39
10^3	89	5.32
10^4	58	5.23
10^5	46	5.22
10^6	41	5.21
10^7	38	5.20
3×10^8	59	5.15
10^{10}	120	5.05

[a] Data from Von Hippel (1954).

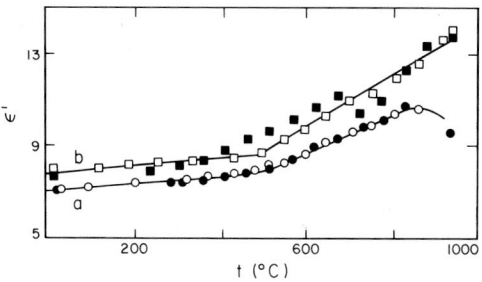

Fig. 14.8. Effect of temperature on the dielectric constant of cesium silicate glasses containing (a) 24 and (b) 26 mole % Cs_2O; frequency 10 GHz; ○, □, heating cycle; ●, ■, cooling cycle. From Amrhein (1963).

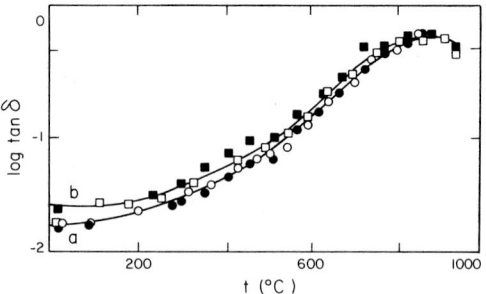

Fig. 14.9. Effect of temperature on the loss tangent of cesium silicate glasses containing (a) 24 and (b) 26 mole % Cs$_2$O; frequency 10 GHz; ○, □, heating cycle; ●, ■, cooling cycle. From Amrhein (1963).

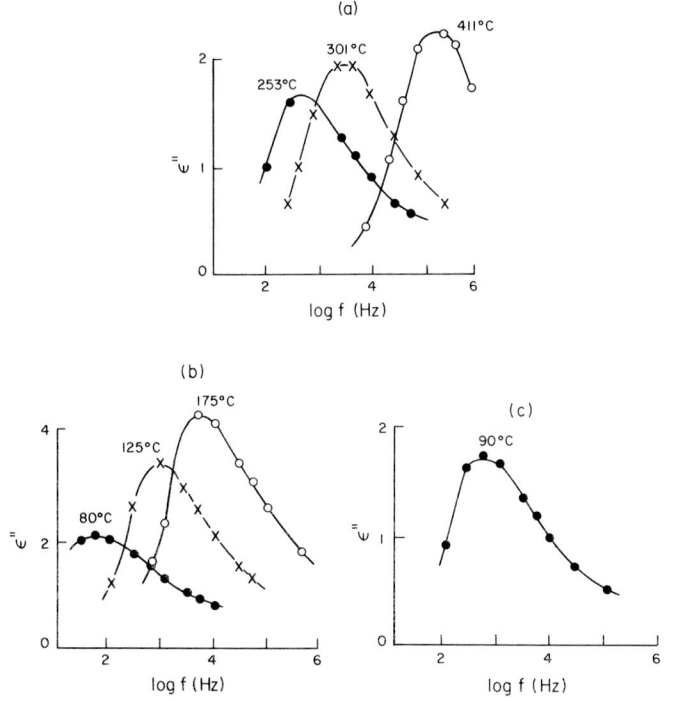

Fig. 14.10. Effect of frequency f on the dielectric loss of cesium silicate glasses at various temperatures. Mole % Cs$_2$O: (a) 5, (b) 15, and (c) 20. From Hakim and Uhlmann (1973).

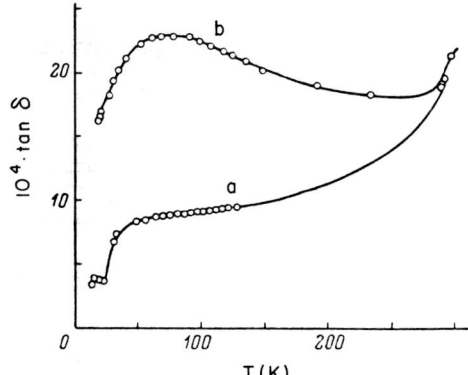

Fig. 14.11. Temperature dependence of the loss tangent of calcium silicate glasses: (a) 1 mole %
CaO, 1 kHz; (b) 50 mole % CaO, 1 kHz. From Volger and Stevels (1956).

TABLE 14.7

Dielectric Parameters of Barium Silicate Glasses[a]

Composition (mole % BaO)	Frequency	$10^4 \tan \delta$	ε'
30	1 MHz	5.4	
	10 GHz	58.6	7.8
40	1 MHz	8.1	
	10 GHz	75.5	9
50	1 MHz	13.5	
	10 GHz	96	9.9

[a] Data read from graph of Mashkovich (1969, 1970).

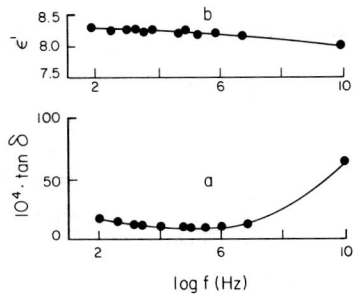

Fig. 14.12. Effect of frequency f on (a) loss
tangent and (b) dielectric constant of BaO–
$2.65SiO_2$ glass at 20°C. From Mashkovich
(1970) and Mashkovich and Udovenko (1965).

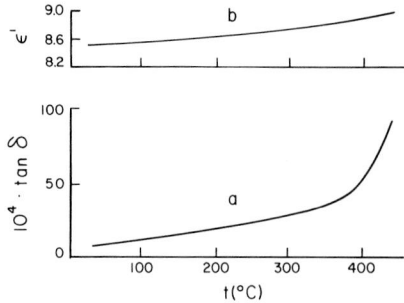

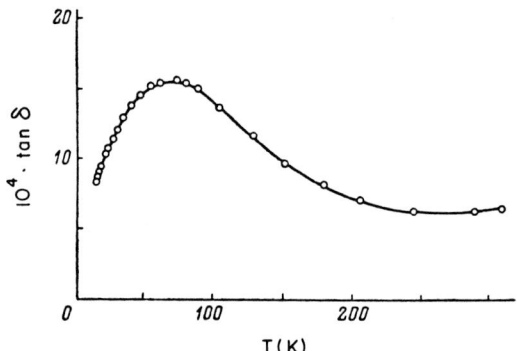

Fig. 14.13. Temperature dependence of (a) loss tangent and (b) dielectric constant of BaO–2.65SiO$_2$ glass. Frequency 0.8 MHz. From Mashkovich (1969, 1970) and Mashkovich and Varshal (1969).

Fig. 14.14. Effect of temperature on loss tangent of 30BaO–70SiO$_2$ (mole %) glass in the low-temperature region. Frequency 32 kHz. From Volger and Stevels (1956).

TABLE 14.8

Effect of Composition on Dielectric Constant and Loss Tangent of Lead Silicate Glasses

Composition (mole % PbO)	f	ε'	Reference[a]	Composition (mole % PbO)	f	$10^4 \tan \delta$	Reference[a]
24.5	1 kHz	8.2	1	38	8.78 GHz	93	2
30		9		45		108	
38		11.7		49		94	
44		14.9		55		124	
50		18		59		136	
60		19.8					
65		20.4					

[a] Data read from graphs of (1) Cohen *et al.* (1973) and (2) Ioffe (1954).

TABLE 14.9

**Temperature Dependence
of the Dielectric Parameters
of 68.6PbO–31.4SiO$_2$ (mole %)
Glass at 1 kHz**[a]

Temperature (°C)	ε'	$-\log \tan \delta$
25	20.9	2.790
70	21.1	2.85
100	21.3	2.82
125	21.5	2.81
150	21.6	2.64
175	21.8	2.36
200	22.0	2.0

[a] Data read from graphs of Schultz and Mizzoni (1973).

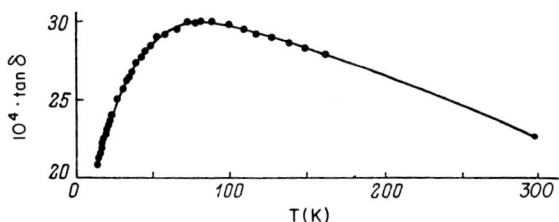

Fig. 14.15. Effect of temperature on the loss tangent of PbO–SiO$_2$ glass at 32 kHz. From Volger and Stevels (1956).

TABLE 14.10

Dielectric Properties of B$_2$O$_3$–SiO$_2$ Glass[a]

Composition (mole % B$_2$O$_3$)	f (GHz)	$10^4 \tan \delta$	ε'
46.3	10	14	3.55

[a] Data from Navias and Green (1946).

TABLE 14.11

Dielectric Properties of Bi_2O_3–SiO_2 Glasses at 20°C[a]

Composition (mole % Bi_2O_3)	10^4 tan δ at given frequency		ε' at given frequency	
	1×10^6 Hz	1.5×10^6 Hz	1×10^6 Hz	1.5×10^6 Hz
25	24.3	36.8	11.0	10.9
30	26.7	42.6	13.6	13.5
35	32.5	46.9	16.7	—
40	37.0	46.9	18.4	17.4
45	42.0	53.1	21.2	20.15
50	49.0	58.5	22.8	22.3
55	48.5	58.5	25.5	23.3
60		58.5		26.3
65		58.5		27.7

[a] Data from Shabanova (1967).

TABLE 14.12

Dielectric Properties of TiO_2–SiO_2 Glass at 20°C[a]

Composition (mole % TiO_2)	f (GHz)	10^4 tan δ	ε'
6.1	1	3.3	3.9

[a] Data from Pavlova and Amatuni (1975).

TABLE 14.13

Dielectric Properties of Li_2O–Na_2O–SiO_2 Glasses

Composition (mole %)			Temperature (°C)	log f (Hz)	10^4 tan δ	ε'	$\varepsilon'/\varepsilon_0$	Reference[a]
Li_2O	Na_2O	SiO_2						
20	16	64	20	8.6532	65	7.78		1
6.4	6.4	87.2	25	2	145		5.15	2
				3	87		5.08	
				4	47		5.05	
				5	28		5.05	
				6	19		5.04	
				7	17		5.03	
				8.48	26		5.00	
				10	52		4.95	

[a] Data from (1) Appen and Bresker (1952) and (2) Von Hippel (1954).

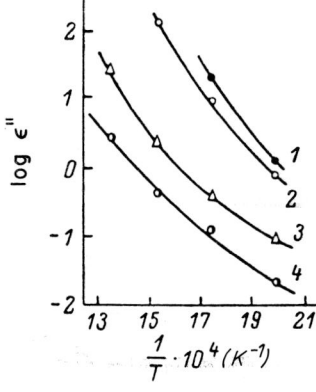

Fig. 14.16. Temperature dependence of dielectric loss of $2.5Li_2O-2.5Na_2O-95SiO_2$ (mole %) glass at (1) 100 Hz, (2) 1 kHz, (3) 100 kHz, and (4) 1 MHz. From Leko (1965, 1967).

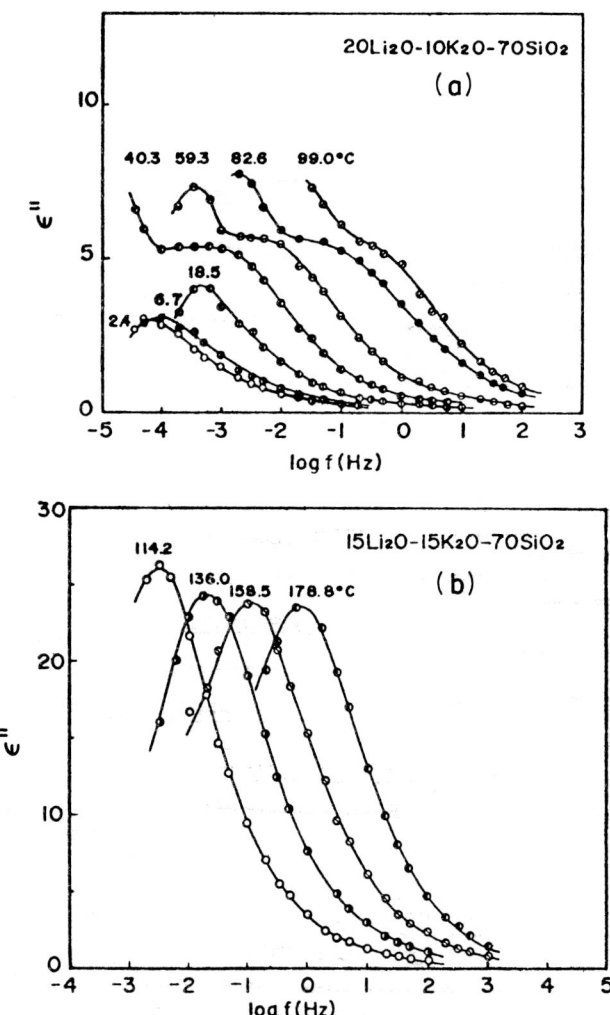

Fig. 14.17. Frequency dependence of the dielectric loss of the glass compositions (mole %): (a) $20Li_2O-10K_2O-70SiO_2$ and (b) $15Li_2O-15K_2O-70SiO_2$ at various temperatures. From Namikawa (1974).

TABLE 14.14

Dielectric Parameters of $Li_2O-K_2O-SiO_2$ Glass[a]

Composition (mole %)			Temperature ($^\circ$C)	log f (Hz)	$10^4 \tan \delta$	$\varepsilon'/\varepsilon_0$
Li_2O	K_2O	SiO_2				
3.3	6.6	90.1	25	2	53	5.23
				3	47	5.19
				4	37	5.17
				5	28	5.15
				6	24	5.14
				7	24	5.10
				8.48	40	5.07
				10	83	5.04

[a] Data from Von Hippel (1954).

TABLE 14.15

Dielectric Properties of $Na_2O-K_2O-SiO_2$ Glass[a]

Composition (mole %)			Temperature ($^\circ$C)	log f (Hz)	$10^4 \tan \delta$	$\varepsilon'/\varepsilon_0$
Na_2O	K_2O	SiO_2				
6.4	6.4	87.2	25	2	102	5.68
				3	75	5.62
				4	42	5.58
				5	31	5.56
				6	25	5.56
				7	23	5.54
				8.48	40	5.51
				10	115	5.50

[a] Data from Von Hippel (1954).

TABLE 14.16

Loss Tangent and Temperature Coefficient of Dielectric Constant of $Na_2O-K_2O-SiO_2$ Glasses at 10 GHz[a]

Composition (mole %)			$-\log \tan \delta$	$\left(\dfrac{1}{\varepsilon'}\right)\dfrac{d\varepsilon'}{dT} \times 10^4$
Na_2O	K_2O	SiO_2		
24	—	76	1.7	5.0
18	6	76	2.1	3.9
12	12	76	2.2	—
6	18	76	2.0	3.1
—	24	76	1.8	3.5

[a] Data read from graph of Amrhein (1963).

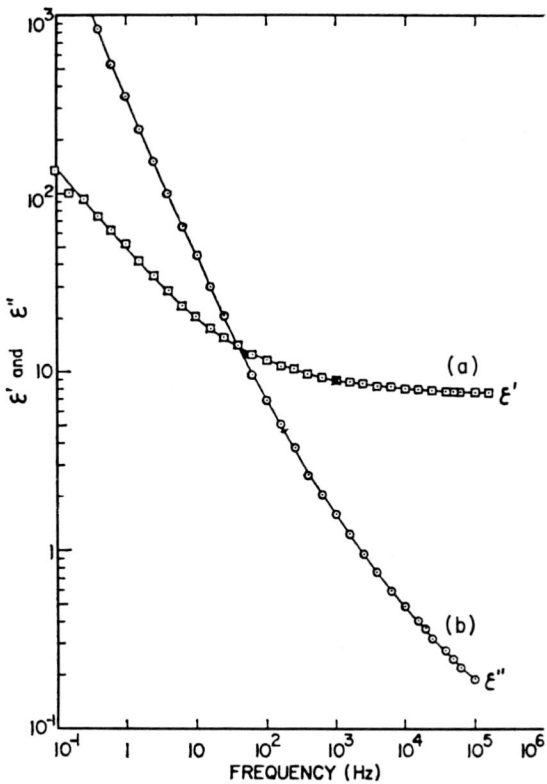

Fig. 14.18. Frequency dependence of (a) dielectric constant and (b) dielectric loss of $12.5Na_2O-12.5K_2O-75SiO_2$ (mole %) glass at 132.5°C. From Higgins *et al.* (1973).

TABLE 14.17

**Loss Tangent and Temperature Coefficient of Dielectric Constant of
$Na_2O-Rb_2O-SiO_2$ Glasses at 10 GHz[a]**

Composition (mole %)			$-\log \tan \delta$	$\left(\dfrac{1}{\varepsilon'}\right)\dfrac{d\varepsilon'}{dT} \times 10^4$
Na_2O	Rb_2O	SiO_2		
24	—	76	1.67	5.01
18	6	76	2.11	3.46
12	12	76	2.11	3.01
6	18	76	2.06	2.34
—	24	76	1.81	2.90

[a] Data read from graph of Amrhein (1963).

<table>
<tr><td colspan="2" align="center">TABLE 14.18</td></tr>
<tr><td colspan="2" align="center">Dielectric Constants of
$x\mathrm{Na_2O}-(1-x)\mathrm{Cs_2O}-5\mathrm{SiO_2}$
Glasses at 1 MHz
and Room Temperature[a]</td></tr>
<tr><td align="center">x</td><td align="center">ε'</td></tr>
<tr><td>0</td><td>8.6</td></tr>
<tr><td>0.2</td><td>8.9</td></tr>
<tr><td>0.4</td><td>7.9</td></tr>
<tr><td>0.6</td><td>7.6</td></tr>
<tr><td>0.75</td><td>8.9</td></tr>
<tr><td>1.0</td><td>8.8</td></tr>
</table>

[a] Data read from graph of Matusita and Sakka (1974).

TABLE 14.19

Loss Tangents of $\mathrm{Li_2O}-\mathrm{MgO}-\mathrm{SiO_2}$ Glasses at 20°C and 24 GHz[a]

Composition (mole %)			
$\mathrm{Li_2O}$	MgO	$\mathrm{SiO_2}$	$10^4 \tan\delta$
30	—	70	269
24	6	70	185
18	12	70	152
12	18	70	137
6	24	70	176
—	30	70	252

[a] Data read from graph of Stevels (1961).

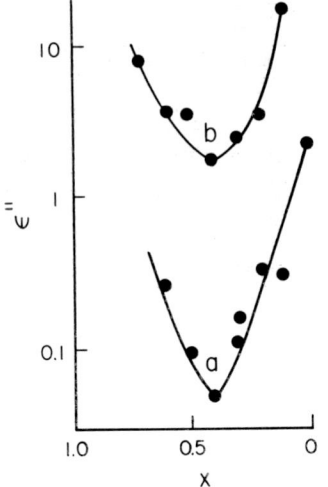

Fig. 14.19. Dielectric loss of $x\mathrm{Na_2O}-(1-x)\mathrm{Cs_2O}-5\mathrm{SiO_2}$ glasses at 1 kHz at (a) 120°C and (b) 260°C. From Terai (1972).

TABLE 14.20

Dielectric Parameters of $Li_2O-MO-SiO_2$ Glasses at 20°C and 450 MHz[a]

Composition (mole %)				
Li_2O	MO	SiO_2	$10^4 \tan \delta$	ε'
	M = Be			
17.4	17.4	65.2	63	7.16
	M = Ca			
17.4	17.4	65.2	57	7.95
25	12.5	62.5	73	8.25
	M = Sr			
17.4	17.4	65.2	58	8.15
	M = Ba			
5.94	19.8	74.26	39	7.45
11.2	18.7	70.1	44	7.80
17.4	17.4	65.2	57	8.45
22.14	16.4	61.46	72	8.92
	M = Zn			
17.4	17.4	65.2	59	7.25
	M = Cd			
17.4	17.4	65.2	49	7.95
	M = Pb			
16	20	64	53	9.23

[a] Data from Appen and Bresker (1952).

TABLE 14.21

Dielectric Parameters of $Na_2O-MO-SiO_2$ Glasses at 20°C and 450 MHz[a]

Composition (mole %)				
Na_2O	MO	SiO_2	$10^4 \tan \delta$	ε'
	M = Be			
16	20	64	52	8.21
17.4	17.4	65.2	55	8.03
17.9	25	57.1	58	8.75
	M = Sr			
16	20	64	48	8.95
17.4	17.4	65.2	53	8.72
17.9	25	57.1	41	9.81
	M = Mn			
14.7	22.1	63.2	56	8.16
18.31	16.52	65.17	48	7.93
	M = Fe			
12	25	63	—	8.44
16	20	64	49	8.46
	M = Co			
12.55	25.45	62.0	49	8.40
15.70	20.77	63.53	44	8.37
	M = Ni			
13.4	23.8	62.8	42	7.75
16.10	18.44	65.46	47	7.86

[a] Data from Appen and Bresker (1952).

TABLE 14.22

Dielectric Properties of $Na_2O–MgO–SiO_2$ Glasses[a]

Composition (mole %)			Temperature (°C)	Frequency (MHz)	$10^4 \tan \delta$	ε'	Reference[a]
Na_2O	MgO	SiO_2					
7.3	23.1	69.6	room temp.	1	90	7.84	1
11.2	23.3	65.5			98	7.81	
11.6	15.0	73.4			104	6.69	
14.7	14.7	70.6			114.5	7.35	
16.7	21.5	61.8			118.5	8.63	
17.5	19.6	62.9			122	7.10	
19.5	7.2	73.3			132	8.91	
22.1	4.9	73.0			141	8.26	
22.2	20.5	57.3			130	8.15	
22.3	20.3	57.4			131	8.43	
22.4	20.8	56.8			123	8.52	
25.1	14.7	60.2			138.5	7.92	
25.5	33.5	41.0			125	8.00	
16	20	64	20	450	50	8.45	2
17.4	17.4	65.2			55	8.36	
17.9	25	57.1			58	8.95	
20	13.3	66.7			62 .	8.00	
20	20	60	21	1		8.75	3
			58			8.9	
			92			9.3	
			137			10.15	
			192			11.8	
10	20	70	69	1		7.3	
			102			7.6	
			138			8.15	
			186			9.1	
			243			11.1	

[a] Data from (1) Botvinkin and Yaroker (1966), (2) Appen and Bresker (1952), and (3) Brailovskaya (1965).

TABLE 14.23

Dielectric Parameters of Soda-Lime Silicate Glasses at 19–21°C as a Function of Frequency[a]

Composition (mole %)			Frequency (kHz)	$10^4 \tan \delta$	ε'
Na$_2$O	CaO	SiO$_2$			
10.0	20.0	70.0	9.3	81	7.62
			52	—	7.58
			131	57.5	7.55
			475	—	7.50
			1690	50.5	7.46
			7650	—	7.42
			38000	51	7.40
			84000	—	7.40
14.8	15.0	70.2	9.2	165	7.85
			52	—	7.78
			130	80	7.72
			470	—	7.66
			1680	65	7.62
			7600	—	7.57
			37500	62	7.57
			83000	—	7.53
19.6	10.0	70.4	9.2	380	8.19
			52	—	8.06
			130	125	7.95
			470	—	7.86
			1680	90	7.80
			7600	—	7.75
			37500	77	7.71
			83000	—	7.70

Composition (mole %)			Frequency (kHz)	$10^4 \tan \delta$	ε'
Na$_2$O	CaO	SiO$_2$			
24.4	5.0	70.6	9.0	360	8.50
			51	—	8.35
			129	240	8.25
			465	—	8.12
			1670	150	8.01
			7500	—	7.95
			37500	115	7.85
			83000	—	7.83
17.6[b]	10.0[b]	72.4[b]	9.1	155	7.75
			52	—	7.69
			130	110	7.63
			465	—	7.57
			1670	82	7.50
			7550	—	7.45
			37500	74	7.40
			83000	81	7.40
11.9	10.0	78.1	9.5	175	6.77
			53.5	—	6.70
			136	95	6.65
			490	—	6.60
			1750	66.5	6.56
			7850	—	6.52
			39500	65	6.50
			87000	—	6.50

[a] Data from Taylor (1959).
[b] Temperature 17–19°C.

TABLE 14.24

Temperature and Frequency Dependence of the Dielectric Constant of Soda–Lime Silicate Glasses[a]

Composition (mole %)			Temperature (°C)	ε' at given frequencies (kHz)											
Na$_2$O	CaO	SiO$_2$		0.1	0.16	0.25	0.4	0.63	1.0	1.6	2.5	4.0	6.3	10	16
10.0	20.0	70.0	19		7.7	7.6	7.6		7.6		7.5		7.5	7.5	
			67		8.1	8.0	7.9		7.8		7.7		7.7	7.7	
			106		8.7	8.6	8.5		8.3		8.1		8.0	7.9	
			151	11.2	10.7		10.0		9.4		9.0		8.7	8.6	
			178	14.4	13.3		11.7		10.6		9.8		9.3	9.1	
			201	18.6	16.9		14.2		12.2		10.9		10.1	9.7	
			225	23.3	21.4		17.7		14.8		12.7		11.1	10.7	
			253	27.5	26.0	24.5	22.7	20.9	19.0	17.2	15.7	14.4	13.3	12.4	11.6
			269	29.2	27.7	26.3	24.8	23.2	21.5	19.6	17.9	16.3	14.9	13.7	12.7
			281	30.7	29.0	27.6	26.2	24.7	23.1	21.3	19.6	17.8	16.2	14.8	13.6
			301	34.5	31.6		28.0		25.3		22.2		18.7	17.0	
			331	43.8	38.4		31.3		28.0		25.4		22.4	20.7	
			362		47.5		37.0		31.2		28.0		25.4	24.0	
			401				43.6		36.4		31.4		28.2	27.1	
			442				50.3		41.4		35.3		31.2	29.7	
14.8	15.0	70.2	139	16.1	14.8		12.7		11.3		10.3		9.6	9.4	
			166	21.0	19.6		16.6		14.0		12.2		10.9	10.4	
			192	23.7	22.8	21.8	20.6	19.2	17.8	16.3	15.0	13.9	12.9	12.0	11.3
			205	24.9	23.9	23.0	22.0	20.9	19.6	18.2	16.8	15.4	14.2	13.2	12.3
			222	26.9	25.2		23.3		21.4		19.0		16.2	14.9	
			252	34.9	30.2		25.4		23.3		21.7		19.5	18.3	
			274		39.0		28.5		24.5		22.7		21.1	20.1	
19.6	10.0	70.4	101	16.8	15.5		13.2		11.6		10.6		9.8	9.6	
			122	20.4	19.3		16.6		14.1		12.2		11.0	10.5	
			140	22.1	21.3	20.4	19.2	18.0	16.6	15.3	14.2	13.1	12.3	11.6	11.0
			154	22.8	22.2	21.5	20.7	19.7	18.6	17.3	16.1	14.8	13.7	12.8	12.0
			176	23.7	23.0		21.8		20.5		18.7		16.3	15.0	
			198	25.0	23.7		22.3		21.2		20.1		18.5	17.4	

24.4	5.0	70.6	227	13.7	26.9		23.4		21.8		20.9		20.0	19.5	
			47	18.5	12.7	18.8	11.3	16.0	10.4	13.6	9.7	11.9	9.3	9.1	10.3
			70	21.5	17.1	20.9	14.4	18.2	12.5	15.6	11.1	13.3	10.2	9.9	11.1
			85	23.3	20.1	21.8	17.3	19.4	14.7	16.7	12.7	14.2	11.3	10.8	11.6
			98	24.2	22.1		19.6		16.9		14.4		12.5	11.7	
			105	26.4	22.9		20.5		18.0		15.4		13.2	12.4	
			121		24.8		22.3		20.1		17.6		15.1	14.0	
			144		26.2		23.9		22.0		20.2		18.0	16.8	
			175				25.4		23.2		21.8		20.3	19.5	
11.9	10.0	78.1	117	11.2	10.5		9.4		8.7		8.2		7.8	7.7	
			137	14.0	12.9		11.0		9.8		8.9		8.3	8.1	
			160	17.4	16.1		13.6		11.6		10.1		9.1	8.8	
			180	19.5	18.6	17.6	16.4	15.2	14.0	12.8	11.9	11.0	10.3	9.8	9.3
			198	20.2	19.7	19.1	18.2	17.2	16.0	14.8	13.7	12.6	11.6	10.8	10.2
			201	20.3	19.9	19.3	18.5	17.6	16.5	15.3	14.1	13.0	12.0	11.1	10.4
			220	20.8	20.4	20.0	19.5	18.9	18.1	17.1	16.1	14.9	13.7	12.7	11.7
			246	21.6	21.0		20.1		19.3		18.0		16.1	15.0	
			268	22.4	21.5		20.4		19.6		18.8		17.6	16.7	
			297	26.3	24.0		21.5		20.3		19.5		18.7	18.2	
17.6	10.0	72.4	14		8.1		7.9		7.8		7.7		7.6	7.6	
			50		8.9		8.6		8.4		8.1		8.0	8.0	
			68	10.3	9.9		9.4		8.9		8.6		8.4	8.3	
			84	12.1	11.4		10.3		9.6		9.1		8.7	8.6	
			103	15.0	13.8		11.9		10.7		9.9		9.3	9.0	
			121	18.2	16.9		14.3		12.3		10.9		10.0	9.7	
			141	20.8	19.9	18.7	17.4	16.0	14.8	13.6	12.7	11.8	11.2	10.6	10.2
			159	22.0	21.3	20.6	19.7	18.6	17.3	16.1	14.8	13.7	12.7	11.9	11.2
			172	22.5	22.0	21.4	20.7	19.9	18.9	17.8	16.5	15.2	14.1	13.0	12.1
			188	22.9	22.3		21.4		20.2		18.4		16.0	14.8	
			209	23.2	22.7		21.8		21.0		19.9		18.2	17.0	
			231	25.2	23.7		22.1		21.3		20.6		19.6	18.8	
			259		27.4		23.4		22.2		21.1		20.6	20.2	
			286				27.5		23.7		22.2		21.4	21.1	

[a] Data from Taylor (1959).

TABLE 14.25

**Loss Tangents of Na_2O–SrO–SiO_2
Glasses at 25.5°C and 1 kHz
(Soda/Silica Ratio = 0.2)[a]**

Composition (mole % SrO)	$10^4 \tan \delta$
0	1055
3.3	449
6.7	268
10	184
13.3	141
16.7	100
20	64
23.2	49

[a] Data read from graph of Rinehart and
Bonino (1959).

TABLE 14.26

Dielectric Parameters of Na_2O–BaO–SiO_2 Glasses

Composition (mole %)			Temperature	Frequency			
Na_2O	BaO	SiO_2	(°C)	(MHz)	$10^4 \tan \delta$	ε'	Reference[a]
5.94	19.8	74.26	20	450	36	7.88	1
11.2	18.7	70.1			41	8.29	
12.5	12.5	75			44	7.56	
15	15	70			50	8.43	
16	20	64			48	9.08	
16.4	22.14	61.46			53	10.05	
17.4	17.4	65.2			54	9.18	
18.7	11.2	70.1			55	8.43	
19.8	5.94	74.26			58	7.82	
20	20	60			59	10.25	
22.14	16.4	61.46			66	9.60	
10	20	70	224	1		8.55	2
			270	1		8.85	
20	20	60	116	1		9.35	
			149			9.65	
			188			10.15	
			235			11.05	
			292			13.25	
10	10	80		1	53		3
				10	30		
				100	58		
10	20	70		1		42	
				10		26	
				100		46	

(continues)

TABLE 14.26 (*Continued*)

Dielectric Parameters of Na$_2$O–BaO–SiO$_2$ Glasses

Composition (mole %)			Temperature (°C)	Frequency (MHz)	$10^4 \tan \delta$	ε'	Reference[a]
Na$_2$O	BaO	SiO$_2$					
20	10	70		1		88	
				10		63	
				100		78	
20	20	60		1		75	
				10		57	
				100		59	

[a] Data from (1) Appen and Bresker (1952), (2) Brailovskaya (1965), and (3) Mazurin (1962).

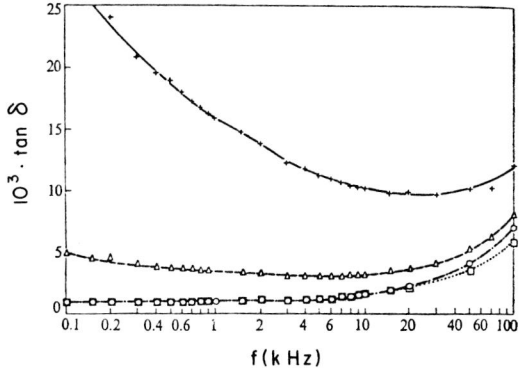

Fig. 14.20. Frequency dependence of the loss tangent of the glass compositions $(1-x)$Na$_2$O–xBaO–2SiO$_2$ at 23°C. Values of x: +, 0.4; △, 0.6; ○, 0.8; □, 1.0. From Graham *et al.* (1969).

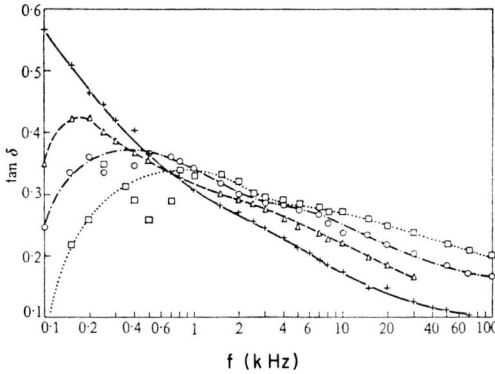

Fig. 14.21. Frequency dependence of loss tangent of 0.8Na$_2$O–0.2BaO–2.0SiO$_2$ glass at various temperatures: +, 94°C; △, 106°C; ○, 118°C; □, 130°C. From Graham *et al.* (1969).

TABLE 14.27

Dielectric Parameters of $Na_2O-ZnO-SiO_2$ Glasses

Composition (mole %)			Temperature (°C)	Frequency (MHz)	$10^4 \tan \delta$	ε'	Reference[a]
Na_2O	ZnO	SiO_2					
10	10	80		1	164		1
				10	131		
				100	132		
10	20	70		1	128		
				10	104		
				100	123		
20	10	70		1	206		
				10	170		
				100	183		
20	20	60		1	169		
				10	114		
				100	128		
16	20	64	20	450	51	8.29	2
17.4	17.4	65.2			52	8.00	
10	20	70	95	1		7.6	3
			130			8.0	
			171			8.8	
			222			10.0	
			286			12.0	
20	20	60	40	1		8.75	
			69			9.05	
			103			9.6	
			145			10.65	
			196			12.8	

[a] Data from (1) Mazurin (1962), (2) Appen and Bresker (1952), and (3) Brailovskaya (1965).

TABLE 14.28

**Loss Tangents of
Na$_2$O–ZnO–SiO$_2$ Glasses
at 25.5°C and 1 kHz
(Soda/Silica Ratio = 0.2)**[a]

Composition (mole % ZnO)	10^4 tan δ
0	1055
1.7	720
3.2	622
6.7	549
10	594
13.3	622
16.7	552
20	441
23.3	326

[a] Data read from graph of Rinehart and Bonino (1959).

TABLE 14.29

Dielectric Parameters of Na$_2$O–CdO–SiO$_2$ Glasses

Composition (mole %)			Temperature (°C)	Frequency (MHz)	10^4 tan δ	ε′	Reference[a]
Na$_2$O	CdO	SiO$_2$					
16	20	64	20	450	45	8.65	1
17.4	17.4	65.2			45	8.37	
20	13.3	66.7			45	8.35	
10	20	70	138	1	—	8.15	2
			177			8.5	
			227			9.05	
			286			10.0	
20	20	60	99	1		8.65	3
			132			8.85	
			174			9.35	
			222			10.1	
			282			12.0	

[a] Data from (1) Appen and Bresker (1952) and (2) Brailovskaya (1965).

TABLE 14.30

**Loss Tangents of $Na_2O-CdO-SiO_2$
Glasses at 25.5°C and 1 kHz
(Soda/Silica Ratio = 0.2)[a]**

Composition (mole % CdO)	$10^4 \tan \delta$
0	1059
3.3	337
6.7	222
10	174
13.2	152
16.7	122
20	107
23.3	88

[a] Data read from graph of Rinehart and
Bonino (1959).

TABLE 14.31

Dielectric Parameters of Soda–Lead Silicate Glasses

Compostion (mole %)			Temperature (°C)	Frequency (MHz)	$10^4 \tan \delta$	ε'	Reference[a]
Na_2O	PbO	SiO_2					
10	10	80		1	65		1
				10	44		
				100	53		
20	10	70		1	80		
				10	59		
				100	70		
20	20	60		1	44		
				10	30		
				100	52		
24.4	—	75.6	25	1	114	6.82	2
25.4	1.4	73.2			64	6.88	
26.4	2.9	70.7			91	7.48	
28.6	6.4	65.0			23	8.42	
34.6	15.4	50.0			80	9.78	
43.7	29.2	27.1			179	12.24	
10	20	70	203	1		8.4	3
			248			8.55	
			298			8.9	
16	20	64	20	450	49	9.63	4

[a] Data from (1) Mazurin (1962), (2) Rao (1963b), (3) Brailovskaya (1965), and (4) Appen and
Bresker (1952).

TABLE 14.32

Dielectric Parameters of $K_2O-MO-SiO_2$ Glasses

Compostion (mole %)			f	Temperature			
K_2O	MO	SiO_2	(MHz)	(°C)	$10^4 \tan \delta$	ε'	Reference[a]
	M = Be						
17.4	17.4	65.2	450	20	51	7.62	1
17.9	25	57.1			58	8.75	
	M = Mg						
17.4	17.4	65.2	450	20	48	7.96	1
20	20	60	1	49		8.6	2
				78		8.8	
				115		9.3	
				158		10.1	
				217		12.05	
	M = Ca						
17.4	17.4	65.2	450	20	48	8.37	1
20	20	60	1	102		8.2	2
				138		8.4	
				182		8.8	
				237		9.8	
				308		12.2	
	M = Sr						
17.4	17.4	65.2	450	20	47	8.45	1
	M = Ba						
5.94	19.8	74.26	450	20	34	7.68	1
11.2	18.7	70.1			38	8.13	
17.4	17.4	65.2			48	8.73	
22.14	16.4	61.46			59	9.05	
20	20	60	1	133		9.23	2
				169		9.75	
				212		9.75	
				265		10.4	
	M = Zn						
17.4	17.4	65.2	450	20	46	7.89	1
20	20	60	1	36		9.0	2
				67		9.2	
				107		9.6	
				158		10.7	
				220		13.2	
	M = Cd						
17.4	17.4	65.2	450	20	42	8.25	1
20	20	60	1	89		8.6	2
				125		8.8	
				168		9.2	
				220		9.9	

[a] Data from (1) Appen and Bresker (1952) and (2) Brailovskaya (1965).

TABLE 14.33
Loss Tangents of K$_2$O–MgO–SiO$_2$ Glasses at 25.5°C and 1 kHz (K$_2$O/SiO$_2$ Ratio = 0.2)[a]

Composition (mole % MgO)	$10^4 \tan \delta$
0	380
3.2	303
6.7	281
10	297
13.3	356
16.7	373
20	289
23.3	226

[a] Data read from graph of Rinehart and Bonino (1959).

TABLE 14.34
Loss Tangents of K$_2$O–ZnO–SiO$_2$ Glasses at 25.5°C and 1 kHz (K$_2$O/SiO$_2$ Ratio = 0.2)[a]

Composition (mole % ZnO)	$10^4 \tan \delta$
0	383
1.7	353
3.2	345
6.7	300
10	304
13.3	304
16.7	255
20	190
23.3	137

[a] Data read from graph of Rinehart and Bonino (1959).

TABLE 14.35

Dielectric Properties of Potassium–Lead Silicate Glasses[a]

Composition (mole %)			f (MHz)	Temperature (°C)	$10^4 \tan \delta$	ε'	Reference[a]
K$_2$O	PbO	SiO$_2$					
16	20	64	450	20	39	9.15	1
20	20	60	1	145		9.5	2
				184		9.7	
				227		10.0	
				282		10.4	
17.5	—	82.5	1	25	114	6.82	3
18.27	1.54	80.19			64	6.88	
19.07	3.22	77.71			91	7.48	
20.90	7.05	72.05			23	8.42	
25.84	17.45	56.71			80	9.78	
33.9	34.3	31.8			179	12.24	

[a] Data from (1) Appen and Bresker (1952), (2) Brailovskaya (1965), and (3) Rao (1963b).

TABLE 14.36

Values of Loss Tangent of Lithium Borosilicate Glasses at 20°C and 1.5 MHz[a]

Composition (mole %)				Composition (mole %)			
Li_2O	B_2O_3	SiO_2	$10^4 \tan \delta$	Li_2O	B_2O_3	SiO_2	$10^4 \tan \delta$
7	83	10	97.5	13	47	40	71.2
14	76	10	70.9	19	41	40	102
16	74	10	28.8	25	35	40	119
18	72	10	52	2.5	38.5	59	47
25	65	10	112	5	36	59	46
5	55	40	111	7.5	33.5	59	40
9	51	40	76.9	10	31	59	85
11	49	40	54.0	12.5	28.9	58.6	127

[a] Data from Stevels (1952).

TABLE 14.37

Dielectric Properties of Lithium Aluminosilicate Glasses at 1 MHz[a]

Composition (wt. %)					Composition (wt. %)				
Li_2O	Al_2O_3	SiO_2	$10^4 \tan \delta$	ε'	Li_2O	Al_2O_3	SiO_2	$10^4 \tan \delta$	ε'
10	10	80	140	7.6	12.5	15	72.5	172	8.1
15	10	75	189	8.5	15	15	70	203	8.4
17.5	10	72.5	205	10.1	17.5	15	67.5	226	10.3
20	10	70	251	10.4	20	15	65	264	10.5

[a] Data from Rauch et al. (1959).

TABLE 14.38

Dielectric Parameters of Sodium Borosilicate Glasses at Room Temperature[a]

Composition (mole %)			$10^4 \tan \delta$ at frequency given		ε' at frequency given	
Na_2O	B_2O_3	SiO_2	1 kHz	1 MHz	1 kHz	1 MHz
4	28	68	142	24	4.6	4.3
12	20	68	337	86	7.3	7.1
16	4	80	540	150	7.2	7.1
16	8	76	600	147	7.9	7.6
16	12	72	546	128	8.2	7.8
16	16	68	454	113	8.3	8.0
16	20	64	360	100	8.0	7.7
16	24	60	340	95	7.25	7.0
16	32	52	—	80	6.1	6.0
20	12	68	715	155	9.8	8.4

[a] Data from Appen and Gan'Fu-Si (1959, 1960).

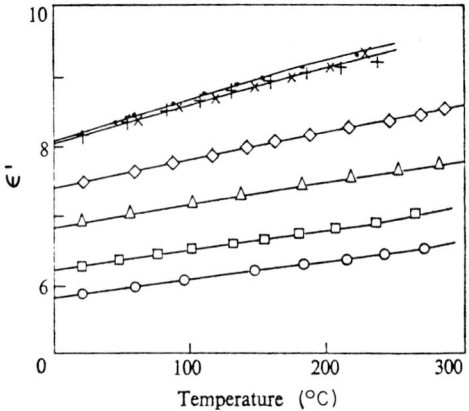

Fig. 14.22. Temperature dependence of dielectric constant of $Na_2O-xAl_2O_3-(5-2x)SiO_2$ glasses at 9.1 GHz. Value of x: $\bigcirc$, 0; $\square$, 0.25; $\triangle$, 0.5; $\diamond$, 0.75; x +, 1.0; $\bullet$, 1.1. From Topping and Isard (1971).

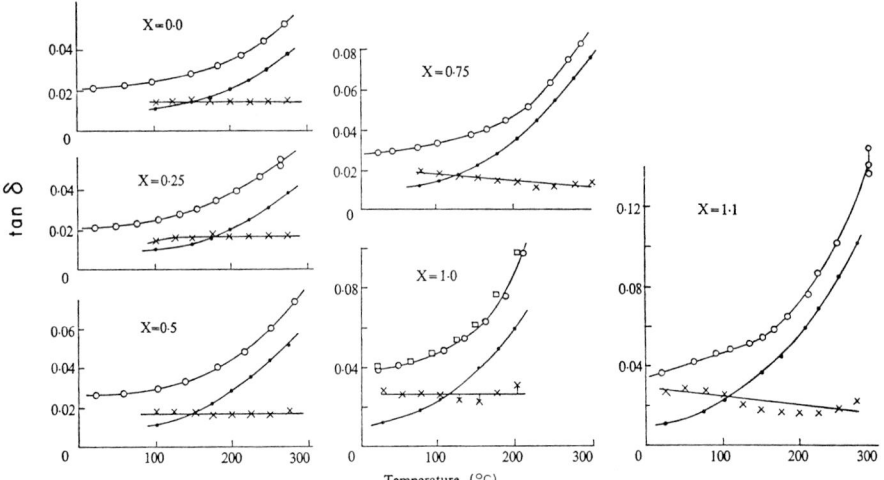

Fig. 14.23. Temperature dependence of loss tangent of $Na_2O-xAl_2O_3-(5-2x)SiO_2$ glasses at 9.1 GHz. $\bigcirc$, measured loss; $\bullet$, calculated contribution from migration loss mechanism; $\times$, residual loss. From Topping and Isard (1971).

TABLE 14.39

Dielectric Parameters of $Na_2O-Bi_2O_3-SiO_2$ Glasses at 1 MHz[a]

Composition (wt. %)				
Na_2O	Bi_2O_3	SiO_2	$10^4 \tan \delta$	ε'
5	60	35	50	12.7
10	50	40	70	15.6
10	60	30	40	14.0
10	65	25	10	15.8
20	50	30	60	13.9

[a] Data from Watanabe *et al.* (1970).

TABLE 14.40

Dielectric Parameters of Potassium Borosilicate Glass at 20°C and 450 MHz[a]

Composition (mole %)				
K_2O	B_2O_3	SiO_2	$10^4 \tan \delta$	ε'
16	20	64	30	6.65

[a] Data from Appen and Bresker (1952).

TABLE 14.41

Loss Tangents of $Li_2O-SnO_2-SiO_2$ Glasses at 1 kHz at Room Temperature[a]

Composition (mole %)			
Li_2O	SnO_2	SiO_2	$10^4 \tan \delta$
33.3	—	66.7	395
33.3	1	65.7	418
33.3	2	64.7	428
33.3	3	63.7	426
33.3	4	62.7	428

[a] Data read from graph of Vakhrameev (1968, 1970).

TABLE 14.42

Dielectric Parameters of $Li_2O-TiO_2-SiO_2$ Glass
at 20°C and 450 MHz[a]

Composition (mole %)				
Li_2O	TiO_2	SiO_2	$10^4 \tan \delta$	ε'
16	20	64	56	9.75

[a] Data from Appen and Bresker (1952).

TABLE 14.43

Loss Tangents of $Na_2O-SnO_2-SiO_2$ Glasses at Room Temperature and 1 kHz[a]

Composition (mole %)				Composition (mole %)			
Na_2O	SnO_2	SiO_2	$10^4 \tan \delta$	Na_2O	SnO_2	SiO_2	$10^4 \tan \delta$
15	—	85	210	25	—	75	365
15	2.5	82.5	169	25	2.5	72.5	310
15	5.0	80	149	25	5.0	70	280
15	7.5	77.5	160	25	7.5	67.5	274
20	—	80	251	25	10.0	65	307
20	2.5	77.5	233	30	—	70	436
20	5.0	75	202	30	5	65	365
20	7.5	72.5	210	30	7.5	62.5	374
20	10.0	70	218	30	10.0	60.0	370

[a] Data read from graph of Vakhrameev (1968, 1970).

TABLE 14.44

Frequency Dependence of Dielectric Properties of Na_2O–TiO_2–SiO_2 Glasses at 29°C[a]

Composition (mole %)			$10^2 \tan \delta$ at given frequency (kHz)					ε' at given frequency (kHz)				
Na_2O	TiO_2	SiO_2	0.06	1	10	100	9.6×10^6	0.06	1	10	100	9.6×10^6
15.5	14.0	70.5	—	—	—	—	1.6	—	—	—	—	8.0
16.5	26.8	56.7	—	—	—	—	1.5	—	—	—	11.7	10.5
21.7	24.9	53.4	20	7.6	3.4	2.1	—	15.9	13.1	12.2	14.2	21.1
22.6	37.4	40.3	17.5	6.6	3.2	1.8	1.3	18.7	15.7	14.7	12.8	11.3
23.0	30.2	46.8	20	7.6	3.6	2.1	1.6	17.5	14.3	13.3	10.0	7.3
23.3	5.4	71.5	44	14.3	6.0	3.0	1.9	17.0	12.0	10.7	11.5	9.6
23.6	21.8	54.8	36	12.0	5.4	2.8	1.6	18.3	13.6	12.2	11.7	—
26.2	12.9	61.0	46	14.9	6.2	3.0	—	22.0	14.2	12.5	12.6	10.6
27.6	25.8	46.6	73.0	20.0	8.2	3.9	2.0	30.0	16.1	13.6	11.7	9.4
27.7	15.8	56.5	111.0	28.0	11.0	5.0	2.2	41.0	16.1	13.0	14.5	12.6
28.0	35.8	36.3	72.0	19.7	7.8	3.6	1.8	34.0	18.3	15.6	13.8	11.8
29.1	31.4	39.7	80.0	22.0	8.4	4.0	2.0	35.0	17.8	15.0	17.4	13.7
25	40	35	82.0	22.0	8.5	3.9	1.8	44.0	22.0	18.9	—	—
30	20	50	—	—	—	—	2.7	—	—	—	—	1.02
30	40	30	—	—	—	—	2.0	—	—	—	—	1.31

[a] Data from Hirayama and Berg (1961).

TABLE 14.45

Temperature Dependence of Dielectric Parameters of Na_2O–TiO_2–SiO_2 Glasses Measured at 100 kHz[a]

Composition (mole %)			$10^2 \tan\delta$ at temperature (°C)				ε' at temperature (°C)			
Na_2O	TiO_2	SiO_2	61	89	125	147	61	89	125	147
21.7	24.9	53.4	4.3	7.7	17.1	27	12.3	13.0	14.7	16.3
22.6	37.4	40.3	3.5	6.7	14.2	23.1	14.7	15.6	17.2	18.8
23.0	30.2	46.8	4.1	7.3	15.6	26	13.4	14.1	15.7	17.5
23.3	26.8	56.7	6.9	12.6	32.0	57.0	10.8	11.7	14.2	18.0
23.6	21.8	54.8	6.0	11.3	26	41	12.3	13.3	15.5	17.7
26.2	12.9	61.0	7.2	13.3	33	57	12.6	13.7	16.8	21
27.6	25.8	46.6	8.7	18.3	42.0	74.0	13.6	15.3	18.8	26.0
27.7	15.8	56.5	12.2	24.0	62.0	110.0	13.0	15.0	22.0	34.0
28.0	35.8	36.3	8.3	17.1	43.0	75.0	15.6	17.3	22.0	31.0
29.1	31.4	39.7	8.9	18.2	45.0	81.0	15.0	16.8	21.0	31.0
25	40	35	120.0	—	—	—	37.0	—	—	—

[a] Data from Hirayama and Berg (1961).

TABLE 14.46

Loss Tangents of Na_2O–ZrO_2–SiO_2 Glasses at 25.5°C and 1 kHz (Soda/Silica Ratio = 0.25)[a]

Composition (mole % ZrO_2)	$10^4 \tan\delta$
0	1230
0.7	1174
1.7	1162
3.3	1204
6.8	1281
10	1414
13.5	1885

[a] Data read from graph of Rinehart and Bonino (1959).

TABLE 14.47

Loss Tangents of K_2O–SnO_2–SiO_2 Glasses at 1 kHz and Room Temperature[a]

Composition (mole %)			
K_2O	SnO_2	SiO_2	$10^4 \tan\delta$
20	—	80	149
20	2.5	77.5	133
20	5.0	75	105
20	7.5	72.5	102

[a] Data read from graph of Vakhrameev (1968, 1970).

TABLE 14.48

Dielectric Parameters of $K_2O-TiO_2-SiO_2$ Glasses at 25°C and 1 MHz Frequency[a]

Composition (wt. %)					Composition (wt. %)				
K_2O	TiO_2	SiO_2	$10^4 \tan \delta$	ε'	K_2O	TiO_2	SiO_2	$10^4 \tan \delta$	ε'
15	5	80	116	6.0	40	5	55	503	9.5
15	12.5	72.5	197	6.9	40	10	50	426	9.7
15	40	45	24	10.2	40	20	40	446	9.9
15	35	50	—	9.4	40	40	20	409	10.9
20	5	75	88	6.8	40	45	15	759	11.5
20	15	65	80	7.6	40	50	10	638	12.0
20	20	60	77	8.2	45	5	50	2133	10.4
20	40	40	30	10.6	45	10	45	2356	10.8
20	45	35	30	10.7	45	20	35	1106	10.7
30	10	60	150	8.2	45	30	25	—	10.5
30	15	55	140	8.3	45	40	15	1386	11.2
30	30	40	60	9.4	45	45	10	—	11.5
30	40	30	110	10.4	45	50	5	7273	—
30	45	25	—	10.9	60	10	30	1238	11.5

[a] Data from Rao (1963c, 1964).

TABLE 14.49

Dielectric Parameters of Na_2O–Nb_2O_5–SiO_2 Glasses at 26°C and Various Temperatures[a]

Composition (mole %)			10^2 tan δ at given frequency (kHz)					ε' at given frequency (kHz)				
Na_2O	Nb_2O_5	SiO_2	0.1	1	10	100	9.6×10^6	0.1	1	10	100	9.6×10^6
15	7.5	77.5	76	9.11	6.29	2.76	1.9	34.7	12.3	11.3	10.7	7.9
19.9	5	75.1	22	11.8	8.92	2.70	1.5	18.9	13.0	11.8	9.2	7.8
12.6	12.5	74.9	14.1	6.96	5.37	2.03	1.1	16.0	14.1	13.3	12.8	9.7
21.9	7	71.1	43	17.7	10.9	3.86	1.6	21.6	15.9	14.0	12.8	9.5
15	15	70	14.7	7.21	6.15	2.35	1.0	18.1	15.9	14.9	14.3	11.1
20	10	70	29	12.5	8.81	3.16	1.7	21.1	16.8	15.3	14.3	9.5
25	5	70	57	24	26	4.87	2.2	22.0	15.3	11.7	9.8	8.4
29.9	5	65.1	143	50	23	7.39	2.3	112	21.5	16.0	13.6	8.9
19.1	19.1	61.8	22	11.0	4.52	1.97	1.3	24.8	21.9	16.5	15.8	14.1
22.1	16	61.9	27	12.1	8.56	3.08	2.0	26.4	21.2	19.3	17.9	12.1
24.9	13	62.1	48	17.7	11.7	3.77	2.0	33.0	20.3	17.8	16.3	11.1
29	9	62	87	30	18.2	5.36	1.7	52	20.8	17.0	15.1	10.1
19.8	26	54.2	12.2	6.17	4.42	2.11	2.1	29.6	26.5	23.9	24.0	18.6
22.9	23	54.1	18.4	8.87	7.00	3.66	2.0	28.8	24.6	22.8	21.5	16.8
26	20	54	30	12.9	8.61	3.09	1.2	30.9	24.4	21.9	20.5	14.8
29.9	16	54.1	73	26	15.3	4.68	1.8	56	25.5	21.2	19.0	12.6
35.1	11	53.9	—	71	38	9.23	2.3	—	39.7	21.5	17.4	11.1
25	25	50	22	10.4	6.75	2.70	1.6	31.7	26.6	24.6	23.4	17.8
30	20	50	56	20	12.6	6.19	1.4	46	26.1	22.5	20.4	14.8
38.1	12	49.9	—	103	55	13.2	2.2	—	68	27.0	18.5	11.9
33.9	19.1	47.0	101	40	19.8	6.09	1.7	89	32.8	23.3	20.4	13.2
40.9	13.6	45.5	—	50	27	8.29	2.0	—	75	46.3	39.3	28.3

[a] Data from Hirayama and Berg (1963).

TABLE 14.50

Dielectric Parameters of $K_2O-Nb_2O_5-SiO_2$ Glasses at 1 MHz Frequency[a]

Composition (mole %)					Composition (mole %)				
K_2O	Nb_2O_5	SiO_2	$10^4 \tan \delta$	ε'	K_2O	Nb_2O_5	SiO_2	$10^4 \tan \delta$	ε'
16.1	19.4	64.5	26	12.7	21.7	13.1	65.2	98	10.3
16.7	16.7	66.6	47	11.7	22.7	9.1	68.2	175	9.3
17.2	13.8	69.0	53	10.2	23.8	4.8	71.4	176	8.7
17.9	10.7	71.4	61	9.5	23.8	28.6	47.6	29	15.5
18.5	7.4	74.1	90	8.9	25	25	50	61	14.0
19.2	3.8	77.0	132	8.2	26.3	21.1	52.6	68	12.3
19.2	23.1	57.7	39	13.7	27.8	16.7	55.5	129	11.2
20	20	60	43	12.3	29.4	11.8	58.8	184	10.7
20.8	16.7	62.5	67	11.1	31.3	6.2	62.5	271	10.1

[a] Data from Rao (1962b).

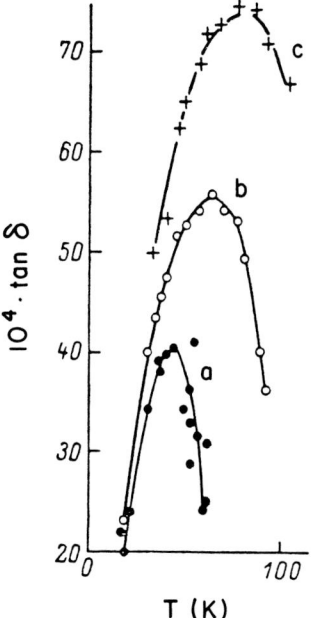

Fig. 14.24. Temperature dependence of the loss tangent of $15MgO-25PbO-60SiO_2$ (mole %) glass at (a) 0.3, (b) 1, and (c) 3 MHz. From Ioffe (1954, 1955).

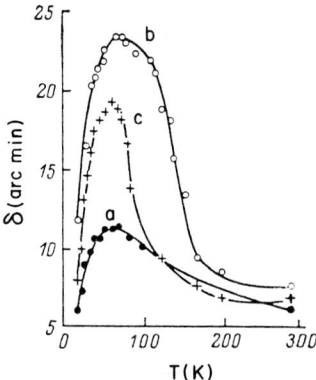

Fig. 14.25. Temperature dependence of the loss angle δ at 1 MHz of the glass compositions (mole %): (a) $5MgO-40PbO-55SiO_2$, (b) $10MgO-40PbO-50SiO_2$, and (c) $15MgO-25PbO-60SiO_2$. From Ioffe (1954, 1955).

TABLE 14.51

Dielectric Parameters of $CaO-PbO-SiO_2$ Glass at 20°C and 450 MHz[a]

Composition (mole %)			$10^4 \tan \delta$	ε'
CaO	PbO	SiO_2		
20	20	60	40	10.44

[a] Data from Appen and Bresker (1952).

TABLE 14.52

Values of Loss Tangent of $CaO-MnO-SiO_2$ Glasses at Different Temperatures[a]

Composition (mole %)			$\log(10^4 \tan \delta)$ at given temperature (°C)		
CaO	MnO	SiO_2	200	300	400
30	10	60	1.55	2.21	3.08
35	10	55	1.70	2.38	3.28
40	10	50	1.62	2.29	3.18
45	10	45	1.84	2.53	3.43
20	20	60	1.70	2.60	3.53
25	20	55	2.20	3.15	4.17
30	20	50	1.90	2.80	3.82
40	20	40	2.12	3.05	4.05
45	20	35	2.58	3.32	4.35
10	30	60	2.17	3.00	3.84
15	30	55	2.45	3.35	4.33
20	30	50	2.22	3.10	4.09
30	30	40	2.65	3.57	4.50
5	40	55	2.40	3.32	4.35
10	40	50	2.53	3.65	4.70
20	40	40	2.62	3.80	4.93
—	50	50	2.95	3.80	4.47

[a] Data from Bezrodn'iy and Kudyshkina (1974).

TABLE 14.53

Dielectric Parameters of SrO–MnO–SiO$_2$ Glasses at 1 kHz[a]

Composition (mole %)					Composition (mole %)				
SrO	MnO	SiO$_2$	$\log(10^4 \tan \delta)$	ε'	SrO	MnO	SiO$_2$	$\log(10^4 \tan \delta)$	ε'
25	10	65	2.25	10.20					
10	30	60	4.12	17.21	40	10	50	2.65	12.81
20	20	60	3.23	13.51	15	40	45	4.23	28.81
30	10	60	2.41	10.91	25	30	45	4.45	22.03
5	40	55	4.01	19.74	35	20	45	4.16	24.21
15	30	55	4.31	19.97	45	10	45	2.82	14.21
25	20	55	3.94	17.12	20	40	40	4.39	30.67
35	10	55	2.50	11.80	30	30	40	4.61	32.13
10	40	50	3.74	17.28	40	20	40	4.82	38.52
20	30	50	3.74	14.94	35	30	35	4.75	34.86
30	20	50	3.71	14.58					

[a] Data from Kuznetsova (1972a,b) and Kuznetsova and Evstrop'ev (1972).

TABLE 14.54

**Dielectric Parameters of BaO–PbO–SiO$_2$ Glasses
at 1 MHz[a]**

Composition (mole %)				
BaO	PbO	SiO$_2$	$10^4 \tan \delta$	ε'
17.5	17.5	65	10.16	12.1
20	20	60	14.32	13.3
22.5	22.5	55	17.95	14.9
23.3	20	56.7	17.82	14.3
25	25	50	22.75	16.2

[a] Data from Sheludyakov (1967).

TABLE 14.55

Dielectric Parameters of BaO–MnO–SiO$_2$ Glasses at 1 kHz[a]

Composition (mole %)				
BaO	MnO	SiO$_2$	$\log(10^4 \tan \delta)$	ε'
25	10	65	2.65	13.01
30	10	60	2.74	13.62
35	10	55	2.91	14.83
40	10	50	3.12	16.81
45	10	45	3.34	17.95
20	20	60	3.25	16.19
25	20	55	3.34	16.83
30	20	50	3.51	19.61
35	20	45	3.64	20.36
40	20	40	3.78	22.19
10	30	60	3.89	14.62
15	30	55	4.05	15.93
20	30	50	4.32	18.61
25	30	45	4.46	25.32
5	40	55	3.32	18.14
10	40	50	3.78	20.10
15	40	45	4.42	27.48
20	40	40	4.83	33.12

[a] Data from Kuznetsova and Evstrop'ev (1972) and Kuznetsova (1972a,b).

TABLE 14.56

Dielectric Parameters of ZnO–MnO–SiO$_2$ Glasses at 1 kHz[a]

Composition (mole %)					Composition (mole %)				
ZnO	MnO	SiO$_2$	$\log(10^4 \tan \delta)$	ε'	ZnO	MnO	SiO$_2$	$\log(10^4 \tan \delta)$	ε'
5	45	50	3.83	21.42	30	20	50	2.85	11.12
10	40	50	3.71	18.63	40	10	50	2.31	9.63
20	30	50	3.16	14.19	45	10	45	2.13	8.61

[a] Data from Kuznetsova (1972a,b) and Kuznetsova and Evstrop'ev (1972).

TABLE 14.57

Dielectric Parameters of CdO–MnO–SiO$_2$ Glasses at 1 kHz[a]

Composition (mole %)				
CdO	MnO	SiO$_2$	$\log(10^4 \tan \delta)$	ε'
10	40	50	3.62	16.81
20	30	50	3.78	20.56
30	20	50	3.61	16.40
40	10	50	3.18	15.61
50	—	50	2.36	14.18
35	10	55	2.25	14.03
45	10	45	2.64	15.21
50	10	40	2.75	16.44
55	10	35	2.84	17.26
60	10	30	3.81	18.02

[a] Data from Kuznetsova (1972a,b) and Kuznetsova and Evstrop'ev (1972).

TABLE 14.58

Loss Tangents and Dielectric Constants of Magnesium Aluminosilicate Glasses at 10 GHz[a]

Composition (mole %)				
MgO	Al$_2$O$_3$	SiO$_2$	$10^4 \tan \delta$	ε'
13	17	70	62	5.3
18	17	65	70	5.7
23	17	60	78	6
28	17	55	86	6.4
18	12	70		5.4
23	12	65		5.7
28	12	60		6.1
33	12	55		6.4

[a] Data read from graph of Mashkovich (1969, 1970).

TABLE 14.59

Temperature Dependence of Dielectric Parameters of 33 MgO–12 Al$_2$O$_3$–55 SiO$_2$ (mole %) Glass at 10 GHz[a]

Temperature (°C)	$10^4 \tan \delta$	ε'	Temperature (°C)	$10^4 \tan \delta$	ε'
−165	66	6.0	100	—	6.4
−135	82	6.2	200	93	6.6
− 15	95	—	300	93	6.7
25	93	6.3	400	—	6.8

[a] Data read from graph of Mashkovich and Smelyanskaya (1965).

TABLE 14.60

Dielectric Parameters of Some MO–Al$_2$O$_3$–SiO$_2$ Glasses at Different Temperatures and Frequencies[a]

Composition (wt. %)			Temperature (°C)	10^4 tan δ at given frequency (kHz)		ε' at given frequency (kHz)	
MO	Al$_2$O$_3$	SiO$_2$		0.1	10	0.1	10
M = Ca							
20.1	36.7	43.2	25	13	16	7.17	7.14
			295	21	19	7.45	7.41
			497	360	58	7.96	7.71
M = Sr							
31.1	31.6	36.6	25	17	18	8.40	8.36
			306.1	21	21	8.75	8.70
			509.0	310	70	9.17	8.99

[a] Data from Bahat (1969).

TABLE 14.61

Dielectric Parameters of Barium Aluminosilicate Glasses

Composition (wt. %)			Temperature (°C)	Frequency (Hz)	10^4 tan δ	ε'	Reference[a]
BaO	Al$_2$O$_3$	SiO$_2$					
30	10	60		1 × 10^6	15.8	6.12	1
40	10	50			16.0	5.16	
50	10	40			12.2	5.7	
40.6 ± 0.1	26.6 ± 0.1	31.83	25	0.1 × 10^3	12	9.46	2
			25	10 × 10^3	14	9.43	
			298.2	0.1 × 10^3	15	9.75	
				10 × 10^3	14	9.71	
			498.7	0.1 × 10^3	340	10.23	
				10 × 10^3	40	9.98	
53	8	39	—	10 × 10^9	68	8.42	3

[a] Data from (1) Naidenov (1963), (2) Bahat (1969), and (3) Navias and Green (1946).

TABLE 14.62

Dielectric Parameters of Cadmium–Bismuth Silicate Glasses at 25°C and 1 MHz Frequency[a]

Composition (wt. %)				
CdO	Bi_2O_3	SiO_2	$10^4 \tan \delta$	ε'
4.8	92.2	3.0	40	33.4
7.5	85	7.5	34	27.9
9.8	88.2	2.0	49	33.5
14.7	83.3	2.0	34	31.9
15	70	15	20	19.4
19.4	77.7	2.9	11	29.1
19.6	78.4	2.0	26	30.8
19.8[b]	79.2	1.0	20	32.4
19.9	79.6	0.5	38	41.6
29.2	58.8	12.0	8	22.4
29.4	68.6	2.0	3	27.4
40[c]	40	20	24	15.3
47.6	47.6	4.8	26	20.3
57.1	38.1	4.8	20	17.3

[a] Data from Rao (1962a).
[b] Composition found by analysis: $19.5\ CdO–78.7Bi_2O_3–1.8SiO_2$ (wt. %).
[c] Composition found by analysis: $37.3CdO–36.5Bi_2O_3–26.4SiO_2$ (wt. %).

TABLE 14.63

Dielectric Parameters of Lead Borosilicate Glasses

Composition (mole %)			Frequency (Hz)	$10^4 \tan \delta$	ε'	Reference[a]
PbO	B_2O_3	SiO_2				
50	5	45	800	15	17.7	1
50	25	25	800	10	16.0	
50	45	5	800	—	15.0	
8.6	27.6	63.8	3×10^9	15	4.76	2
			10×10^9	20	4.64	

[a] Data from (1) Bischoff (1955) and (2) Navias and Green (1946).

TABLE 14.64

**Temperature Dependence of
Dielectric Constant and Loss Tangent
of $39.2PbO-2Al_2O_3-58.8SiO_2$ (mole %)
Glass at 10 GHz[a]**

Temperature (°C)	$10^4 \tan \delta$	ε'
−175	127	—
−145	141	11.6
− 70	147	11.9
− 15	149	12.2
25	144	12.5
100	147	12.7
200	147	—
300	148	—
400	145	13.1

[a] Data read from graph of Mashkovich and Smelyanskaya (1965).

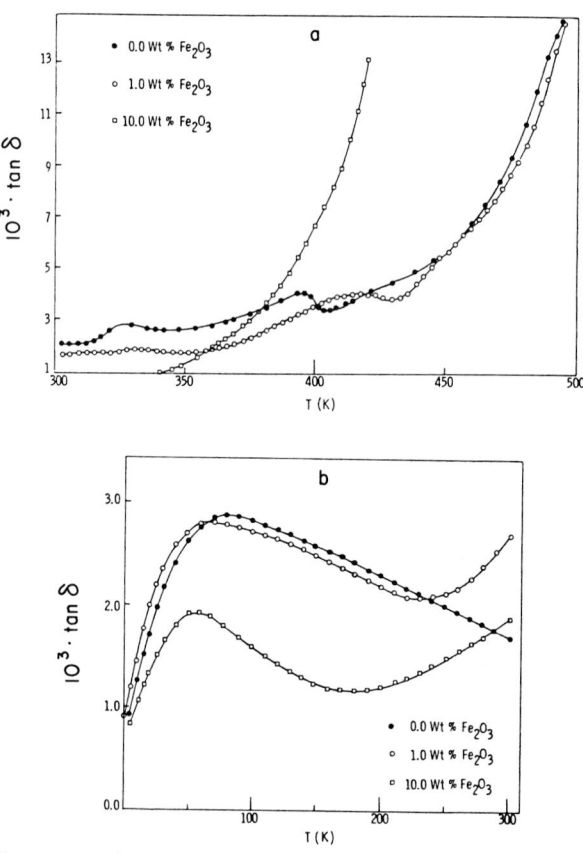

Fig. 14.26. Temperature dependence of loss tangent of $PbO-Fe_2O_3-SiO_2$ glasses at 1 kHz in (a) high-temperature and (b) low-temperature regions. From El-Bayoumi and MacCrone (1976).

TABLE 14.65

**Loss Tangent of BaO–TiO$_2$–SiO$_2$ Glass
at 20°C and 10 GHz[a]**

Composition (mole %)			
BaO	TiO$_2$	SiO$_2$	10^4 tan δ
30	30	40	35

[a] Data from Mashkovich (1970).

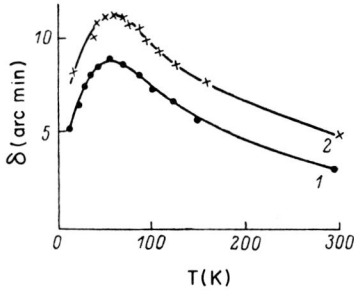

Fig. 14.27. Temperature dependence of the loss angle δ of the glass compositions (mole %):
(1) 40PbO–12TiO$_2$–48SiO$_2$ and (2) 50PbO–6TiO$_2$–44SiO$_2$ at 1 MHz. From Ioffe (1954).

Chapter **15**

Ionic Diffusion

I MEASUREMENT OF IONIC DIFFUSION

Ionic diffusion can be studied with radioactive tracers; such diffusion, called "tracer diffusion," takes place without chemical changes or ionic gradients in the sample. Either the tracer can be deposited on the sample surface from a solution to provide a concentrated source of tracer, or the tracer concentration at the sample surface can be maintained constant by a large dilute source such as a fused salt.

The diffusion coefficient D is defined by the equation

$$J = -D\, \partial c/\partial x, \tag{1}$$

where J is the flux of diffusing species and $\partial c/\partial x$ is its gradient of concentration c in the x direction. From the equation of continuity,

$$\frac{\partial c}{\partial t} = \frac{\partial(-J)}{\partial x} = D\,\frac{\partial^2 c}{\partial x^2}, \tag{2}$$

where t is the time and D is independent of concentration.

If at time zero an amount Q of substance per unit area is added to one face of a semi-infinite solid containing none of this substance, then the solution of Eq. (2) is

$$c = \frac{Q}{\sqrt{\pi Dt}} \exp\left(-\frac{x^2}{4Dt}\right), \tag{3}$$

where c is the concentration of diffusion substance at a distance x from the face after diffusion for a time t. To calculate the diffusion coefficient D from Eq. (3), only the slope of the log concentration against x^2 plot for a certain time of diffusion is needed, since Q and t are constants for these conditions. The boundary condition of Eq. (3) applies when material containing a radioactive or isotropic tracer of high specific activity is deposited on a solid. The concentration profile in the solid after diffusion for a certain time can be determined by removing successive layers of the solid and analyzing for the tracer with a counter or mass spectrometer.

If the solute concentration at the face of a semi-infinite solid that contains no solute initially is maintained at c_i, then with D constant,

$$c = c_i \operatorname{erfc}\left(\frac{x}{2\sqrt{Dt}}\right), \tag{4}$$

where

$$\operatorname{erfc}(z) = \frac{2}{\sqrt{\pi}} \int_z^\infty e^{-y^2} \, dy.$$

The total amount of material M that has diffused into the solid after a time t is

$$M = 2c_i \left(\frac{Dt}{\pi}\right)^{1/2}. \tag{5}$$

TEMPERATURE DEPENDENCE II

The diffusion coefficient D follows an Arrhenius equation as a function of temperature:

$$D = D_0 \exp(-E_D/RT), \tag{6}$$

where D_0 is the pre-exponential factor with the same units as D, usually square centimeters per second; and E_D is the activation energy, usually given in kilocalories per mole or in SI units in kilojoules per mole. These parameters are related to atomic mechanisms in the theory section.

INTERIONIC DIFFUSION III

If glass is placed in contact with a medium containing monovalent cations, such as a fused salt or an aqueous solution, these cations can exchange with monovalent cations in the glass and interdiffuse with them into the glass. The glass can be considered as a matrix of immobile negative groups with associated mobile cations. An exchange cation normally has a different mobility from the original ion; therefore, as interdiffusion proceeds, one ion tends to outrun the other and an electrical charge is built up. However, accompanying this charge is a gradient in electrical potential that slows down the fast ion and speeds up the slow one. To preserve electrical neutrality the fluxes of the two ions must be equal and opposite, and the electrical potential ensures this condition in spite of the difference in mobility of the two ions.

For two exchanging ions A and B this process can be described by an interdiffusion coefficient $\tilde{D}$:

$$\tilde{D} = \frac{D_A D_B}{N_A D_A + N_B D_B}, \tag{7}$$

where D_A and D_B are the tracer diffusion coefficients of the ions and N_A and N_B their mole fractions; $N_A + N_B = 1$. Ratios of diffusion coefficients D_A/D_B for monovalent ions in different glasses are given in Table 15.1.

Ion exchange is important in strengthening glass by exchanging a large ion such as potassium for a smaller ion such as sodium already in the glass. This

TABLE 15.1

Ratios of Diffusion Coefficients of Monovalent Ions in Rigid Silicate Glasses[a]

Glass type	Mole % M_2O	Temperature (°C)	Ions A–B	D_A/D_B
Soda-lime	15.2 Na_2O	354	Na–Ag	4
			Na–Ag	12
			Na–K	500
	14.0 Na_2O	250	Ag–Li	2
	13.0 Na_2O	350	Na–H	1800
			Na–K	1400
			Na–Rb	3.4×10^5
Potash-lime	12.5 K_2O	350	K–H	2
			K–Rb	190
Sodium binary	13 Na_2O	415	Na–K	25
	20 Na_2O	415	Na–K	39
	25 Na_2O	415	Na–K	48
Potassium binary	13 K_2O	415	Na–K	0.5
	20 K_2O	415	Na–K	0.5
	25 K_2O	415	Na–K	0.6
Sodium aluminosilicate (hydrated)	12.5 Na_2O	400	Na–K	~ 300
	27 Na_2O	25	Na–K	3; 7
Lithium aluminosilicate	5.2 Li_2O	415	Na–Li	1
Pyrex borosilicate	4 Na_2O	335	Na–Ag	1
		373	Na–K	~ 30
Vitreous silica (G.E. 204)	1.1×10^{-4} Na_2O	380	Na–Li	6.7
	3.4×10^{-4} Li_2O	380	Na–Ag	12
		380	Na–K	>500
(Spectrosil)	?	380	Na–Li	33
		380	Na–K	1400
(Vitreosil)	$\sim 5 \times 10^{-4}$ Na_2O	1000	Na–H	$\sim 10^4$

[a] From Doremus (1973).

exchange gives rise to a compressive surface stress, so that a high applied tensile stress is required to propagate surface flaws. Ion exchange rates also determine the chemical durability of silicate glasses (see Chapter 18) because of the reaction and exchange between water and alkali ions:

$$Na^+ \text{ (glass)} + 2H_2O = H_3O^+ \text{ (glass)} + NaOH \qquad (8)$$

THEORIES OF IONIC TRANSPORT   IV

The diffusion coefficient D of Eq. (1) can be described in terms of a random walk of molecules in discrete steps. From the probability of finding a molecule at a certain place after a fixed number of steps

$$D = g\overline{\lambda^2}\Gamma, \qquad (9)$$

where g is a geometrical constant, $\overline{\lambda^2}$ is the average of the squares of the step lengths, and Γ is the average number of steps made in unit time. The jump rate Γ_m for ionic motion can be calculated from various models; its usual form is

$$\Gamma_m = v \exp\left(\frac{\Delta S_m/R - \Delta H_m}{RT}\right), \qquad (10)$$

where v is the vibration frequency of the diffusing atom and ΔS_m and ΔH_m are the entropy and enthalpy of activation for the motion of an atom from one side to another.

The diffusion coefficients of monovalent cations in oxide glasses cannot be described by considering the glass to be a viscous liquid; the Stokes–Einstein equation predicts much too low a diffusion coefficient. Furthermore, these cations diffuse much more rapidly than the lattice elements, such as silicon and oxygen, so the glass behaves like a rigid solid with mobile interstitial ions.

The activation energy for molecular diffusion in glass is a strong function of the size of the diffusing molecule, as described in Chapter 17. However, the activation energy for ionic diffusion in fused silica is about the same at low temperatures for the following ions: hydrogen, lithium, sodium, and potassium. The activation energies for transport in binary alkali silicate glasses increase with increasing size at lower ionic concentrations, but are all about the same at higher concentrations. Crudely, one can think of two different processes being involved in ionic diffusion: a "squeezing" of the ions through the glass network, which should be harder the larger the ion, and a tearing away of the cation from the anionic site, which should be easier for the larger ions, since they are less tightly bound. Thus, with these two competing effects one might expect a maximum in diffusion coefficient at some intermediate size, which is the result shown in Table 15.1 with a maximum rate of transport of sodium ion in various silicate glasses.

V TABLES AND FIGURES

Tables 15.2–15.10 and Figs. 15.1–15.9 list data on the ionic diffusion of binary glasses. Data for ternary glasses are given in Tables 15.11 through 15.43 and Figs. 15.10 through 15.18.

BIBLIOGRAPHY

Barr, L. W., and Mundy, J. N. (1972). *In* "Amorphous Materials" (R. W. Douglas and B. Ellis, eds.), p. 234. Wiley (Interscience), New York.
Barton, J. L., and Morain, M. (1970). *J. Non-Cryst. Solids* **3**, 115.
Beier, W., and Frischat, G. H. (1980). *J. Non-Cryst. Solids* **38/39**, 569.
Beier, W., and Frischat, G. H. (1984). *Glastech. Ber.* **57**, 71.
Borom, M. P., and Pask, J. A. (1968). *J. Am. Ceram. Soc.* **51**, 490.
deBerg, K. C., and Lauder, I. (1978). *Phys. Chem. Glasses* **19**, 78.
DiMarcello, F. V. (1966). *Am. Ceram. Soc. Bull.* **45**, 420.
Doremus, R. H. (1962). *In* "Modern Aspects of the Vitreous State" (J. D. Mackenzie, ed.), Vol. 2, p. 1. Butterworth, London.
Doremus, R. H. (1973). "Glass Science." Wiley, New York.
Ershov, G. S., and Popova, E. A. (1964). *Russ. J. Inorg. Chem. (Engl. Transl.)* **9**, 361.
Evstrop'ev, K. K. (1960). *Struct. Glass* **2**, 237.
Evstrop'ev, K. K. (1963a). *Opt.-Mekh. Prom.* No. 5, p. 38.
Evstrop'ev, K. K. (1963b). *Opt.-Mekh. Prom.* No. 6, p. 32.
Evstrop'ev, K. K. (1966). Investigation of the processes of ionic diffusion and conductivity in glass. Avtoref., Dokt. Diss., Leningrad.
Evstrop'ev, K. K. (1970a). *In* "Stekloobraznoe Sostoyanie," p. 25. Erevan.
Evstrop'ev, K. K. (1970b). "Diffusion Processes in Glass," Leningrad.
Evstrop'ev, K. K., and Kharyuzov, V. A. (1961). *Dokl. Akad. Nauk SSSR* **136**, 140.
Evstrop'ev, K. K., and Pavlovskii, V. K. (1961). *Opt.-Mekh. Prom.* No. 8, p. 46.
Evstrop'ev, K. K., and Pavlovskii, V. K. (1966). *J. Appl. Chem. USSR (Engl. Transl.)* **39**, 2219.
Evstrop'ev, K. K., and Pavlovskii, V. K. (1967). *Inorg. Matter. (Engl. Transl.)* **3**, 592.
Evstrop'ev, K. K., Pavlova, G. A., and Pavlovskii, V. K. (1966). *J. Appl. Chem. USSR (Engl. Transl.)* **39**, 416.
Fleming, J. W., and Day, D. E. (1972). *J. Am. Ceram. Soc.* **55**, 186.
Frischat, G. H. (1967). *Glastech. Ber.* **40**, 382.
Frischat, G. H. (1971a). *Glastech. Ber.* **44**, 93.
Frischat, G. H. (1971b). *J. Mater. Sci.* **6**, 1229.
Frischat, G. H. (1972). *React. Solids, Proc. Int. Symp., 7th, Bristol* p. 199.
Frischat, G. H. (1975a). *Cent. Glass Ceram. Res. Inst. Bull.* **22**, 129.
Frischat, G. H. (1975b). *In* "Ionic Diffusion in Oxide Glasses," Trans. Tech. Publ., Bay Village, Ohio.
Frischat, G. H., and Oel, H. J. (1966a). *Glastech. Ber.* **39**, 50.
Frischat, G. H., and Oel, H. J. (1966b). *Glastech. Ber.* **39**, 524.
Garfinkel, H. M., and Rauscher, H. E. (1966). *J. Appl. Phys.* **37**, 2169.
Gupta, Y. P., and King, T. B. (1967). *Trans. Metall. Soc. AIME* **239**, 1701.
Hagel, W. C., and Mackenzie, J. D. (1964). *Phys. Chem. Glasses* **5**, 113.
Hahnert, M., Kolitsch, A., and Richter, E. (1984). *Z. Chem.* **21**, 229.
Haider, Z., and Roberts, G. J. (1970). *Glass Technol.* **11**, 158.
Haven, Y., and Verkerk, B. (1965). *Phys. Chem. Glasses* **6**, 38.
Hayami, R., and Terai, R. (1972). *Phys. Chem. Glasses* **13**, 102.
Hayami, R., and Terai, R. (1973). *Osaka Kogyo Gijutsu Shikensho Kiho* **24**, 151.

Heckman, R. W., Ringlien, J. A., and Williams, E. L. (1967). *Phys. Chem. Glasses* **8,** 145.

Henderson, J., Yang, L., and Derge, G. (1961). *Trans. Metall. Soc. AIME* **221,** 56.

Hlavac, J., and Matousek, J. (1971). *Silikaty* **15,** 333.

Ivanov, I. A., and Pavlov, A. B. (1975). *Sov. Electrochem.* (*Engl. Transl.*) **11,** 1276.

Jain, H., and Peterson, N. L. (1983). *J. Am. Ceram. Soc.* **66,** 174.

Jain, H., Peterson, N. L., and Downing, H. L. (1983). *J. Non-Cryst. Solids* **55,** 283.

Johnson, J. R. (1950). Thesis, Ohio State Univ., Columbus.

Johnson, J. R., Bristow, R. H., and Blau, H. H. (1951). *J. Am. Ceram. Soc.* **34,** 165.

Kahl, L., and Schiewer, E. (1971). *Atomwirtschaft* **16,** 434.

Kaps, C. (1984). *J. Non-Cryst. Solids* **65,** 189.

Kelly, J. E., III, and Tomozawa, M. (1982). *J. Non-Cryst. Solids* **51,** 345.

King, T. B., and Koros, P. J. (1959). *In* "Kinetics of High Temperature Processes (W. D. Kingery, ed.), p. 80. MIT Press, Cambridge, Massachusetts.

Kingery, W. D., and Lecron, J. A. (1960). *Phys. Chem. Glasses* **1,** 87.

Köhler, W., and Frischat, G. H. (1978). *Phys. Chem. Glasses* **19,** 103.

Kolitsch, A., and Richter, E. (1981a). *Z. Chem.* **21,** 376.

Kolitsch, A., and Richter, E. (1981b). *Z. Chem.* **21,** 377.

Kolitsch, A., Kuchler, R., Richter, E., and Hinz, W. (1978). *Silikattechnik* **29,** 369.

Kolitsch, A., Richter, E., and Hinz, W. (1979). *Silikattechnik* **30,** 172.

Kolitsch, A., Richter, E., and Hinz, W. (1980a). *Silikattechnik* **31,** 41.

Kolitsch, A., Richter, E., and Hinz, W. (1980b). *Z. Chem.* **20,** 423.

Kolitsch, A., Richter, E., Gehrke, E., Hinz, W., and Muller, W. (1980c). *Silikattechnik* **31,** 136.

Levi, H. W., Lutze, W., Malow, G., and Sedighi, N. (1971). *Phys. Status Solidi A* **5,** 617.

Lim, C., and Day, D. E. (1977a). *J. Am. Ceram. Soc.* **60,** 198.

Lim, C., and Day, D. E. (1977b). *J. Am. Ceram. Soc.* **60,** 473.

Lim, C., and Day, D. E. (1978). *J. Am. Ceram. Soc.* **61,** 329.

Lindner, R., Hassenteufels, W., Kotera, Y., and Matzke, H. (1960). *Z. Phys. Chem. N. F.* **23,** 408.

McGrail, B. P., Kumar, A., and Day, D. E. (1984). *J. Am. Ceram. Soc.* **67,** 463.

McVay, G. L., and Day, D. E. (1970). *J. Am. Ceram. Soc.* **53,** 508.

Malinin, V. R., and Evstrop'ev, K. K. (1972). *Sov. Radiochem.* (*Engl. Transl.*) **14,** 169.

Malinin, V. R., Evstrop'ev, K. K., and Tsekhomskii, V. A. (1972). *Zh. Prikl. Khim.* **45,** 184.

Malkin, V. I., and Mogutnov, B. M. (1961). *Dokl. Akad. Nauk SSSR* **141,** 1127.

Matousek, J. (1967). *Silikaty* **12,** 89.

Moiseev, V. V., and Zhabrev, V. A. (1965). *Silic. Ind.* **30,** 495.

Moiseev, V. V., and Zhabrev, V. A. (1969). *Inorg. Mater.* (*Engl. Transl.*) **5,** 793.

Müller-Warmuth, W., Krämer, F., and Dutz, H. (1971). *Int. Congr. Glass, Versailles* **2,** 303.

Muller, W., Hahnert, M., and Kolitsch, A. (1981). *Silikattechnik* **32,** 55.

Negodaev, G. D. (1973). Investigation of the diffusion process of alkaline cations in alkaline silicate glass in narrow temperature range. Autoref., Candidate's Dissertation, Leningrad.

Negodaev, G. D., and Malinin, V. R. (1974). *In* "Stekloobraznoe Sostoyanie," p. 146. Erevan.

Negodaev, G. D., Ivanov, I. A., and Evstrop'ev, K. K. (1971). *Radiokhimiya* **13,** 651.

Negodaev, G. D., Ivanov, I. A., and Evstrop'ev, K. K. (1972a). *Inorg. Mater.* (*Engl. Transl.*) **8,** 298.

Negodaev, G. D., Ivanov, I. A., and Evstrop'ev, K. K. (1972b). *Sov. Electrochem.* (*Engl. Transl.*) **8,** 227.

Negodaev, G. D., Ivanov, I. A., and Evstrop'ev, K. K. (1973). *Sov. Electrochem.* (*Engl. Transl.*) **9,** 1554.

Niwa, K. (1957). *Nippon Kinzoku Gakkaishi* **21,** 304.

Obarara, R. A. (1973). M. S. Thesis, Rensselaer Polytech. Inst., Troy, New York.

Pavlovskii, V. K. (1970). *Stekloobraznoe Sostoyanie* **5,** 148.

Perron, P. O., and Bell, H. B. (1967). *Trans. Br. Ceram. Soc.* **66,** 347.

Rawal, B. S., and Cooper, A. R. (1979). *J. Mater, Sci.* **14,** 1425.

Richter, E., Kolitsch, A., Dresden., and Hahnert, M. (1982). *Glastech. Ber.* **55,** 171.

Riebling, E. F. (1969). *Am. Ceram. Soc. Bull.* **48,** 766.

Saito, T., and Maruya, S. (1957). *Nippon Kinzoku Gakkaishi* **21,** 728.

Schaeffer, H. A. (1971). *Int. Congr. Glass, Versailles* **1,** 133.

Schaeffer, H. A. (1974). *Phys. Status Solidi A* **22,** 281.

Schaeffer, H. A., and Oel, H. J. (1969). *Glastech. Ber.* **42,** 493.

Schafer, H., and Maywald, H. (1967). *Glastech. Ber.* **40,** 253.

Schlichting, J. (1982). *Glastech. Ber.* **55,** 167.

Souquet, J. L., Deportes, C., and Besson, J. (1968). *Silic. Ind.* **23,** 75.

Sviridov, S. I., Roskova, G. P., Moiseev, V. V., and Zhabrev, V. A. (1976). *Sov. J. Glass Phys. Chem. (Engl. Transl.)* **2,** 514.

Takizawa, K., Sakai, T., and Ohishi, K. (1981). *Nippon Kagaku Kaishi* **13,** 1522.

Taylor, R. H., Robertson, J., Morris, S. B., Williamson, J., and Atkinson, A. (1980). *J. Mater. Sci.* **15,** 670.

Terai, R. (1968). *Yogyo Kyokaishi* **76,** 189.

Terai, R. (1969a). *Phys. Chem. Glasses* **10,** 146.

Terai, R. (1969b). *Yogyo Kyokaishi* **77,** 318.

Terai, R. (1970). *Osaka Kogyo Gijutsu Shikensho Kiho* **21,** 64.

Terai, R. (1971a). *J. Non-Cryst. Solids* **6,** 121.

Terai, R. (1971b). *Osaka Kogyo Gijutsu Shikensho Kiho* **22,** 73.

Terai, R., and Hayami, R. (1975). *J. Non-Cryst. Solids* **18,** 217.

Terai, R., and Kitaoka, T. (1968). *Yogyo Kyokaishi* **76,** 393.

Terai, R., and Kitaoka, T. (1969). *Osaka Kogyo Gijutsu Shikenso Kiho* **20,** 58.

Terai, R., and Oishi, Y. (1977). *Glastech. Ber.* **50,** 68.

Terai, R., and Okawa, E. (1974). *Int. Congr. Glass* [*Pap.*], *10th, Kyoto* **8,** 23.

Terai, R., and Okawa, E. (1977). *Yogyo Kyokaishi* **85,** 294.

Terai, R., Kitaoka, T., and Ueno, T. (1969a). *Yogyo Kyokaishi* **77,** 88.

Terai, R., Kitaoka, T., and Ueno, T. (1969b). *Osaka Kogyo Gijutsu Shikensho Kiho* **20,** 198.

Towers, H., and Chipman, J. (1957). *J. Met.* **9,** 769.

Truhlarova, M., and Veprek, O. (1969). *Glastech. Ber.* **42,** 9.

Truhlarova, M., and Veprek, O. (1970a) *Silikaty* **14,** 1.

Truhlarova, M., and Veprek, O. (1970b). *Glastech. Ber* **43,** 191.

Ueda, H., and Oishi, Y. (1970). *Kenyko Hokoku—Asahi Garasu Kogyo Gijutsu Shoreikai* **16,** 201.

Varshneya, A. K., and Cooper, A. R. (1972a). *J. Am. Ceram. Soc.* 55, 220.

Varshneya, A. K., and Cooper, A. R. (1972b). *J. Am. Ceram. Soc.* **55,** 312.

Wakabayashi, H., and Hayami, R. (1974). *Int. Congr. Glass* [*Pap*], *10th, Kyoto,* Sect. 8, p. 36.

Williams, E. L., and Heckman, R. W. (1964). *Phys. Chem. Glasses* **5,** 166.

Wilson, C. G., and Carter, A. C. (1964). *Phys. Chem. Glasses* **5,** 111.

Winchell, P. (1969). *High Temp. Sci.* **1,** 200.

Wollast, R., Lauder, I., and May, H. (1973). *Silic. Ind.* **38,** 169.

Zhabrev, V. A. (1970). Dissertation, Leningrad.

Zhabrev, V. A., and Moiseev, V. V. (1965). *In* "Stekloobraznoe Sostoyanie," p. 288. Moscow.

Zhabrev, V. A., and Moiseev, V. V. (1970). *Stekloobraznoe Sostoyanie* **5,** 143.

TABLE 15.2

**Values of Arrhenius Parameters for the Diffusion of
Various Species in Lithium Silicate Glasses**

Composition (mole % Li_2O)	Diffusing species	Temperature range (°C)	D_0 ($cm^2\ s^{-1}$)	E_D (kcal $mole^{-1}$)	Reference[a]
10	Na^+	$< T_g$	0.468	27.05	1
17	Na^+	$< T_g$	0.759	27.2	1
18	Li^+	285–420	2.71×10^{-3}	15.7	2
25	Na^+	$< T_g$	0.692	26.65	1
25	Li^+	270–440	2.77×10^{-3}	15.3	2
33	Li^+	290–450	2.35×10^{-2}	17.9	2
33.3	Na^+	$< T_g$	0.490	26.15	1
33.33	^{18}O	350–450	$\left(1.33 \begin{smallmatrix} +2.17 \\ -0.83 \end{smallmatrix}\right) \times 10^{-10}$	20.6 ± 1.3	3

[a] Data from (1) Pavlovskii (1970), (2) Beier and Frischat (1980, 1984), and (3) Takizawa *et al.* (1981).

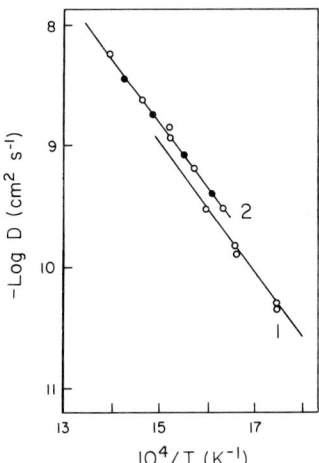

Fig. 15.1. Temperature dependence of diffusion coefficient of ^{110}Ag in lithium silicate glasses containing (1) 15 and (2) 20 mole % Li_2O. From Malinin *et al.* (1972).

TABLE 15.3

Values of Arrhenius Parameters for Na Self-Diffusion in Sodium Silicate Glasses

Composition (mole % Na_2O)	Temperature range (°C)	D_0 ($cm^2 s^{-1}$)	E_D (kcal $mole^{-1}$)	References[a]
5.0	300–500	8.71×10^{-3}	26.2	1, 2
7.5	300–500	8.91×10^{-3}	20.6	1, 2
10.0	300–500	7.94×10^{-4}	17.45	1, 2
10.0	270–470	1.03×10^{-2}	21.1	3, 4
12.6	245–405	2.50×10^{-3}	18.7	5
13.0	300–500	1.23×10^{-3}	17.15	2, 6
13.0	384–473	1.434	26.88	7
14.8	245–405	3.49×10^{-3}	18.5	5
15	300–450	1.2×10^{-2}	20.2	8
16.7	200–500	2.92×10^{-2}	21.3	3, 4
16.7	350–500	3.80×10^{-2}	21.6	9, 10
17.4	348–446	$(8 \pm 3) \times 10^{-4}$	16.3	11
17.4	363–459	0.204	23.6	7
18.9	382–399	1.176×10^{-2}	19.94	7
20.0	300–500	1.29×10^{-3}	17.35	1, 2
20.0	270–440	6.65×10^{-3}	19.0	3, 4
20.0	300–415	—	19.3	6
22.0	245–405	6.61×10^{-4}	16.5	5
23.6	850–1500	4.67×10^{-3}	16.1	12
25.0	300–500	1.95×10^{-3}	16.55	1, 2
25.0	220–420	2.63×10^{-3}	16.9	3, 4
25.0	350–500	2.10×10^{-2}	19.5	13
25.0	355–495	2.05×10^{-2}	19.4	14

(*continues*)

TABLE 15.3 (*Continued*)

Composition (mole % Na_2O)	Temperature range (°C)	D_0 ($cm^2\ s^{-1}$)	E_D (kcal mole^{-1})	References[a]
25.0	250–450	8.19×10^{-4}	15.4	15
25.0	300–450	$\left(7.9^{+4.0}_{-1.6}\right) \times 10^{-4}$	15.2	16
25.0	300–450	7.94×10^{-4}	15.2	17
25.0	250–450	7.94×10^{-4}	15.0	18
25.0	450–750	6.31×10^{-1}	25.0	18
25.0	750–1150	3.16×10^{-3}	14.0	18
25.0	850–1150	3.02×10^{-3}	14.1	19
26.39	300–430	1.19×10^{-2}	18.7	20
30.0	300–500	1.995×10^{-3}	16.15	1, 2
30.0	352–436	0.232	22.15	7
33.3	100–250	1.58×10^{-4}	13.2	16, 21
33.3	100–400	2.46×10^{-3}	15.7	22
33.3	220–400	6.67×10^{-3}	16.3	3, 4, 23
33.3	300–450	3.2×10^{-2}	18.9	8
33.3	354–481	0.268	21.4	24
33.3	$< T_g\ (465°C)$	8.2×10^{-3}	17.1	
33.3	$> T_g\ (465°C)$	16	27.7	
33.3	850–1210	1.8×10^{-3}	11.9	25
33.3	850–1050	1.91×10^{-3}	12.3	19
33.3	850–1500	2.29×10^{-3}	12.4	12
33.3	1027–1417	5.79×10^{-4}	7.008	26
33.4	925–1220	3.5×10^{-3}	13.1	27
33.9	100–427	2.91×10^{-3}	15.8	28

[a] Data from (1) Evstrop'ev and Pavlovskii (1967), (2) Evstrop'ev (1963a), (3) Terai (1971a,b), (4) Terai (1968). (5) Frischat (1975b), (6) Evstrop'ev (1960), (7) Haven and Verkerk (1965), (8) Kolitsch *et al.* (1978), (9) Hayami and Terai (1972), (10) Terai (1971), (11) Barr *et al.* (1972), (12) Gupta and King (1967), (13) McVay and Day (1970), (14) Obarara (1973), (15) Fleming and Day (1972), (16) Moiseev and Zhabrev (1969), (17) Malinin and Evstrop'ev (1972), (18) Negodaev *et al.* (1973) and Negodaev and Malinin (1974), (19) Negodaev *et al.* (1971, 1972a,b), (20) Lim and Day (1977a), (21) Moiseev and Zhabrev (1965), (22) Frischat (1971a), (23) Terai (1969a), (24) Schafer and Maywald (1967), (25) Malkin and Mogutnov (1961), (26) Perron and Bell (1967), (27) Hlavac and Matousek (1971d), and (28) Frischat (1967).

TABLE 15.4

Values of Arrhenius Parameters for the Diffusion of Various Species in Sodium Silicate Glasses

Composition (mole % Na_2O)	Species	Temperature range (°C)	D_0 ($cm^2 s^{-1}$)	E_D (kcal mole^{-1})	Reference[a]
13.14	^{18}O	550–700	5×10^{-10}	24	1
14.3	K^+		5.3×10^{-4}	21	2
15.0	K^+	350–510	4.6×10^{-3}	23.7	3
	Rb^+	350–510	0.32	34.6	3
	Cs^+	350–510	0.51	43.5	3
16.7	Cs^+	400–530	94.2	51.5	4
25.0	K^+	250–450	0.159	27.6	5
	K^+	300–450	0.50	29.7	6
	K^+	330–750	0.631	28.0	7
	K^+	850–1050	7.94×10^{-3}	19	7
	Rb^+	300–450	1.2	38.7	6
	Rb^+	350–500	0.02	29.4	8
	Rb^+	370–750	10	38	7
	Rb^+	950–1150	3.98×10^{-2}	27	7
	Cs^+	300–450	2.7	49.7	6
	Cs^+	460–750	125.89	47	7
	Cs^+	950–1250	19.95×10^{-2}	34	7
	Ca^{2+}	550–750	1.0	39	7
	Ca^{2+}	750–1050	19.95×10^{-3}	31	7
	^{89}Sr	790–1140	7.94×10^{-3}	28	9
	^{140}Ba	760–1150	12.59×10^{-3}	28	9
33.3	K^+	350–510	4.57	30.6	3
	Rb^+	350–510	13.8	36.6	3
	Cs^+	350–510	2.8×10^3	67.0	3

[a] Data from (1) DiMarcello (1966), (2) Wakabayashi and Hayami (1974), (3) Kolitsch *et al.* (1978), (4) Hayami and Terai (1972), (5) Fleming and Day (1972), (6) Malinin and Evstrop'ev (1972), (7) Negodaev *et al.* (1973), and Negodaev and Malinin (1974), (8) McVay and Day (1970), and (9) Ivanov and Pavlov (1975).

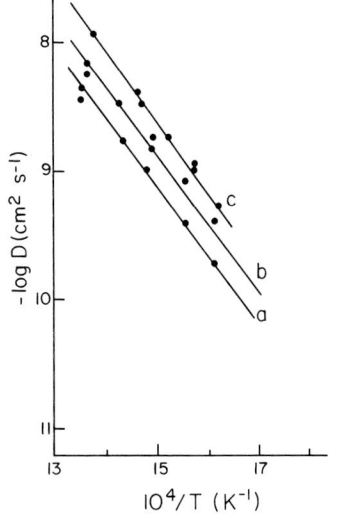

Fig. 15.2. Temperature dependence of ^{110}Ag diffusion coefficient in sodium silicate glasses containing (a) 15, (b) 20, and (c) 25 mole % Na_2O. From Malinin *et al.* (1972).

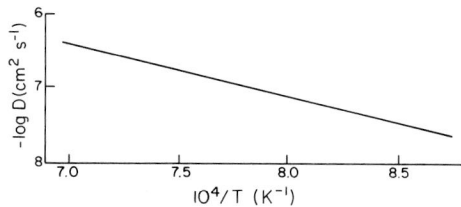

Fig. 15.3. Temperature dependence of diffusion coefficient of Fe^{2+} in $Na_2O-2SiO_2$ glass. From Borom and Pask (1968).

TABLE 15.5

Diffusion Coefficients of SiO_2 in Sodium Silicate Glass Melts[a]

Composition (mole % Na_2O)	Temperature (°C)	D (cm^2 s^{-1})	Composition (mole % Na_2O)	Temperature (°C)	D (cm^2 s^{-1})
33.4	1050	1.64×10^{-8}	35	1100	2.63×10^{-8}
	1100	2.58×10^{-8}		1114	1.74×10^{-8}
	1150	3.76×10^{-8}		1169	3.02×10^{-8}
	1200	4.55×10^{-8}		1175	1.45×10^{-7}
	1250	7.55×10^{-8}		1182	1.86×10^{-7}
				1232	3.02×10^{-7}
				1250	5.75×10^{-7}
				1310	2.16×10^{-6}
			52	1185	7.94×10^{-6}

[a] Data from Truhlarova and Veprek (1970a,b) and Souquet *et al.* (1968).

TABLE 15.6

Values of Arrhenius Parameters for the Diffusion of Various Ions in Potassium Silicate Glasses

Composition (mole % K_2O)	Species	Temperature range (°C)	D_0 (cm^2 s^{-1})	E_D (kcal mole^{-1})	References[a]
5	^{22}Na	300–500	0.55	35.25	1
7.5	^{22}Na	300–500	0.158	31.95	1
10	^{22}Na	300–500	5.01×10^{-2}	28.7	1
	K$^+$	300–500	1.995×10^{-4}	19.4	2
15	^{22}Na	300–500	6.61×10^{-2}	27.3	1
	K$^+$	300–500	5.25×10^{-4}	19.25	2
16	^{86}Rb	415–510	4.133	32.8	3
17.8	^{86}Rb	415–510	0.423	28.8	3
20	^{22}Na	300–500	6.31×10^{-2}	25.95	1
	K$^+$	300–500	3.80×10^{-4}	18.45	2
25	Na$^+$	250–450	0.109	25.5	8
	Na$^+$	320–700	0.316	26	4
	Na$^+$	300–450	0.30	25.8	9
	Na$^+$	800–1150	3.98×10^{-3}	17	4
	K$^+$	250–450	3.71×10^{-4}	15.7	8
	^{42}K	270–480	1.58×10^{-3}	17	4
	K$^+$	300–450	1.51×10^{-3}	17.1	9
	^{42}K	480–800	1.58×10^{-1}	24	4
	^{42}K	850–1050	5.01×10^{-3}	17	4
	Rb$^+$	300–450	0.24	25.7	9
	Rb$^+$	320–800	19.95×10^{-2}	26	4
	Rb$^+$	950–1150	7.94×10^{-3}	19	4
	Cs$^+$	300–450	1.26	31.4	9
	Cs$^+$	380–800	0.398	30	4
	Cs$^+$	950–1250	2.51×10^{-2}	24	4
	^{45}Ca	880–1190	0.501	45	5
	^{89}Sr	810–1180	7.94×10^{-3}	33	5
	^{140}Ba	850–1200	1.58×10^{-2}	32	5
25.08	Na$^+$	300–430	0.10	25.1	7
26.4	Si	600–1300	1.9×10^{-2}	38.9	6
27	K$^+$	300–500	2.40×10^{-3}	18.15	2
30	^{22}Na	300–500	0.209	25.45	1
33.3	K$^+$	300–500	9.33×10^{-4}	15.95	2

[a] Data from (1) Pavlovskii (1970), (2) Evstrop'ev and Pavlovskii (1967), (3) Evstrop'ev and Pavlovskii (1961), (4) Negodaev *et al.* (1973) and Negodaev and Malinin (1974), (5) Ivanov and Pavlov (1975), (6) Wollast *et al.* (1973), (7) Lim and Day (1977a), (8) Fleming and Day (1972), and (9) Malinin and Evstrop'ev (1972).

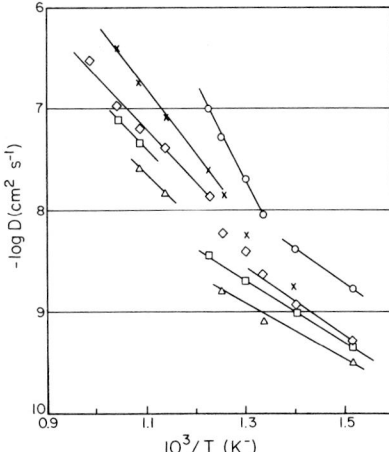

Fig. 15.4. Temperature dependence of K self-diffusion coefficient in potassium silicate glasses below and above T_g, as measured by Johnson (1950) and plotted by Doremus (1962). Weight percent of K_2O: $\bigcirc$, 43.7; x, 33.8; $\diamond$, 27.7; $\square$, 24.2; $\triangle$, 20.4.

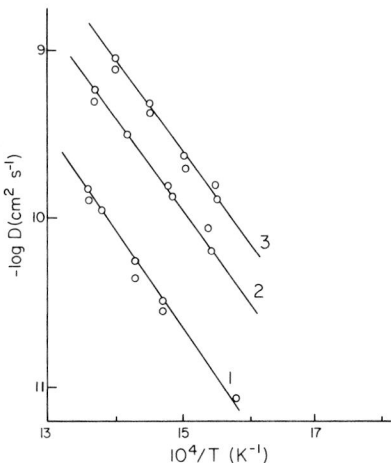

Fig. 15.5. Temperature dependence of diffusion coefficient of ^{110}Ag in potassium silicate glasses containing (1) 15, (2) 20, and (3) 25 mole % K_2O. From Malinin *et al.* (1972).

TABLE 15.7

Values of Arrhenius Parameters for the Diffusion of Various Ions in Rubidium Silicate Glasses

Composition (mole % Rb_2O)	Species	Temperature range (°C)	D_0 (cm^2 s^{-1})	E_D (kcal mole^{-1})	References[a]
20	Rb^+	300–500	3.39×10^{-4}	18.1	1
	Na^+	350–500	6.15	34.4	3
25	Na^+	320–850	0.794	30	2
	Na^+	300–450	0.89	30.2	4
	Na^+	900–1200	1.26×10^{-2}	20	2
	K^+	350–850	3.98×10^{-2}	23	2
	K^+	300–450	1.58×10^{-2}	22	4
	K^+	900–1200	1.26×10^{-2}	19	2
	Rb^+	280–500	5.01×10^{-3}	19	2
	Rb^+	350–500	5.31×10^{-2}	24.4	3
	Rb^+	500–850	5.01×10^{-2}	23	2
	Rb^+	300–450	5.01×10^{-3}	19.2	4
	Rb^+	900–1200	7.94×10^{-3}	18	2
	Cs^+	320–850	0.251	27	2
	Cs^+	300–450	0.26	27	4
	Cs^+	900–1200	1.26×10^{-2}	20	2

[a] Data from (1) Evstrop'ev and Pavlovskii (1967), (2) Negodaev and Malinin (1974), Negodaev et al. (1973), (3) McVay and Day (1970), and (4) Malinin and Evstrop'ev (1972).

TABLE 15.8

Values of Arrhenius Parameters for the Diffusion of Various Ions in Cesium Silicate Glasses

Composition (mole % Cs_2O)	Species	Temperature range (°C)	D_0 (cm^2 s^{-1})	E_D (kcal mole^{-1})	References[a]
12.5	Cs^+	350–470	1.89×10^{-5}	18.9	3
14.3	Cs^+	350–470	4.0×10^{-5}	18.7	3
16.7	Na^+	350–500	83.6	42	2
	Cs^+	400–530	8.7×10^{-3}	24.5	2
	Cs^+	350–470	1.3×10^{-4}	18.7	3
25	Na^+	300–450	15.85	35.0	4
	Na^+	340–900	1.995	33	1
	Na^+	950–1200	1.8	22	1
	Rb^+	300–450	0.158	26	4
	Rb^+	320–900	12.6×10^{-2}	26	1
	Cs^+	300–500	6.31×10^{-3}	21	1
	Cs^+	300–450	1.38×10^{-2}	21.8	4
	Cs^+	500–800	12.59×10^{-2}	25	1
	Cs^+	950–1200	1.0×10^{-2}	20	1

[a] Data from (1) Negodaev et al. (1973) and Negodaev and Malinin (1974), (2) Hayami and Terai (1972), (3) Terai (1969b, 1970), and (4) Malinin and Evstrop'ev (1972).

TABLE 15.9

Diffusion Coefficient of SiO₂ in Calcium Silicate Glass Melt[a]

Composition (mole % CaO)	Temperature (°C)	$D \times 10^6$ (cm² s⁻¹)
51.8	1700	0.50

[a] Data from Ershov and Popova (1964).

TABLE 15.10

Diffusion Coefficients of ²²Na and ¹⁴⁰Ba in BaSiO₃ Glass[a]

Diffusing Species	Temperature (°C)	$D \times 10^{12}$ (cm² s⁻¹)
²²Na	655	2 ± 1
¹⁴⁰Ba	655	2 ± 1

[a] From Evstrop'ev (1963b, 1966) and Evstrop'ev and Kharyuzov (1961).

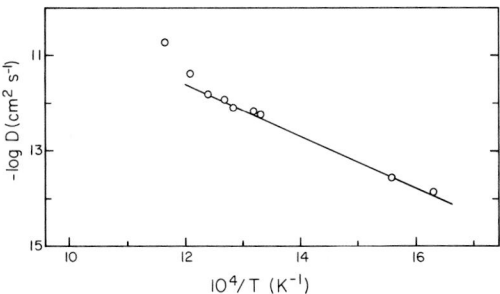

Fig. 15.6. Temperature dependence of Pb self-diffusion coefficient in PbSiO₃ glass. From Lindner *et al.* (1960).

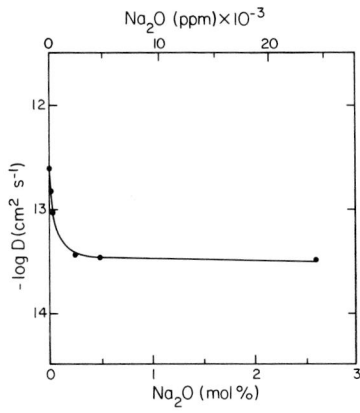

Fig. 15.7. Diffusion coefficients of ²²Na in 70B₂O₃–30SiO₂ (mole %) glass at 350°C as a function of sodium concentration. From Kelly and Tomozawa (1982).

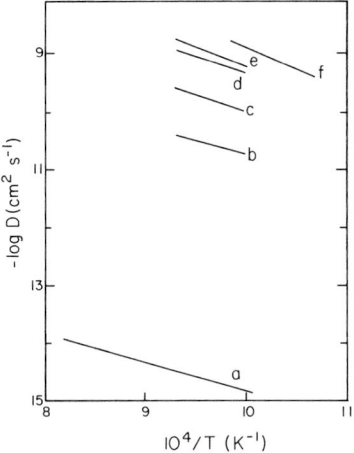

Fig. 15.8. Temperature dependence of oxygen self-diffusion coefficients in B_2O_3–SiO_2 glasses containing (a) 0, (b) 1, (c) 5, (d) 20, (e) 30, and (f) 100% B_2O_3. From Schlichting (1982).

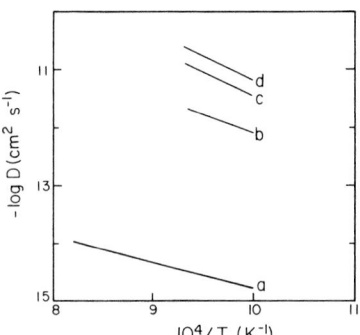

Fig. 15.9. Temperature dependence of oxygen self-diffusion coefficients in GeO_2–SiO_2 glasses containing (a) 0, (b) 5, (c) 20, and (d) 30% GeO_2. From Schlichting (1982).

TABLE 15.11

Values of Arrhenius Parameters for the Diffusion of Various Ions in Li_2O–Na_2O–SiO_2 Glasses

Composition (mole %)			Species	Temperature range (°C)	D_0 ($cm^2\ s^{-1}$)	E_D (kcal mole^{-1})	Reference[a]
Li_2O	Na_2O	SiO_2					
5.3	21.0	73.7	Na^+	200–450	3.14×10^{-2}	21.4	1
16.5	16.5	67	^7Li	$< T_g$ (392°C)	—	20–22	2
16.7	16.7	66.6	^{24}Na	850–1200	2.82×10^{-3}	13.4	3

[a] Data from (1) Terai et al. (1969a), (2) Müller-Warmuth et al. (1971), and (3) Malkin and Mogutnov (1961).

TABLE 15.12

Values of Arrhenius Parameters for the Diffusion of Various Ions in $Li_2O-K_2O-SiO_2$ Glasses

Composition (mole %)							
Li_2O	K_2O	SiO_2	Species	Temperature range (°C)	D_0 (cm^2 s^{-1})	E_D (kcal mole^{-1})	Reference[a]
16.5	16.5	67	^7Li	$< T_g$ (481)	—	25–30	1
0	20	80	^{22}Na	$< T_g$	6.31×10^{-2}	26.0	2
5	15	80			0.776	29.9	
10	10	80			8.13	32.7	
15	5	80			7.41	31.1	
20	0	80			0.63	27.1	
0	30	70	^{22}Na	$< T_g$	0.209	25.5	2
5	25	70			0.603	28.6	
10	20	70			2.51	31.8	
15	15	70			9.33	33.8	
20	10	70			3.63	31.3	
25	5	70			2.04	29.3	
30	0	70			0.63	26.5	
16.7	16.7	66.6	^{42}K	850–1200	8.53×10^{-3}	18.2	3

[a] Data from (1) Müller-Warmuth et al. (1971), (2) Malinin and Evstrop'ev (1972), and (3) Malkin and Mogutnov (1961).

TABLE 15.13

Arrhenius Parameters for the Self-Diffusion of ^7Li in $Li_2O-Cs_2O-SiO_2$ Glass[a]

Composition (mole %)					
Li_2O	Cs_2O	SiO_2	Species	Temperature range (°C)	E_D (kcal mole^{-1})
16.5	16.5	67	^7Li	$< T_g$ (480°C)	22–25

[a] Data from Müller-Warmuth et al. (1971).

TABLE 15.14

Arrhenius Parameters for Diffusion of Na and K Ions in Na_2O–K_2O–SiO_2 Glasses

Na₂O	K₂O	SiO₂	Species	Temperature range (°C)	D_0 (cm² s⁻¹)	E_D (kcal mole⁻¹)	Reference[a]
12.5	12.5	75	^{22}Na	300–700	0.398	26 ± 2	1
				850–1050	6.31×10^{-3}	17 ± 2	
			^{42}K	350–700	0.50	27 ± 3	
				850–1050	6.31×10^{-3}	18 ± 1	
16.7	16.7	66.6	^{24}Na	850–1200	6.37×10^{-3}	16.1	2
			^{42}K	850–1200	5.08×10^{-3}	15.8	
21	5.3	73.7	^{22}Na	200–450	2.55×10^{-2}	21.3	3
0	25	75	^{22}Na	250–450	0.109	25.5	4
			^{42}K	250–450	3.71×10^{-4}	15.7	
7.5	17.5	75	^{22}Na	250–450	0.214	25.9	
			^{42}K	250–450	4.77×10^{-2}	24.2	
15	10	75	^{22}Na	250–450	0.13	24.3	
			^{42}K	250–450	2.43	30.4	
25	0	75	^{22}Na	250–450	8.19×10^{-4}	15.4	
			^{42}K	250–450	0.159	27.6	
8.0	12.17	79.83	^{22}Na	300–430	2.18×10^{-2}	24.1	7
13.64	5.72	80.64	^{22}Na	300–430	1.53×10^{-2}	22.2	
21.18	1.36	77.46	^{22}Na	300–430	6.14×10^{-2}	22.1	

Na₂O	K₂O	SiO₂	Species	D(cm² s⁻¹) at given temperature (°C)			Reference[a]
				300	415	500	
10	2.5	87.5	^{22}Na	5.75×10^{-11}		12.88×10^{-9}	5
10	5	85		3.09×10^{-11}		12.30×10^{-9}	
10	10	80		3.55×10^{-11}		21.38×10^{-9}	
10	15	75		5.50×10^{-11}		5.25×10^{-8}	
10	20	70		9.55×10^{-11}		10.0×10^{-8}	
2.5	10	87.5		2.45×10^{-12}		4.57×10^{-9}	
5	10	85		6.17×10^{-12}		8.71×10^{-9}	
15	10	75		8.51×10^{-11}		4.48×10^{-8}	
20	10	70		19.05×10^{-11}		6.76×10^{-8}	
5	15	80		1×10^{-11}		2.09×10^{-9}	
12.5	7.5	80		2.29×10^{-11}		5.62×10^{-9}	
15	5	80		4.68×10^{-11}		9.77×10^{-9}	
2.5	17.5	80	^{42}K		4.6×10^{-10}		6
5	15	80			3.3×10^{-10}		
7.5	12.5	80			2.6×10^{-10}		
10	10	80			2.2×10^{-10}		
12.5	7.5	80			1.9×10^{-10}		
15	5	80			1.74×10^{-10}		
17.5	2.5	80			1.6×10^{-10}		

[a] Data from (1) Negodaev (1973), Negodaev et al. (1973) and Negodaev and Malinin (1974), (2) Malkin and Mogutnov (1961), (3) Terai et al. (1969a,b) and Terai (1971b), (4) Fleming and Day (1972), (5) Evstrop'ev (1970a,b), (6) Evstrop'ev and Pavlovskii (1961), and (7) Lim and Day (1977b).

TABLE 15.15

Arrhenius Parameters for Self-Diffusion of Na and Rb Ions in $Na_2O-Rb_2O-SiO_2$ Glasses[a]

Composition (mole %)			Species	Temperature range (°C)	D_0 $(cm^2\,s^{-1})$	E_D $(kcal\,mole^{-1})$
Na_2O	Rb_2O	SiO_2				
0	25	75	Na^+	350–500	6.15	34.4
			Rb^+	350–500	5.31×10^{-2}	24.4
6.25	18.75	75	Na^+	350–500	4.45	32.6
			Rb^+	350–500	4.34	32.4
12.5	12.5	75	Na^+	350–500	2.23	30.2
			Rb^+	350–500	32.4	36.9
18.75	6.25	75	Na^+	350–500	1.31	27.4
			Rb^+	350–500	27.1	38.0
25	0	75	Na^+	350–500	2.10×10^{-2}	19.5
			Rb^+	350–500	2.01×10^{-2}	29.4

[a] Data from McVay and Day (1970).

TABLE 15.16

Self-Diffusion Coefficients of Na and Rb Ions in $Na_2O-Rb_2O-SiO_2$ Glasses[a]

Composition (mole %)			Species	$D(cm^2\,s^{-1})$ at given temperature (°C)	
Na_2O	Rb_2O	SiO_2		300	500
5	15	80	^{22}Na	3.47×10^{-12}	2.45×10^{-9}
			^{86}Rb	1.51×10^{-12}	9.33×10^{-10}
10	10	80	^{22}Na	7.08×10^{-12}	5.89×10^{-9}
			^{86}Rb	2.95×10^{-13}	3.16×10^{-10}
15	5	80	^{22}Na	2.51×10^{-11}	7.08×10^{-9}
			^{86}Rb	6.46×10^{-14}	3.16×10^{-10}
10	2.5	87.5	^{22}Na	7.59×10^{-11}	8.51×10^{-9}
10	5	85	^{22}Na	3.24×10^{-11}	5.13×10^{-9}
0	20	80	^{22}Na	7.94×10^{-13}	4.47×10^{-10}

[a] Data from Evstrop'ev (1970a,b).

TABLE 15.17

Arrhenius Parameters for Self-Diffusion of Na and Cs Ions in $Na_2O-Cs_2O-SiO_2$ Glasses

Composition (mole %)				Temperature range	D_0	E_D	
Na_2O	Cs_2O	SiO_2	Species	(°C)	$(cm^2 s^{-1})$	(kcal mole^{-1})	Reference[a]
—	16.7	83.3	Na^+	350–500	83.6	42.0	1
			Cs^+	400–530	0.0087	24.5	
1.67	15.03	83.3	Na^+	370–500	281	43.6	1
			Cs^+	400–500	12.8	37.5	
3.34	13.36	83.3	Na^+	400–500	50.0	40.8	1
			Cs^+	400–500	172	42.7	
4.175	12.525	83.3	Na^+	350–500	2.75	36.2	1
			Cs^+	400–500	653	45.7	
5.01	11.69	83.3	Na^+	370–500	1.08	34.6	1
			Cs^+	400–500	680	46.2	
6.68	10.02	83.3	Na^+	370–500	3.43	35.8	1
			Cs^+	400–500	2780	49.5	
8.35	8.35	83.3	Na^+	350–500	1.66	33.9	1
			Cs^+	400–480	3240	50.5	
10.02	6.68	83.3	Na^+	370–500	1.70	32.6	1
			Cs^+	400–500	94.6	46.5	
12.525	4.175	83.3	Na^+	350–500	0.241	27.6	1
			Cs^+	400–500	45.8	47.1	
16.7	0	83.3	Na^+	350–500	0.038	21.6	1
			Cs^+	400–530	94.2	51.5	
12.5	12.5	75	^{22}Na	300–700	0.50	29 ± 3	2
				1000–1200	0.01259	21 ± 2	
			^{134}Cs	380–700	3.16	36 ± 4	
				1000–1200	0.02	23 ± 2	

[a] Data from (1) Hayami and Terai (1972, 1973) and (2) Negodaev (1973), Negodaev *et al.* (1973), and Negodaev and Malinin (1974).

TABLE 15.18

Diffusion Coefficients of Various Ions in $Na_2O-Cs_2O-SiO_2$ Glasses

Composition (mole %)				Temperature	D	
Na_2O	Cs_2O	SiO_2	Species	(°C)	(cm^2 s^{-1})	Reference[a]
—	25	75	^{22}Na	395.5	1.89×10^{-11}	2
			^{86}Rb	396.5	7.36×10^{-10}	1
			^{137}Cs	395.5	9.66×10^{-10}	2
6.425	18.575	75	^{22}Na	396.5	2.28×10^{-11}	2
			^{86}Rb	396.5	6.4×10^{-11}	1
			^{137}Cs	396.5	5.71×10^{-11}	2
14.875	10.125	75	^{22}Na	396.5	9.58×10^{-11}	2
			^{86}Rb	396.5	5.33×10^{-12}	1
			^{137}Cs	396.5	8.78×10^{-13}	2
16.3	8.7	75	^{22}Na	305.8	1.20×10^{-11}	2
			^{22}Na	345.8	4.43×10^{-11}	2
			^{22}Na	396.5	2.13×10^{-10}	2
			^{86}Rb	396.5	3.31×10^{-12}	1
			^{137}Cs	396.5	3.26×10^{-13}	2
18.95	6.05	75	^{22}Na	396.5	6.35×10^{-10}	2
			^{86}Rb	396.5	1.94×10^{-12}	1
			^{137}Cs	396.5	6.08×10^{-14}	2
19.55	5.45	75	^{22}Na	396.5	1.05×10^{-9}	2
			^{137}Cs	396.5	2.49×10^{-12}	2
22.325	2.675	75	^{22}Na	198.0	5.19×10^{-12}	2
			^{22}Na	396.5	2.88×10^{-9}	2
			^{86}Rb	396.5	2.08×10^{-12}	1
			^{137}Cs	396.5	1.40×10^{-14}	2
25	—	75	^{22}Na	396.5	8.25×10^{-9}	2
			^{86}Rb	396.5	1.99×10^{-12}	1
			^{137}Cs	396.5	4.46×10^{-15}	2

[a] Data from (1) Jain and Peterson (1983), and (2) Jain et al. (1983).

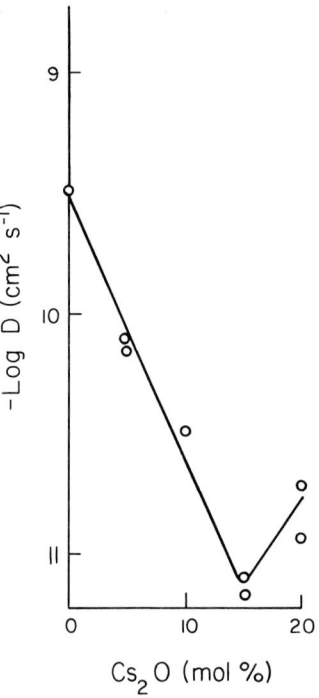

Fig. 15.10. Diffusion coefficients of ^{22}Na in $20(K_2O + Cs_2O)-80SiO_2$ (mole %) glass at $415°C$ as a function of mole % of Cs_2O. From Evstrop'ev and Pavlovskii (1961).

TABLE 15.19

Diffusion Coefficients of Na in Na$_2$O–BeO–SiO$_2$ Glass[a]

Composition (mole %)				Temperature	D
Na$_2$O	BeO	SiO$_2$	Species	(°C)	(cm^2 s^{-1})
20	10	70	^{22}Na	300	1.41×10^{-10}
				500	1.26×10^{-8}

[a] Data from Evstrop'ev (1963a,b).

TABLE 15.20

Arrhenius Parameters for the Diffusion of Various Species in Soda-Lime Silicate Glasses

Composition (mole %)							
Na_2O	CaO	SiO_2	Species	Temperature range (°C)	D_0 (cm^2 s^{-1})	E_D (kcal mole^{-1})	Reference[a]
14.0	12.0	74.0	Na^+	347–565	1.6×10^{-3}	20.6	1, 2
15.5	10.7	73.8		220–510	8.31×10^{-3}	22.8	3
15.5	12.8	71.7		300–560	1.14×10^{-2}	23.4	4
15.5	12.8	71.7		350–450	2.82×10^{-2}	15.6	5
15.5	12.8	71.7		350–530	0.5×10^{-2}	19.6	6
15.9	11.9	72.2		200–550	7.58×10^{-3}	22.0	7, 8
16.7	16.7	66.6		315–560	12.6×10^{-3}	23.3	9
16.7	16.7	66.6		850–1200	11.66×10^{-3}	19.9	10
16.7	33.3	50		200–600	3.76×10^{-2}	27.3	3, 11
20.0	10.0	70		200–450	5.71×10^{-3}	20.6	12
20.0	10.0	70		250–450	5.37×10^{-3}	20.5	9
25.0	12.5	62.5		260–460	8.16×10^{-3}	20.1	9
14.0	12.0	74.0	^{42}K	347–565	1.25×10^{-2}	30.1	1, 2
15.5	12.8	71.7	Rb^+	450–535	4×10^{-7}	20.7	13
15.5	12.8	71.7	Cs^+	450–535	3×10^{-8}	20.7	13
15.5	12.8	71.7	Sr^{2+}	450–535	9×10^{-3}	39.2	13
15.5	10.7	73.8	^{45}Ca	474–545	39.7	53.7	3
16.7	33.3	50.0	^{45}Ca	460–650	$3.39 \times 10^{+2}$	56.2	3, 14
15.5	12.8	71.7	^{18}O	460–525	$2.1 \times 10^{+3}$	66.5	15
15.5	12.8	71.7	^{18}O	800–1470	4.5×10^{-2}	38.6	15

[a] Data from (1) Richter *et al.* (1982), (2) Kolitsch and Richter (1981b), (3) Frischat (1971a), (4) Levi *et al.* (1971), (5) Wilson and Carter (1964), (6) Johnson *et al.* (1951), (7) Williams and Heckman (1964), (8) Heckman *et al.* (1967), (9) Terai *et al.* (1969a), (10) Malkin and Mogutnov (1961), (11) Frischat and Oel (1966a), (12) Terai and Kitaoka (1968), (13) Kahl and Schiewer (1971), (14) Frischat and Oel (1966b), and (15) Kingery and Lecron (1960).

TABLE 15.21

Diffusion Coefficients of ^{110}Ag and ^{18}O in Soda-Lime Silicate Glasses

Composition (mole %)						
Na_2O	CaO	SiO_2	Species	Temperature (°C)	D (cm^2 s^{-1})	Reference[a]
14.6	6.68	78.72	^{110}Ag	340	2.1×10^{-10}	1
				385	3.7×10^{-10}	
				450	7.3×10^{-10}	
				480	8.2×10^{-10}	
				510	6.3×10^{-9}	
				550	8.8×10^{-9}	
				600	1.6×10^{-8}	
15.5	12.8	71.7	^{18}O	515	3.71×10^{-16}	2
				540	1.77×10^{-15}	
				553	2.11×10^{-14}	
				635	3.41×10^{-12}	
				808	2.03×10^{-10}	

[a] Data from (1) Matousek (1967) and (2) Terai and Oishi (1977).

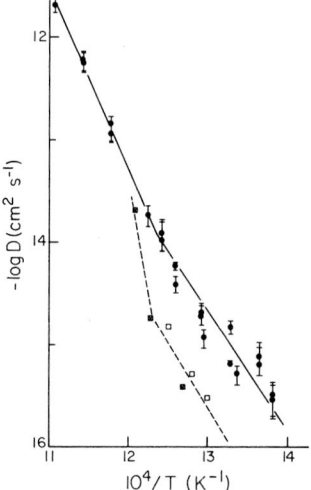

Fig. 15.11. Temperature dependence of diffusion coefficients of calcium and oxygen in $16Na_2O-12CaO-72SiO_2$ (wt. %) glass: ●, calcium; □, ■, oxygen. From Terai and Okawa (1977).

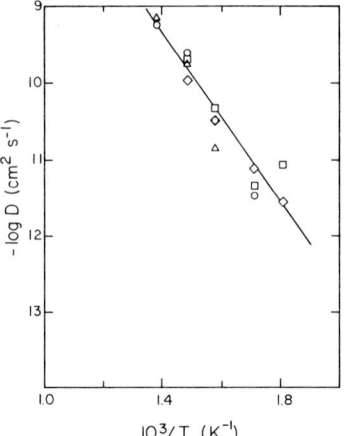

Fig. 15.12. Temperature dependence of diffusion coefficient of Ag^+ as a function of its concentration in $18Na_2O-10CaO-72SiO_2$ (mole %) glass; ppm of Ag: △, 50; ○, 80; □, 360; ◇, 700. From Barton and Morain (1970).

TABLE 15.22

Arrhenius Parameters for Na Diffusion in $Na_2O-MO-SiO_2$ Glasses

Composition (mole %)			Species	Temperature range (°C)	D_0 (cm² s⁻¹)	E_D (kcal mole⁻¹)	Reference[a]
Na_2O	MO	SiO_2					
	M = Mg						
20	10	70	^{22}Na	200–450	28.8×10^{-4}	19.0	1
16.7	16.7	66.6	^{24}Na	850–1200	15.42×10^{-3}	20.5	2
	M = Sr						
20	10	70	^{22}Na	200–450	80.4×10^{-4}	22.2	1
16.7	16.7	66.6	^{24}Na	850–1200	115.8×10^{-3}	26.2	2
	M = Ba						
20	10	70	^{22}Na	200–450	29.7×10^{-3}	23.8	1
16.7	16.7	66.6	^{24}Na	850–1200	2.333	34.8	2
	M = Zn						
20	10	70	^{22}Na	200–450	55.3×10^{-4}	20.2	1
	M = Pb						
20	10	70	^{22}Na	200–450	3.54×10^{-2}	23.7	1
	M = Cd						
20	10	70	^{22}Na	200–450	12.3×10^{-3}	22.2	1

[a] Data from (1) Terai and Kitaoka (1968, 1969) and (2) Malkin and Mogutnov (1961).

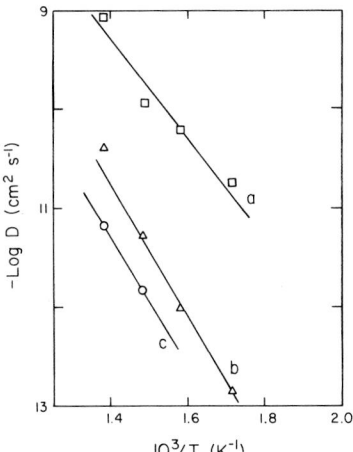

Fig. 15.13. Temperature dependence of Ag^+ diffusion coefficients in $13Na_2O-15MO-72SiO_2$ (mole %) glasses where M is (a) Mg, (b) Sr, (c) Ba. From Barton and Morain (1970).

TABLE 15.23

Arrhenius Parameters for ^{42}K Diffusion in $K_2O-MO-SiO_2$ (M = Mg, Ba) Glasses[a]

Composition (mole %)					
K_2O	MO	SiO_2	Temperature range	D_0 (cm² s⁻¹)	E_D (kcal mole⁻¹)
	M = Mg				
16.7	16.7	66.6	850–1200	9.33×10^{-3}	21.9
	M = Ba				
16.7	16.7	66.6	850–1200	56.23	45.9

[a] Data from Malkin and Mogutnov (1961) and Winchell (1969).

TABLE 15.24

Arrhenius Parameters for the Diffusion of ^{42}K and ^{22}Na Ions in $K_2O-CaO-SiO_2$ Glasses

Composition (mole %)							
K_2O	CaO	SiO_2	Species	Temperature range (°C)	D_0 (cm² s⁻¹)	E_D (kcal mole⁻¹)	Reference[a]
16.7	16.7	66.6	^{42}K	850–1200	169.82×10^{-3}	29.8	1
14	12	74	^{42}K	347–565	1.0×10^{-4}	21	2
14	12	74	^{22}Na	347–565	1.3×10^{-2}	26.8	2

[a] Data from (1) Malkin and Mogutnov (1961) and (2) Richter et al. (1982).

TABLE 15.25

Diffusion Coefficients of ^{110}Ag and ^{45}Ca Ions in $K_2O-CaO-SiO_2$ Glasses

Glass composition (mole %)						
K_2O	CaO	SiO_2	Species	Temperature (°C)	$D \times 10^{10}$ (cm² s⁻¹)	Reference[a]
14.6	6.68	78.72	^{110}Ag	340	0.11	1
				385	0.22	
				440	0.46	
				500	1.0	
				550	3.2	
				560	4.6	
				580	9.8	
				605	23	
				630	29	
18	8	74	^{45}Ca	595	$D = 7.78 \times 10^{-14}$	2

[a] Data from (1) Matousek (1967) and (2) Frischat (1971a, 1972).

TABLE 15.26

Arrhenius Parameters for Self-Diffusion Coefficients of Various Ions in $K_2O-SrO-SiO_2$ Glasses

Composition (mole %)							
K_2O	SrO	SiO_2	Species	Temperature range (°C)	D_0 (cm^2 s^{-1})	E_D (kcal mole^{-1})	Reference[a]
15.15	13.75	71.10	^{42}K	530–830	36.5	40.8	1
			^{85}Sr	530–830	0.17	42.7	1
			O	T_g^b–727	7.6×10^{14}	119	2
			O	327–T_g^b	1.0×10^{-12}	10	2
16.7	16.7	66.6	^{42}K	850–1200	0.479	32.7	3

[a] Data from (1) Varshneya and Cooper (1972a), (2) Rawal and Cooper (1979), and (3) Malkin and Mogutnov (1961).

[b] $T_g = 600°C$.

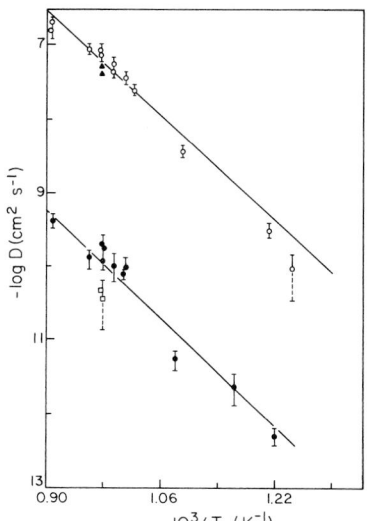

Fig. 15.14. Temperature dependence of self-diffusion coefficients of various ions in $15.15K_2O-13.75SrO-71.10SiO_2$ (mole %) glass: ○, ^{42}K; ●, ^{85}Sr; □, ^{31}Si; ▲, ^{24}Na. From Varshneya and Cooper (1972a,b).

TABLE 15.27

Arrhenius Parameters for the Self-Diffusion Coefficients of Oxygen in $K_2O-PbO-SiO_2$ Glasses

Composition (mole %)							
K_2O	PbO	SiO_2	Species	Temperature range (°C)	D_0 (cm^2 s^{-1})	E_D (kcal mole^{-1})	Reference[a]
8.7	33	58.3	^{18}O	275–425	1×10^{-10}	12	1
10	38	52	O	578–678	$(3.2^{+9.8}_{-2.4}) \times 10^{+6}$	71.6 ± 2.5	2

[a] Data from (1) Schaeffer and Oel (1969) and deBerg and Lauder (1978).

TABLE 15.28

Arrhenius Parameters for Diffusion of ^{22}Na in Sodium Borosilicate Glasses

Composition (mole %)			Temperature range (°C)	D_0 (cm^2 s^{-1})	E_D (kcal mole^{-1})	Reference[a]
Na$_2$O	B$_2$O$_3$	SiO$_2$				
0.37	4	95.67	375–550	0.26	30.5	1
13.8	8.8	77.4	250–400	1.4×10^{-4}	17.8	2
18.5	7.4	74.1	381–565	3.5×10^{-3}	20.6	4
28.6	14.3	57.1	100–250	3.98×10^{-5}	13.1	3
30.3	9.10	60.6	100–250	6.31×10^{-6}	11.7	3
31.3	6.25	62.45	100–250	5.01×10^{-6}	11.5	3
31.7	4.76	63.54	100–250	3.98×10^{-6}	12.1	3
32.3	3.22	64.48	100–250	1.59×10^{-4}	13.0	3

[a] Data from (1) Zhabrev and Moiseev (1965, 1970) and Moiseev and Zhabrev (1965, 1969), (2) McGrail, et al. (1984), (3) Lim and Day (1978), and (4) Hahnert et al. (1984).

TABLE 15.29

Diffusion Coefficients of ^{22}Na in Sodium Borosilicate Glasses[a]

Composition (mole %)			Temperature (°C)	$D \times 10^{10}$
Na$_2$O	B$_2$O$_3$	SiO$_2$		
20	0	80	380	32
20	10	70	380	7.5
20	15	65	380	6.1
20	20	60	380	3.1
20	30	50	380	2.0
20	40	40	380	1.2
20	60	20	380	0.88

[a] Data from Evstrop'ev (1963a).

TABLE 15.30

Arrhenius Parameters for Diffusion of K Ions in Sodium Borosilicate Glasses[a]

Composition (mole %)			Species	Temperature range (°C)	D_0 (cm^2 s^{-1})	E_D (kcal mole^{-1})
Na$_2$O	B$_2$O$_3$	SiO$_2$				
18.5	7.4	74.1	^{42}K	381–565	5.4×10^{-2}	29.2

[a] Data from Hahnert et al. (1984).

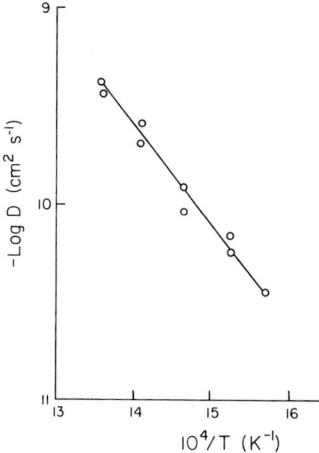

Fig. 15.15. Temperature dependence of the diffusion coefficient of ^{110}Ag in $20Na_2O-15B_2O_3-65SiO_2$ (mole %) glass. From Malinin *et al.* (1972).

TABLE 15.31

Arrhenius Parameters for the Self-Diffusion of Na⁺ Ions in Sodium Aluminosilicate Glasses

Composition (mole %)			Temperature range (°C)	D_0 (cm² s⁻¹)	E_D (kcal mole⁻¹)	Reference[a]
Na₂O	Al₂O₃	SiO₂				
0.18	0.24	99.58	375–550	0.05	26.6	1
11.01	16.08	72.90	200–600	2.09×10^{-3}	15.8	2, 3
11.0	18.0	71.0	352–716	5.02×10^{-4}	14.5	4
13.76	14.24	72.02	200–600	1.52×10^{-3}	14.6	2, 3
15.7	12.1	72.2	200–600	3.23×10^{-3}	17.2	2, 3
18.5	7.4	74.1	380–520	4.7×10^{-3}	18.7	5, 6
18.9	5.66	75.44	100–250	7.95×10^{-5}	14.3	7
21.0	5.3	73.7	200–450	3.03×10^{-3}	18.0	8–10
23.8	28.6	47.6	200–760	1.31×10^{-3}	15.0	11, 12
24.4	26.8	48.8	200–730	1.33×10^{-3}	14.9	11, 12
25.0	25.0	50.0	200–730	1.01×10^{-3}	14.2	11, 12
25.6	23.1	51.3	200–740	1.20×10^{-3}	14.8	11, 12
25.6	5.26	69.14	100–250	4.67×10^{-5}	12.5	7
27.0	18.9	54.1	200–700	1.73×10^{-3}	15.6	11, 12
27.0	3.0	70.0	228–436	2.96×10^{-4}	13.7	4
28.6	14.3	57.1	200–530	3.10×10^{-3}	16.6	11, 12
28.6	14.3	57.1	100–250	8.13×10^{-4}	14.9	13
29.5	4.92	65.58	100–250	9.85×10^{-6}	10.1	7
29.85	10.45	59.70	354–490	2.0×10^{-3}	16.0	14, 15
30.3	9.1	60.6	100–250	6.32×10^{-6}	11.1	13
30.3	9.1	60.6	200–430	3.61×10^{-3}	16.5	11, 12
31.1	8.75	60.15	100–470	2.78×10^{-3}	16.4	16, 17
31.3	6.25	62.45	100–250	3.98×10^{-5}	10.5	13
31.6	1.26	67.14	100–430	4.16×10^{-3}	16.4	16, 17
31.7	4.76	63.54	100–250	6.02×10^{-6}	8.8	7, 13

(continues)

TABLE 15.31 (*Continued*)

Composition (mole %)			Temperature range (°C)	D_0 (cm^2 s^{-1})	E_D (kcal mole^{-1})	Reference[a]
Na$_2$O	Al$_2$O$_3$	SiO$_2$				
31.7	4.76	63.54	90–300	1.95×10^{-4}	12.7	18
31.7	4.76	63.54	90–300	3.63×10^{-4}	12.5	18
31.8	5.8	62.4	100–460	1.78×10^{-3}	15.8	16, 17, 19
31.8	5.8	62.4	100–400	1.78×10^{-3}	15.8	20
31.8	5.8	62.4	354–473	9.7×10^{-3}	17.9	21
31.8	5.8	62.4	200–575	37.6×10^{-3}	27.3	20
32.1	5.4	62.4	200–480	1.8×10^{-3}	15.7	22
32.3	3.22	64.48	100–250	6.31×10^{-4}	14.2	13
32.3	3.2	64.5	200–430	7.81×10^{-3}	16.8	11, 12
32.5	4.48	63.02	100–450	2.64×10^{-3}	15.9	16, 17
32.7	2.53	64.77	100–440	2.20×10^{-3}	15.9	16, 17
32.8	1.62	65.58	100–250	2.51×10^{-4}	14.2	13
33.0	1.00	66.0	100–250	8.13×10^{-3}	17.2	13
33.2	0.33	66.47	100–250	7.76×10^{-2}	17.9	13
33.9	1.69	64.41	100–250	2.54×10^{-4}	13.4	7
34.5	3.45	62.05	100–250	4.0×10^{-4}	12.7	7
35.1	5.26	59.64	100–250	8.91×10^{-5}	12.0	7
35.7	7.14	57.16	100–250	9.1×10^{-5}	12.9	7
36.8	4.42	58.78	100–250	1.58×10^{-5}	10.2	7
37.0	11.1	51.9	100–250	2.04×10^{-5}	14.1	7

[a] Data from (1) Lim and Day (1978), (2) Williams and Heckman (1964), (3) Heckman *et al.* (1967), (4) Garfinkel and Rauscher (1966), (5) Muller *et al.* (1981), (6) Kolitsch *et al.* (1980b), (7) Moiseev and Zhabrev (1969), (8) Terai *et al.* (1969a), (9) Terai *et al.* (1969b), (10) Terai (1971b), (11) Terai (1968), (12) Terai (1969a), (13) Moiseev and Zhabrev (1965), (14) Kolitsch *et al.* (1980a), (15) Kolitsch *et al.*, (1980c), (16) Frischat (1967), (17) Frischat (1971a), (18) Evstrop'ev and Pavlovskii (1966), (19) Frischat (1971b), (20) Frischat (1972), (21) Kolitsch *et al.* (1979) and (22) Köhler and Frischat (1978).

TABLE 15.32

Arrhenius Parameters for the Diffusion of Various Ions in Sodium Aluminosilicate Glasses

Composition (mole %)			Species	Temperature range (°C)	D_0 (cm^2 s^{-1})	E_D (kcal mole^{-1})	Reference[a]
Na$_2$O	Al$_2$O$_3$	SiO$_2$					
18.52	7.41	74.07	K$^+$	347–490	6.9×10^{-3}	25.6	1
29.85	10.45	59.70	Ag$^+$	350–490	2.2×10^{-3}	18.2	2
29.85	10.45	59.70	K$^+$	354–473	6.8×10^{-3}	23.1	3
31.8	5.8	62.4	K$^+$	354–473	0.78	30.2	4
32.2	5.4	62.4	^{59}Fe	455	5.5×10^{-15} [b]	—	5

[a] Data from (1) Kolitsch *et al.* (1980b), (2) Kolitsch and Richter (1981a), (3) Kolitsch *et al.* (1980a), (4) Kolitsch *et al.* (1979), and (5) Köhler and Frischat (1978).
[b] Diffusion coefficient at 455°C.

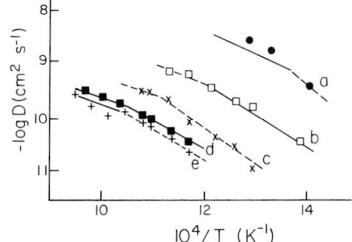

Fig. 15.16. Temperature dependence of the diffusion coefficient of H_2O in $20Na_2O-xAl_2O_3-(80-x)SiO_2$ (mole %) glasses with different x values: (a) 0, (b) 5, (c) 10, (d) 15, (e) 20. From Haider and Roberts (1970).

TABLE 15.33

Arrhenius Parameters for Na Diffusion in Sodium Gallosilicate Glasses[a]

Glass composition (wt. %)			Species	Temperature range (°C)	D_0 (cm^2 s^{-1})	E_D (kcal mole^{-1})
Na$_2$O	Ga$_2$O$_3$	SiO$_2$				
25.70	22.80	51.36	^{22}Na	100–250	1.00×10^{-5}	12.0
28.83	16.50	53.92	^{22}Na	100–250	7.94×10^{-5}	12.6
29.30	12.63	57.08	^{22}Na	100–250	11.67×10^{-5}	13.0
30.48	9.90	57.78	^{22}Na	100–250	16.75×10^{-5}	13.4

[a] Data from Zhabrev (1970), Zhabrev and Moiseev (1965, 1970), and Moiseev and Zhabrev (1965, 1969).

TABLE 15.34

Arrhenius Parameters for ^{22}Na Diffusion in Na$_2$O–La$_2$O$_3$–SiO$_2$ Glasses[a]

Glass composition (mole %)			Temperature range (°C)	D_0 (cm^2 s^{-1})	E_D (kcal mole^{-1})
Na$_2$O	La$_2$O$_3$	SiO$_2$			
21.5	5.3	73.2	220–400	3.81×10^{-3}	19.6
32.3	3.22	64.48	220–400	1.59×10^{-4}	13.0
31.7	4.76	63.54	220–400	3.98×10^{-6}	12.1

[a] Data from Terai *et al.* (1969a).

TABLE 15.35

Diffusion Coefficients of ^{22}Na and ^{59}Fe in Na$_2$O–Fe$_2$O$_3$–SiO$_2$ Glasses

Glass composition (mole %)			Species	Temperature (°C)	$D \times 10^9$ (cm^2 s^{-1})	Reference[a]
Na$_2$O	Fe$_2$O$_3$	SiO$_2$				
13	—	87	^{22}Na	415	3.6	1
13	1	86	^{22}Na	415	2.0	1
13	5	82	^{22}Na	415	1.2	1
13	14.5	72.5	^{22}Na	415	2.1	1
31.2	5.8	62.4	^{59}Fe	450	1.0	2

[a] Data from (1) Evstrop'ev (1963a) and (2) Frischat (1975a).

TABLE 15.36

Arrhenius Parameters for Diffusion of Ions in Potassium Borosilicate Glass[a]

Composition (mole %)				Temperature range	D_0	E_D
K_2O	B_2O_3	SiO_2	Species	(°C)	$(cm^2 s^{-1})$	(kcal mole^{-1})
18.4	7.4	74.1	^{22}Na	381–565	3.7×10^{-3}	22
			^{42}K	381–565	7.4×10^{-4}	20.6

[a] Data from Hahnert et al. (1984).

TABLE 15.37

Arrhenius Parameters for Diffusion of Various Ions in Potassium Aluminosilicate Glasses

Composition (mole %)				Temperature range	D_0	E_D	
K_2O	Al_2O_3	SiO_2	Species	(°C)	$(cm^2 s^{-1})$	(kcal mole^{-1})	Reference[a]
18.52	7.41	74.07	Na^+	350–490	5.2×10^{-3}	22.9	1
18.52	7.41	74.07	K^+	350–490	3.3×10^{-4}	18.4	1
29.85	10.45	59.70	Na^+	354–473	2.5×10^{-3}	19.9	2
29.85	10.45	59.70	K^+	354–490	0.7×10^{-3}	15.9	2, 3
29.85	10.45	59.70	Ag^+	350–490	2.8×10^{-3}	22.0	4
31.8	5.8	62.4	Na^+	354–473	3.0×10^{-3}	19.9	5
31.8	5.8	62.4	^{22}Na	200–400	$(3.59 \pm 1.08) \times 10^{-1}$	26.7 ± 0.3	6, 7
31.8	5.8	62.4	K^+	354–473	2.0×10^{-3}	17.3	5

[a] Data from (1) Kolitsch et al. (1980b), (2) Kolitsch et al. (1980a), (3) Kolitsch et al. (1980c), (4) Kolitsch and Richter (1981a), (5) Kolitsch et al. (1979), (6) Frischat (1972), and (7) Frischat (1971b).

TABLE 15.38

Arrhenius Parameters for ^{22}Na Diffusion in Na_2O–SnO_2–SiO_2 Glasses[a]

Composition (mole %)			Temperature range	D_0	E_D
Na_2O	SnO_2	SiO_2	(°C)	$(cm^2 s^{-1})$	(kcal mole^{-1})
32.04	3.24	64.72	150–350	8.45×10^{-4}	14.1
31.96	4.75	63.29	150–350	12.39×10^{-4}	14.5
30.70	6.01	63.29	150–350	9.89×10^{-4}	14.35
31.02	8.73	60.24	150–350	8.72×10^{-4}	14.2

[a] Data from Sviridov et al. (1976).

TABLE 15.39

Arrhenius Parameters for ^{22}Na Diffusion in $Na_2O–TiO_2–SiO_2$ Glasses[a]

Composition (mole %)			Temperature range (°C)	D_0 (cm^2 s^{-1})	E_D (kcal mole^{-1})
Na_2O	TiO_2	SiO_2			
32.04	3.24	64.72	150–350	8.09×10^{-4}	14.3
32.50	5.00	62.50	150–350	2.98×10^{-3}	16.2
31.68	6.21	62.11	150–350	9.22×10^{-4}	15.1
30.30	9.09	60.61	150–350	1.82×10^{-3}	15.8

[a] Data from Sviridov et al. (1976).

TABLE 15.40

Arrhenius Parameters for Self-Diffusion of ^{24}Na and ^{32}P in $Na_2O–P_2O_5–SiO_2$ Glasses[a]

Composition (mole %)			Species	Temperature range (°C)	$D_0 \times 10^4$ (cm^2 s^{-1})	E_D (kcal mole^{-1})
Na_2O	P_2O_5	SiO_2				
33.3	—	66.7	^{22}Na	1300–1690	5.76	7.05 ± 0.35
36.1	1.6	62.3	^{24}Na	1370–1635	27.7	11.50 ± 0.5
			^{32}P	1370–1635	42.7	15.85 ± 0.75
38.7	3.2	58.1	^{24}Na	1340–1530	15.6	9.60 ± 0.55
			^{32}P	1340–1530	520	22.4 ± 0.2
41.3	4.7	54.0	^{24}Na	1370–1675	13.4	9.05 ± 0.3
			^{32}P	1370–1675	3.76	8.4 ± 0.4
43.7	6.3	50.0	^{24}Na	1380–1675	21.9	11.3 ± 0.35
			^{32}P	1380–1675	92.7	18.55 ± 0.2

[a] Data from Perron and Bell (1967).

TABLE 15.41

Diffusion Coefficient of ^{22}Na in Magnesium Aluminosilicate Glass[a]

Composition (wt. %)			Species	Temperature (°C)	D (cm^2 s^{-1})
MgO	Al_2O_3	SiO_2			
20	18	62	^{22}Na	670	5.5×10^{-11}

[a] Data from Evstrop'ev et al. (1966).

TABLE 15.42

Arrhenius Parameters for Self-Diffusion Coefficients of Various Species in Calcium Aluminosilicate Glasses

Composition (mole %)							
CaO	Al_2O_3	SiO_2	Species	Temperature range	D_0 (cm^2 s^{-1})	E_D (kcal mole^{-1})	Reference[a]
43.75	12.5	43.75	^{26}Al	1400–1485	5.4 ± 0.2	60 ± 10	1
47	6	47	^{26}Al	1440–1520	$(4.3 \pm 0.3) \times 10^4$	85 ± 17.5	1
45[b]	13.3	41.7	^{30}Si	660–760	10^0–10^{-3}	55 ± 10	2
45[b]	13.3	41.7	^{18}O	625–830	$(2.79^{+2.27}_{-1.25}) \times 10^{-3}$	57.7	3
45.33	12.44	42.23	^{18}O	765–845	4	69	4

[a] Data from (1) Henderson et al. (1961), (2) Schaeffer (1971, 1974), (3) Hagel and Mackenzie (1964), and (4) Kingery and Lecron (1960).

[b] Glass compositions approximate.

TABLE 15.43

Diffusion Coefficients of Various Species in Calcium Aluminosilicate Glass Melts

Composition (wt. %)				Temperature	D	
CaO	Al_2O_3	SiO_2	Species	(°C)	(cm^2 s^{-1})	Reference[a]
39.4	21.2	38.8	^{32}P	1260	1.1×10^{-5}	1
				1320	5.7×10^{-5}	
				1400	12×10^{-5}	
				1500	18×10^{-5}	
38.5	20.9	40.5	^{31}Si	1365	0.47×10^{-7}	1
				1430	1.05×10^{-7}	
11	18	71	^{45}Ca	1350	0.39×10^{-7}	2
				1400	3.2×10^{-7}	
39.8	19.0	41.2	^{45}Ca	1350	3.9×10^{-7}	3
				1395	6.9×10^{-7}	
				1440	10×10^{-7}	
43.0	20.2	36.8	^{45}Ca	1440	8.0×10^{-7}	3
				1510	19×10^{-7}	
45.2	18.6	36.2	^{45}Ca	1510	8.5×10^{-7}	3
				1530	9.9×10^{-7}	
45.6	20.3	34.1	^{45}Ca	1440	3.8×10^{-7}	3
				1485	7.1×10^{-7}	
				1510	8.4×10^{-7}	
				1530	10.3×10^{-7}	
				1575	13.0×10^{-7}	
48.4	20.3	31.3	^{45}Ca	1540	5.5×10^{-7}	3
				1565	8.1×10^{-7}	
48.9	11.7	39.4	^{45}Ca	1440	11.5×10^{-7}	3
				1510	17×10^{-7}	

(continues)

TABLE 15.43 (*Continued*)

Composition (wt. %)				Temperature	D	
CaO	Al$_2$O$_3$	SiO$_2$	Species	(°C)	(cm^2 s^{-1})	Reference[a]
48.7	51.3	—	^{45}Ca	1420	3.3×10^{-7}	3
				1440	5.0×10^{-7}	
				1485	7.8×10^{-7}	
55.2	—	44.8	^{45}Ca	1485	7.1×10^{-7}	3
				1510	8.7×10^{-7}	
				1530	11×10^{-7}	
35.0	15.0	50.0	Sn(IV)	1307	5.6×10^{-15}	4
				1410	5.3×10^{-13}	
				1415	3.5×10^{-13}	
				1455	2.5×10^{-12}	
				1463	1.5×10^{-12}	
				1505	$(2.1 \pm 0.5) \times 10^{-11}$	
10	50	40	SiO$_2$	1700	0.97×10^{-6}	5
20	50	30		1700	1.07×10^{-6}	
30	50	20		1700	1.05×10^{-6}	
35	35	30		1600	1.30×10^{-6}	
40	40	20		1600	1.41×10^{-6}	
40	40	20		1700	1.51×10^{-6}	
40	50	10		1700	1.48×10^{-6}	
45	45	10		1600	1.48×10^{-6}	
50	10	40		1700	1.12×10^{-6}	
50	20	30		1700	1.51×10^{-6}	
50	30	20		1700	3.16×10^{-6}	
50	40	10		1700	3.89×10^{-6}	

[a] Data from (1) Towers and Chipman (1957), (2) Niwa (1957), (3) Saito and Maruya (1957), (4) Taylor *et al.* (1980), and (5) Ershov and Papova (1964).

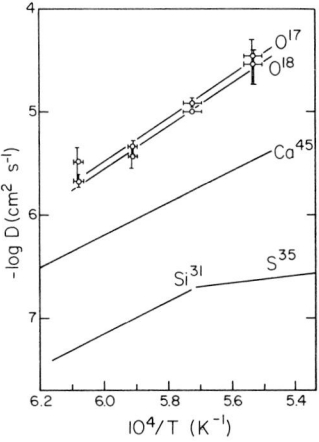

Fig. 15.17. Temperature variation of ^{17}O, ^{18}O, ^{45}Ca, ^{31}Si, and ^{35}S diffusion coefficients in 40CaO–20Al$_2$O$_3$–40SiO$_2$ (wt. %) glass melt. From King and Koros (1959).

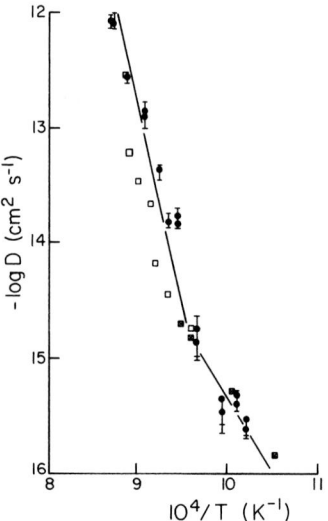

Fig. 15.18. Temperature variation of the diffusion coefficients of Ca and O in 40CaO–20Al$_2$O$_3$–40SiO$_2$ (wt. %) glass melt; ●, calcium; □, ■, oxygen. From Terai and Okawa (1974, 1977).

Part V
Other Properties

Chapter 16

Refractive Index and Dispersion

Many uses of glass take advantage of its optical transparency. The use of glass as an optical component requires knowledge of the refractive index and its dependence on the wavelength of light.

The refractive index n of a transparent material is the ratio of the velocity of light in a vacuum, c, to the velocity in the medium, v:

$$n = c/v. \tag{1}$$

The refractive index is a function of wavelength, so the wavelength at which it is measured must be specified. The usual wavelength is the sodium D line at 0.5893 μm wavelength, and n_D is the value of n at this wavelength. The difference between n values at the hydrogen F (0.4861 μm) and hydrogen C (0.6563 μm) lines is used as a measure of the wavelength dependence of the refractive index and is called the dispersion $n_F - n_C$. The Abbe number v is another measure of wavelength dependence of n:

$$v = (n_D - 1)/(n_F - n_C). \tag{2}$$

Values of n_D and v for a number of optical glasses are given in Table 3.10.

A simple method for estimating the refractive index of a glass involves the use of a series of organic liquids of different refractive index. A small glass sample, which can be of irregular shape, is immersed in different liquids and observed in the microscope. The liquid in which the sample is least visible is the one with refractive index closest to the glass. This result follows from the equation for reflection R from surface of glass with index n_G in contact with a liquid of index n_L:

$$R = \frac{(n_L - n_G)^2}{(n_L + n_G)^2}, \tag{3}$$

where R is a minimum when $n_L - n_G$ is smallest.

A Pulfrich or Abbe refractometer can be used to measure the refractive index of a glass to about four decimal places. A glass plate with polished faces is placed on a prism, with a liquid film of intermediate index between the plate and prism. The critical angle of refraction at the interface is measured, from which n can be calculated. A variety of other techniques are available for measurements of higher precision (Dickson, 1969).

With these techniques and much care, the refractive index can be measured to a precision of six decimal places or better. However, it takes great care in the melting and processing of glass to achieve uniformity in the refractive index to this precision. It is also extremely difficult to reproduce values of n_D to better than four decimal places from one melt to another. Small differences in impurities, heat treatment, and melting conditions change n_D from one batch to another. Thus the values of n_D found in tables in this chapter are usually given to four decimal places only.

II TABLES AND FIGURES

Tables 16.1–16.27 present data on the refractive index and dispersion of binary glasses; and Tables 16.28–16.121 and Figs. 16.1–16.6 those of ternary glasses.

BIBLIOGRAPHY

Akimov, V. V. (1960a). *J. Appl. Chem. USSR (Engl. Transl.)* **33**, 2160.
Akimov, V. V. (1960b). In "Stekloobraznaye Sostoyaniye," p. 488. Moscow.
Akimov, V. V. (1973). *Zh. Prikl. Khim.* **46**, 2178.
Aleksandrov, V. I., Borik, M. A., Dechev, G. K., Markov, N. I., Myzina, V. A., Osiko, V. V., Tatarintsev, V. M., and Khodakovskaya, R. Y. (1980). *Sov. J. Glass Phys. Chem. (Engl. Transl.)* **6**, 117.
Alekseeva, Z. D., and Polozok, N. V. (1972). *Inorg. Mater. (Engl. Transl.)* **8**, 135.
Ammar, M. M., El-Badry, K., Moussa, M. R., Gharib, S., and Halawa, M. (1975). *Cent. Glass Ceram. Res. Inst. Bull.* **22**, 10.
Ammar, M. M., Gharib, S., Halawa, M. M., El-Badry, K., Ghoneim, N. A., and ElBatal, H. A. (1982). *J. Non-Cryst. Solids* **53**, 165.
Appen, A. A. (1953). *Zh. Prikl. Khim.* **26**, 9.
Appen, A. A. (1954). *Steklo Keram.* **3**, 7.
Appen, A. A. (1958). *Trans. Congr. Int. Verre, 4th, Paris* p. 103.
Appen, A. A., and Gan' Fu-Si (1959a). *J. Appl. Chem. USSR (Engl. Transl.)* **32**, 1006.
Appen, A. A., and Gan' Fu-Si (1959b). *J. Appl. Chem. USSR (Engl. Transl.)* **32**, 1013.
Appen, A. A., and Gan' Fu-Si (1959c). *J. Appl. Chem. USSR (Engl. Transl.)* **32**, 1239.
Aramaki, S., and Roy, R. (1962). *J. Am. Ceram. Soc.* **45**, 229.
Bezborodov, M. A., and Bobkova, N. M. (1957). *Dokl. Akad. Nauk SSSR* **116**, 652.
Bezborodov, M. A., and Bobkova, N. M. (1958). "Influence of Cesium on Some Properties of Silicate and Borate Glasses," Minsk.
Bezborodov, M. A., and Bobkova, N. M. (1959). *Silikattechnik* **10**, 584.
Bezborodov, M. A., and Mazurenko, V. D. (1960). *Dokl. Akad. Nauk BSSR* **4**, 58.

Bezborodov, M. A., and Mazurenko, V. D. (1961a). *Izv. Vuzov* **4**, 261.

Bezborodov, M. A., and Mazurenko, V. D. (1961b). *Steklo Keram.* **12**, 161.

Bihuniak, P. P., and Condrate, R. A. (1981). *J. Non-Cryst. Solids* **44**, 331.

Bobkova, N. M. (1958). *Tr. Beloruss. Politekh. Inst.* **63**, 16.

Bogatyreva, V. V., Bogatyrev, Y. Z., and Solov'yeva, T. I. (1973). *Sov. J. Opt. Technol. (Engl. Transl.)* **40**, 495.

Bollin, P. L. (1972). *J. Am. Ceram. Soc.* **55**, 483.

Botvinkin, O. K., and Demichev, S. A. (1964). *Steklo* No. 2, p. 1.

Botvinkin, O. K., and Yaroker, K. G. (1966). *Steklo* No. 3, p. 1.

Botvinkin, O. K., and Yaroker, K. G. (1967). *Inorg. Mater. (Engl. Transl.)* **3**, 1427.

Bowen, N. L., and Schairer, J. F. (1935). *Am. J. Sci.* **29**, 151.

Bowen, N. L., Schairer, J. F., and Willems, H. W. V. (1930). *Am. J. Sci.* **20**, 405.

Bowen, N. L., Schairer, J. F., and Posnjak, E. (1933). *Am. J. Sci.* **26**, 193.

Brekhovskikh, S. M., and Sesorova, V. N. (1960). *In* "The Structure of Glass," p. 444. Moscow-Leningrad.

Bruckner, R., and Navarro, J. F. (1966). *Glastech. Ber.* **39**, 283.

Chakrabarty, M. R. (1969). *Am. Ceram. Soc. Bull.* **48**, 1076.

Chang, Ying-hua, and Ying, Chin-wen (1965). *J. Chin. Silic. Soc.* **4**, 1.

Cleek, G. W., and Babcock, C. L. (1973). "Properties of Glasses in Some Ternary Systems Containing BaO and SiO_2." *Natl. Bur. Stand., Monogr.* **135**, U.S. Govt. Printing Office, Washington, D.C.

Cleek, G. W., and Hamilton, E. (1956). *J. Res. Natl. Bur. Stand. (U.S.)* **57**, 317.

Danilova, N. P. (1967). *Steklo* **1**, 89.

Danilova, N. P., and Dubrovo, S. K. (1965). *In* "Issledovaniya u Oblasti Khimii Silikatovi Okislov," p. 18. Moscow.

Danilova, N. P., and Dubrovo, S. K. (1967). *J. Appl. Chem. USSR (Engl. Transl.)* **40**, 959.

Danilova, N. P., and Dubrovo, S. K. (1971). "Glass Formation of Silicates of Barium," 3509-71 Dep. VINITI.

Demkina, L. I. (1958). "Issledovaniye Zavisimosti Svoystu Stekol at ikh Sostava, Moscow, Prilozheniye," p. 7.

Dickson, J. H. (1969). "Optical Instruments and Techniques." Oriel Press, London.

Dietzel, A., and Florke, O. W. (1955). *Glastech. Ber.* **28**, 423.

Dietzel, A., and Scholze, H. (1955). *Glastech. Ber.* **30**, 135.

Dubrovo, S. K. (1960). *In* "Stekloobraznaye Sostoyaniye," p. 418. Moscow.

Dubrovo, S. K., and Kasymova, S. S. (1964). *Uzb. Khim. Zh.* No. 8, p. 14.

Dubrovo, S. K., and Shmidt, Y. A. (1959). *J. Appl. Chem. USSR (Engl. Transl.)* **32**, 767.

Dubrovo, S. K., and Shnypikov, A. D. (1966). *Inorg. Mater. (Engl. Transl.)* **2**, 1417.

Dubrovo, S. K., and Tsekhomskaya, T. S. (1964a). *J. Appl. Chem. USSR (Engl. Transl.)* **37**, 1224.

Dubrovo, S. K., and Tsekhomskaya, T. S. (1964b). *In* "Kimiya Redkikh Elementov," p. 91. Leningrad.

Dubrovo, S. K., Danilova, N. P., and Tsekhomskaya, T. S. (1965). *In* "Issledovaniya v Oblasti Khim. Silikat. i Okislov," p. 11. Moscow.

Ellern, G. A., and Pavlushkin, N. M. (1969). *Tr. Mosk. Khim.-Tekhnol. Inst., Silik.* **59**, 30.

Eskola, P. (1922). *Am. J. Sci.* **4**, 331.

Evstrop'ev, K. S., and Zopin, A. P. (1967). *Opt.-Mekh. Prom.* No. 2, p. 38.

Faick, C. A., and Finn, A. N. (1931a). *J. Res. Natl. Bur. Stand. (U.S.)* **6**, 993.

Faick, C. A., and Finn, A. N. (1931b). *J. Am. Ceram. Soc.* **14**, 518.

Faick, C. A., Young, J. C., Hubbard, D., and Finn, A. N. (1935). *J. Res. Natl. Bur. Stand. (U.S.)* **14**, 133.

Faust, G. T., and Peck, A. B. (1938). *J. Am. Ceram. Soc.* **21**, 320.

Fleming, J. W. (1976). *J. Am. Ceram. Soc.* **59**, 503.

Fleming, J. W. (1978). *Electron. Lett.* **14**, 326.

Galant, E. I. (1980). *Fiz. Khim. Stekla* **6**, 121.

Galant, E. I., Makarova, T. M., Malchanov, V. S., and Tsekhomskii, V. A. (1966). *Opt.-Mekh. Prom.* No. 4, p. 32.

Gambaryan, S. G., Batanova, A., and Chetverikov, S. D. (1967). *Prom. Arm., Sov. Nar. Khoz. Arm. SSR, Tekh.-Ekon. Byull.* No. 3, p. 19.

Geller, R. F., and Bunting, E. N. (1939). *J. Res. Natl. Bur. Stand. (U.S.)* **23**, 275.

Geller, R. F., and Bunting, E. N. (1943). *J. Res. Natl. Bur. Stand. (U.S.)* **31**, 255.

Graham, P. W. L., and Rindone, G. E. (1967). *Phys. Chem. Glasses* **8**, 160.

Greene, K. T., and Morgan, W. R. (1941). *J. Am. Ceram. Soc.* **24**, 111.

Gunawardane, R. P., and Glasser, F. P. (1974). *J. Am. Ceram. Soc.* **57**, 201.

Hamilton, E. H., and Cleek, G. W. (1958). *J. Res. Natl. Bur. Stand. (U.S.)* **61**, 89.

Hamilton, E. H., Cleek, G. W., and Grauer, O. H. (1958). *J. Am. Ceram. Soc.* **41**, 209.

Hamilton, E. H., Waxler, R. M., and Nivert, J. M. (1959). *J. Res. Natl. Bur. Stand. (U.S.)* **62**, 59.

Hirayama, C., and Berg, D. (1963). *J. Am. Ceram. Soc.* **46**, 85.

Horn, W. F., and Hummel, F. A. (1955). *J. Soc. Glass Technol.* **39**, 113.

Huang, Y. Y., Sarkar, A., and Schultz, P. C. (1978). *J. Non-Cryst. Solids* **27**, 29.

Hubbard, D., and Cleek, G. W. (1952). *J. Res. Natl. Bur. Stand. (U.S.)* **49**, 267.

Hurt, I. C., and Phillips, C. J. (1970). *J. Am. Ceram. Soc.* **53**, 269.

Iskhakov, K. S. (1971). *Uzb. Khim. Zh.* No. 2, p. 79.

Jabra, R., Pelous, J., and Phalippou, J. (1980). *J. Non-Cryst. Solids* **37**, 349.

Kanazawa, T., Kawazoe, H., and Ikeda, M. (1970). *Yogyo Kyokaishi* **78**, 121.

Karlsson, K. (1970). *Suom. Kemistil.* **43**, 479.

Kasymova, S. S. (1964a). *Steklo* No. 3, p. 87.

Kasymova, S. S. (1964b). Physico-chemical properties of sodium–strontium–silicate glasses. Candidate's Dissertation, Tashkent.

Kim, K. H. (1968). *Taehan Hwahakhoe Chi* **12**, 65.

Kleine, R., Maksimova, O. S., Rasmane, D. N., and Kanunnikova, T. D. (1970). *In* "Stekla Steklovidnye Pokrytiya" (J. R. Eiduks, ed.), p. 24. Riga.

Konijnendijk, W. L. (1975). The structure of borosilicate glasses. Thesis, Eindhoven.

Kordes, E. (1939). *Z. Anorg. Allg. Chem.* **241**, 1.

Kracek, F. C. (1930). *J. Phys. Chem.* **34**, 2641.

Krakau, K. A. (1949). *In* "Physiochemical Properties of the Ternary System of Sodium Oxide, Lead Oxide, and Silica," pp. 123–131. Izd. Akad. Nauk, Moscow–Leningrad.

Lai, C. F., and Silverman, A. (1928). *J. Am. Ceram. Soc.* **11**, 535.

Lai, C. F., and Silverman, A. (1930). *J. Am. Ceram. Soc.* **13**, 393.

Larsen, E. S. (1909). *Am. J. Sci.* **28**, 263.

Mazurenko, V. D. (1962). *In* "Steklo i Silikatnye Materialy," p. 90. Minsk.

Mochida, N., Takahashi, K., and Shibusawa, S. (1980). *Yogyo Kyokaishi* **88**, 583.

Morey, G. W. (1954). "The Properties of Glass." Reinhold, New York.

Morey, G. W., and Merwin, H. E. (1932). *J. Opt. Soc. Am.* **22**, 632.

Moriya, T., Akao, Y., and Hatano, N. (1960). *Yogyo Kyokaishi* **68**, 145.

Murthy, M. K., and Topping, J. A. (1975). *J. Am. Ceram. Soc.* **58**, 460.

Peddle, C. T. (1920). *J. Soc. Glass Technol.* **4**, 3.

Phillips, B., and Scroger, M. G. (1965). *J. Am. Ceram. Soc.* **48**, 398.

Polukhin, V. N. (1960). *Opt.-Mekh. Prom.* No. 11, p. 18.

Rao, B. V. J. (1962a). *J. Am. Ceram. Soc.* **45**, 555.

Rao, B. V. J. (1962b). *J. Sci. Ind. Res., Sect. B* **21**, 108.

Rao, B. V. J. (1963a). *Phys. Chem. Glasses* **4**, 22.

Rao, B. V. J. (1963b). *J. Am. Ceram. Soc.* **46**, 107.

Rao, B. V. J. (1965). *C. R. Congr. Int. Verre, 7th, Brussels* **1**, 104.

Riegel, E. R., and Sharp, D. (1934). *J. Am. Ceram. Soc.* **17**, 88.

Roedder, E. W. (1951). *Am. J. Sci.* **249**, 81.

Schairer, J. F., and Bowen, N. L. (1955). *Am. J. Sci.* **253**, 681.

Schairer, J. F., and Bowen, N. L. (1956). *Am. J. Sci.* **254**, 129.

Schairer, J. F., and Yagi, K. (1952). *Am. J. Sci.* Bowen vol., p. 2, 471.

Schroeder, J. (1980). *J. Non-Cryst. Solids* **40**, 549.

Schultz, P. C. (1976). *J. Am. Ceram. Soc.* **59**, 214.

Schultz, P. C., and Dumbaugh, W. H. (1980). *J. Non-Cryst. Solids* **38/39,** 33.

Sedmalis, U., Bol'shii, Y. Y., Gudkina, N. N., and Eiduks, J. (1974). *Uch. Zap. Latv. un-ta* **203,** 138.

Sedykh, T. S., Pustil'nik, A. I., and Mikheikin, V. I. (1975). *Inorg. Mater.* (*Engl. Transl.*) **11,** 987.

Shchavelev, O. S., Kasymova, S. S., and Petrovskii, G. T. (1974). *J. Appl. Chem. USSR* (*Engl. Transl.*) **47,** 13.

Shelby, J. E. (1978). *J. Appl. Phys.* **49,** 5885.

Shelby, J. E. (1979). *J. Appl. Phys.* **50,** 8010.

Shelby, J. E., and Day, D. E. (1969). *J. Am. Ceram. Soc.* **52,** 169.

Shibata, N. and Edahiro, T. (1982). *Trans. IECE Japan* **65-E,** 166.

Shmidt, Y. A., and Alekseeva, Z. D. (1964). *J. Appl. Chem. USSR* (*Engl. Transl.*) **37,** 2266.

Shnypikov, A. D. (1967). *Steklo* No. 1, p. 85.

Simpson, H. E. (1959). *Glass Ind.* **40,** 454.

Simpson, H. E. (1961). *Glass Ind.* **42,** 222.

Snow, R. B. (1943). *J. Am. Ceram. Soc.* **26,** 11.

Sun, K.-H., and Silverman, A. (1941). *J. Am. Ceram. Soc.* **45,** 182.

Takahashi, K., Mochida, N., and Hatta, G. (1975). *Yogyo Kyokaishi* **83,** 103.

Takahashi, K., Mochida, N., Matsui, H., Takeuchi, S., and Gohshi, Y. (1976). *Yogyo Kyokaishi* **84,** 482.

Takahashi, K., Mochida, N., and Yoshida, Y. (1977). *Yogyo Kyokaishi* **85,** 330.

Tiwari, A. N., and Das, A. R. (1972). *Am. Ceram. Soc. Bull.* **51,** 695.

Topol, L. E., Hengstenberg, D. H., Blander, M., Happe, R. A., Richardson, N. L., and Nelson, R. S. (1973). *J. Non-Cryst. Solids* **12,** 377.

Topping, J. A., Fuchs, P., and Murthy, M. K. (1974). *J. Am. Ceram. Soc.* **57,** 205.

Tsekhomskaya, T. S. (1966). Stekloobraznaye gallosilikaty litiya i produkty ikh kristallizatsii. Avtoref., Candidate's Dissertation, Leningrad.

Urusovskaya, L. N. (1960). *J. Appl. Chem. USSR* (*Engl. Transl.*) **33,** 1971.

Van Uitert, L. G., Pinnow, D. A., Williams, J. C., Rich, T. C., Jaefer, R. E., and Grodkiewicz, W. H. (1973). *Mater. Res. Bull.* **8,** 469.

Vargin, V. V., Zasolotskaya, M. V., Kind, N. E., Kondrat'ev, Y. N., Milyukov, E. M., and Tudorovskaya, N. A. (1971). "Catalyzed Crystallization of Glasses of $Li_2O-Al_2O_3-SiO_2$ System," Part 2. Leningrad.

Vargin, V. V., Dzhavuktsyan, S. G., Mishel', V. E., and Pevzhev, B. Z. (1972). *J. Appl. Chem. USSR* (*Engl. Transl.*) **45,** 1228.

Varshal, B. G. (1972). *Inorg. Mater.* (*Engl. Transl.*) **8,** 812.

Verweij, H., Buster, J. H. J. M., and Remmers, G. F. (1979). *J. Mater. Sci.* **14,** 931.

Watanabe, K., Suniyoshi, Y., and Anbo, E. (1970). *Yogyo Kyokaishi* **78,** 165.

Wemple, S. H., Pinnow, D. A., Rich, T. C., Jaeger, R. E., and Van Uitert, L. G. (1973). *J. Appl. Phys.* **44,** 5432.

Williamson, J., and Glasser, F. P. (1964a). *Phys. Chem. Glasses* **5,** 52.

Williamson, J., and Glasser, F. P. (1964b). *Nature* (*London*) **201,** 286.

TABLE 16.1

Refractive Index of Lithium Silicate Glasses[a]

Composition (mole % Li_2O)	n_D	Composition (mole % Li_2O)	n_D
10.3	1.474	33.3	1.532
14.0	1.484	37.1	1.542
16.7	1.491	39.8	1.547
20.0	1.500	41.7	1.549
25.0	1.511	42.8	1.553
27.8	1.518	45.4	1.555
30.0	1.524	50.0	1.559

[a] Values were derived from a data base consisting of eight data sets of different workers. The useful studies are: Dubrovo and Shmidt (1959); Vargin *et al.* (1971); Kracek (1930); Hubbard and Cleek (1952); Verweij *et al.* (1979).

TABLE 16.2

Refractive Index of Sodium Silicate Glasses[a]

Composition (mole % Na_2O)	n_D	Composition (mole % Na_2O)	n_D
5	1.4655	30	1.5027
10	1.475	35	1.5072
15	1.482	40	1.511
20	1.4906	45	1.5142
25	1.4977	50	1.517

[a] Values derived from a data base consisting of 10 data sets of different workers. The useful studies are: Urusovskaya (1960); Akimov (1960a); Faick and Finn (1931a,b); Peddle (1920); Morey and Merwin (1932); Dubrovo and Shmidt (1959); Bezborodov and Bobkova (1958); Verweij *et al.* (1979).

TABLE 16.3

Refractive Index of Potassium Silicate Glasses[a]

Composition (mole % K_2O)	n_D	Composition (mole % K_2O)	n_D
10	1.480	30	1.5048
15	1.4867	35	1.5093
20	1.4937	40	1.5136
25	1.4995	45	1.5175

[a] Values derived from a data base consisting of nine data sets of different workers. The useful studies are: Rao (1963a,b); Faust and Peck (1938); Urusovskaya (1960); Shelby and Day (1969); Shmidt and Alekseeva (1964).

TABLE 16.4

Refractive Index of Rubidium Silicate Glasses[a]

Composition (mole % Rb_2O)	n_D	Composition (mole % Rb_2O)	n_D
8.4	1.480	20.0	1.504
11.8	1.487	25.0	1.511
13.5	1.490	26.6	1.514
15.3	1.495	33.1	1.521

[a] Data from Shmidt and Alekseeva (1964); and Shelby and Day (1969).

TABLE 16.5

Refractive Indices of Cesium Silicate Glasses

Composition (mole % Cs_2O)	n_D	Reference[a]	Composition (mole % Cs_2O)	n_D	Reference[a]
4.0	1.477	1	10.5	1.499	2
5.0	1.481		11.1	1.499	
5.8	1.488		11.8	1.501	
8.0	1.493		12.5	1.504	
12.2	1.510		13.3	1.505	
20.0	1.530		14.3	1.507	
26.7	1.548		15.4	1.511	
			16.7	1.515	
8.0	1.490	2	18.2	1.521	
9.1	1.494		20.0	1.522	
10.0	1.498				

[a] Data from (1) Shmidt and Alekseeva (1964) and (2) Bezborodov and Bobkova (1957, 1958, 1959).

TABLE 16.6

Refractive Index of Magnesium Silicate Glass[a]

Composition (mole % MgO)	n_D
50	1.5802

[a] Data from Larsen (1909).

TABLE 16.7

Optical Constants of Calcium Silicate Glasses

Composition (mole % CaO)	n_D	$(n_F - n_C) \times 10^4$	Reference[a]
39.0	1.5905	103.0	1
44.6	1.612	109.5	
50.0	1.6295	115.5	
52.9	1.635	116.0	
57.5	1.6455	121.5	
40	1.580		2[b]
45	1.610		
48	1.617		
52	1.629		
55.5	1.636		
50	1.628		3

[a] Data from (1) Morey and Merwin (1932), (2) Shelby (1979), and (3) Dietzel and Florke (1955).

[b] Data from this reference read from a graph.

TABLE 16.8

Refractive Indices of Strontium Silicate Glasses

Composition (mole % SrO)	n_D	Reference[a]	Composition (mole % SrO)	n_D	Reference[a]
35	1.586	1[b]	33.25	1.584	2
40	1.601		36.70	1.5915	
45	1.617		46.50	1.624	
			49.90	1.632	
			54.10	1.644	

[a] Data from (1) Shelby (1979) and (2) Eskola (1922).

[b] Values from this reference were read from a graph.

TABLE 16.9

Refractive Indices of Barium Silicate Glasses

Composition (mole % BaO)	n_D	Reference[a]	Composition (mole % BaO)	n_D	Reference[a]
15	1.535	1[b]	26.9	1.5816	4
22	1.561		33.3	1.6176	
25	1.573		40.2	1.6405	
30	1.591		24.3	1.567	5
33	1.611		28.1	1.585	
38	1.626		33.3	1.609	
40	1.632		42.1	1.645	
33.33	1.608	2	50	1.646	6
33.33	1.610	3			

[a] Data from (1) Shelby (1979), (2) Graham and Rindone (1967), (3) Greene and Morgan (1941), (4) Hamilton *et al* (1958), (5) Eskola (1922), and (6) Dietzel and Florke (1955).

[b] Data from this reference were read from a graph.

TABLE 16.10

Refractive Indices of Zinc Silicate Glasses[a]

Composition (mole % ZnO)	n_D
45.0	1.6353
50.2	1.6540
60	1.697[b]
66	1.722[b]

[a] Data from Williamson and Glasser (1964a,b).

[b] Values read from graph.

TABLE 16.11

Refractive Index of Lead Silicate Glasses[a]

Composition (mole % PbO)	n_D	Composition (mole % PbO)	n_D
20.8	1.617	49.5	1.906
22.8	1.633	49.7	1.913
24.9	1.651	50.5	1.919
27.4	1.672	52.5	1.941
29.7	1.695	53.5	1.955
31.2	1.711	54.5	1.964
33.0	1.727	56.4	1.989
34.7	1.744	59.4	2.026
36.6	1.763	59.9	2.025
38.6	1.786	61.4	2.046
40.8	1.809	63.4	2.075
41.7	1.819	66.0	2.103
42.5	1.827	66.6	2.096
43.6	1.839	69.9	2.144
44.1	1.846	74.2	2.182
45.6	1.859		
47.8	1.887		

[a] Values were derived from a data base consisting of seven data sets of different workers. The useful studies are Bogatyreva *et al.* (1973), Demkina (1958), and Kordes (1939).

TABLE 16.12

Refractive Indices of B_2O_3–SiO_2 Glasses[a]

Composition (mole % B_2O_3)	n_D	n (at 0.589 μm)
13.5	1.458	—
14.29	—	1.4559
20.0	—	1.4575
32.0	1.456	—
33.33	—	1.4588
94.26	1.453	—

[a] Data from Murthy and Topping (1975), Fleming (1976), Appen and Gan'Fu-Si (1959a), and Wemple et al. (1973).

TABLE 16.13

Refractive Indices of B_2O_3–SiO_2 (13.5 mole % B_2O_3) Glass at Various Wavelengths[a]

Wavelength λ (μm)	n	Wavelength λ (μm)	n
0.435833	1.4665	0.734620	1.4537
0.479991	1.4632	0.807902	1.4523
0.508582	1.4615	0.852111	1.4515
0.546073	1.4596	0.917224	1.4504
0.589263	1.4579	1.002439	1.4493
0.643847	1.4560	1.08303	1.4481

[a] Data from Fleming (1976).

TABLE 16.14

Refractive Index of B_2O_3–SiO_2 Glasses with Different Water Contents[a]

Composition (mole %) B_2O_3	SiO_2	Water content (ppm OH)	$n_{5145 Å}$
5	95	320	1.4604
10	90	9	1.4592
15	85	75	1.4584
20	80	258	1.4582
30	70	620	1.4588
50	50	2155	1.4604
75	25	3408	1.4612
90	10	4496	1.4617

[a] Data from Jabra et al. (1980).

TABLE 16.15

Refractive Indices of Borosilicate Glasses with Varied Water Content[a]

Composition (mole %)			
B_2O_3	SiO_2	H_2O	n_D
39.26	60.69	0.044	1.4616
44.55	55.40	0.048	1.4612
51.58	48.36	0.059	1.4608
56.38	43.57	0.053	1.4603
64.61	35.35	0.041	1.4579
70.56	29.39	0.049	1.4573
77.65	22.27	0.082	1.4599
84.07	15.84	0.091	1.4598
89.19	10.72	0.089	1.4600
90.38	9.55	0.074	1.4593
93.97	5.95	0.082	1.4623
95.31	4.59	0.097	1.4628
97.69	2.17	0.139	1.4630

[a] Data from Bruckner and Navarro (1966).

TABLE 16.16

Coefficients of the Sellmeier Equation $n^2 = 1 + \sum_{i=1}^{3} A_i \lambda^2 / (\lambda^2 - l_i^2)$ for Refractive Index (n) of B_2O_3–SiO_2 Glasses[a]

Composition (mole %)		Coefficients					
B_2O_3	SiO_2	A_1	l_1	A_2	l_2	A_3	l_3
13.3[b]	86.7	0.690618	0.061900	0.401996	0.123662	0.898817	9.098960
5.2[c]	94.8	0.6910021	0.004981838	0.4022430	0.01375664	0.9439644	97.93353

[a] Data from Fleming (1978) and Shibata and Edahiro (1982).
[b] Wavelength (λ) range 0.8–1.5 μm.
[c] Wavelength range 0.4047–2.0581 μm.

TABLE 16.17

Refractive Indices of Al_2O_3–SiO_2 Glasses

Composition (mole % Al_2O_3)	n_D	Reference[a]	Composition (mole % Al_2O_3)	n_D	Reference[a]
10	1.478	1	70.21	1.629	2
20	1.504		76.95	1.634	
33	1.535		84.13	1.720	
50	1.574		91.80	1.728	
60	1.597				

[a] Data from (1) Aramaki and Roy (1962) and (2) Topal et al. (1973).

TABLE 16.18

Refractive Index of Bismuth Silicate Glasses[a]

Composition (mole % Bi_2O_3)	n^b	Composition (mole % Bi_2O_3)	n^b
26.25	1.75	41.28	1.97
31.00	1.83	43.40	1.98
36.38	1.89	46.55	1.99
39.15	1.93	49.57	2.07

[a] Data from Tiwari and Das (1972).
[b] Calculated values.

TABLE 16.19

Refractive Index of GeO_2–SiO_2 Glasses[a]

Composition[b] (mole % GeO_2)	n_D	Composition[b] (mole % GeO_2)	n_D
0.0	1.458	27.0	1.499
1.0	1.460	48.0	1.516
2.0	1.460	59.0	1.556
5.5	1.466	61.0	1.552
11.0	1.482	78.5	1.576
17.5	1.484	100.0	1.603
24.1	1.490		

[a] Data from Huang et al. (1978).
[b] Glasses prepared by vapor-phase oxidation process.

TABLE 16.20

Refractive Indices of GeO_2–SiO_2 Glasses[a]

Composition (mole % GeO_2)	n_D	Composition (mole % GeO_2)	n_D
0	1.458	12.0	1.474
2.5	1.463	16.0	1.479
6.8	1.469	64.3^b	1.556
8.0	1.471	100	1.609

[a] Data from Galant (1980).
[b] Glass prepared by flame spherulization; rest by melting in molybdenum crucible.

TABLE 16.21

Refractive Index of GeO_2–SiO_2 Glasses at Various Wavelengths[a]

Wavelength λ	n at given mole % GeO_2		
(μm)	4.1	7.0	13.5
0.435833	1.4738	1.4787	1.4902
0.479991	1.4704	1.4752	1.4864
0.508582	1.4686	1.4733	1.4845
0.546073	1.4667	1.4713	1.4825
0.589262	1.4650	1.4695	1.4803
0.643847	1.4632	1.4675	1.4784
0.734620	1.4609	1.4653	1.4758
0.807902	1.4595	1.4638	1.4744
0.852111	1.4588	1.4629	1.4737
0.917224	1.4578	1.4621	1.4726
1.002439	1.4566	1.4610	1.4717
1.08303	1.4556	1.4600	1.4706

[a] Data from Fleming (1976).

TABLE 16.22

Coefficients of the Sellmeier Dispersion Equation $n^2 = 1 + \sum_{i=1} A_i\lambda^2/(\lambda^2 - l_i^2)$ for Refractive Index (n) of GeO_2–SiO_2 Glasses in the Wavelength (λ) Range 0.4047 to 2.0581 μm[a]

Composition (mole %)		Coefficients					
GeO_2	SiO_2	A_1	A_2	A_3	l_1	l_2	l_3
6.3	93.7	0.7083952	0.4203993	0.8663412	0.007290464	0.01050294	97.93428
8.7	91.3	0.7133103	0.4250904	0.8631980	0.006910297	0.01165674	97.93434
11.2	88.8	0.7186243	0.4301997	0.8543265	0.004026394	0.01632475	97.93440
15.0	85.0	0.7249180	0.4381220	0.8221368	0.007596374	0.01162396	97.93472
19.3	80.7	0.7347008	0.4461191	0.8081698	0.005847345	0.01552717	97.93484

[a] Data from Shibata and Edahiro (1982).

TABLE 16.23
Refractive Indices of TiO$_2$–SiO$_2$ Glasses[a]

Composition (mole % TiO$_2$)	$n_{6328.2\text{Å}}$
2.4	1.46494
7.4	1.48149
9.0	1.48408

[a] Data from Schroeder (1980).

TABLE 16.24
Refractive Indices[a] of Some MO$_2$–SiO$_2$ Glasses[b]

Composition (mole % MO$_2$)	n_D	Composition (mole % MO$_2$)	n_D
M = Ti		M = Hf	
0.46	1.462	0.08	1.460
1.86	1.467	0.30	1.462
5.28	1.482	0.70	1.463
		0.84	1.464
M = Zr			
0.15	1.460		
0.43	1.462		
0.62	1.463		
0.93	1.465		

[a] Data from Bihuniak and Condrate (1981).
[b] Prepared by the sol–gel technique.

TABLE 16.25
Refractive Index of P$_2$O$_5$–SiO$_2$ Glasses[a]

Composition (mole % PO$_{5/2}$)	n_D	Composition (mole % PO$_{5/2}$)	n_D
17.4	1.471	49.6	1.525
28.5	1.480	50.2	1.533
32.8	1.481	57.9	1.523
42.8	1.507	59.8	1.537
45.8	1.498	64.2	1.545

[a] Data from Takahashi et al. (1976).

TABLE 16.26

Coefficients of the Sellmeier Equation $n^2 = 1 + \sum_{i=1}^{3} A_i \lambda^2 / (\lambda^2 - l_i^2)$
For Refractive Index (n) of P$_2$O$_5$–SiO$_2$ Glasses[a]

Composition (mole %)		Coefficients					
P$_2$O$_5$	SiO$_2$	A_1	l_1	A_2	l_2	A_3	l_3
9.1[b]	90.1	0.695790	0.061568	0.452497	0.119921	0.712513	8.656641
10.5[c]	89.5	0.7058489	0.005202431	0.4176021	0.01287730	0.8952753	97.93401

[a] Data from Fleming (1978) and Shibata and Edahiro (1982).
[b] Wavelength range 0.8–1.5 μm.
[c] Wavelength range 0.4047–2.0581 μm.

TABLE 16.27

Refractive Index of Some $M_2O_5-SiO_2$ Glasses[a]

Composition (mole % M_2O_5)	n_D
M = Nb	
47.49	2.11
81.12	2.19
M = Ta	
43.52	2.04

[a] Data from Topol *et al.* (1973).

TABLE 16.28

Refractive Index of $Li_2O-Na_2O-SiO_2$ Glasses[a]

Composition (mole %)			
Li_2O	Na_2O	SiO_2	n_D
0.5	24.5	75.0	1.499
1.25	23.75	75.0	1.500
3.75	21.25	75.0	1.503
8.75	16.25	75.0	1.506
12.5	12.5	75.0	1.508
16.25	8.75	75.0	1.510
21.25	3.75	75.0	1.512
23.75	1.25	75.0	1.511
24.5	0.5	75.0	1.512
24.875	0.125	75.0	1.513

[a] Data from Shelby and Day (1969).

TABLE 16.29

Refractive Index of $Li_2O-K_2O-SiO_2$ Glasses

Composition (mole %)					
Li_2O	K_2O	SiO_2	n_D	$n_{6328.2Å}$	Reference[a]
—	33	67		1.50498	1
8	25	67		1.51216	
16.5	16.5	67		1.51648	
25	8	67		1.52169	
33	—	67		1.53304	
0.125	24.875	75.0	1.499		2
0.5	24.5	75.0	1.500		
1.25	23.75	75.0	1.501		
3.75	21.25	75.0	1.503		
8.75	16.25	75.0	1.505		
12.5	12.5	75.0	1.506		
16.25	8.75	75.0	1.507		
21.25	3.75	75.0	1.510		
23.75	1.25	75.0	1.512		
24.5	0.5	75.0	1.513		

[a] Data from (1) Schroeder (1980) and (2) Shelby and Day (1969).

TABLE 16.30

**Refractive Index of
Li$_2$O–Cs$_2$O–SiO$_2$ Glasses[a]**

Composition (mole %)			
Li$_2$O	Cs$_2$O	SiO$_2$	n
20.0	—	80.0	1.501
19.1	0.9	80.0	1.503
18.2	1.8	80.0	1.505
15.0	5.0	80.0	1.511
10.0	10.0	80.0	1.522
5.0	15.0	80.0	1.527
—	20.0	80.0	1.530

[a] Data from Alekseeva and Polozok (1972).

TABLE 16.31

Refractive Index of Na$_2$O–K$_2$O–SiO$_2$ Glasses

Composition (mole %)				Composition (mole %)			
Na$_2$O	K$_2$O	SiO$_2$	n_D[a]	Na$_2$O	K$_2$O	SiO$_2$	$n_{6328.2 Å}$[b]
0.5	24.5	75.0	1.499	—	25	75.0	1.48752
1.25	23.75	75.0	1.499	0.5	24.5	75.0	1.49761
2.5	22.5	75.0	1.500	2.5	22.5	75.0	1.49918
5	20	75.0	1.500	6.25	18.75	75.0	1.49864
10	15	75.0	1.500	9	16	75	1.48783
12.5	12.5	75.0	1.501	12.5	12.5	75.0	1.49778
15	10	75.0	1.501	18.75	6.25	75.0	1.49711
20	5	75.0	1.499	20	5	75	1.49656
22.5	2.5	75.0	1.499	25	—	75	1.49539
23.75	1.25	75.0	1.498				
24.5	0.5	75.0	1.498				
24.875	0.125	75.0	1.498				

[a] Data from Shelby and Day (1969).
[b] Data from Schroeder (1980).

TABLE 16.32

Refractive Index of Na$_2$O–K$_2$O–SiO$_2$ Glasses[a]

Composition (mole %)				Composition (mole %)			
Na$_2$O	K$_2$O	SiO$_2$	n_D	Na$_2$O	K$_2$O	SiO$_2$	n_D
8	16	76	1.4989	16	4	80	1.4922
12	12	76	1.4990	4	15	81	1.4927
16	8	76	1.4990	8	11	81	1.4923
20	4	76	1.4980	12	7	81	1.4912
4	19	77	1.4985	16	3	81	1.4902
8	15	77	1.4980	4	14	82	1.4911
16	7	77	1.4965	8	10	82	1.4910
20	3	77	1.4962	12	6	82	1.4895
4	18	78	1.4970	16	2	82	1.4883
16	6	78	1.4959	4	13	83	1.4902
20	2	78	1.4944	8	9	83	1.4891
4	17	79	1.4958	12	5	83	1.4878
8	13	79	1.4964	16	1	83	1.4864
12	9	79	1.4961	4	12	84	1.4881
16	5	79	1.4946	8	8	84	1.4875
18	3	79	1.4937	12	4	84	1.4860
20	1	79	1.4929	4	11	85	1.4870
4	16	80	1.4946	12	3	85	1.4841
8	12	80	1.4938	4	10	86	1.4848
12	8	80	1.4936	12	2	86	1.4813

[a] Data from Urusovskaya (1960).

TABLE 16.33

Refractive Index of Na$_2$O–Rb$_2$O–SiO$_2$ Glasses[a]

Composition (mole %)			
Na$_2$O	Rb$_2$O	SiO$_2$	n_D
12.5	12.5	75.0	1.506
23.75	1.25	75.0	1.500
24.5	0.5	75.0	1.499

[a] Data from Shelby and Day (1969).

TABLE 16.34

Refractive Index of Na$_2$O–Cs$_2$O–SiO$_2$ Glasses[a]

Composition (mole %)			
Na$_2$O	Cs$_2$O	SiO$_2$	n
1.5	18.5	80.0	1.523
5.0	15.0	80.0	1.518
8.0	12.0	80.0	1.514
10.0	10.0	80.0	1.511
11.4	8.6	80.0	1.507
13.3	6.7	80.0	1.520
16.0	4.0	80.0	1.495
18.2	1.8	80.0	1.491

[a] Data from Alekseeva and Polozok (1972).

TABLE 16.35

Optical Constants of $Na_2O-Tl_2O-SiO_2$ Glasses with Na_2O-SiO_2 Ratio of 0.25[a]

Tl_2O (wt. %)	n_D	v_D	$(n_F - n_C) \times 10^5$
3	1.49526	57.76	857.5
6	1.50050	55.87	895.9
10	1.50902	53.07	959.2
12	1.51150	52.21	979.8
21	1.53034	46.29	1130.2
30	1.54812	42.48	1284.4
39	1.57664	37.32	1527.9
51	1.61600	31.58	1950.9
60	1.70989	22.43	3165.0

[a] Data from Polukhin (1960).

TABLE 16.36

Refractive Index of $K_2O-Rb_2O-SiO_2$ Glasses[a]

Composition (mole %)			
K_2O	Rb_2O	SiO_2	n_D
12.5	12.5	75.0	1.506
21.25	3.75	75.0	1.504
23.75	1.25	75.0	1.501
24.5	0.5	75.0	1.500

[a] Data from Shelby and Day (1969).

TABLE 16.37

Refractive Index of $Li_2O-BaO-SiO_2$ Glasses[a]

Composition (mole %)				Composition (mole %)			
Li_2O	BaO	SiO_2	n_D	Li_2O	BaO	SiO_2	n_D
35	5	60	1.572	35	30	35	1.672
40	5	55	1.580	5	35	60	1.628
45	5	50	1.594	10	35	55	1.639
35	10	55	1.591	15	35	50	1.650
45	10	45	1.613	20	35	45	1.660
25	15	60	1.591	25	35	40	1.671
30	15	55	1.600	30	35	35	1.681
35	15	50	1.613	35	35	30	1.694
40	15	45	1.623	5	40	55	1.647
45	15	40	1.634	15	40	45	1.669
25	20	55	1.609	25	40	35	1.671
35	20	45	1.632	5	45	50	1.668
45	20	35	1.654	10	45	45	1.679
15	25	60	1.611	15	45	40	1.690
20	25	55	1.620	20	45	35	1.700
25	25	50	1.632	25	45	30	1.712
30	25	45	1.641	5	50	45	1.687
35	25	40	1.653	15	50	35	1.709
40	25	35	1.663	5	55	40	1.707
45	25	30	1.674	10	55	35	1.719
15	30	55	1.628	15	55	30	1.729
25	30	45	1.650				

[a] Data from Bezborodov and Mazurenko (1960, 1961a,b) and Mazurenko (1962).

TABLE 16.38

Refractive Indices of $Li_2O-ZnO-SiO_2$ Glasses

Composition (mole %)					Composition (mole %)				
Li_2O	ZnO	SiO_2	n_D	Reference[a]	Li_2O	ZnO	SiO_2	n_D	Reference[a]
15	20	65	1.558	1	25	25	50	1.583	1
15	25	60	1.573		25	30	45	1.598	
15	30	55	1.589		30	5	65	1.536	
15	35	50	1.602		30	10	60	1.551	
20	5	75	1.518		30	15	55	1.563	
20	10	70	1.532		30	20	50	1.577	
20	15	65	1.549		35	5	60	1.544	
20	20	60	1.563		35	10	55	1.557	
20	25	55	1.578		35	15	50	1.570	
20	30	50	1.592						
25	5	70	1.528		34.19	6.28	59.53	1.553	2
25	10	65	1.542		40.59	2.91	56.50	1.548	
25	15	60	1.556		26.80	6.27	66.63	1.548	
25	20	55	1.570		33.84	3.03	63.15	1.535	

[a] Data from (1) Vargin et al. (1972) and (2) Ammar et al. (1982). Values from ref. (1) were read from a graph.

TABLE 16.39

Refractive Index of Na$_2$O–BeO–SiO$_2$ Glasses[a]

Composition (mole %)				Composition (mole %)			
Na$_2$O	BeO	SiO$_2$	n_D	Na$_2$O	BeO	SiO$_2$	n_D
15.4	19.2	65.4	1.5210	17.4	17.4	65.2	1.5206
16	16	68	1.5173	17.4	21.7	60.9	1.5265
16	20	64	1.5237	18.2	9.1	72.7	1.5107
16.7	12.5	70.8	1.5121	18.2	13.6	68.2	1.5165
16.7	16.7	66.6	1.5193	18.2	18.2	63.6	1.5222
16.7	20.8	62.5	1.5251	19	14.3	66.7	1.5177
17.4	8.7	73.9	1.5081	19.1	9.5	71.4	1.5122
17.4	13	69.6	1.5141	20	10	70	1.5136

[a] Data from Lai and Silverman (1928, 1930).

TABLE 16.40

Refractive Index of Na$_2$O–MgO–SiO$_2$ Glasses[a]

Composition (mole %)				Composition (mole %)			
Na$_2$O	MgO	SiO$_2$	n_D	Na$_2$O	MgO	SiO$_2$	n_D
7.1	22.9	70	1.5142	11	17.2	71.8	1.506
10	20	70	1.5137	11	13.5	65.5	1.510
10.5	14.5	75	1.5112	11	27.3	61.7	1.5152
11.8	18.2	70	1.5092	11	29.6	59.4	1.5156
15	15	70	1.5084	11	32.3	56.7	1.5191
18.8	11.2	70	1.506	11	34.3	54.7	1.5234
11.3	23.7	65	1.5157	16	14.8	69.2	1.5084
11.7	23.3	65	1.5151	16	18.5	65.5	1.5115
13.2	21.8	65	1.5144	16	22.5	61.5	1.5148
15.9	19.1	65	1.5136	16	24.4	59.6	1.5176
16.6	18.4	65	1.5131	16	25.7	58.3	1.5175
21.9	13.1	65	1.513	16	26.3	57.7	1.520
25.8	9.2	65	1.5133	16	28	56	1.5206
10.6	32.4	57	1.5191	16	30.5	53.5	1.5253
17.2	25.8	57	1.5196	22	13.1	64.9	1.5168
17.9	25.1	57	1.5182	22	15.3	62.7	1.5185
21.5	21.5	57	1.523	22	17.7	60.3	1.5165
22.2	20.8	57	1.5206	22	19.9	58.1	1.5213
23.2	19.8	57	1.5188	22	21.4	56.6	1.5207
27	16	57	1.5217	22	23.4	54.6	1.522
11	11	78	1.5004	22	24.8	53.2	1.5245
11	12.8	76.2	1.5029				

[a] Data from Botvinkin and Yaroker (1966, 1967).

TABLE 16.41

Refractive Index of Na$_2$O–MgO–SiO$_2$ Glasses[a]

Composition (mole %)			
Na$_2$O	MgO	SiO$_2$	$n_{6328.2 Å}$
20	10	70	1.50467
20	20	70	1.51370
20	30	70	1.52825
37.5	19.9	42.6	1.61605

[a] Data from Schroeder (1980).

TABLE 16.42

Refractive Index and Optical Dispersion of Na$_2$O–CaO–SiO$_2$ Glasses[a]

Composition (wt. %)					Composition (wt. %)				
Na$_2$O	CaO	SiO$_2$	n_D	$(n_F - n_C) \times 10^4$	Na$_2$O	CaO	SiO$_2$	n_D	$(n_F - n_C) \times 10^4$
3.90	5.70	90.40	1.483	—	5.73	20.02	74.25	1.5325	89
6.00	4.00	90.00	1.4805	75.5	12.82	12.98	74.20	1.524	88.5
10.89	3.01	86.10	1.486	78	18.48	7.34	74.18	1.5135	88.5
8.00	6.00	86.00	1.4905	78	20.73	5.12	74.15	1.5085	87.5
9.21	7.58	83.21	1.4985	80.5	13.88	12.02	74.10	1.522	88.5
6.55	11.65	81.80	1.507		15.04	11.03	73.93	1.521	89
16.98	5.01	78.01	1.5035	85	7.32	19.04	73.64	1.535	90
12.00	10.00	78.00	1.511	85	12.95	13.44	73.61	1.526	88
14.93	8.01	77.06	1.510	85.5	13.30	13.13	73.57	1.526	89.5
12.85	11.03	76.12	1.517	87	16.42	10.02	73.56	1.5195	88
8.90	14.99	76.11	1.523	87	23.48	3.07	73.45	1.505	87
18.89	5.01	76.10	1.506	85.5	21.71	5.00	73.24	1.5095	88
6.18	17.96	75.86	1.5275	88	13.81	13.02	73.17	1.5255	88
18.64	5.76	75.60	1.509	—	22.84	4.01	73.15	1.508	87.5
4.48	20.23	75.29	1.533	—	17.91	9.01	73.08	1.518	88.5
17.25	7.54	75.21	1.512	88	11.82	16.03	72.15	1.533	90.5
17.50	7.50	75.00	1.5125	87	8.92	19.02	72.06	1.5385	92
15.04	9.98	74.98	1.517	87.5	21.66	6.54	71.80	1.514	89
13.40	11.64	74.96	1.519	88	18.52	10.10	71.38	1.522	91
10.08	14.97	74.95	1.525	87	21.35	7.54	71.11	1.5155	90
15.20	10.10	74.70	1.517	87	10.91	18.02	71.07	1.5375	92
15.28	10.03	74.69	1.5175	86.5	5.46	23.66	70.88	1.549	93
5.46	20.25	74.29	1.5345	88.5	9.11	20.19	70.70	1.544	93.5
24.02	5.32	70.66	1.513	90	23.45	10.07	66.48	1.5265	94
13.78	15.66	70.56	1.5335	91	32.20	1.34	66.46	1.5065	91
15.68	14.04	70.28	1.5315	92	18.69	14.91	66.40	1.537	94.5
14.77	15.00	70.23	1.5335	92.5	17.80	16.04	66.16	1.540	94.5
23.26	6.52	70.22	1.516	91	18.95	15.06	65.99	1.538	—
24.86	4.94	70.20	1.513	91	19.15	15.00	65.85	1.538	—
15.91	14.08	70.01	1.531	92	9.07	25.28	65.65	1.559	97

(continues)

TABLE 16.42 (*Continued*)

Composition (wt. %)					Composition (wt. %)				
Na$_2$O	CaO	SiO$_2$	n_D	$(n_F - n_C) \times 10^4$	Na$_2$O	CaO	SiO$_2$	n_D	$(n_F - n_C) \times 10^4$
24.71	5.31	69.98	1.5135	91.5	16.64	17.83	65.53	1.5425	95.5
25.55	4.61	69.84	1.514	90.5	21.80	13.09	65.11	1.5335	95
17.23	13.00	69.77	1.5285	90.5	10.53	24.38	65.09	1.559	97.5
25.79	5.01	69.20	1.5135	91.5	30.87	4.27	64.86	1.515	94
5.56	25.40	69.04	1.5525	94	32.82	2.40	64.78	1.511	93.5
20.54	10.78	68.68	1.5265	93	32.43	3.02	64.55	1.513	94
20.90	10.84	68.26	1.527	92.5	27.20	8.33	64.47	1.524	96
13.83	18.04	68.13	1.5415	94	14.15	21.47	64.38	1.553	97
24.95	7.00	68.05	1.5185	92.5	8.22	28.51	63.27	1.5695	100
22.10	10.14	67.76	1.5255	—	5.11	32.59	62.30	1.5805	—
19.64	13.03	67.33	1.532	93	9.71	28.18	62.11	1.5705	100.5
3.87	28.84	67.29	1.5645	97	28.60	10.04	61.36	1.5295	97
17.45	15.57	66.98	1.538	94	10.42	28.46	61.12	1.571	100.5
5.98	27.24	66.78	1.563	97	7.44	32.17	60.39	1.5815	102.5
12.44	20.79	66.77	1.549	94.5	16.81	23.04	60.15	1.560	100
19.99	20.00	60.01	1.5535	100	39.91	7.52	52.57	1.531	105
5.01	35.70	59.29	1.591	105	20.80	27.10	52.10	1.5745	108
34.08	6.66	59.26	1.5235	100	31.87	16.08	52.05	1.5505	106
9.97	31.14	58.89	1.580	104.5	10.23	37.98	51.79	1.5995	110.5
26.22	15.00	58.78	1.5425	100	10.61	37.63	51.76	1.5985	111
17.14	24.24	58.62	1.563	102	23.49	24.95	51.56	1.5675	—
28.47	14.31	57.22	1.5425	101	18.00	30.65	51.35	1.584	108
15.79	27.07	57.14	1.571	104	39.63	9.16	51.21	1.538	—
33.91	9.04	57.05	1.531	101	14.68	34.13	51.19	1.5895	109.5
23.06	20.15	56.79	1.555	102.5	9.88	39.03	51.09	1.603	111.5
23.50	21.14	55.36	1.5595	105.5	17.38	31.65	50.97	1.584	109
9.97	34.72	55.31	1.590	108.5	36.37	12.66	50.97	1.5435	107
16.41	28.31	55.28	1.577	106.5	23.99	25.10	50.91	1.5685	109
33.87	10.87	55.26	1.5365	102	37.44	11.68	50.88	1.5405	105.5
6.13	38.75	55.12	1.6025	109	9.60	39.64	50.76	1.604	112
4.56	40.42	55.02	1.6065	110.5	21.40	27.90	50.70	1.576	—
41.94	3.33	54.73	1.5215	101.5	33.47	15.98	50.55	1.550	107
38.32	7.50	54.18	1.529	101.5	29.77	20.18	50.05	1.5605	109
31.33	14.71	53.96	1.5435	102.5	34.42	15.54	50.04	1.5495	108
21.06	24.97	53.97	1.567	104.5	26.81	23.57	49.62	1.566	109
4.44	42.04	53.52	1.6105	111	5.30	45.44	49.26	1.6205	113.5
5.18	41.50	53.32	1.610	109.5	28.42	22.46	49.12	—	109.5
28.44	18.29	53.27	1.5525	107	30.00	21.00	49.00	1.561	—
7.40	44.30	48.30	1.6165	114.5	14.47	40.24	45.29	1.6065	113
16.98	35.54	47.48	1.5945	112.5	4.79	50.06	45.15	1.6305	120
40.76	12.32	46.92	1.5445	109.5	34.91	20.02	45.07	1.562	111.5
7.98	45.16	46.86	1.6165	115	22.60	32.50	44.90	1.5875	114.5
36.50	17.27	46.23	1.555	110	29.00	27.18	43.82	1.5785	114.5
44.77	9.05	46.18	1.539	111	20.88	38.16	40.96	1.599	119
44.56	10.07	45.37	1.5405	111.5	26.86	32.24	40.90	1.587	—
9.27	45.38	45.35	1.619	118	38.91	23.43	37.66	1.5745	—

^{a} Data from Morey and Merwin (1932).

TABLE 16.43

Refractive Index of Na$_2$O–SrO–SiO$_2$ Glasses[a]

Composition (mole %)				Composition (mole %)			
Na$_2$O	SrO	SiO$_2$	n_D	Na$_2$O	SrO	SiO$_2$	n_D
10	10	80	1.512	10	30	60	1.581
15	5	80	1.500	15	25	60	1.566
10	15	75	1.530	17	23	60	1.563
12.5	12.5	75	1.527	20	20	60	1.557
20	5	75	1.511	22	18	60	1.553
10	20	70	1.545	23.4	16.6	60	1.550
15	15	70	1.536	24.3	15.7	60	1.547
20	10	70	1.526	25	15	60	1.544
25	5	70	1.515	10	40	50	1.604
25	10	65	1.534	25	25	50	1.574
5	35	60	1.594	40	10	50	1.534

[a] Data from Dubrovo and Kasymova (1964) and Kasymova (1964a,b).

TABLE 16.44

Refractive Index of Na$_2$O–BaO–SiO$_2$ Glasses[a]

Composition (wt. %)				Composition (wt. %)			
Na$_2$O	BaO	SiO$_2$	n_D	Na$_2$O	BaO	SiO$_2$	n_D
3.4	50.5	46.1	1.597	20.4	22.4	57.2	1.539
6.8	44.9	48.3	1.583	23.8	16.8	59.4	1.529
10.2	39.3	50.5	1.571	27.2	11.2	61.6	1.520
13.6	33.7	52.7	1.560	30.6	5.6	63.8	1.513
17.0	28.0	55.0	1.549				

[a] Data from Greene and Morgan (1941).

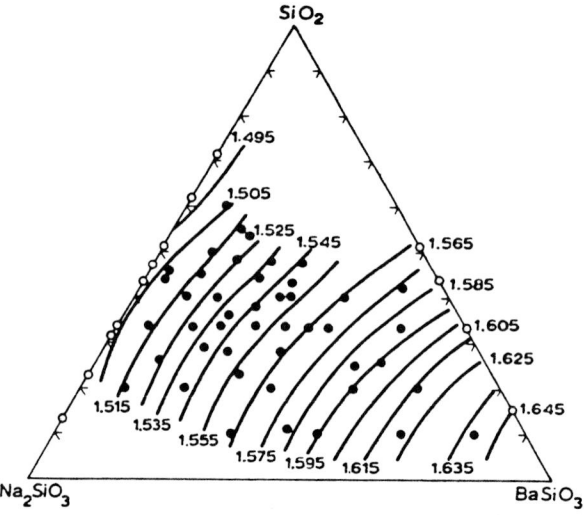

Fig. 16.1. Isofracts showing refractive indices of quenched $Na_2O-BaO-SiO_2$ glasses, measured at 589 nm. Data from Gunawardane and Glasser (1974).

TABLE 16.45

Refractive Index of $Na_2O-ZnO-SiO_2$ Glasses[a]

Composition (mole %)				Composition (mole %)			
Na_2O	ZnO	SiO_2	n_D	Na_2O	ZnO	SiO_2	n_D
10	15	75	1.523	30	5	65	1.516
12.5	12.5	75	1.521	10	30	60	1.584
15	10	75	1.516	15	25	60	1.568
20	5	75	1.508	20	20	60	1.554
25	—	75	1.501	25	15	60	1.542
10	20	70	1.541	30	10	60	1.532
15	15	70	1.535	15	35	50	1.608
20	10	70	1.522	20	30	50	1.589
25	5	70	1.514	25	25	50	1.572
10	25	65	1.563	30	20	50	1.562
15	20	65	1.551	35	15	50	1.551
20	15	65	1.539	40	10	50	1.539
25	10	65	1.527				

[a] Data from Hurt and Phillips (1970).

TABLE 16.46

Optical Constants of $Na_2O-CdO-SiO_2$ Glass[a]

Composition (mole %)				
Na_2O	CdO	SiO_2	n_D	$(n_F - n_C) \times 10^5$
16	20	64	1.5728	1180

[a] Data from Appen (1953, 1954, 1958).

TABLE 16.47
Optical Constants of Na$_2$O–PbO–SiO$_2$ Glasses[a]

| Composition (mole %) | | | | | |
Na$_2$O	PbO	SiO$_2$	n_D	$(n_F - n_C) \times 10^5$	ν
7.5	29	63.5	1.7070	2338	30.2
9.37	37.38	53.25	1.7958	3210	24.79
10.77	33.05	56.18	1.7534	2856	26.4
11	25	64	1.6787	2164	31.4
11.65	24.56	63.79	1.6766	2155	31.4
12.71	30.21	57.08	1.7200	2588	27.84
14.3	28.6	57.1	1.7177	2551	28.13
15.8	19.7	64.5	1.6432	1933	33.3
16	26	58	1.6984	2359	29.60
17.75	17.5	64.75	1.6327	1837	34.5
18.7	16.5	64.8	1.6223	1752	35.6
18.74	24.27	56.99	1.6880	2333	29.5
20.22	15.41	64.37	1.6105	1670	36.7

| Composition (mole %) | | | | | |
Na$_2$O	PbO	SiO$_2$	n_D	$(n_F - n_C) \times 10^5$	ν
20.29	19.83	59.88	1.6543	2038	32.1
11.54	21.86	66.60	1.653		
12.46	22.02	65.52	1.653		
16.68	19.72	63.60	1.642		
17.97	17.66	64.37	1.626		
18.63	22.80	58.57	1.671		
19.47	15.95	64.58	1.617		
19.82	18.85	61.33	1.640		
20.75	13.81	65.44	1.598		
21.31	19.62	59.07	1.650		
21.66	15.19	63.15	1.613		
23.45	19.56	56.99	1.650		

[a] Data from Krakau (1949).

TABLE 16.48

Refractive Index of
$K_2O-BeO-SiO_2$ Glasses[a]

Composition (mole %)			
K_2O	BeO	SiO_2	n_D
15.4	26.9	57.7	1.5163
16	24	60	1.5152
16	28	56	1.5187
16.7	20.8	62.5	1.5146
16.7	25	58.3	1.5173
16.7	29.1	54.2	1.5197
17.4	21.7	60.9	1.5192
17.4	26.1	56.5	1.5184
17.4	30.4	52.2	1.5216
18.2	22.7	59.1	1.5170
18.2	27.3	54.5	1.5201
19	23.8	57.2	1.5183

[a] Data from Lai and Silverman (1928, 1930).

TABLE 16.49

Refractive Indices of $K_2O-MgO-SiO_2$ Glasses[a]

Composition (wt. %)				Composition (wt. %)			
K_2O	MgO	SiO_2	n	K_2O	MgO	SiO_2	n
19.03	8.15	72.82	1.499	12.00	9.00	79.00	1.492
17.60	15.07	67.33	1.512	16.00	12.00	72.00	1.499
16.36	21.01	62.63	1.528	13.00	16.00	71.00	1.507
15.29	26.19	58.52	1.534	2.00	26.00	72.00	1.534
14.35	30.72	54.93	1.547	26.00	2.00	72.00	1.493
				27.18	2.00	70.82	1.496
				31.00	2.00	67.00	1.497
36.99	15.83	47.18	1.527	28.46	15.00	56.54	1.515
31.94	27.34	40.72	1.554	23.27	9.96	66.77	1.500
26.53	22.71	50.76	1.531	4.00	31.00	65.00	1.551
29.93	12.81	57.26	1.510	31.68	16.00	52.32	1.517
18.24	12.00	69.76	1.505	40.12	5.00	54.88	1.508
37.36	8.60	54.04	1.512	32.00	6.00	62.00	1.503
26.57	5.68	67.75	1.501	31.50	4.00	64.50	1.499
25.14	10.76	64.10	1.506	27.50	8.50	64.00	1.500
18.65	10.00	71.35	1.502	25.00	9.50	65.50	1.501
8.00	23.00	69.00	1.524	24.48	7.50	68.02	1.497
33.09	14.16	52.75	1.517	22.50	9.50	68.00	1.498
7.00	18.00	75.00	1.512	23.99	7.50	68.51	1.497
13.00	10.00	77.00	1.495	20.02	10.50	69.48	1.498
40.17	8.60	51.23	1.515	19.50	6.00	74.50	1.493

(*continues*)

TABLE 16.49 (*Continued*)

Composition (wt. %)				Composition (wt. %)			
K$_2$O	MgO	SiO$_2$	n	K$_2$O	MgO	SiO$_2$	n
26.00	14.00	60.00	1.512	17.00	6.50	76.50	1.490
27.87	6.41	65.72	1.503	22.91	7.50	69.59	1.497
35.00	5.00	60.00	1.507	25.49	5.45	69.06	1.498
21.67	9.27	69.06	1.498	19.83	16.97	63.20	1.515
22.69	19.43	57.88	1.522	13.17	11.27	75.56	1.498
10.00	10.00	80.00	1.494	11.00	8.50	80.50	1.490
7.00	15.00	78.00	1.502	15.00	7.00	78.00	1.490
11.00	13.00	76.00	1.503	24.00	2.50	73.50	1.493
19.00	4.00	77.00	1.490	28.61	3.20	68.19	1.497
41.76	2.86	55.38	1.510	22.53	11.46	66.00	1.503
34.00	3.00	63.00	1.506	21.00	12.00	67.00	1.503
29.49	9.61	60.90	1.509	19.72	12.05	68.23	1.502
				5.50	12.50	82.00	1.496
7.00	11.00	82.00	1.494	9.00	8.50	82.50	1.488
14.38	12.30	73.32	1.504	16.00	6.50	77.50	1.490
16.00	8.00	76.00	1.493	18.50	5.50	76.00	1.490
11.00	5.00	84.00	1.483	24.50	3.50	72.00	1.496
7.00	9.00	84.00	1.487	25.71	3.50	70.79	1.496
4.00	17.00	79.00	1.505	27.00	3.50	69.50	1.497
15.00	5.00	80.00	1.485	17.50	9.00	73.50	1.497
22.00	4.00	74.00	1.495	16.50	11.00	72.50	1.498
21.00	8.00	71.00	1.497	17.00	11.50	71.50	1.498
24.00	5.00	71.00	1.497	11.55	14.83	73.62	1.503
10.00	7.00	83.00	1.485	37.78	5.64	56.78	1.508
13.00	7.00	80.00	1.487	34.14	8.97	56.89	1.508
26.00	7.50	66.50	1.500	30.00	2.00	68.00	1.497
24.00	13.50	62.50	1.508	30.00	4.00	66.00	1.499
23.00	13.00	64.00	1.506	39.06	11.14	49.80	1.516
19.84	13.31	66.85	1.505	41.54	5.50	52.96	1.510
20.45	11.00	68.55	1.500	43.03	13.00	43.97	1.522
9.27	19.83	70.90	1.517	46.08	9.86	44.06	1.522
22.08	9.45	68.47	1.498				
22.90	9.81	67.29	1.499	48.41	20.72	30.87	1.547
23.73	10.16	66.11	1.500	54.01	11.56	34.43	1.527
24.56	10.51	64.93	1.501				
26.21	11.22	62.57	1.504	37.80	14.00	48.20	1.519
30.00	5.81	64.19	1.501	41.10	6.50	52.40	1.512
23.50	12.00	64.50	1.504	36.39	11.63	51.98	1.514
14.50	8.00	77.50	1.490	35.00	14.00	51.00	1.517
21.50	5.00	73.50	1.492	46.78	13.12	40.10	1.525
13.00	8.00	79.00	1.490				
16.28	6.97	76.75	1.490				
22.00	3.00	75.00	1.491	44.81	4.00	51.19	1.515
25.00	1.00	74.00	1.491	40.00	20.00	40.00	1.531
20.77	11.25	67.98	1.501	44.07	16.95	38.98	1.528
23.75	11.25	65.00	1.504	11.50	16.00	72.50	1.509

[a] Data from Roedder (1951).

TABLE 16.50

Refractive Index of $K_2O-CaO-SiO_2$ Glasses[a]

Composition (wt. %)				Composition (wt. %)			
K_2O	CaO	SiO_2	n_D	K_2O	CaO	SiO_2	n_D
5	—	95	1.463	34.1	—	65.9	1.502
10	—	90	1.468	3.9	30.4	65.7	1.567
5.1	9.8	85.1	1.495	34.6	—	65.4	1.501
17.5	—	82.5	1.482	10.5	24.4	65.1	1.551
11.6	9.1	79.3	1.503	12.7	22.6	64.7	1.547
15.0	5.0	80.0	1.493	16.6	19.8	63.6	1.541
13.7	10.5	75.8	1.508	28.3	8.4	63.3	1.518
19.8	5.0	75.2	1.500	—	37.4	62.6	1.590
24.8	—	75.2	1.491	22.1	15.4	62.5	1.533
26.3	—	73.7	1.493	32.6	4.9	62.5	1.513
5.4	22.4	72.2	1.510	8.6	31.7	59.7	1.574
12.6	15.0	72.4	1.521	24.7	15.4	59.9	1.534
17.3	10.2	72.5	1.510	25.4	14.1	60.5	1.532
10.5	16.9	72.6	1.526	32.2	7.2	60.6	1.518
27.3	1.0	71.7	1.495	31.4	9.3	59.3	1.523
28.3	—	71.7	1.495	41.0	—	59.0	1.506
29.1	—	70.9	1.495	41.1	—	58.9	1.506
18.3	11.0	70.7	1.516	27.7	14.6	57.7	1.533
22.4	8.1	69.5	1.511	34.3	7.2	58.5	1.519
24.3	7.6	68.1	1.512	20.0	22.3	57.7	1.511
32.1	—	67.9	1.498	—	42.9	57.1	1.612
38.3	4.7	57.0	1.518	27.7	24.2	48.1	1.565
43.3	—	56.7	1.508	25.1	26.5	48.4	1.569
5.3	39.9	54.8	1.597	46.0	5.8	48.2	1.527
15.2	29.7	55.1	1.570	24.3	29.0	46.7	1.575
28.3	17.0	54.7	1.544	34.5	20.7	44.8	1.560
—	48.3	51.7	1.629	—	55.8	44.2	1.645
24.6	24.9	50.5	1.585	32.8	23.5	43.7	1.564
34.0	15.6	50.4	1.547	17.9	39.6	42.5	1.603
9.6	40.6	49.8	1.601	16.5	41.4	42.1	1.603
30.6	19.7	49.7	1.555	47.9	10.6	41.5	1.544
—	51.2	48.8	1.635				

[a] Data from Morey (1954).

TABLE 16.51
Optical Constants of K_2O–SrO–SiO_2 Glasses[a]

Composition (mole %)						Composition (mole %)					
K_2O	SrO	SiO_2	n_D	$(n_F - n_C) \times 10^5$	v_D	K_2O	SrO	SiO_2	n_D	$(n_F - n_C) \times 10^5$	v_D
14.9	—	85.1	1.4870	784	62.1	20.5	14.4	65.1	1.5360[b]	—	—
10.9	27.1	62.0	1.5642	988	57.1	21.4	10.6	68.0	1.5229	892	58.6
11.6	22.1	66.3	1.5461	935	58.4	22.3	6.9	70.8	1.5164[b]	—	—
12.6	15.4	72.0	1.5255	877	59.9	23.1	3.4	73.5	1.5066[b]	—	—
13.5	9.6	76.9	1.5100	841	60.7	29.1	—	70.9	1.5060	—	—
14.2	4.5	81.3	1.5001	816	61.3	5.4	10.1	84.5	1.5058	805	62.8
19.1	—	80.9	1.4940	—	—	6.9	8.3	84.8	1.5030	805	62.5
13.4	30.0	56.6	1.5755	1039	55.4	7.6	14.2	78.2	1.5208	854	61.0
14.0	26.4	59.6	1.5658	1008	56.1	8.7	6.1	85.2	1.4999	804	62.3
15.5	18.7	65.8	1.5413	938	57.7	9.9	18.7	71.4	1.5384	909	59.2
16.8	11.8	71.4	1.5231	888	58.9	10.8	7.4	81.8	1.5060	824	61.4
18.0	5.6	76.4	1.5088	850	59.9	11.6	3.6	84.8	1.4950	797	62.1
24.0	—	76.0	1.4980[b]	—	—	14.7	4.6	80.7	1.5027	823	61.1
16.8	30.0	53.2	1.5765	—	—	21.3	26.4	52.3	1.5714	—	—
17.6	26.6	55.8	1.5678	1027	55.3	24.0	17.2	58.8	1.5458	978	55.8
18.6	22.4	59.0	1.5560	991	56.1	26.5	8.5	65.0	1.5250	916	57.3
19.6	18.3	62.1	1.5446	956	57.0	34.7	10.8	54.5	1.5380	—	—

[a] Data from Shchavelev et al. (1974).

[b] Measured by Olireimov's method; rest by Pulfrich method.

TABLE 16.52

Refractive Indices of $K_2O-ZnO-SiO_2$ Glasses[a]

Composition (mole %)				Composition (mole %)			
K_2O	ZnO	SiO_2	n_D	K_2O	ZnO	SiO_2	n_D
10	5	85	1.485	20	15	65	1.528
10	10	80	1.498	20	20	60	1.539
10	15	75	1.513	20	25	55	1.551
10	20	70	1.527	20	30	50	1.563
10	25	65	1.542	20	35	45	1.577
10	30	60	1.555	25	5	70	1.512
10	35	55	1.569	25	10	65	1.524
10	40	50	1.585	25	15	60	1.534
10	45	45	1.598	25	20	55	1.544
15	5	80	1.495	25	25	50	1.556
15	10	75	1.508	25	30	45	1.567
15	15	70	1.520	30	5	65	1.518
15	20	65	1.534	30	10	60	1.528
15	25	60	1.546	30	15	55	1.540
15	30	55	1.559	30	20	50	1.550
15	35	50	1.573	30	25	45	1.561
15	40	45	1.589	35	5	60	1.522
20	5	75	1.504	35	10	55	1.532
20	10	70	1.517	35	15	50	1.544
				35	20	45	1.554

[a] Data from graph in Vargin *et al.* (1972).

TABLE 16.53

Refractive Index of $Rb_2O-CaO-SiO_2$ Glasses[a]

Composition (mole %)				Composition (mole %)			
Rb_2O	CaO	SiO_2	n_D	Rb_2O	CaO	SiO_2	n_D
3.3	44.6	52.1	1.610	11.7	39.3	49.0	1.595
3.3	55.3	41.4	1.610	11.8	26.4	61.8	1.560
3.4	22.6	74.0	1.555	12	13.4	74.6	1.535
3.4	33.6	63.0	1.585	17.2	42.8	40.0	1.600
7.3	24.4	68.3	1.560	17.3	28.8	53.9	1.575
7.3	36.2	56.5	1.595	17.4	14.6	68.0	1.545
7.3	48	44.7	1.615	23.8	31.8	44.4	1.580
7.4	12.3	80.3	1.520	24	16	60	1.550
11.6	52	36.4	1.615	32.1	17.9	50.0	1.570

[a] Data from Simpson (1959, 1961).

TABLE 16.54

Refractive Index of $Rb_2O-BaO-SiO_2$ Glasses[a]

Composition (mole %)				Composition (mole %)			
Rb_2O	BaO	SiO_2	n_D	Rb_2O	BaO	SiO_2	n_D
4.6	22.8	72.6	1.575	15.6	19.2	65.2	1.590
5.1	31.2	63.7	1.600	17.4	28.3	54.3	1.600
8.6	10.6	80.8	1.535	19.3	5.8	74.9	1.530
9.4	17.3	73.3	1.550	21.2	12.9	65.9	1.590
11.5	35	53.5	1.605	23.6	21.5	54.9	1.595
14.2	11.7	74.1	1.540	26.7	6.5	66.8	1.570

[a] Data from Simpson (1959, 1961).

TABLE 16.55

Refractive Index of $Rb_2O-PbO-SiO_2$ Glasses[a]

Composition (mole %)				Composition (mole %)			
Rb_2O	PbO	SiO_2	n_D	Rb_2O	PbO	SiO_2	n_D
5.6	23.7	70.7	1.710	15.1	37.9	47.0	1.720
6.5	32.7	60.8	1.710	16.6	14	69.4	1.595
10	12.5	77.5	1.575	19.1	21.3	59.6	1.650
11.2	18.8	70.0	1.660	19.6	4.1	76.3	1.520
12.8	27	60.2	1.635	22	9.3	68.7	1.590
14.8	8.2	77.0	1.545	22.3	31.3	46.4	1.680

[a] Data from Simpson (1959, 1961).

TABLE 16.56

Refractive Index of $Cs_2O-CaO-SiO_2$ Glasses[a]

Composition (mole %)				Composition (mole %)			
Cs_2O	CaO	SiO_2	n_D	Cs_2O	CaO	SiO_2	n_D
10	15	75	1.531	15	7.5	77.5	1.524
10	12.5	77.5	1.524	15	5	80	1.518
10	10	80	1.518	17.5	12.5	70	1.540
12.5	15	72.5	1.540	17.5	10	72.5	1.537
12.5	12.5	75	1.531	17.5	7.5	75.0	1.532
12.5	10	77.5	1.524	17.5	5	77.5	1.524
12.5	7.5	80.0	1.518	20	10	70	1.542
15	15	70	1.542	20	7.5	72.5	1.537
15	12.5	72.5	1.537	20	5	75	1.530
15	10	75	1.533				

[a] Data from Bezborodov and Bobkova (1957) and Bobkova (1958).

TABLE 16.57

Refractive Index of Lithium Borosilicate Glasses[a]

Composition (mole %)				Composition (mole %)			
Li_2O	B_2O_3	SiO_2	n_D	Li_2O	B_2O_3	SiO_2	n_D
11.3	11.3	77.4	1.493	13.3	33.3	53.4	1.513
12.5	12.5	75	1.493	14	30	56	1.508
13	13	74	1.503	14.3	28.6	57.1	1.513
15	15	70	1.501	15	25	60	1.512
16.7	16.7	66.6	1.510	16.5	17.5	66	1.507
17.2	17.2	65.6	1.514	17	15	68	1.507
24	24	52	1.537	17.1	14.3	68.6	1.508
27.5	27.5	45	1.558	18.2	9.1	72.7	1.505
30	30	40	1.558	18.7	6.3	75.0	1.505
33.3	33.3	33.4	1.565	19	5	76	1.505
35	35	30	1.569	19.4	3.1	77.5	1.504
40	40	20	1.570	23.3	30	46.7	1.544
45	45	10	1.569	24.2	27.5	48.3	1.546
33.3	—	66.7	1.536	25.5	23.5	51.0	1.547
10	33.3	56.7	1.497	26.7	20	53.3	1.547
10	50	40	1.505				
10.4	30	59.6	1.501	27.3	18.2	54.5	1.546
10.5	29.4	60.1	1.502	28	16	56	1.545
11.2	25	63.8	1.500	28.3	15	56.7	1.548
11.4	23.5	65.1	1.494	30	10	60	1.546
12	20	68	1.493	31.7	5	63.3	1.542
12	40	48	1.505				
12.3	17.7	70	1.494				

[a] Data from Akimov (1973).

TABLE 16.58

Refractive Index of Lithium Aluminosilicate Glasses[a]

Composition (mole %)				Composition (mole %)			
Li_2O	Al_2O_3	SiO_2	n_D	Li_2O	Al_2O_3	SiO_2	n_D
12.5	12.5	75.0	1.509	25	20	55	1.538
16.7	16.7	66.6	1.516	25	25	50	1.543
20	10	70	1.516	28.6	14.3	57.1	1.540
20	15	65	1.519	30.3	9.1	60.6	1.538
20	20	60	1.525	31.7	4.8	63.5	1.536
25	15	60	1.535	32.8	1.6	65.6	1.533

[a] Data from Dubrovo and Shmidt (1959).

TABLE 16.59

Refractive Index of Lithium Gallosilicate Glasses[a]

Composition (mole %)				Composition (mole %)			
Li_2O	Ga_2O_3	SiO_2	n_D	Li_2O	Ga_2O_3	SiO_2	n_D
12	22	66	1.578	20	25	55	1.603
12.5	12.5	75	1.537	20	27.5	52	1.614
12.5	17.5	70	1.550	25	5	70	1.539
13.3	21.7	65	1.574	25	10.7	64.3	1.562
16	12	72	1.542	25	15	60	1.575
16.7	16.7	66.6	1.562	25	25	50	1.611
16	21	63	1.581	27.5	10.4	62.1	1.563
16	28	56	1.613	27.5	18.1	54.4	1.595
20	2.5	77.5	1.510	27.5	27.5	45	1.623
20	5	75	1.525	30	5	65	1.551
20	10	70	1.546	30	10	60	1.570
20	12.5	67.5	1.556	30	14	56	1.584
20	15	65	1.564	30	20	50	1.604
20	20	60	1.582	35	5	60	1.561
20	22.5	57.5	1.591				

[a] Data from Tsekhomskaya (1966) and Dubrovo and Tsekhomskaya (1964a,b).

TABLE 16.60

Refractive Index and Optical Dispersion of Sodium Borosilicate Glasses[a]

Composition (mole %)[b]					Composition (mole %)[b]				
Na_2O	B_2O_3	SiO_2	n_D	$(n_F - n_C) \times 10^5$	Na_2O	B_2O_3	SiO_2	n_D	$(n_F - n_C) \times 10^5$
5	15	80	1.4726	707	12.3	10.2	77.5	1.5055	775
5	20	75	1.4711	716	12.3	17.7	70	1.5085	769
5	21.1	73.9	1.4703	718	12.5	12.5	75	1.5087	774
5	25	70	1.4695	722	12.5	25	62.5	1.5040	765
5	28.3	66.7	1.4699	727	12.9	19.4	67.7	1.5105	774
5	30	65	1.4705	729	13	13	74	1.5100	777
7.5	15	77.5	1.4850	728	13.3	20	66.7	1.5122	777
8.3	16.7	75	1.4879	734	13.3	33.3	53.4	1.5014	770
8.8	8.7	82.5	1.4933	746	14	7	79	1.5012	784
10	5	85	1.4873	753	14	21	65	1.5139	781
10	6.7	83.3	1.4929	753	14	30	56	1.5093	779
10	8.3	81.7	1.4962	755	14.5	3.6	81.9	1.4927	789
10	10	80	1.4983	753	15	5	80	1.4975	788
10	12.5	77.5	1.4989	753	15	7.5	77.5	1.5048	790
10	15	75	1.4980	752	15	10	75	1.5092	791
10	20	70	1.4942	750	15	12.5	72.5	1.5136	790
10	23.3	66.7	1.4917	747	15	15	70	1.5162	789
10	25	65	1.4904	746	15	18.3	66.7	1.5170	787
10	27.5	62.5	1.4890	746	15	18.7	66.3	1.5171	788
10	30	60	1.4875	745	15	20	65	1.5170	791
10	33.3	56.7	1.4854	748	15	25	60	1.5156	788
10	50	40	1.4843	759	15	27.5	57.5	1.5141	787
11.3	11.2	77.5	1.5036	764	15	30	55	1.5128	784
12	8	80	1.5002	767	15	40	45	1.5052	783
12	15	73	1.5079	766	15.4	23.1	61.5	1.5176	790

12	20	68	1.5047	766	16	20	64	1.5197	795
16.7	33.3	50	1.5165	793	16.4	13.6	70	1.5171	803
17	15	68	1.5188	803	16.7	8.3	75	1.5087	809
17.1	14.3	68.6	1.5185	806	16.7	16.6	66.7	1.5199	802
17.5	12.5	70	1.5165	808	22.2	33.3	44.5	1.5279	819
17.5	17.5	65	1.5216	807	22.5	10	67.5	1.5167	844
17.5	30	52.5	1.5208	794	22.5	15	62.5	1.5224	836
17.8	22.2	60	1.5230	804	22.5	20	57.5	1.5264	830
18	10	72	1.5129	814	22.5	22.5	55	1.5273	824
18	12	70	1.5165	812	22.9	25	52.5	1.5280	825
18.3	20	62	1.5231	808	23	17.1	60	1.5244	835
19	26.7	55	1.5235	806	23.3	23	54	1.5276	828
19	5	76	1.5037	821	24	11.7	65	1.5190	840
19	19	62	1.5239	815	24.2	24	52	1.5284	833
19.4	23.8	57.2	1.5250	807	25	27.5	48.3	1.5291	830
20	3.1	77.5	1.4993	826	25	5	70	1.5093	859
20	5	75	1.5052	826	25	8.3	66.7	1.5151	855
20	10	70	1.5154	828	25	10	65	1.5173	853
20	13.3	66.7	1.5199	824	25	12.5	62.5	1.5203	850
20	15	65	1.5217	820	25	15	60	1.5227	850
20	16.7	63.3	1.5233	822	25	16.7	58.3	1.5250	844
20	17.8	62.2	1.5239	820	25	20	55	1.5267	842
20	20	60	1.5249	818	25	20.8	54.2	1.5268	841
20	25	55	1.5263	812	25	25	50	1.5286	840
20.2	30	50	1.5254	809	25	27.5	47.5	1.5289	832
21	40	40	1.5214	806	25	30	45	1.5292	829
21.7	19	60.8	1.5250	819	25	33.3	41.7	1.5296	830
22	21	58	1.5262	822	25	37.5	37.5	1.5288	825
22.2	13.3	65	1.5205	831	26.7	13.3	60	1.5211	859
	22	56	1.5269	825	27.5	25	47.5	1.5277	850
	11.1	66.7	1.5179	835	27.5	27.5	45	1.5285	846

(continues)

TABLE 16.60 (Continued)

Composition (mole %)[b]

Na$_2$O	B$_2$O$_3$	SiO$_2$	n_D	$(n_F - n_C) \times 10^5$
28	16	56	1.5231	866
28.3	5	66.7	1.5103	880
30	5	65	1.5117	888
30	10	60	1.5172	880
30	15	55	1.5222	875
30	15.5	54.5	1.5224	872
30	20	50	1.5255	864
30	25	45	1.5279	855
30	27.5	42.5	1.5274	859

Composition (mole %)[b]

Na$_2$O	B$_2$O$_3$	SiO$_2$	n_D	$(n_F - n_C) \times 10^5$
27.5	30	42.5	1.5291	843
30	30	40	1.5281	855
30	33.3	36.7	1.5282	853
30	37.5	32.5	1.5281	843
31.7	5	63.3	1.5119	899
33.3	33.3	33.4	1.5257	863
36	30	34	1.5233	885
37.5	25	37.5	1.5211	902
40	13.3	46.7[c]	1.5163	944
40	20	40[c]	1.5175	925

[a] Data from Akimov (1960a,b).
[b] Chemical analysis of 38 glasses revealed the presence of certain impurities: hundredths of 1% of MgO, CaO, Fe$_2$O$_3$ and As$_2$O$_3$ and about 0.15% Al$_2$O$_3$. On an average, the glasses were found to contain, 0.8% more SiO$_2$, 0.35% less Na$_2$O, and 0.45% less B$_2$O$_3$ than the batch composition.
[c] Glasses not subjected to fine annealing.

TABLE 16.61

Coefficients of the Sellmeier Equation $n^2 = 1 + \sum\limits_{i=1}^{3} A_0\lambda^2/(\lambda^2 - l_i^2)$ for Refractive Index (n) of

$Na_2O-B_2O_3-SiO_2$ Glass at $23.5 \pm 0.5°C$ in the Wavelength (λ) Range 0.8 to 1.5 μm [a]

Composition (mole %)			Coefficients					
Na_2O	B_2O_3	SiO_2	A_1	l_1	A_2	l_2	A_3	l_3
16.9	32.5	50.6	0.796468	0.094359	0.497614	0.093386	0.358924	5.999652

[a] Data from Fleming (1978).

TABLE 16.62

Refractive Index of Sodium Aluminosilicate Glasses[a]

Composition (wt. %)				Composition (wt. %)			
Na_2O	Al_2O_3	SiO_2	n_D	Na_2O	Al_2O_3	SiO_2	n_D
21.36	0.93	77.71	1.4926	31.67	2.41	65.92	1.5047
21.54	2.85	75.61	1.4949	31.23	2.87	65.90	1.5052
19.70	4.82	75.48	1.4935	24.69	9.43	65.88	1.5025
19.67	5.12	75.21	1.4934	29.55	4.75	65.70	1.5049
23.02	2.00	74.98	1.4952	24.94	9.96	65.10	1.5032
23.98	1.07	74.95	1.4960	34.20	1.02	64.78	1.5059
22.12	3.00	74.88	1.4955	32.46	6.57	60.97	1.5073
24.27	0.98	74.75	1.4970	29.53	9.59	60.88	1.5070
21.50	4.69	73.81	1.4957	38.13	1.09	60.78	1.5094
25.87	0.98	73.15	1.4984	36.42	2.82	60.76	1.5092
19.60	9.32	71.08	1.4970	34.46	4.86	60.68	1.5088
22.63	6.49	70.88	1.4988	29.71	9.84	60.45	1.5072
26.31	2.85	70.84	1.5007	39.23	4.83	55.94	1.5124
24.45	4.92	70.63	1.5000	41.32	2.88	55.80	1.5130
24.84	4.86	70.30	1.5003	34.66	9.68	55.66	1.5110
27.86	2.03	70.11	1.5015	37.50	6.94	55.56	1.5120
29.04	0.97	69.99	1.5020	34.92	9.83	55.25	1.5116
25.45	5.77	68.78	1.5006	44.17	4.88	50.95	1.5157
25.15	6.75	68.10	1.5018	39.54	9.57	50.89	1.5150
29.30	2.95	67.75	1.5030	42.43	6.71	50.86	1.5155
28.34	4.45	67.21	1.5026	42.43	7.00	50.57	1.5154
31.61	2.03	66.36	1.5050	46.82	2.86	50.32	1.5162

[a] Data from Faick et al. (1935).

TABLE 16.63

Refractive Index of $Na_2O-Al_2O_3-SiO_2$ Glasses[a]

Composition (mole %)				Composition (mole %)			
Na_2O	Al_2O_3	SiO_2	$n_{6328.2 Å}$	Na_2O	Al_2O_3	SiO_2	$n_{6328.2 Å}$
25	—	75	1.49540	33	15	52	1.15229
25	5	70	1.49949	40	—	60	1.50850
25	10	65	1.50151	40	5	55	1.51139
33	—	67	1.50285	40	10	50	1.51456
33	5	62	1.50627	40	15	45	1.51791
33	10	57	1.50921	40	20	40	1.52139

[a] Data from Schroeder (1980).

TABLE 16.64

Refractive Index of Sodium Gallosilicate Glasses[a]

Composition (mole %)				Composition (mole %)			
Na_2O	Ga_2O_3	SiO_2	n_D	Na_2O	Ga_2O_3	SiO_2	n_D
12.5	12.5	75.0	1.519	20	20	60	1.556
12.5	17.5	70.0	1.538	20	28	52	1.581
13.3	21.7	65.0	1.553	21	9	70	1.529
15	5	80	1.503	21	11.3	67.7	1.531
15	10	75	1.515	24.2	10.8	65	1.531
15	15	70	1.527	25	15	60	1.545
15	20	65	1.542	25	25	50	1.574
16	21	63	1.55	28.6	14.3	57.1	1.545
16.66	16.66	66.68	1.537	30	5	65	1.519
16.66	12.5	70.84	1.530	30	10	60	1.533
16.66	8.33	75.01	1.514	30.3	9.1	60.6	1.535
18.3	11.7	70.0	1.524	31.7	4.8	63.5	1.52
20	10	70	1.527				

[a] Data from Dubrovo (1960), Danilova and Dubrovo (1965), Dubrovo et al. (1965), and Takahashi et al. (1975).

TABLE 16.65

Refractive Index of Some Na$_2$O–M$_2$O$_3$–SiO$_2$ Glasses[a]

Composition (mole %)			
Na$_2$O	M$_2$O$_3$	SiO$_2$	n_D
	M = Sc		
21	9	70	1.578
30	5	65	1.546
30	10	60	1.580
	M = V		
21	9	70	1.573
30	5	65	1.543
30	10	60	1.581
	M = In		
21	9	70	1.594
30	5	65	1.549
30	10	60	1.597
	M = Sb		
21	9	70	1.600
30	5	65	1.555
30	10	60	1.587

[a] Data from Takahashi *et al.* (1975).

TABLE 16.66

Refractive Index of Na$_2$O–La$_2$O$_3$–SiO$_2$ Glasses[a]

Composition (mole %)				Composition (mole %)			
Na$_2$O	La$_2$O$_3$	SiO$_2$	n_D	Na$_2$O	La$_2$O$_3$	SiO$_2$	n_D
6	7	87	1.550	20	14	66	1.640
12	1	87	1.491	21	5	74	1.545
12	5	83	1.538	21	9	70	1.598
12	12	76	1.613	25	5	70	1.550
12	16	72	1.657	27	13	60	1.622
15	10	75	1.598	30	5	65	1.556
15	15	70	1.648	30	10	60	1.604
16	11	73	1.598	35	5	60	1.559
20	5	75	1.547	39	1	60	1.526

[a] Data from Shnypikov (1967), Dubrovo and Shnypikov (1966), and Takahashi *et al.* (1975).

TABLE 16.67

Optical Constants of $Na_2O-Sb_2O_3-SiO_2$ Glasses with Na_2O-SiO_2 Ratio of 0.25[a]

Sb_2O_3 (wt. %)	n_D	$(n_F - n_C) \times 10^5$	ν_D
0	1.49090	828.0	59.29
3	1.50116	887.8	56.45
6	1.50778	926.3	54.82
9	1.51662	979.2	52.76
12	1.52530	1030.2	50.99
15	1.53478	1090.4	49.04
18	1.54409	1146.1	47.48
21	1.55397	1209.2	45.82
24	1.56306	1260.4	44.67
27	1.57382	1330.9	43.11
30	1.58459	1402.7	41.67
50	1.64524	1871.9	34.47

[a] Data from Polukhin (1960).

TABLE 16.68

Refractive Index of $Na_2O-Bi_2O_3-SiO_2$ Glasses

Composition (wt. %)				Composition (wt. %)			
Na_2O	Bi_2O_3	SiO_2	$n_D{}^a$	Na_2O	Bi_2O_3	SiO_2	$n_D{}^b$
13.55	52.87	33.58	1.712	5	60	35	>1.741
14.24	44.92	40.84	1.6668	10	50	40	1.6873
14.48	45.75	39.77	1.670	10	60	30	>1.741
14.99	34.51	50.50	1.6109	10	65	25	>1.741
18.23	34.34	47.43	1.624	15	40	45	1.6398
19.18	29.66	51.16	1.5948	20	30	50	1.6023
20.83	29.52	49.65	1.600	20	50	30	1.6897
21.78	22.00	56.22	1.5697	25	28.6	46.4	1.5651
22.73	20.67	56.60	1.565	25	40	35	1.6242
23.78	12.40	63.82	1.5352	30	30	40	1.6142
24.30	12.71	62.99	1.535	32	9	59	1.5307

[a] Data from Riegel and Sharp (1934).
[b] Data from Watanabe et al. (1970).

TABLE 16.69

Refractive Index of $Na_2O-Fe_2O_3-SiO_2$ Glasses[a]

Composition (wt. %)				Composition (wt. %)			
Na_2O	Fe_2O_3	SiO_2	n	Na_2O	Fe_2O_3	SiO_2	n
20.51	—	79.49	1.491	18.1	11.75	70.15	1.540
45.71	—	54.29	1.513	17.95	12.5	69.55	1.544
25.28	1.25	73.47	1.503	24	12.5	63.5	1.552
25.09	2	72.91	1.507	29.77	12.5	57.73	1.558
22.78	2.5	74.72	1.504	16.51	14	69.49	1.545
24.83	3	72.17	1.510	12.48	15	72.52	1.543
24.45	4.5	71.05	1.517	17.44	15	67.56	1.553
19.49	5	80.01	1.512	23.31	15	61.69	1.564
26.06	5	68.94	1.519	28.92	15	56.08	1.569
30.88	5	64.12	1.525	31.52	15	53.48	1.574
36.23	5	58.77	1.529	32.62	16.81	50.57	1.583
24.06	6	69.94	1.523	28.25	17	54.75	1.579
21.61	7.5	70.89	1.527	7.62	20	72.38	1.548
23.68	7.5	68.82	1.529	11.74	20	68.26	1.563
25.23	8	66.77	1.532	18.69	20	61.31	1.581
42.06	8	49.94	1.548	26	20	54	1.591
13.21	10	76.79	1.522	32.51	20	47.49	1.602
21.03	10	68.97	1.538	40.63	20	39.37	1.612
24.68	10	65.32	1.541	9.9	22.5	67.6	1.566
30.62	10	59.38	1.548	26.38	22.5	51.12	1.606
36.57	10	53.43	1.554	31.49	22.5	46.01	1.613
45.71	10	44.29	1.562	11.19	23.75	65.06	1.576
38.21	11.5	50.29	1.562	11.01	25	63.99	1.584
15.38	25	59.62	1.595	17.28	32.5	50.22	1.641
25.53	25	49.47	1.617	20.57	32.5	46.93	1.651
30.48	25	44.52	1.626	25.03	32.5	42.47	1.661
29.46	27.5	43.04	1.639	16.64	35	48.36	1.653
14.67	28.5	56.83	1.611	23.11	35	41.89	1.677
14.51	29.25	56.24	1.614	26.41	35	38.59	1.682
14.36	30	55.64	1.617	33.01	35	31.99	1.701
21.33	30	48.67	1.636	19.05	37.5	43.45	1.679
25.95	30	44.05	1.647	31.74	37.5	30.76	1.716
32	30	38	1.657	20.43	40	39.57	1.698
35.55	30	34.45	1.667	27.43	40	32.57	1.719
24.18	32	43.82	1.659	30.48	40	29.52	1.732

[a] Data from Bowen *et al.* (1930).

TABLE 16.70

Refractive Index of Potassium Borosilicate Glasses

Composition (mole %)				Composition (mole %)			
K_2O	B_2O_3	SiO_2	$n_D{}^a$	K_2O	B_2O_3	SiO_2	$n_D{}^b$
1	34	65	1.4580	4	28	68	1.465
5	30	65	1.4729	8	24	68	1.481
20	15	65	1.5099	16	4	80	1.4980
3	82	15	1.4647	16	8	76	1.5065
10	75	15	1.4829	16	12	72	1.5125
14.5	70.5	15	1.4860	16	16	68	1.5153
20	65	15	1.4931	16	20	64	1.5154
30	55	15	1.5050	16	24	60	1.5140
				16	32	52	1.5086
				24	8	68	1.5097
				28	4	68	1.5076

[a] Data from Konijnendijk (1975).
[b] Data from Appen and Gan'Fu-Si (1959a,b,c).

TABLE 16.71

Refractive Index of Potassium Aluminosilicate Glasses[a]

Composition (mole %)				Composition (mole %)			
K_2O	Al_2O_3	SiO_2	$n_{6328.2Å}$	K_2O	Al_2O_3	SiO_2	$n_{6328.2Å}$
25	—	75	1.49752	33	5	62	1.50739
25	5	70	1.49971	33	10	57	1.50991
25	10	65	1.50251	33	15	52	1.5161
25	15	60	1.50445	40	10	50	1.51777
33	—	67	1.50498				

[a] Data from Schroeder (1980).

TABLE 16.72

Refractive Index of $K_2O-Bi_2O_3-SiO_2$ Glasses[a]

Composition (wt. %)			
K_2O	Bi_2O_3	SiO_2	n_D
20.15	47.72	32.13	1.6915
22.87	40.66	36.47	1.6420
26.46	31.35	42.19	1.6110
28.71	25.50	45.79	1.5470
31.38	18.58	50.04	1.5295

[a] Data from Riegel and Sharp (1934).

TABLE 16.73

Refractive Index of K$_2$O–Fe$_2$O$_3$–SiO$_2$ Glasses[a]

Composition (wt. %)				Composition (wt. %)			
K$_2$O	Fe$_2$O$_3$	SiO$_2$	n_D	K$_2$O	Fe$_2$O$_3$	SiO$_2$	n_D
27.04	4	68.96	1.510	18.65	10	71.35	1.524
23.10	18	58.90	1.564	17.61	15	67.39	1.543
21.69	23	55.31	1.585	17.09	17.5	65.41	1.552
21.13	25	53.87	1.599	16.58	20	63.42	1.562
19.06	32.32	48.62	1.621	16.47	20.5	63.03	1.562
23.16	3	75.84	1.502	16.37	21	62.63	1.563
22.45	6	71.55	1.515	15.95	23	61.05	1.571
21.49	10	68.51	1.528	15.54	25	59.46	1.579
19.10	20	60.90	1.566	15.33	26	58.67	1.582
18.39	23	58.61	1.578	16.11	12	71.89	1.528
20.31	2	77.69	1.493	15.56	15	69.44	1.541
19.48	6	74.52	1.509	15.01	18	66.99	1.551
19.17	7.5	73.33	1.513	14.64	20	65.36	1.559

[a] Data from Faust and Peck (1938).

TABLE 16.74

Refractive Index of Rubidium Borosilicate Glasses[a]

Composition (mole %)				Composition (mole %)			
Rb$_2$O	B$_2$O$_3$	SiO$_2$	n_D	Rb$_2$O	B$_2$O$_3$	SiO$_2$	n_D
12.4	10.8	76.8	1.505	13	46.3	40.7	1.500
7.9	20.7	71.4	1.495	8.3	53.8	37.9	1.480
17.8	11.9	70.3	1.515	3.5	61	35.5	1.470
12.5	22.7	64.8	1.515	26.2	41.8	32.0	1.490
8	31.4	60.6	1.480	18.9	51.4	29.7	1.485
25	13	62	1.515	13.3	59.2	27.5	1.495
17.9	24.8	57.3	1.520	8.5	66.1	25.4	1.485
12.7	34.1	53.2	1.495	3.6	72.5	23.9	1.470
8.1	42.2	49.7	1.470	26.7	56.5	16.8	1.480
25.5	27.4	47.1	1.515	19.3	65.1	15.6	1.480
18.4	37.7	43.9	1.490				

[a] Data from Simpson (1961).

TABLE 16.75

Refractive Index of Rb$_2$O–Bi$_2$O$_3$–SiO$_2$ Glasses[a]

Composition (mole %)			
Rb$_2$O	Bi$_2$O$_3$	SiO$_2$	n_D
22	71.7	6.3	1.98
36.2	53.3	10.5	1.96
67.4	24.7	7.9	1.90

[a] Data from Rao (1965).

TABLE 16.76

Refractive Index of Cesium Aluminosilicate Glasses[a]

Composition (mole %)			
Cs$_2$O	Al$_2$O$_3$	SiO$_2$	n_D
5	5	90	1.499
5	10	85	1.509
10	5	85	1.521
7.5	12.5	80	1.543
20	5	75	1.565
10	25	65	1.581

[a] Data from Bollin (1972).

TABLE 16.77

Refractive Index of Cs$_2$O–Bi$_2$O$_3$–SiO$_2$ Glasses[a]

Composition (mole %)			
Cs$_2$O	Bi$_2$O$_3$	SiO$_2$	n_D
14.3	78.8	6.9	1.98
27	66.4	6.6	1.98
58.8	35.6	5.6	1.96

[a] Data from Rao (1965).

TABLE 16.78

Refractive Index of Thallium Borosilicate Glasses[a]

Composition (mole %)				Composition (mole %)			
Tl$_2$O	B$_2$O$_3$	SiO$_2$	n_D	Tl$_2$O	B$_2$O$_3$	SiO$_2$	n_D
1.4	45.8	52.8	1.465	18.4	38.2	43.4	1.66–1.67
3.9	44.2	51.9	1.49–1.50	25.8	33.9	40.3	1.75
6.1	43.5	50.4	1.54–1.55	38	28	34	1.90–1.95
8.8	42.2	49.0	1.539	58.3	19.5	22.0	>2.11
13.3	40	46.7	1.62				

[a] Data from Kim (1968).

TABLE 16.79

Refractive Index of $Li_2O-ZrO_2-SiO_2$ Glasses[a]

Composition (mole %)			
Li_2O	ZrO_2	SiO_2	n_D
20	2.5	77.5	1.5165
20	5	75	1.5346
20	7	73	1.5462
20	9.5	70.5	1.5626
20	12.5	67.5	1.5813

[a] Data read from graph of Ellern and Pavlushkin (1969).

TABLE 16.80

Refractive Index of $Na_2O-GeO_2-SiO_2$ Glasses[a]

Composition (mole %)				Composition (mole %)			
Na_2O	GeO_2	SiO_2	n	Na_2O	GeO_2	SiO_2	n
20	10	70	1.519	30	10	60	1.522
20	20	60	1.540	30	20	50	1.541
20	30	50	1.569	30	30	40	1.557
20	40	40	1.592	30	40	30	1.581
20	50	30	1.617	30	50	20	1.598
20	60	20	1.638	30	60	10	1.614
20	70	10	1.662				

[a] Data from Takahashi et al. (1977).

TABLE 16.81

Optical Constants of $Na_2O-SnO_2-SiO_2$ Glasses[a]

Composition (mole %)				
Na_2O	SnO_2	SiO_2	n_D	$(n_F - n_C) \times 10^5$
13	1	86	1.4857	790
13	2	85	1.4941	819
13	5	82	1.5036	837
13	7	80	1.5163	866
13	8.5	78.5	1.5241	888
13	10	77	1.5307	905

[a] Data from Galant et al. (1966).

TABLE 16.82

Refractive Index of Some
$Na_2O-MO_2-SiO_2$ Glasses[a]

Composition (mole %)			
Na_2O	MO_2	SiO_2	n
	M = Sn		
20	5	75	1.518
20	10	70	1.539
30	5	65	1.525
30	10	60	1.543
	M = Th		
20	5	75	1.532
30	5	65	1.540
	M = Hf		
18	12	70	1.549

[a] Data from Takahashi *et al.* (1977) and Brekhovskikh and Sesorova (1960).

TABLE 16.83

Optical Constants of $Na_2O-TiO_2-SiO_2$ Glasses[a]

Composition (mole %)					Composition (mole %)				
Na_2O	TiO_2	SiO_2	n_D	v	Na_2O	TiO_2	SiO_2	n_D	v
15	5	80	1.5184	51.5	20	20	60	1.6424	33.5
20	5	75	1.5272	50.4	25	20	55	1.6402	34.1
25	5	70	1.5318	49.7	30	20	50	1.6362	34.5
10	10	80	1.5481	43.5	35	20	45	1.6348	34.2
15	10	75	1.5574	43.8	40	20	40	1.6358	33.3
20	10	70	1.5635	43.7	15	25	60	1.6817	29.0
25	10	65	1.5666	43.5	20	25	55	1.6837	29.7
30	10	60	1.5672	43.2	25	25	50	1.6783	30.6
35	10	55	1.5688	42.6	30	25	45	1.6739	30.9
10	15	75	1.5868	36.4	35	25	40	1.6709	30.8
15	15	70	1.5982	37.8	15	30	55	1.7249	26.0
20	15	65	1.6019	38.2	20	30	50	1.7255	26.7
25	15	60	1.6016	38.5	25	30	45	1.7184	27.6
30	15	55	1.6006	38.6	30	30	40	1.7109	28.1
35	15	50	1.6007	38.2	35	30	35	1.7061	28.2
40	15	45	1.6020	37.1	25	35	40	1.7595	25.1
23	17	60	1.6185	36.4	30	35	35	1.7496	25.7
10	20	70	1.6257	31.4	35	35	30	1.7413	26.0
15	20	65	1.6412	32.7					

[a] Data from Hamilton and Cleek (1958).

TABLE 16.84

Refractive Index of Na_2O–ZrO_2–SiO_2 Glasses[a]

Composition (mole %)			n_D		Composition (mole %)			n_D	
Na_2O	ZrO_2	SiO_2	Refractometer method	Obreimova's method	Na_2O	ZrO_2	SiO_2	Refractometer method	Obreimova's method
13.5	11.0	75.5	1.5851	1.584	15.4	5.2	79.4	1.5327	1.525
16.0	9.4	74.6	—	1.572	15.8	8.0	76.2	1.552	1.55
18.5	8.0	73.5	1.5564	1.557	16.0	9.4	74.6	1.5608	1.56
23.2	5.1	71.7	1.5382	1.539	16.5	12.5	71.0	1.5851	1.5848
25.4	3.8	70.8	1.529	1.529	16.6	7.2	76.2	1.5509	1.55
27.6	2.5	69.9	1.5222	—	21.1	8.0	70.9	1.5542	1.554
29.7	1.2	69.1	1.5003	1.5007	26.4	8.0	65.6	1.562	1.563
14.8	1.2	84.0	1.5	1.4902	31.8	8.0	60.2	1.5692	1.569
15.0	2.5	82.5	1.502	1.51	37.2	8.0	54.8	1.5785	1.5775
15.2	3.8	81.0	1.5139	1.5132	39.8	8.0	52.2	1.5799	1.5807

[a] Data from Botvinkin and Demichev (1964).

TABLE 16.85

Refractive Index of $Na_2O-TeO_2-SiO_2$ Glasses[a]

Composition (mole %)				Composition (mole %)			
Na_2O	TeO_2	SiO_2	n_D	Na_2O	TeO_2	SiO_2	n_D
25	4.8	70.2	1.525	30	20.2	49.8	1.607
25	9.8	65.2	1.550	30	24.9	45.1	1.629
25	19.6	55.4	1.601	35	5	60	1.532
25	24.9	50.1	1.625	35	10	55	1.554
30	4.8	65.2	1.530	40	5	55	1.534
30	10	60	1.553	45	5	50	1.533
30	15	55	1.578	50	4.8	45.2	1.536

[a] Data from Mochida *et al.* (1980).

TABLE 16.86

Refractive Index of $K_2O-TiO_2-SiO_2$ Glasses[a]

Composition (wt. %)				Composition (wt. %)			
K_2O	TiO_2	SiO_2	n_D	K_2O	TiO_2	SiO_2	n_D
15	12.5	72.5	1.542	40	40	20	1.678
15	20	65	1.552	40	45	15	1.705
20	20	60	1.584	40	50	10	1.730
20	30	50	1.610	40	55	5	1.750
20	40	40	1.720	40	58	2	1.770
20	45	35	1.755	45	10	45	1.544
30	10	60	1.534	45	20	35	1.584
30	20	50	1.580	45	30	25	1.628
30	30	40	1.624	45	40	15	1.668
30	40	30	1.680	45	45	10	1.692
40	10	50	1.540	45	50	5	1.710
40	20	40	1.582	60	10	30	1.536
40	30	30	1.632	60	20	20	1.584

[a] Data from Rao (1963a,b).

TABLE 16.87

Refractive Index of Na$_2$O–Nb$_2$O$_5$–SiO$_2$ Glasses[a]

Composition (mole %)				Composition (mole %)			
Na$_2$O	Nb$_2$O$_5$	SiO$_2$	n_D	Na$_2$O	Nb$_2$O$_5$	SiO$_2$	n_D
15	7.5	77.5	1.596	19.1	19.1	61.8	1.760
19.9	5	75.1	1.567	35.1	11	53.9	1.656
25	5	70	1.572	29.9	16	54.1	1.720
21.9	7	71.1	1.592	26	20	54	1.770
20	10	70	1.634	22.9	23	54.1	1.805
15	15	70	1.698	19.8	26	54.2	1.840
29.9	5	65.1	1.576	38.1	12	49.9	1.666
29	9	62	1.625	30	20	50	1.765
24.9	13	62.1	1.680	25	25	50	1.825
22.1	16	61.9	1.728	33.9	19.1	47	1.755

[a] Data from Hirayama and Berg (1963).

TABLE 16.88

Refractive Index of K$_2$O–Nb$_2$O$_5$–SiO$_2$ Glasses[a]

Composition (mole %)				Composition (mole %)			
K$_2$O	Nb$_2$O$_5$	SiO$_2$	n_D	K$_2$O	Nb$_2$O$_5$	SiO$_2$	n_D
16.1	19.4	64.5	1.722	21.7	13.1	65.2	1.643
16.7	16.7	66.6	1.692	22.7	9.1	68.2	1.600
17.2	13.8	69.0	1.658	23.8	4.8	71.4	1.550
17.9	10.7	71.4	1.622	23.8	28.6	47.6	1.800
18.5	7.4	74.1	1.582	25	25	50	1.765
19.2	3.8	77.0	1.536	26.3	21.1	52.6	1.725
19.2	23.1	57.7	1.758	27.8	16.7	55.5	1.674
20	20	60	1.725	29.4	11.8	58.8	1.626
20.8	16.7	62.5	1.682	31.3	6.2	62.5	1.574

[a] Data from Rao (1962b).

TABLE 16.89

Refractive Index of K$_2$O–Ta$_2$O$_5$–SiO$_2$ Glasses[a]

Composition (wt. %)				Composition (wt. %)			
K$_2$O	Ta$_2$O$_5$	SiO$_2$	n_D	K$_2$O	Ta$_2$O$_5$	SiO$_2$	n_D
36.1	5.3	58.6	1.508	31.4	18.1	50.5	1.529
33.4	5.6	61.0	1.506	28.6	19.2	52.2	1.531
30.1	5.9	64.0	1.507	25.6	20	54.4	1.530
26.6	6.1	67.3	1.503	34.8	20.5	44.7	1.541
22.4	6.4	71.2	1.498	22.3	20.9	56.8	1.530
29.2	8.6	62.2	1.510	25	22	53	1.534
34.5	10	55.5	1.515	30.1	22	47.9	1.539
31.2	10.5	58.3	1.516	37.5	22.2	40.3	1.552
28.5	10.9	60.6	1.514	27.3	22.9	49.8	1.538
25	11.4	63.6	1.510	24.4	23.8	51.8	1.537
21	12.3	66.7	1.510	28.7	25.2	46.1	1.548
24.3	13.9	61.8	1.515	26.1	26.3	47.6	1.544
32.9	14.2	52.9	1.520	23.2	27.1	49.7	1.541
30.1	15.1	54.8	1.523	32.3	28.5	39.2	1.560
26.6	15.5	57.9	1.521	26.6	30.5	42.9	1.561
23.6	16.6	59.8	1.519	24.5	35.9	39.6	1.574
29.3	17.2	53.5	1.526				

[a] Data from Sun and Silverman (1941).

TABLE 16.90

Refractive Index of Na$_2$O–U$_3$O$_8$–SiO$_2$ Glasses[a]

Composition (wt. %)				Composition (wt. %)			
Na$_2$O	U$_3$O$_8$	SiO$_2$	n_D	Na$_2$O	U$_3$O$_8$	SiO$_2$	n_D
11	50	39	1.644	27	20	53	1.545
11.2	52	36.8	1.611	30.5	10	59.5	1.523
14	41	45	1.573	34	—	66	1.505
16.8	29	54.2	1.545	35.5	30	34.5	1.576
20	15	65	1.520	40.5	20	39.5	1.553
20	40	40	1.596	45.5	10	44.5	1.531
23.5	30	46.5	1.573	50.8	—	49.2	1.513
23.7	—	76.3	1.495				

[a] Data from Chakrabarty (1969).

TABLE 16.91

Refractive Index of MgO–CaO–SiO$_2$ Glasses[a]

Composition (mole %)				Composition (mole %)			
MgO	CaO	SiO$_2$	n_D	MgO	CaO	SiO$_2$	n_D
2.8	47.2	50.0	1.6262	25	25	50	1.6073
4.9	50.9	44.2	1.6410	31.7	18.3	50.0	1.6007
8.4	41.6	50.0	1.6223	32.7	23.3	44.0	1.6225
14.4	35.6	50.0	1.6175	36.5	13.5	50.0	1.5960
19.7	30.3	50.0	1.6122	45.6	4.4	50.0	1.5852
20.3	38.9	40.8	1.6392	47.8	2.2	50.0	1.5821
21.7	28.3	50.0	1.6105				

[a] Data from Larsen (1909).

TABLE 16.92

**Refractive Indices of MgO–FeO–SiO$_2$ Glasses
at 25°C Using White Light**[a]

Composition (wt. %)				
MgO	FeO	Fe$_2$O$_3$	SiO$_2$	n
3.07	57.73	1.38	37.82	1.750
5.66	52.59	0.95	41.80	1.725
9.72	50.25	1.13	38.90	1.730
9.40	46.18	0.93	43.29	1.702
10.80	44.80	1.11	43.29	1.700
14.98	39.05	1.04	44.93	1.687
14.85	44.30	1.25	39.60	1.716
14.55	37.30	0.61	47.54	1.668
21.55	30.66	0.79	47.00	1.661
24.30	26.51	0.58	48.61	1.648
19.45	26.44	0.59	53.52	1.633
19.30	28.35	0.69	51.66	1.642
11.87	40.05	0.63	47.45	1.672
4.78	51.36	0.89	42.97	1.713
15.91	29.83	0.59	53.67	1.638

[a] Data from Bowen and Schairer (1935).

TABLE 16.93

**Refractive Index of
SrO–CaO–SiO₂ Glasses**[a]

Composition (mole %)			
SrO	CaO	SiO₂	n
4.6	45.4	50	1.6265
14.9	35.1	50	1.627
23.8	26.2	50	1.628
27.1	22.9	50	1.628

[a] Data from Eskola (1922).

TABLE 16.94

**Refractive Index of BaO–CaO–SiO₂
Glasses**[a]

Composition (mole %)			
BaO	CaO	SiO₂	n
7.7	42.3	50	1.6345
13.3	36.7	50	1.6395
17.6	32.4	50	1.644
22.4	27.6	50	1.649
31	19	50	1.657

[a] Data from Eskola (1922).

TABLE 16.95

Refractive Index of FeO–CaO–SiO₂ Glasses[a] **at 25°C Using White Light**[b]

Composition (wt. %)					Composition (wt. %)				
FeO	Fe₂O₃	CaO	SiO₂	n	FeO	Fe₂O₃	CaO	SiO₂	n
6.50	0.30	42.68	50.52	1.636	28.74	2.19	32.56	36.51	1.725
8.62	0.58	40.35	50.45	1.639	31.12	1.74	29.62	37.52	1.725
11.94	0.64	37.59	49.83	1.648	32.14	1.88	29.56	36.42	1.735
14.44	0.46	35.27	49.83	1.650	27.61	1.26	28.70	42.43	1.699
17.78	0.62	32.47	49.13	1.656	29.09	1.00	25.24	44.67	1.690
19.71	0.75	30.44	49.10	1.657	34.27	1.31	24.02	40.40	1.717
25.32	0.96	25.11	48.61	1.669	31.68	1.56	28.90	37.86	1.725
28.96	0.74	22.31	47.99	1.675	32.23	2.03	29.88	35.86	1.735
31.73	1.16	19.25	47.86	1.678	33.69	1.63	28.86	35.82	1.734
34.15	0.84	17.01	48.00	1.680	38.23	1.61	23.67	36.49	1.740
36.71	1.18	14.55	47.56	1.685	41.55	0.79	15.58	42.08	1.720
41.76	1.15	9.77	47.32	1.691	41.27	1.19	19.26	38.28	1.737
15.00	0.63	27.79	56.58	1.622	45.04	1.33	13.37	40.26	1.734
19.73	0.76	24.85	54.66	1.632	43.95	1.05	12.00	43.00	1.715
27.61	0.74	20.99	50.66	1.668	5.21	0.50	51.61	42.68	1.662
45.22	1.01	10.36	43.41	1.715	16.43	1.50	43.51	38.56	1.688
52.04	0.90	5.51	41.55	1.735	20.86	1.36	40.88	36.90	1.712
5.40	0.35	34.72	59.53	1.602	33.69	1.63	28.86	35.82	1.734
10.75	0.44	34.54	54.27	1.628	45.97	1.92	18.04	34.07	1.77
14.92	1.14	40.49	43.45	1.673	48.09	1.52	14.41	35.98	1.76
17.05	1.69	42.10	39.16	1.688	54.42	1.73	9.72	34.13	1.79
20.84	0.99	33.51	44.66	1.681	48.73	1.49	11.19	38.59	1.745
22.55	1.50	36.67	39.28	1.706	52.96	1.09	9.01	36.94	1.765
23.60	2.11	37.15	37.14	1.714	53.30	0.98	5.96	39.76	1.745
23.83	1.09	35.04	40.04	1.708					

[a] Unannealed glasses.
[b] Data from Bowen et al. (1933).

TABLE 16.96

Optical Constants of ZnO–BaO–SiO₂ Glasses[a]

Composition (mole %)			n_C	n_D	n_F	v
ZnO	BaO	SiO₂				
14	40	46	1.67453	1.67848	1.68817	49.7
6	36	58	1.63224	1.63568	1.64406	53.8
10	36	54	1.65006	1.65371	1.66261	52.1
14	36	50	1.66165	1.66544	1.67474	50.8
18	36	46	1.67210	1.67605	1.68574	49.6
10	34	56	1.64106	1.64461	1.65327	52.8
14	34	52	1.64960	1.65327	1.66222	51.7
16	34	50	1.65815	1.66192	1.67112	51.1
18	34	48	1.66512	1.66898	1.67841	50.3
20	34	46	1.66629	1.67017	1.67968	50.0
3	33	64	1.60761	1.61084	1.61862	55.5
16	32	52	1.65227	1.65598	1.66504	51.4
18	32	50	1.65849	1.66228	1.67151	50.9
6	30	64	1.61047	1.61369	1.62156	55.3
8	30	62	1.61841	1.62172	1.62980	54.6
10	30	60	1.62461	1.62800	1.63625	54.0
12	30	58	1.62890	1.63232	1.64071	53.5
14	30	56	1.63785	1.64141	1.65002	52.7
16	30	54	1.64489	1.64852	1.65735	52.0
18	30	52	1.65060	1.65430	1.66330	51.5
20	30	50	1.65650	1.66028	1.66949	50.8

Composition (mole %)			n_C	n_D	n_F	v
ZnO	BaO	SiO₂				
10	28	62	1.61549	1.61880	1.62682	54.6
12	28	60	1.62199	1.62540	1.63360	53.9
14	28	58	1.63151	1.63501	1.64354	52.8
16	28	56	1.63674	1.64029	1.64892	52.6
20	28	52	1.64842	1.65212	1.66112	51.4
24	28	48	1.66336	1.66725	1.67672	50.0
26	28	46	1.67040	1.67438	1.68412	49.2
4	26	70	1.58630	1.58933	1.59672	56.6
6	26	68	1.59460	1.59772	1.60528	56.0
8	26	66	1.60183	1.60495	1.61272	55.5
10	26	64	1.60618	1.60941	1.61723	55.2
12	26	62	1.61447	1.61779	1.62581	54.5
14	26	60	1.62000	1.62336	1.63164	53.6
16	26	58	1.63005	1.63354	1.64201	53.0
18	26	56	1.63488	1.63842	1.64705	52.5
20	26	54	1.64365	1.64729	1.65617	51.7
24	26	50	1.65731	1.66112	1.67044	50.4
28	26	46	1.66964	1.67361	1.68336	49.1
6	24	70	1.58480	1.58785	1.59519	56.6
10	24	66	1.59904	1.60228	1.60990	55.5
14	24	62	1.61290	1.61622	1.62426	54.2

(continues)

TABLE 16.96 (*Continued*)

| Composition (mole %) | | | | | | | Composition (mole %) | | | | | | |
ZnO	BaO	SiO$_2$	n_C	n_D	n_F	ν	ZnO	BaO	SiO$_2$	n_C	n_D	n_F	ν
24	30	46	1.66957	1.67352	1.68316	49.6	18	24	58	1.62586	1.62932	1.63772	53.0
2	28	70	1.59088	1.59392	1.60134	56.8	20	24	56	1.63743	1.64102	1.64977	52.0
8	28	64	1.60711	1.61035	1.61816	55.2	22	24	54	1.64219	1.64584	1.65473	51.5
24	24	52	1.64742	1.65114	1.66017	51.0	34	14	52	1.64298	1.64675	1.65593	49.9
26	24	50	1.65830	1.66216	1.67157	49.9	38	14	48	1.65546	1.65940	1.66906	48.5
14	20	66	1.59986	1.60307	1.61084	54.9	40	14	46	1.66423	1.66826	1.67816	48.0
20	20	60	1.62306	1.62651	1.63488	53.0	32	12	56	1.62320	1.62677	1.63543	51.3
22	20	58	1.61816	1.62158	1.62985	53.2	34	12	54	1.63434	1.63803	1.64699	50.4
28	20	52	1.64567	1.64940	1.65849	50.7	38	12	50	1.64908	1.65292	1.66235	49.2
30	20	50	1.65172	1.65553	1.66482	50.0	40	12	48	1.65487	1.65882	1.66847	48.4
30	18	52	1.64598	1.64976	1.65891	50.2	36	10	54	1.63301	1.63671	1.64572	50.1
34	18	48	1.66042	1.66436	1.67400	48.9	38	10	52	1.64045	1.64424	1.65348	49.4
24	16	60	1.61432	1.61774	1.62604	52.7	42	10	48	1.65416	1.65810	1.66776	48.4
30	16	54	1.63462	1.63832	1.64727	50.5	42	8	50	1.64709	1.65098	1.66047	48.7
34	16	50	1.65131	1.65516	1.66454	49.5	42	6	52	1.63757	1.64134	1.65066	49.0

^{a} Data from Cleek and Babcock (1973).

TABLE 16.97

Refractive Index of Beryllium Aluminosilicate Glass[a]

Compsition (mole %)			
BeO	Al$_2$O$_3$	SiO$_2$	n_D
25	25	50	1.583

[a] Data from Appen (1954, 1958).

TABLE 16.98

Refractive Index of Magnesium Aluminosilicate Glasses[a]

Composition (wt. %)				Composition (wt. %)			
MgO	Al$_2$O$_3$	SiO$_2$	n_D	MgO	Al$_2$O$_3$	SiO$_2$	n_D
20.3	18.3	61.4	1.5447	14	26	60	1.5380
10	23.5	66.5	1.521	26	18	56	1.5620
16.1	34.8	49.1	1.556	22	22	56	1.5570
25.7	22.8	51.5	1.5695	18	26	56	1.5520
25	21	54	1.5640	14	30	56	1.5450
10	22	68	1.519	26	22	52	1.5690
18	18	64	1.5375	22	26	52	1.5630
14	22	64	1.532	18	30	52	1.5575
10	26	64	1.525	30	22	48	1.582
22	18	60	1.5500	26	26	48	1.575
18	22	60	1.5440	22	30	48	1.570

[a] Data from Gambaryan et al. (1967).

TABLE 16.99

Refractive Index of Calcium Aluminosilicate Glasses[a]

Composition (mole %)				Composition (mole %)			
CaO	Al$_2$O$_3$	SiO$_2$	n_D	CaO	Al$_2$O$_3$	SiO$_2$	n_D
22.2	11.1	66.7	1.5497	29.3	21.9	48.8	1.5819
22.7	13.6	63.7	1.5534	33.3	19.1	47.6	1.5900
23.2	16.3	60.5	1.5578	37.2	16.3	46.5	1.5985
23.8	19.1	57.1	1.5640	22.8	26.6	50.6	1.5713
24.4	22	53.6	1.5697	27.2	23.4	49.4	1.5778
25	25	50	1.5739	29.4	29.4	41.2	1.5907
25.3	26.6	48.1	1.5777	26.3	26.3	47.4	1.5792
25.6	28.2	46.2	1.5813	22.7	22.7	54.6	1.5664
26.3	31.6	42.1	1.5875	20.8	20.8	58.4	1.5573
20.5	28.2	51.3	1.5700	20	20	60	1.5549
				16.7	16.7	66.6	1.545
				12.5	12.5	75	1.533

[a] Data from Chang and Ying (1965) except last two entries, right column, which are from Varshal (1972).

TABLE 16.100

Refractive Index of Calcium Gallosilicate Glasses[a]

Composition (mole %)				Composition (mole %)			
CaO	Ga_2O_3	SiO_2	n_D	CaO	Ga_2O_3	SiO_2	n_D
15	10	75	1.542	30	10	60	1.602
15	15	70	1.570	30	30	40	1.684
15	25	60	1.616	40	10	50	1.633
20	10	70	1.561	40	20	40	1.664
20	15	65	1.583	40	30	30	1.708
20	20	60	1.610	40	40	20	1.745
20	25	55	1.630	40	45	15	1.757
20	30	50	1.653	50	10	40	1.664
25	15	60	1.604	50	40	10	1.757
25	25	50	1.645				
25	35	40	1.688				

[a] Data from Danilova (1967) and Danilova and Dubrovo (1967).

TABLE 16.101

Refractive Index of Strontium Aluminosilicate Glasses[a]

Composition (mole %)				Composition (mole %)			
SrO	Al_2O_3	SiO_2	n_D	SrO	Al_2O_3	SiO_2	n_D
30	5	65	1.563	35	20	45	1.591
25	10	65	1.551	55	5	40	1.614
35	5	60	1.584	50	10	40	1.614
30	10	60	1.583	45	15	40	1.611
40	5	55	1.592	40	20	40	1.606
35	10	55	1.589	35	25	40	1.596
45	5	50	1.603	60	5	35	1.611
40	10	50	1.604	55	10	35	1.605
35	15	50	1.588	50	15	35	1.608
50	5	45	1.611	45	20	35	1.608
45	10	45	1.611	40	25	35	1.608
40	15	45	1.604	35	30	35	1.598

[a] Data from Iskhakov (1971).

TABLE 16.102

Optical Constants of Barium Borosilicate Glasses[a]

BaO	B₂O₃	SiO₂	n_D	v	BaO	B₂O₃	SiO₂	n_D	v
16.7	75.4	7.9	1.5472	62.7	28.6	55.8	15.6	1.6130	63.0
20.6	71.4	8.0	1.5666	63.2	28.6	59.6	11.8	1.6104	63.6
20.7	63.6	15.7	1.5711	63.7	28.8	63.2	8.0	1.6078	62.5
24.0	39.9	36.1	1.6032	63.6	30.1	20.3	49.6	1.6238	60.0
24.2	47.6	28.2	1.6003	63.5	32	3	65	1.6122	—
24.3	42.9	32.8	1.6022	63.6	32.0	20.4	47.6	1.6303	59.3
24.3	59.8	15.9	1.5931	63.2	32.1	22.6	45.3	1.6319	59.9
24.4	51.4	24.2	1.5980	63.6	32.1	24.4	43.5	1.6322	59.9
24.4	55.5	20.1	1.5952	63.7	32.1	50.4	17.5	1.6286	61.9
24.4	55.5	20.1	1.5961	63.8	32.2	28.0	39.8	1.6330	60.0
24.5	63.5	12.0	1.5906	63.2	32.2	32.1	35.7	1.6333	60.8
24.5	67.5	8.0	1.5875	63.1	32.3	6.5	61.2	1.6195	57.1
24.5	71.4	4.1	1.5850	63.4	32.3	10.3	57.4	1.6236	57.9
28.1	12.5	59.4	1.6099	59.5	32.3	30.1	37.6	1.6335	60.4
28.2	23.9	47.9	1:6184	61.5	32.3	53.6	14.1	1.6268	62.1
28.2	32.0	39.8	1.6203	62.3	32.3	58.8	8.9	1.6249	62.0
28.2	36.2	35.6	1.6200	62.3	32.3	59.4	8.3	1.6242	62.4
28.2	39.7	32.1	1.6197	62.7	32.4	16.1	51.5	1.6280	59.0
28.3	27.6	44.1	1.6197	61.6	32.4	33.8	33.8	1.6333	60.9
28.4	35.5	36.1	1.6204	62.2	32.4	43.5	24.1	1.6316	62.0
28.5	47.4	24.1	1.6172	63.0	32.4	45.7	21.9	1.6308	62.0
28.6	16.0	55.4	1.6139	60.3	32.4	49.5	18.1	1.6289	61.8
28.6	40.0	31.4	1.6205	62.3	32.4	49.9	17.7	1.6290	62.0
32.5	35.5	32.0	1.6332	61.0	36.6	59.4	4.0	1.6351	61.1
32.5	51.6	15.9	1.6282	62.1	36.9	48.2	14.9	1.6419	60.7
32.5	61.7	5.8	1.6220	61.8	37	48	15	1.6429	60.3
32.6	25.5	41.9	1.6327	60.1	37.2	24.1	38.7	1.6458	58.3
32.6	41.4	26.0	1.6322	61.6	37.2	27.7	35.1	1.6464	58.8
32.6	64.2	3.2	1.6205	62.2	37.7	37.1	25.2	1.6468	59.2
32.7	47.8	19.5	1.6305	61.6	38.2	51.9	9.9	1.6433	60.1
34.2	32.6	33.2	1.6375	60.1	39.8	55.1	5.1	1.6433	60.0
34.3	35.8	29.9	1.6390	60.6	40.0	16.5	43.5	1.6516	56.3
35.9	16.7	47.4	1.6409	57.7	40.1	26.0	33.9	1.6527	57.5
36.0	8.5	55.5	1.6358	56.7	40.2	20.0	39.8	1.6527	56.8
36.1	24.3	39.6	1.6435	58.7	40.3	23.8	35.9	1.6530	56.8
36.3	28.0	35.7	1.6441	59.5	40.3	27.6	32.1	1.6532	57.7
36.4	31.9	31.7	1.6443	59.7	40.3	31.7	28.0	1.6527	58.1
36.4	35.5	28.1	1.6440	60.1	40.3	35.8	23.9	1.6522	58.6
36.4	39.7	23.9	1.6432	60.2	40.3	43.7	16.0	1.6501	59.2
36.4	43.5	20.1	1.6423	60.5	40.4	39.8	19.8	1.6513	58.7
36.5	43.5	20.0	1.6425	60.4	40.5	47.6	11.9	1.6488	59.4
36.5	55.7	7.8	1.6376	61.1	40.6	51.5	7.9	1.6470	59.2
36.6	39.4	24.0	1.6435	60.0	40.7	55.5	3.8	1.6447	59.7
36.6	47.3	16.1	1.6413	60.5	44.0	20.4	35.6	1.6610	55.2
36.6	51.5	11.9	1.6402	60.8	44.1	32.1	23.8	1.6592	56.4
44.2	28.1	27.7	1.6604	56.3	44.9	23.7	31.4	1.6607	55.4
44.2	28.2	27.6	1.6604	55.8	47.2	22.9	29.9	1.6665	54.7
44.3	24.1	31.6	1.6610	55.7	47.2	26.5	26.3	1.6657	54.7
44.3	32.0	23.7	1.6598	56.8	47.5	30.7	21.8	1.6648	55.0
44.7	39.7	15.6	1.6576	56.9	50.0	26.5	23.5	1.6712	53.4

[a] Data from Hamilton et al. (1958).

TABLE 16.103

Optical Constants of Barium Aluminosilicate Glasses[a]

Composition (mole %)						
BaO	Al_2O_3	SiO_2	n_C	n_D	n_F	v
30	6	64	1.60438	1.60758	1.61537	55.3
32	6	62	1.60887	1.61207	1.61991	55.5
34	6	60	1.61158	1.61483	1.62264	55.6
36	6	58	1.62514	1.62862	1.63672	54.3
38	6	56	1.62831	1.63170	1.63993	54.3
40	6	54	1.63238	1.63586	1.64415	54.0
34	8	58	1.61694	1.62023	1.62827	54.8
36	8	56	1.62142	1.62473	1.63280	54.9
38	8	54	1.62491	1.62822	1.63626	55.4
42	8	50	1.63861	1.64207	1.65063	53.4
28	10	62	1.59091	1.59397	1.60132	57.0
38	10	52	1.62786	1.63127	1.63952	54.2
40	10	50	1.62626	1.62964	1.63784	54.4
42	10	48	1.63339	1.63684	1.64523	53.8
36	12	52	1.62024	1.62357	1.63163	54.7
40	12	48	1.63315	1.63660	1.64501	53.7
40	14	46	1.63264	1.63611	1.64452	53.6
42	14	44	1.64402	1.64765	1.65635	52.5

[a] Data from Cleek and Babcock (1973).

TABLE 16.104

Refractive Index of Barium Gallosilicate Glasses[a]

Composition (mole %)				Composition (mole %)			
BaO	Ga_2O_3	SiO_2	n_D	BaO	Ga_2O_3	SiO_2	n_D
20	10	70	1.577	30	30	40	1.682
20	15	65	1.591	40	10	50	1.661
20	20	60	1.610	40	20	40	1.688
20	25	55	1.633	40	30	30	1.710
20	30	50	1.656	40	40	20	1.736
30	10	60	1.619	40	45	15	1.746

[a] Data from Danilova and Dubrovo (1971).

TABLE 16.105

Optical Constants of BaO–La$_2$O$_3$–SiO$_2$ Glasses[a]

Composition (mole %)						
BaO	La$_2$O$_3$	SiO$_2$	n_C	n_D	n_F	v
28	2	70	1.60655	1.60974	1.61743	56.0
30	2	68	1.61357	1.61685	1.62465	55.6
33	2	65	1.62518	1.62853	1.63662	54.9
35	2	63	1.63521	1.63865	1.64698	54.3
38	2	60	1.64261	1.64611	1.65462	53.8
23	5	72	1.61743	1.62076	1.62864	55.4
25	5	70	1.62824	1.63162	1.63978	54.8
28	5	67	1.63952	1.64299	1.65139	54.2
33	5	62	1.65768	1.66126	1.67007	53.3
35	5	60	1.66330	1.66700	1.67598	52.6
37	5	58	1.67280	1.67660	1.68584	51.9
23	7	70	1.64081	1.64428	1.65270	54.2
25	7	68	1.64872	1.65227	1.66086	53.8
30	7	63	1.66779	1.67152	1.68057	52.6
35	7	58	1.68199	1.68584	1.69530	51.5
38	7	55	1.69098	1.69496	1.70467	50.8
20	10	70	1.66660	1.67032	1.67942	52.3
23	10	67	1.67071	1.67445	1.68352	52.7
25	10	65	1.67861	1.68243	1.69172	52.0
23	12	65	1.69550	1.69950	1.70927	50.8

[a] Data from Cleek and Babcock (1973).

TABLE 16.106

Refractive Index of Zinc Borosilicate Glasses[a]

Composition (mole %)				Composition (mole %)			
ZnO	B$_2$O$_3$	SiO$_2$	n_D	ZnO	B$_2$O$_3$	SiO$_2$	n_D
50	45	5	1.6409	57.5	27.5	15	1.6614
51.5	38.5	10	1.6439	57.5	32.5	10	1.6614
51.5	43.5	5	1.6449	57.5	37.5	5	1.6617
52	38	10	1.6455	59	23	18	1.6662
52.5	37.5	10	1.6465	60	20	20	1.6721
52.5	42.5	5	1.6479	60	25.2	14.8	1.6698
53	37	10	1.6485	60	30	10	1.6691
53.5	36.5	10	1.6499	60	32	18	1.6687
55.5	34.5	10	1.6541	60	35	15	1.6684
55	40	5	1.6544	62.5	22.5	15	1.6798
57.5	22.5	20	1.6619	62.5	27.5	10	1.6772

[a] Data from Hamilton et al. (1959).

TABLE 16.107

Refractive Index of Zinc Aluminosilicate Glasses[a]

Composition (mole %)			
ZnO	Al_2O_3	SiO_2	n_D
14.3	14.3	71.4	1.536
16.7	16.7	66.6	1.544
20	20	60	1.565

[a] Data from Varshal (1972).

TABLE 16.108

Refractive Index of $CdO-Bi_2O_3-SiO_2$ Glasses[a]

Composition (wt. %)			
CdO	Bi_2O_3	SiO_2	n_D
19.6	78.4	2	2.2261
19.5[b]	78.7[b]	1.8[b]	>1.98
57.1	38.1	4.8	1.98
37.3[b]	36.5[b]	26.4[b]	1.88

[a] Data from Rao (1962a).
[b] Found by analysis; others are batch compositions.

TABLE 16.109

Refractive Index of Lead Borosilicate Glasses[a]

Composition (wt. %)				Composition (wt. %)			
PbO	B_2O_3	SiO_2	n_D	PbO	B_2O_3	SiO_2	n_D
8.9	29.1	62	1.48	50.5	29.2	20.3	1.65
10.0	9.7	80.3	1.485	60.1	10.3	29.6	1.72
10.0	68.9	21.1	1.485	60.1	29.9	10.0	1.72
10.1	48.9	41.0	1.48	65	20	15	1.76
10.2	79.9	9.9	1.485	69.9	10.2	19.9	1.79
19.4	28.9	51.7	1.505	70	5	25	1.80
19.8	10.0	70.2	1.495	70	15	15	1.78
20.2	69.9	9.9	1.51	70	25	5	1.80
20.6	49.0	30.4	1.505				
29.5	9.9	60.6	1.535	75	5	20	1.85
30.1	59.9	10.0	1.54	75	20	5	1.85
30.2	29.3	40.5	1.54	80	5	15	1.92
30.6	49.1	20.3	1.54	80	15	5	1.92
39.9	10.0	50.1	1.585	85	5	10	2.02
40.2	29.1	30.7	1.59	85	10	5	2.02
40.2	49.9	9.9	1.59	90	5	5	2.11
50.2	10.2	39.6	1.63				
50.2	39.8	10.0	1.65				

[a] Data from Geller and Bunting (1939).

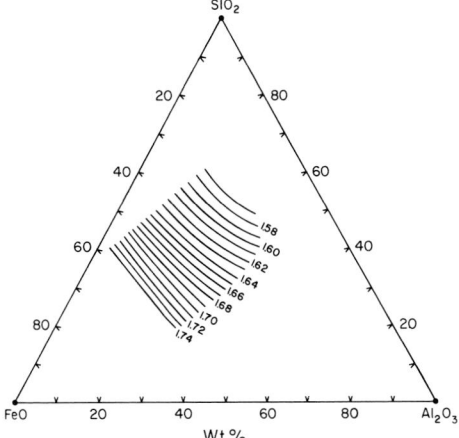

Fig. 16.2. Refractive indices of FeO-Al$_2$O$_3$–SiO$_2$ glasses at 25°C. Data from Schairer and Yagi (1952).

TABLE 16.110

Refractive Index of SrO–TiO$_2$–SiO$_2$ Glasses[a]

Composition (mole %)				Composition (mole %)			
SrO	TiO$_2$	SiO$_2$	n_D	SrO	TiO$_2$	SiO$_2$	n_D
30	20	50	1.74	35	10	55	1.684
30	25	45	1.79	35	15	50	1.716
30	30	40	1.86	35	20	45	1.75
30	35	35	1.90	35	25	40	1.80
30	40	30	1.96	35	30	35	1.86
30	45	25	2.00	40	10	50	1.712
30	50	20	2.03	40	15	45	1.728
35	—	65	1.580	40	20	40	1.75
35	5	60	1.622				

[a] Data from Evstrop'ev and Zopin (1967).

TABLE 16.111

Optical Constants of BaO–TiO₂–SiO₂ Glasses[a]

Composition (mole %)						
BaO	TiO₂	SiO₂	n_C	n_D	n_F	v
25	20	55	1.72414	1.73021	1.74562	34.0
25	25	50	1.77046	1.77760	1.79589	30.6
25	30	45	1.81406	1.82236	1.84382	27.6
25	35	40	1.85733	1.86682	1.89161	25.3
30	5	65	1.62772	1.63139	1.64037	49.9
30	10	60	1.66807	1.67250	1.68347	43.7
30	15	55	1.70750	1.71276	1.72596	38.6
30	20	50	1.74790	1.75412	1.76986	34.3
30	25	45	1.78860	1.79585	1.81434	30.9
30	30	40	1.82865	1.83697	1.85847	28.1
35	5	60	1.63026	1.63399	1.64310	49.4
35	10	55	1.68703	1.69148	1.70306	43.1
35	15	50	1.72494	1.73037	1.74401	38.3

[a] Data from Cleek and Hamilton (1956) and Cleek and Babcock (1973).

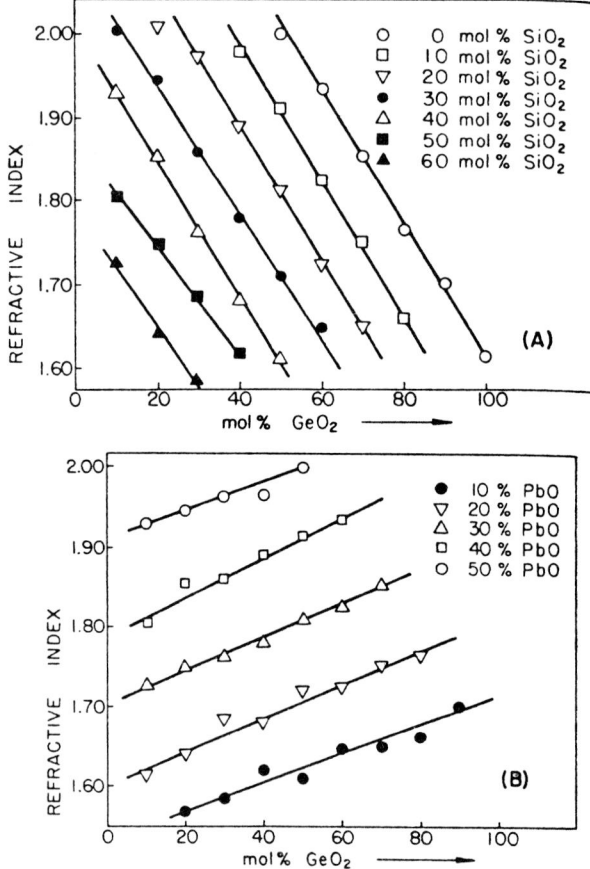

Fig. 16.3. Variation of refractive index (determined using white light) with composition for constant amounts of (a) SiO₂ and (b) PbO in the PbO–GeO₂–SiO₂ system. Data from Topping *et al.* (1974).

TABLE 16.112

**Refractive Index of PbO–TiO$_2$–SiO$_2$
Glasses**[a]

Composition (mole %)			
PbO	TiO$_2$	SiO$_2$	n_D
40	30	30	2.07
50	20	30	2.05
50	30	20	1.98

[a] Data from Kleine *et al.* (1970).

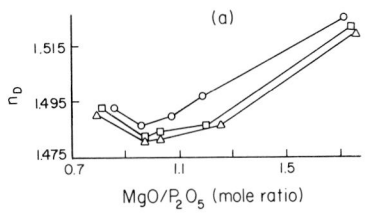
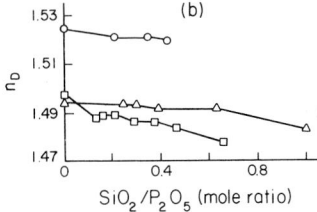

Fig. 16.4. Refractive indices of MgO–P$_2$O$_5$–SiO$_2$ glasses at 25°C: (a) SiO$_2$–P$_2$O$_5$ ratio: ○, 0.0; □, 0.3; △, 0.4. (b) MgO–P$_2$O$_5$ ratio: ○, 1.7; □, 1.2; △, 0.8. Data from Kanazawa *et al.* (1970).

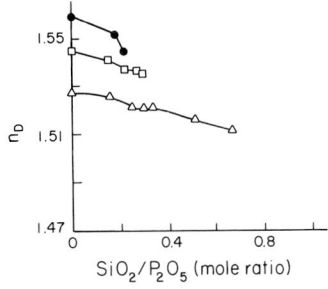

Fig. 16.5. Refractive indices of CaO–P$_2$O$_5$–SiO$_2$ Glasses with CaO–P$_2$O$_5$ (mole ratio) of: △, 0.8–0.9; □, 1.2; ●; 1.7. Data from Kanazawa *et al.* (1970).

TABLE 16.113

Optical Constants of BaO–Nb$_2$O$_5$–SiO$_2$ Glasses[a]

Composition (mole %)			n_C	n_D	n_F	v
BaO	Nb$_2$O$_5$	SiO$_2$				
32	2	66	1.63111	1.63471	1.64352	51.1
36	2	62	1.64421	1.64794	1.65704	50.5
40	2	58	1.66006	1.66394	1.67345	49.6
32	6	62	1.68287	1.68738	1.69856	43.8
36	6	58	1.69921	1.70391	1.71558	43.0
40	6	54	1.71062	1.71543	1.72739	42.7
44	6	50	1.72775	1.73290	1.74573	40.8
32	10	58	1.73419	1.73977	1.75373	37.9
36	10	54	1.74802	1.75372	1.76803	37.7
40	10	50	1.75390	1.75959	1.77385	38.1
44	10	46	1.76825	1.77413	1.78884	37.6
32	14	54	1.78176	1.78844	1.80532	33.5
36	14	50	1.79500	1.80176	1.81887	33.6
40	14	46	1.80584	1.81269	1.83002	33.6
44	14	42	1.81450	1.82140	1.83888	33.7
48	14	38	1.82084	1.82777	1.84529	33.8
32	18	50	1.82988	1.83792	1.85805	29.8
36	18	46	1.84130	1.84922	1.86945	30.2
40	18	42	1.84890	1.85680	1.87702	30.5
44	18	38	1.85714	1.86508	1.88533	30.7
48	18	34	1.86176	1.86967	1.88983	31.0
36	22	42	1.88213	1.89116	1.91444	27.6
40	22	38	1.89258	1.90166	1.92498	27.8

[a] Data from Cleek and Babcock (1973).

TABLE 16.114

Optical Constants of BaO–Ta$_2$O$_5$–SiO$_2$ Glasses[a]

BaO	Ta$_2$O$_5$	SiO$_2$	n_C	n_D	n_F	v
Composition (mole %)						
28	2	70	1.60786	1.61119	1.61928	53.5
30	2	68	1.61803	1.62145	1.62972	53.2
33	2	65	1.63352	1.63706	1.64571	52.3
35	2	63	1.63998	1.64360	1.65237	51.9
38	2	60	1.65926	1.66309	1.67244	50.3
40	2	58	1.65740	1.66117	1.67037	51.0
27	5	68	1.63942	1.64325	1.65258	48.9
30	5	65	1.65577	1.65974	1.66942	48.4
32	5	63	1.66787	1.67160	1.68066	52.5
35	5	60	1.67544	1.67959	1.68972	47.6
37	5	58	1.68377	1.68799	1.69834	47.2
40	5	55	1.69318	1.69750	1.70809	46.8
45	5	50	1.70802	1.71249	1.72349	46.0
47	5	48	1.70870	1.71306	1.72431	45.7
33	7	60	1.69368	1.69814	1.70907	45.4
35	7	58	1.69592	1.70039	1.71136	45.4
38	7	55	1.70816	1.71274	1.72403	44.9
40	7	53	1.71379	1.71844	1.72991	44.6
43	7	50	1.72379	1.72848	1.74023	44.3
45	7	48	1.72115	1.72595	1.73749	44.4
46	7	47	1.72156	1.72624	1.73778	44.8
48	7	45	1.73675	1.74166	1.75380	43.5
45	10	45	1.74781	1.75293	1.76564	42.2
48	10	42	1.76542	1.77082	1.78424	41.0

[a] Data from Cleek and Babcock (1973).

TABLE 16.115

Refractive Index of B$_2$O$_3$–Al$_2$O$_3$–SiO$_2$ Glasses[a]

B$_2$O$_3$	Al$_2$O$_3$	SiO$_2$	n_D	B$_2$O$_3$	Al$_2$O$_3$	SiO$_2$	n_D
Composition (wt. %)				Composition (wt. %)			
10.4	5.3	84.6	1.461	7.6	45.0	47.6	1.541
12.5	9.5	78.3	1.467	24.1	6.5	68.8	1.461
9.0	20.2	70.8	1.485	19.6	11.6	69.2	1.470
17.5	5.7	76.5	1.461	20.9	20.9	57.9	1.490
17.5	9.5	72.6	1.468	9.9	34.9	55.6	1.519
15.8	20.1	64.5	1.488	30.2	6.0	63.4	1.461
13.4	30.9	55.3	1.498	28.2	11.5	60.1	1.471

[a] Data from Dietzel and Scholze (1955).

TABLE 16.116

Refractive Indices of Al_2O_3–La_2O_3–SiO_2 Glasses[a]

Composition (mole %)				Composition (mole %)			
Al_2O_3	La_2O_3	SiO_2	n_D	Al_2O_3	La_2O_3	SiO_2	n_D
20.5	24.5	55	1.762	37.5	7.5	55	1.635
22.5	22.5	55	1.754	19.5	11.5	69	1.652
23.5	21.5	55	1.742	21	10	69	1.635
25.5	19.5	55	1.728	21.5	9.5	69	1.624
27	18	55	1.711	23	8	69	1.609
29	16	55	1.703	24.5	6.5	69	1.590
33	12	55	1.663	27	4	69	1.570
35.5	9.5	55	1.646				

[a] Data read from graph of Aleksandrov *et al.* (1980).

TABLE 16.117

Refractive Index of Nd_2O_3–Al_2O_3–SiO_2 Glass[a]

Composition (wt. %)			
Nd_2O_3	Al_2O_3	SiO_2	n_D
38.5	28.5	33.0	1.636

[a] Data from Karlsson (1970).

TABLE 16.118

Refractive Index of B_2O_3–GeO_2–SiO_2 Glasses at Various Wavelengths[a]

Wavelength λ (μm)	n for given composition[b]		
	A	B	C
0.435833	1.4678	1.4722	1.4796
0.479991	1.4645	1.4688	1.4760
0.508582	1.4628	1.4670	1.4742
0.546073	1.4608	1.4651	1.4722
0.589262	1.4592	1.4633	1.4703
0.643847	1.4575	1.4615	1.4684
0.734620	1.4552	1.4591	1.4660
0.807902	1.4538	1.4578	1.4645
0.852111	1.4531	1.4569	1.4638
0.917224	1.4521	1.4559	1.4627
1.002439	1.4509	1.4547	1.4615
1.08303	1.4499	1.4535	1.4605

[a] Data from Fleming (1976).
[b] Composition (mole %): A, $5.4B_2O_3$–$0.1GeO_2$–$94.5SiO_2$; B, $9.7B_2O_3$–$4.03GeO_2$–$86.27SiO_2$; and C, $7.7B_2O_3$–$9.1GeO_2$–$83.2SiO_2$.

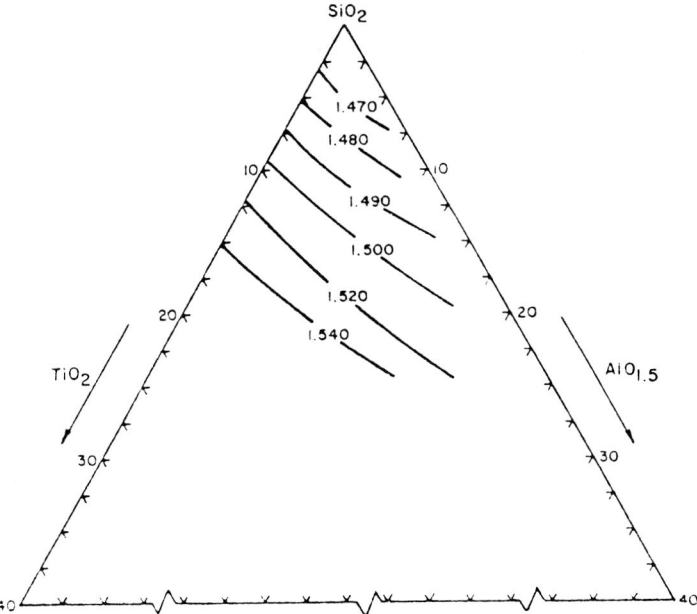

Fig. 16.6. Refractive index n_D of silica-rich $Al_2O_3-TiO_2-SiO_2$ glasses; composition in cation %. Data from Schultz and Dumbaugh (1980).

TABLE 16.119

Refractive Index of $B_2O_3-P_2O_5-SiO_2$ Glasses[a]

Composition (wt. %)				Composition (wt. %)			
B_2O_3	P_2O_5	SiO_2	n^b	B_2O_3	P_2O_5	SiO_2	n^b
1.36	1.20	97.44	1.458	17.20	23.81	58.99	1.486
5.55	1.41	93.04	1.460	16.01	31.89	52.10	1.494
3.17	5.23	91.60	1.462	25.41	32.74	41.85	1.503
6.14	6.53	87.33	1.464	27.31	45.21	27.48	1.517
7.35	9.57	83.08	1.466	30.01	50.27	19.72	1.527
10.41	18.07	71.52	1.472	41.64	48.96	9.40	1.536

[a] Data from Horn and Hummel (1955).

[b] Wavelength not given.

TABLE 16.120

Refractive Index of
$Al_2O_3-P_2O_5-SiO_2$ Glasses[a]

Composition (wt. %)			
Al_2O_3	P_2O_5	SiO_2	n_D
19.9	78.9	0.7	1.527
21.2	78.6	0.9	1.523
23.9	74.5	0.8	1.515

[a] Data from Moriya et al. (1960).

TABLE 16.121

Coefficients of the Sellmeier Dispersion Equation $n^2 = 1 + \sum_{i=1}^{3} A_i\lambda^2/(\lambda^2 - l_i^2)$ for Refractive Index
(n) of $GeO_2-P_2O_5-SiO_2$ Glass in the Wavelength (λ) Range 0.4047 to 2.0581 μm[a]

Composition (mole %)			Coefficients					
GeO_2	P_2O_5	SiO_2	A_1	A_2	A_3	l_1	l_2	l_3
0.8	3.6	95.6	0.7026425	0.4143438	0.8952753	0.006640930	0.01094264	97.93401

[a] Data from Shibata and Edahiro (1982).

Chapter 17

Solubility, Permeability, and Diffusion of Gases in Glass

INTRODUCTION I

Gases dissolve molecularly in silicate glasses because of their open structure. Small molecules diffuse rapidly in these glasses at higher temperatures. The molecular solubility is not highly dependent on temperature, and becomes smaller as the molecules become larger. Diffusion of gases in glass increases strongly at higher temperatures. [For more details see Doremus (1973) and Shelby (1979a) for reviews.]

MOLECULAR SOLUBILITY II

The most common method of measuring molecular solubility in glass is to combine measurements, described below, of the permeation coefficient K and the diffusion coefficient D of gas across a glass membrane. The solubility $\alpha_{Bu} = K/D$ is dimensionless. The most direct method of measuring solubility is to saturate the glass, usually in the form of fibers or a powder, with gas at a certain temperature and then to pump out the gas at a higher temperature and measure the volume of gas given off. The pressure drop of a gas in a vessel containing glass samples also gives a measure of the solubility.

The solubility α_{Bu} (Bunsen coefficient) is defined as the volume of gas at standard temperature and pressure (0°C and 1 atm) that dissolves in unit volume of glass per unit external partial pressure of the gas. The solubility α_{Os} (Ostwald coefficient) is the ratio of the concentration of gas dissolved in the glass C_i to the concentration of gas C_g in the gas phase:

$$\alpha_{Os} = C_i/C_g = T\alpha_{Bu}/273. \tag{1}$$

Other solubilities are expressed as the volume of gas dissolved per mole of glass per atmosphere partial pressure, and the weight fraction or percentage of gas dissolved in the glass. The Ostwald coefficient has the advantage that it does not depend directly on the external partial pressure of the gas.

III   MEASUREMENT OF MOLECULAR PERMEATION AND DIFFUSION IN GLASS

The permeation and diffusion of a gas through glass have usually been measured in two different ways: with a glass membrane (in the form of a tube or a bulb) or in a powder. Thin membranes of many different glasses can be made by blowing or polishing, but sometimes it is difficult to seal them into a vacuum system, in which case the powder technique is useful.

Adsorption of a gas on the glass surface has been considered to be a necessary prelude to its solution and diffusion in the glass; however, the amount of gas dissolved in glass is proportional to the gas pressure at pressures far above that required for saturation of surface adsorption sites. Thus it seems likely that the gas molecules can dissolve directly in the glass without surface adsorption. Equilibrium between the gas and gas dissolved at the glass surface occurs rapidly for molecular solubility, and surface adsorption and reactions are not involved.

A glass membrane is mounted in a vacuum system to measure permeation and diffusion of a gas through it. First the spaces on both sides of the membrane are pumped out, and then the permeating gas is introduced on one side at the desired pressure. The amount of gas that appears on the other side of the membrane is then measured with time. The amount of gas that permeates the glass is kept so low that the drop in pressure on the inlet side is negligible and the pressure on the outlet side is small. Under these conditions the concentrations of gas dissolved in the membrane at its two surfaces are constant, being equal to some value C_i on the inlet side and zero on the outlet side. Since thin-walled tubes or bulbs of comparatively large diameter are used in these experiments, the diffusion equation for a slab applies without appreciable error. The flux J of diffusing gas, which passes through unit surface area of the membrane of thickness L in unit time under the boundary conditions $C = C_i$ when $x = 0$, $C = 0$ when $x = L$, and $C = 0$ when $t = 0$, is

$$J = \frac{DC_i}{L}\left[1 + 2\sum_{n=0}^{\infty}(\cos n\pi)\exp\left(\frac{-Dn^2\pi^2t}{L^2}\right)\right] \tag{2}$$

where t is the time after emission of the gas. This equation can be transformed to

$$J = 2C_i \frac{D^{1/2}}{t} \sum_{n=0}^{\infty} \exp\left[\frac{-L^2(2n+1)^2}{4Dt}\right] \tag{3}$$

where the series converges rapidly for small t. For long times ($t \gg D/L^2$) these equations reduce to $J = DC_i/L$, which is the "steady-state" solution and implies a constant gradient throughout the membrane.

The measured pressure p of gas on the outlet side is related to J by the equation

$$J = \left(\frac{V}{ART_p}\right)\left(\frac{dp}{dt}\right)$$

where V is the volume on the outlet side where p is measured, A is the surface area of the glass membrane, R is the gas constant, and T_p is the temperature at which the pressure is measured. The results of permeation measurements in the steady state are usually expressed in terms of a permeation velocity K, which is defined as

$$K = \left(\frac{VL}{\Delta pA}\right)\left(\frac{dp}{dt}\right) = \frac{LRT_pJ}{\Delta p}$$

where Δp is the pressure difference across the membrane. The units of K generally used are cubic centimeters of gas at STP per square centimeter of glass area per sec per atm of gas pressure difference per cm of glass thickness. If the outlet pressure is negligible, $p = C_gRT$, where C_g is the concentration of gas on the inlet side in moles per cubic centimeter and T is the temperature of the diffusion experiment. Then, since $J = DC_i/L$, $K = DT_pC_i/C_gT$, where T_p is 273 K. The solubility α_{Bu} is defined as the volume of gas (at STP) dissolved in unit volume of glass per atmosphere of external gas pressure:

$$\alpha_{Bu} = \frac{C_iT_p}{C_gT} \quad \text{and} \quad K = D\alpha_{Bu}. \tag{4}$$

The units of K are also cubic centimeters per second, since α_{Bu} is dimensionless.

In the early stages of permeation D can be calculated from the measured change in pressure with time (dp/dt) in the outlet side. For this purpose Eq. (3) is useful, since for short times only the first term of the series is needed, so that a plot of $\log(t^{1/2}\,dp/dt]$ against $1/t$ is linear, because J is proportional to dp/dt; the slope of the line is equal to $L^2/4D$, from which D can be calculated since L is known. This method was suggested by Rogers et al. (1954) and gives the most accurate values of D from permeation data. The intercept of this line with the $\log(t^{1/2}\,dp/dt)$ axis ($1/t \to 0$) can also be used to calculate the solubility α_{Bu} if D is known. When the outlet pressure p_o is plotted against time, the curve becomes linear at long times (that is also the steady-state region), and the diffusion coefficient can as well be found from the intercept t_c of this line with the pressure axis from the relation $D = L^2/6t_e$, as shown by Barrer (1941).

If the gas on the inlet side is pumped out rapidly after the steady state is attained, the diffusion coefficient can be calculated from the rate at which the dissolved gas diffuses out of the membrane. The initial concentration for this desorption is given by $C = C_i(1 - x/L)$, and the boundary conditions are that $C = 0$ at both surfaces. The flux at the outlet is then

$$J = \frac{2C_iD}{L} \sum_{n=0}^{\infty} (\cos n\pi) \exp\left(\frac{-Dn^2\pi^2t}{L^2}\right). \tag{5}$$

Thus the diffusion coefficient and solubility can be calculated from permeability data in a variety of ways, and the comparison between the various calculations should give some of the reliability of the results.

In the powder method the diffusion coefficient is calculated from the amount of gas absorbed or desorbed from a powdered sample, and the solubility is found from the total amount of gas absorbed or desorbed per unit weight of glass. For desorption from spheres of radius R, the amount of gas desorbed Q is

$$\frac{Q}{Q_\infty} = \frac{6(Dt)^{1/2}}{\pi R} - \frac{3Dt}{R^2} \tag{6}$$

where Q_∞ is the total amount of gas desorbed. A comparison of the experimental curve with Eq. (6) gives a measure of D when t, R, and Q_∞ are known.

IV TABLES AND FIGURES

Data for binary glasses are presented in Tables 17.1–17.28 and Figs. 17.1–17.7, and data for ternary glasses in Tables 17.29–17.52 and Figs. 17.8–17.19.

BIBLIOGRAPHY

Barrer, R. M. (1941). "Diffusion in and Through Solids," Cambridge Univ. Press, Cambridge, England.
Barton, J. L., and Morain, M. (1970). J. Non-Cryst. Solids 3, 115.
Doremus, R. H. (1973). "Glass Science," Chap. 8. Wiley, New York.
Eschbach, H. L., Jaeckel, R., and Müller, D. (1963). Z. Naturforsch., 18A, 434.
Faile, S. P., and Roy, D. M. (1966). J. Am. Ceram. Soc. 49, 638.
Franz, H., and Scholze, H. (1963). Glastech. Ber. 36, 347.
Garbe, S. (1961). Glastech. Ber. 34, 413.
Haider, Z., and Roberts, G. J. (1970). Glass Technol. 11, 158.
Kröger, C., and Goldmann, N. (1962). Glastech. Ber. 35, 459.
Kurkjian, C. R., and Russell, L. E. (1958) J. Soc. Glass Technol. 42, 130.
May, H. B., Lauder, I., and Wollast, R. (1974). J. Am. Ceram. Soc. 57, 197.

Mulfinger, H.-O., and Scholze, H. (1962a). *Glastech. Ber.* **35,** 466.

Mulfinger, H.-O., and Scholze, H. (1962b). *Glastech. Ber.* **35,** 495.

Mulfinger, H.-O., Dietzel, A., and Navarro, J. M. F. (1972). *Glastech. Ber.* **45,** 389.

Nair, K. M., White, W. B., and Roy, R. (1965). *J. Am. Ceram. Soc.* **48,** 52.

Newkirk, T. F., and Tooley, F. V. (1949). *J. Am. Ceram. Soc.* **32,** 272.

Norton, F. J. (1953). *J. Am. Ceram. Soc.* **36,** 90.

Oishi, Y., Terai, R., and Ueda, H. (1975). *In* "Mass Transport Phenomena in Ceramics" (A. R. Cooper and A. H. Heuer, eds.), p. 297. Plenum, New York.

Orlova, G. P., and Rudnitskaya, E. S. (1965). *In* "Stekloobraznoe Sostoyanie," p. 161. Moscow.

Papadopoulos, K. (1973). *Phys. Chem. Glasses* **14,** 60.

Pashkeev, I. Y., Antonenko, V. I., and Kozheurov, V. A. (1969). *Russ. J. Phys. Chem.* (*Engl. Transl.*) **43,** 1772.

Pevzner, B. Z., Nyunin, G. I., and Appen, A. A. (1975). *Fiz. Khim. Stekla* **1,** 318.

Rogers, W. A., Burlitz, R. S., and Alpert, D. (1954). *J. Appl. Phys.* **25,** 868.

Roy, D. M., Faile, S. P., and Tuttle, O. F. (1964). *Phys. Chem. Glasses* **5,** 176.

Russell, L. E. (1957). *J. Soc. Glass Technol.* **41,** 304.

Scholze, H., and Mulfinger, H.-O. (1959). *Glastech. Ber.* **32,** 381.

Scholze, H., Mulfinger, H.-O., and Franz, H. (1962). *Adv. Glass Technol., Tech. Pap. Int. Congr. Glass, 6th, Washington, D.C.,* 230.

Shelby, J. E. (1972). *J. Am. Ceram. Soc.* **55,** 195.

Shelby, J. E. (1973a). *J. Am. Ceram. Soc.* **56,** 263.

Shelby, J. E. (1973b). *J. Am. Ceram. Soc.* **56,** 340.

Shelby, J. E. (1974). *J. Am. Ceram. Soc.* **57,** 260.

Shelby, J. E. (1975). *In* "Mass Transport Phenomena in Ceramics" (A. R. Cooper and A. H. Heuer, eds.), p. 367. Plenum, New York.

Shelby, J. E. (1977a). *Phys. Non-Cryst. Solids, Int. Conf., 4th, Clausthal-Zellerfeld, Ger., 1976* p. 509.

Shelby, J. E. (1977b). *J. Appl. Phys.* **48,** 1497.

Shelby, J. E. (1978). *J. Appl. Phys.* **49,** 2748.

Shelby, J. E. (1979a). Molecular solubility and diffusion. *In* "Glass II" (M. Tomozawa and R. H. Doremus, eds.), Treatise on Materials Science and Technology, Vol. 17, p. 1. Academic Press, New York.

Shelby, J. E. (1979b). *J. Appl. Phys.* **50,** 8010.

Shelby, J. E. (1981). *J. Non-Cryst. Solids* **45,** 411.

Shelby, J. E., and Eagan, R. J. (1976). *J. Am. Ceram. Soc.* **59,** 420.

Shelby, J. E., and Keeton, S. C. (1974). *J. Appl. Phys.* **45,** 1458.

Shelby, J. E., and McVay, G. L. (1975). *J. Am. Ceram. Soc.* **58,** 147.

Wollast, R., Lauder, I., and May, H. (1973). *Silic. Ind.* **38,** 169.

Woods, K. N., and Doremus, R. H. (1971). *Phys. Chem. Glasses* **12,** 69.

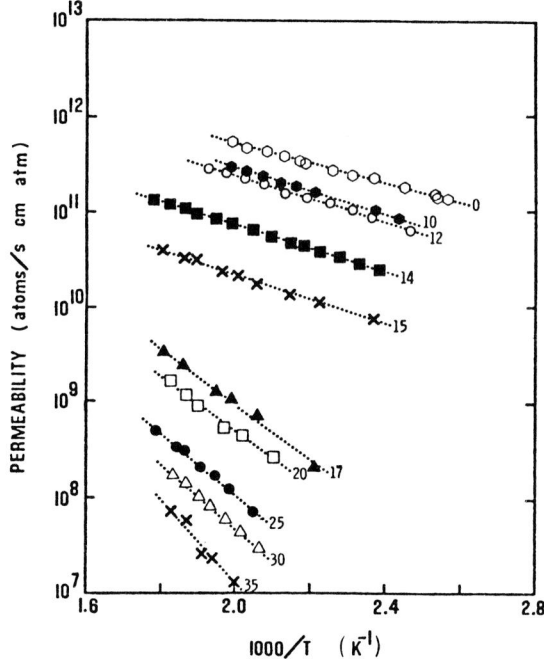

Fig. 17.1. Permeability of helium in Li_2O-SiO_2 glasses as a function of temperature. The number on each curve is the mole % Li_2O in the glass. From Shelby (1977b).

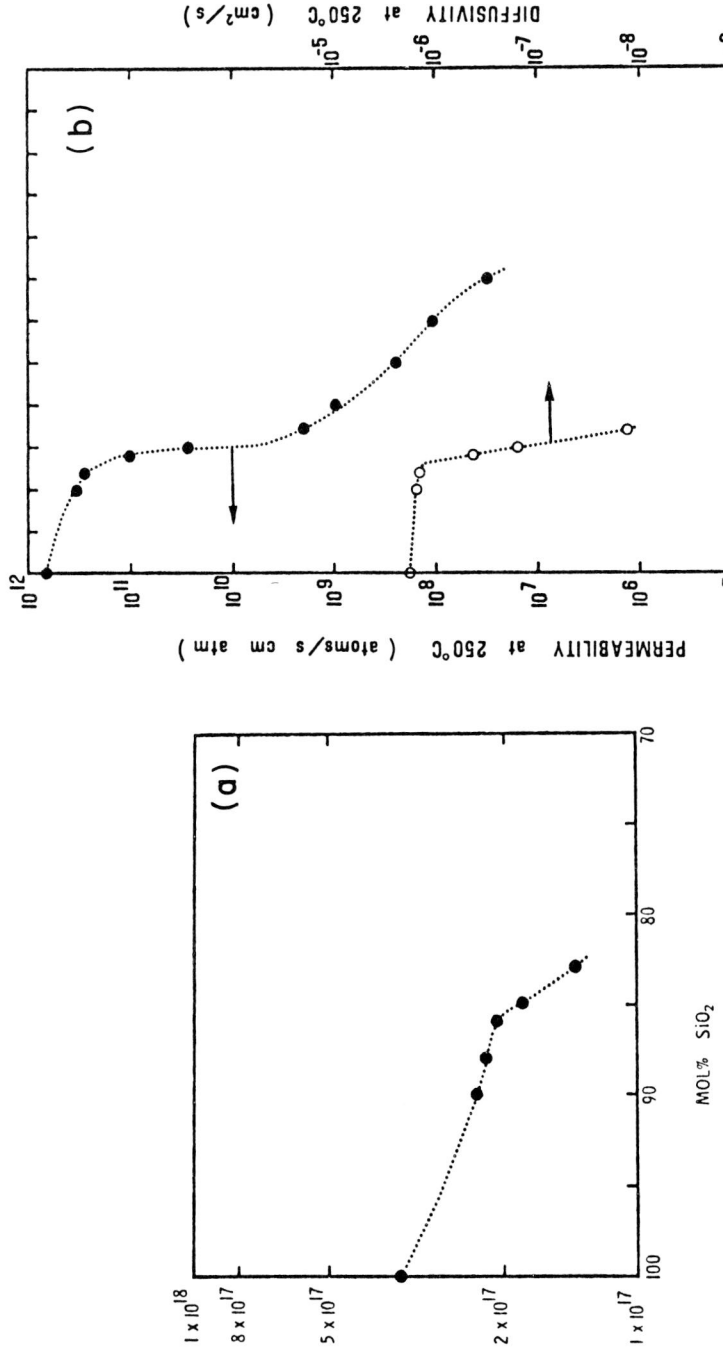

Fig. 17.2. Composition dependence of the isothermal (a) solubility and (b) permeability and diffusivity of helium in $Li_2O–SiO_2$ glasses at 250°C. From Shelby (1977b).

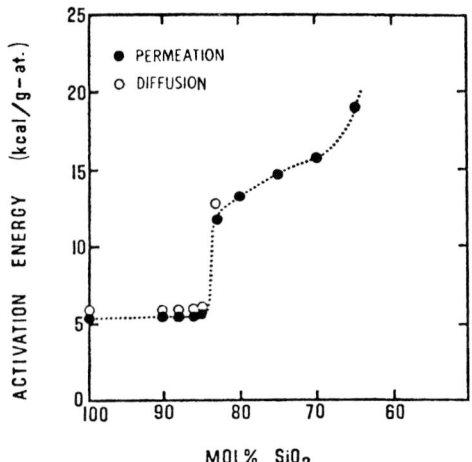

Fig. 17.3. Composition dependence of the Arrhenius activation energies for permeability and diffusivity of helium in Li_2O-SiO_2 glasses. From Shelby (1977b).

TABLE 17.1

Solubilities of Various Gases in Lithium Silicate Glass Melts

			Solubility[a]			
Species	Composition (mole % Li_2O)	Temperature (°C)	$S \times 10^3$ (N cm^3 mole^{-1} atm^{-1})	$\alpha_{Bu} \times 10^3$	$\alpha_{Os} \times 10^3$	Reference[b]
He	20	1400	102.63	4.12	25.24	1, 2
	20	1480	102.46	4.11	26.38	
	25	1400	79.58	3.26	19.97	
	25	1480	82.70	3.33	21.63	
Ne	20	1480		2.32		2
	25	1480		1.79		
	30	1480		1.63		
N_2	25	1480		1.17		2
	30	1480		1.35		

[a] α_{Bu}, Bunsen coefficient; α_{Os}, Ostwald coefficient.
[b] Data from (1) Mulfinger and Scholze (1962a) and (2) Mulfinger et al. (1972).

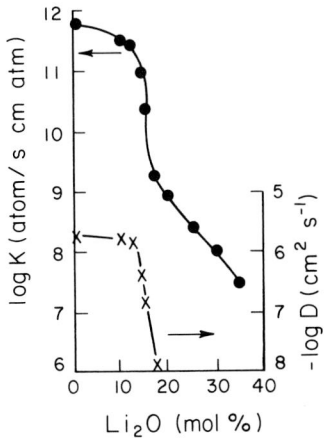

Fig. 17.4. Composition dependence of gas permeability and diffusivity of helium in lithium silicate glasses at 250°C. From Shelby (1977a).

TABLE 17.2

Diffusion Coefficients of Helium and Water in Lithium Silicate Glass Melts

Species	Composition (mole % Li_2O)	Temperature (°C)	$D \times 10^6$ ($cm^2\,s^{-1}$)	References[a]
He	20	1400	115	1
		1480	143	
	25	1400	115	
		1480	173	
H_2O	25.1	1300	4.1	2
		1400	6.2	
	29.7	1100	2.0	
		1200	3.1	
		1300	4.2	
		1400	6.7	

[a] Data from (1) Mulfinger and Scholze (1962b) and (2) Scholze and Mulfinger (1959).

TABLE 17.3

**Solubilities of Water in Lithium Silicate Glass Melts at
Water Vapor Pressure of 760 torr**[a]

Composition (mole % Li_2O)	Temperature (°C)	Solubility of H_2O	
		mole % H_2O	Ostwald coefficient
2.0	1700	0.152	9.2
4.5	1700	0.170	10.4
9.6	1600	0.222	13.2
13.8	1600	0.252	15.2
17.9	1480	0.308	17.7
27.5	1480	0.400	23.6
27.7	1400	0.391	22.3
27.9	1320	0.382	20.8
28.0	1250	0.377	19.8
37.0	1480	0.475	29.4
37.5	1400	0.461	27.4
37.8	1320	0.454	25.8
37.9	1250	0.445	24.4

[a] Data from Franz and Scholze (1963) and Scholze et al. (1962).

TABLE 17.4

Parameters of the Equations $S = S_0 \exp(-E_s/RT)$ and $K = K_0 T \exp(-E_k/RT)$ for the Solubility and Permeability of Helium in Sodium Silicate Glasses[a]

Composition (mole % Na_2O)	Temperature range (°C)	$S_0 \times 10^{-17}$ (atoms cm^{-3} atm^{-1})	E_s (cal g-atom^{-1})	$K_0 \times 10^{-10}$ (atoms s^{-1} cm^{-1} K^{-1} atm^{-1})	E_k (cal g-atom^{-1})
4.32	64–194	3.43	−78 ± 89	5.7	4623 ± 86
8.30	141–202	3.31	−15 ± 240	2.9	4755 ± 186
9.88	121–219	2.46	−68 ± 141	2.3	5003 ± 188
12.26	146–224	2.91	251 ± 226	3.9	5971 ± 174
13.72	157–251	5.23	874 ± 100	12.3	7327 ± 33
15.58	148–224	2.15	225 ± 165	3.6	6844 ± 121
18.32	146–207	1.37	−118 ± 158	7.2	7878 ± 184
19.00	165–216	0.85	15 ± 1210	9.8	8646 ± 286
22.50	172–250	8.73	2255 ± 529	13.7	9609 ± 433
24.81	184–256	4.75	1843 ± 384	18.8	10,318 ± 140
29.59	217–284	15.2	3658 ± 520	48.5	12,399 ± 233
33.05	222–276	18.2	4031 ± 525	36.8	12,566 ± 585
36.07	229–298	16.3	4589 ± 387	55.0	13,617 ± 911
39.74	254–271	—	—	~117	~14,975

[a] Data from Shelby (1973a).

TABLE 17.5

Permeability of Helium in Sodium Silicate Glasses[a]

Composition (mole % Na$_2$O)	Temperature (°C)	$K \times 10^{13}$ (g mm cm^{-2} s^{-1} atm^{-1})
20	300	5.3
25	300	3.1
30	300	1.4
	373	2.28[b]
	403	2.38[b]

[a] Data source: graph in Newkirk and Tooley (1949).

TABLE 17.6

Solubilities of Oxygen in Na$_2$O–2SiO$_2$ at 900°C as a Function of Oxygen Pressure[a]

P_{O_2} (MPa)	S (wt. %)
17	0.117
21	0.267
29	0.237

[a] Data read from graph of Nair et al. (1965).

TABLE 17.7

Solubilities of Helium and Neon in Sodium Silicate Glass Melts[a]

Species	Composition (mole % Na$_2$O)	Temperature (°C)	$\alpha_{Bu} \times 10^4$ [b]
He	20	1400	43.2
		1480	45.8
	25	1400	32.7
		1480	35.7
Ne	20	1480	22.6
	25	1480	21.6
	30	1480	18.7

[a] Data from Mulfinger et al. (1972).
[b] α_{Bu}, Bunsen coefficient.

TABLE 17.8

Solubility of Water in Sodium Silicate Glass Melts under the Water Vapor Pressure of 760 torr[a]

Composition (mole % Na_2O)	Temperature (°C)	Solubility of H_2O	
		(mole % H_2O)	Ostwald coefficient
2.6	1700	0.164	9.8
5.9	1600	0.213	12.1
9.5	1600	0.275	15.6
13.7	1480	0.334	17.8
17.5	1480	0.383	20.3
17.75	1400	0.375	19.1
17.9	1320	0.370	17.9
18.0	1250	0.365	17.0
25.6	1480	0.515	27.0
25.8	1400	0.500	25.0
25.9	1320	0.488	23.4
26.0	1250	0.477	22.1
26.65	1480	0.525	27.6
27.2	1400	0.509	25.5
27.6	1320	0.495	23.8
27.9	1250	0.483	22.4
36.2	1480	0.670	34.6
37.0	1400	0.655	32.6
37.5	1320	0.635	30.2
37.8	1250	0.621	28.5

[a] Data from Franz and Scholze (1963) and Scholze *et al.* (1962).

TABLE 17.9

Parameters of the Equation $D = D_0 T \exp(-E_D/RT)$ for the Diffusivity of Helium in Sodium Silicate Glasses[a]

Composition (mole % Na_2O)	Temperature range (°C)	$D_0 \times 10^7$ (cm^2 s^{-1} K^{-1})	E_D (cal g-atom^{-1})
4.32	64–194	6.6	5482 ± 79
8.30	141–202	4.8	5825 ± 54
9.88	121–219	4.2	5946 ± 113
12.26	146–224	6.2	6628 ± 125
13.72	157–251	11.1	7396 ± 87
15.58	148–224	7.4	7480 ± 85
18.32	146–207	23.4	8886 ± 280
19.00	165–216	23.6	9097 ± 82
22.50	172–250	16.3	9075 ± 271
24.81	184–256	19.4	9452 ± 287
29.59	217–284	16.6	9718 ± 448
33.05	222–276	10.5	9571 ± 440
36.07	229–298	17.9	10,090 ± 546
39.74	254–271	15.0	10,156 ± 195

[a] Data from Shelby (1973a).

TABLE 17.10

Arrhenius Parameters for the Diffusion of Oxygen-18 in Sodium Silicate Glass Melt[a]

Composition (mole % Na_2O)	Temperature range (°C)	$D_0 \times 10^3$ (cm² s⁻¹)	E_D (kcal mole⁻¹)
20	1061–1395	79	44.5

[a] Data from Oishi *et al.* (1975).

TABLE 17.11

Diffusion Coefficients of Various Gases in Sodium Silicate Glass Melts

Species	Composition (mole % Na_2O)	Temperature (°C)	$D \times 10^6$ (cm² s⁻¹)	References[a]
H_2O	20.2	1200	1.2	1
		1300	2.4	
		1400	3.3	
	24.9	1000	0.85	
		1100	1.3	
		1200	2.6	
		1300	3.4	
		1400	5.3	
	27.2	1000	1.0	
		1100	1.7	
		1200	2.6	
		1300	3.7	
		1400	6.2	
He	20	1400	205	2
		1480	261	
	25	1400	229	
		1480	301	
CO_2	25	1300	10	3

[a] Data from (1) Scholze and Mulfinger (1959), (2) Mulfinger and Scholze (1962b), and (3) Kröger and Goldmann (1962).

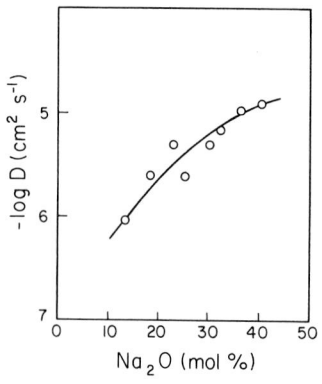

Fig. 17.5. Composition dependence of the diffusion coefficient of water in sodium silicate glasses at 900°C. From Garbe (1961).

TABLE 17.12

Parameters of the Equations $S = S_0 \exp(-E_s/RT)$ and $K = K_0 T \exp(-E_k/RT)$ for the Solubility and Permeability of Helium in Potassium Silicate Glasses[a]

Composition (mole % K_2O)	Temperature range (°C)	$S_0 \times 10^{-17}$ (atoms cm^{-3} atm^{-1})	E_s (cal g-atom^{-1})	$K_0 \times 10^{-10}$ (atoms s^{-1} cm^{-1} K^{-1} atm^{-1})	E_k (cal g-atom^{-1})
10	147–226	1.4	−34	9.0	6160
12	139–213	1.3	−11	10.1	6490
16	139–222	0.89	+32	10.7	7230
20	145–227	0.42	−443	10.2	7690
24	152–231	0.26	−558	7.2	7980
28	145–227	0.50	+289	27.3	9670
32	170–234	0.41	+600	17.4	10,024

[a] Data from Shelby (1974).

TABLE 17.13

Effect of Pressure on the Solubility of Various Gases in $K_2O-4SiO_2$ Glass[a]

Gas	Temperature (°C)	Pressure (GPa)	Solubility (mole % gas)
Ar	750	0.20	0.3
		0.37	1.3
		0.50	2.4
	800	0.02	1.5
		0.09	1.95
		0.19	3.1
		0.94	6.9
N_2	800	0.25	1.5
		0.37	2.1
CO_2	800	0.05	1.0
		0.19	4.0
		0.22	5.0
		0.39	7.7

[a] Data read from graph of Faile and Roy (1966) and Roy et al. (1964).

TABLE 17.14

Solubilities of Helium, Neon, and Nitrogen in Potassium Silicate Glass Melts[a]

Species	Composition (mole % K_2O)	Temperature (°C)	$\alpha_{Bu} \times 10^4$ [b]
He	16.5	1400	51.8
		1480	54.8
	20	1400	45.5
		1480	49.9
	24	1400	37.5
		1480	43.4
Ne	16.5	1480	31.5
	24	1480	27.6
	30	1480	27.6
N_2	16.5	1480	12.0
	20	1480	12.1
	23	1480	11.7

[a] Data from Mulfinger et al. (1972).
[b] α_{Bu}, Bunsen coefficient.

TABLE 17.15

Solubility of Water in Potassium Silicate Glass Melts under the Water Vapor Pressure of 760 torr[a]

Composition (mole % K_2O)	Temperature (°C)	Solubility of H_2O	
		(mole % H_2O)	Ostwald coefficient
1.3	1700	0.149	8.8
4.5	1600	0.219	12.3
5.1	1480	0.241	12.6
7.5	1480	0.294	15.1
10.7	1480	0.363	18.2
14.4	1480	0.485	23.7
16.0	1400	0.485	22.5
17.0	1320	0.482	21.2
17.6	1250	0.480	20.2
26.7	1480	0.805	36.1
27.25	1400	0.781	33.7
27.6	1320	0.760	31.3
27.85	1250	0.745	29.3
35.6	1480	1.075	45.6
36.65	1400	1.048	42.5
37.25	1320	1.010	39.6
37.7	1250	0.984	37.9

[a] Data from Franz and Scholze (1963) and Scholze et al. (1962).

TABLE 17.16

Parameters of the Equation $D = D_0 T \exp(-E_D/RT)$ for the Diffusivity of Helium in Potassium Silicate Glasses[a]

Composition (mole % K_2O)	Temperature range (°C)	$D_0 \times 10^7$ (cm^2 s^{-1} K^{-1})	E_D (cal g-atom^{-1})
10	147–226	6.3	6200
12	139–213	7.8	6500
16	139–222	12.0	7200
20	145–227	24.0	8130
24	152–231	28.1	8540
28	145–227	54.8	9390
32	170–234	42.5	9420

[a] Data from Shelby (1974).

TABLE 17.17

Arrhenius Parameters for the Diffusion of Oxygen-18 in Potassium Silicate Glass Melts[a]

Composition (mole % K_2O)	Temperature range (°C)	D_0 (cm^2 s^{-1})	E_D (kcal mole^{-1})
17.5	750–1000	40^{+5}_{-4}	61.7
26.4	700–1000	$0.24^{+0.05}_{-0.04}$	46.0

[a] Data from Wollast et al. (1973) and May et al. (1974).

TABLE 17.18

Diffusion Coefficients of Helium and Water in Potassium Silicate Glass Melts

Species	Composition (mole % K_2O)	Temperature (°C)	$D \times 10^6$ (cm^2 s^{-1})	References[a]
He	16	1400	219	1
		1480	262	
	19	1400	228	
		1480	303	
	23	1400	281	
		1480	359	
H_2O	19.9	1000	0.18	2
		1100	0.36	
		1200	0.79	
		1300	1.5	
		1400	2.7	
	23.2	1000	0.23	
		1100	0.60	
		1200	1.5	
		1300	2.4	
		1400	4.5	
	29.1	1000	0.64	
		1100	1.2	
		1200	2.0	
		1300	3.8	
		1400	5.5	

[a] Data from (1) Mulfinger and Scholze (1962b) and Scholze and Mulfinger (1959).

TABLE 17.19

Diffusion Coefficients of Water in Rubidium Silicate Glass Melts[a]

Composition (mole % Rb_2O)	Temperature (°C)	$D \times 10^6$ (cm^2 s^{-1})
17.1	1000	0.07
	1100	0.17
	1200	0.36
	1300	0.65
	1400	1.3
19.1	1000	0.17
	1100	0.37
	1200	0.77
	1300	1.5
	1400	2.5

[a] Data from Scholze and Mulfinger (1959).

TABLE 17.20

Solubility of H$_2$O in Strontium (33 mole % SrO) Silicate Glass Melts at 1 atm Partial Pressure of Water Vapor[a]

Temperature (°C)	Solubility of H$_2$O[b] (wt. %)
1400	0.038
1500	0.059
1600	0.091

[a] Data from Russell (1957).
[b] Values read from graph.

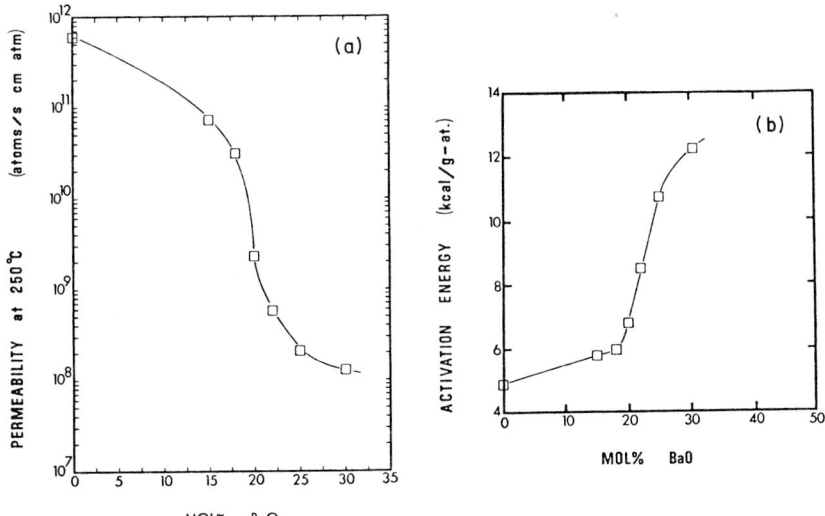

Fig. 17.6. Effect of glass composition on helium (a) permeability at 250°C and (b) activation energy for permeation for BaO–SiO$_2$ system. From Shelby (1979b).

TABLE 17.21

Solubility of H$_2$O in (33 mole % BaO) Barium Silicate Glass Melt at 1 atm Partial Pressure of Water Vapor[a]

Temperature (°C)	Solubility of H$_2$O[b] (wt. %)
1420	0.04
1510	0.043
1600	0.054

[a] Data from Russell (1957).
[b] Values read from graph.

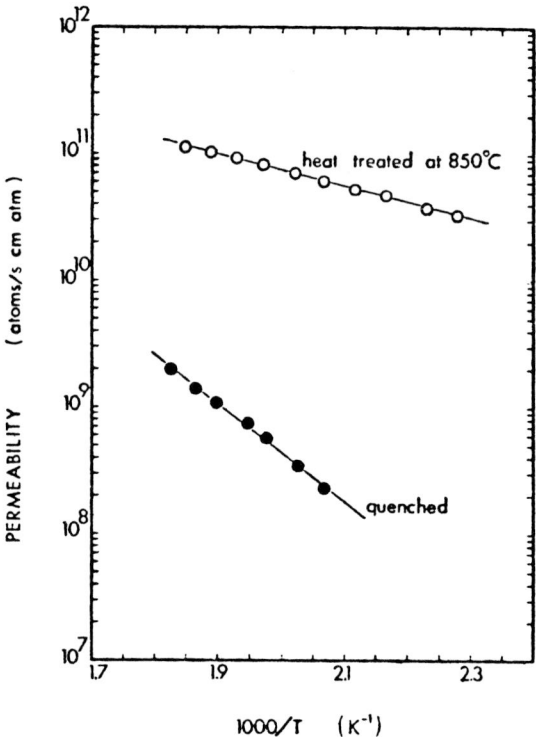

Fig. 17.7. Effect of heat treatment (1 hr at 850°C) on helium permeability in 15BaO–85SiO$_2$ glass. From Shelby (1979b).

TABLE 17.22

Solubility of CO$_2$ in B$_2$O$_3$–SiO$_2$ Glass (75 mole % B$_2$O$_3$)[a]

Temperature (°C)	Time of pretreatment with CO$_2$ (min)	Pressure (atm)	Solubility of CO$_2$ [N (cm^3 CO$_2$) (g glass)$^{-1}$]
—	0	—	0.33×10^{-2}
600	30	25	0.83×10^{-2}
600	30	50	5.38×10^{-2}

[a] Data from Kröger and Goldmann (1962).

TABLE 17.23

Solubility of CO_2 in B_2O_3–SiO_2 Glass Melts[a]

Composition (mole % B_2O_3)	Temperature (°C)	Time (min)	Solubility of CO_2 $[(g\ CO_2)\ (g\ melt)^{-1}]$
50.0	1100	90	2.0×10^{-7}
	1300	90	2.0×10^{-7}
	1500	90	9.7×10^{-7}
66.7	900	90	1.0×10^{-6}
	900	360	2.1×10^{-6}
	1100	45	9.3×10^{-7}
	1100	90	1.0×10^{-6}
	1100	180	1.7×10^{-6}
	1100	360	1.9×10^{-6}
	1300	90	1.3×10^{-6}
75.0	1100	90	2.5×10^{-7}
	1300	45	9.0×10^{-8}
	1300	90	1.7×10^{-7}
	1300	180	1.9×10^{-7}
	1500	90	1.3×10^{-7}

[a] Data from Kröger and Goldmann (1962).

TABLE 17.24

Permeability of Helium at 40°C
in B_2O_3–SiO_2 Glasses[a]

Composition (mole % B_2O_3)	$K \times 10^{-11}$ (atoms s^{-1} cm^{-1} atm^{-1})
40	2
75	2.27
80	2
94.5	1.24
97.5	1.13
100	0.6

[a] Data read from graph of Shelby (1975).

TABLE 17.25

Diffusion Coefficient of CO_2 in B_2O_3–SiO_2 Glass Melt[a]

Composition (mole % B_2O_3)	Species	Temperature (°C)	$D \times 10^5$ (cm^2 s^{-1})
66.7	CO_2	1100	1.5

[a] Data from Kröger and Goldmann (1962).

TABLE 17.26

Parameters of the Equations $S = S_0 \exp(-E_s/RT)$ and $K = K_0 T \exp(-E_k/RT)$ for the Solubility and Permeability of Helium and Neon in TiO_2–SiO_2 Glasses

Species	Composition (mole % TiO_2)	Temperature range (°C)	$S_0 \times 10^{-17}$ (atoms cm^{-3} atm^{-1})	E_s (cal g-atom^{-1})	$K_0 \times 10^{-10}$ (atoms cm^{-1} atm^{-1} °C^{-1} s^{-1})	E_k (cal g-atom^{-1})	References[a]
He	2.3	219–648	2.1 ± 0.3	667 ± 84	7.1 ± 1.0	4162 ± 55	1
	5.1	219–691	2.4 ± 0.2	598 ± 105	8.0 ± 0.7	4075 ± 66	
	6.6	148–734	2.3 ± 0.2	455 ± 80	7.6 ± 0.4	4153 ± 60	
	6.6	282–619	1.6 ± 0.2	713 ± 152	5.5 ± 0.5	3955 ± 102	
	10.3	210–681	1.9 ± 0.1	674 ± 117	6.3 ± 0.3	3867 ± 75	
	10.3	229–619	1.6 ± 0.2	674 ± 138	5.7 ± 0.4	3928 ± 116	
Ne	2.3	498–834	0.87	1382 ± 271	0.76	7760 ± 101	2
	6.6	481–876	0.92	1294 ± 315	0.84	7693 ± 137	
	10.3	438–850	1.55	613 ± 202	0.99	7602 ± 120	

[a] Data from (1) Shelby (1972) and (2) Shelby (1973b).

TABLE 17.27

Parameters of the Equation $D = D_0 T \exp(-E_D/RT)$ for the Diffusivity of Helium and Neon in TiO_2-SiO_2 Glasses

Species	Composition (mole % TiO_2)	Temperature range (°C)	$D_0 \times 10^7$ (cm² s⁻¹ K⁻¹)	E_D (cal g-atom⁻¹)	References[a]
He	2.3	219–648	3.45 ± 0.11	4829 ± 29	1
	5.1	219–691	3.27 ± 0.09	4673 ± 39	
	6.6	148–734	3.24 ± 0.09	4608 ± 20	
	6.6	282–619	3.40 ± 0.10	4668 ± 50	
	10.3	210–681	3.31 ± 0.10	4541 ± 42	
	10.3	229–619	3.52 ± 0.12	4602 ± 22	
Ne	2.3	498–834	0.87	9142 ± 170	2
	6.6	481–876	0.91	8987 ± 178	
	10.3	438–850	0.64	8215 ± 82	

[a] Data from (1) Shelby (1972) and (2) Shelby (1973b).

TABLE 17.28

Parameters of the Equation $D = D_0 \exp(\sigma^2/4R^2T^2) \exp(-\varepsilon/RT)$ for the Diffusivity of Helium in TiO_2-SiO_2 Glasses[a]

Composition (mole % TiO_2)	Temperature range (°C)	$D_0 \times 10^4$ (cm² s⁻¹)	ε (cal g-atom⁻¹)	σ (cal g-atom⁻¹)
2.3	219–759	10.4	7490	1860
5.1	219–691	11.9	7820	2180
6.6	148–734	10.6	7440	1960
8.8	201–709	9.6	7360	1840
10.3	210–681	8.6	6760	1520

[a] Data from Shelby and Keeton (1974).

TABLE 17.29

Solubility of Water in $Li_2O-K_2O-SiO_2$ Glass Melts under the Partial Pressure of Water at 760 torr[a]

Composition (mole %)			S_{H_2O} (mole %) at given temperature (°C)			
Li_2O	K_2O	SiO_2	1250	1320	1400	1480
5	15	80	0.513	0.521	0.529	0.540
10	10	80	0.450	0.455	0.461	0.468
15	5	80	0.378	0.383	0.390	0.398

[a] Data from Franz and Scholze (1963).

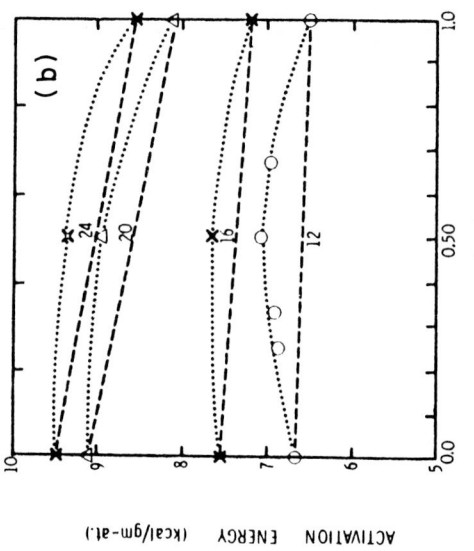

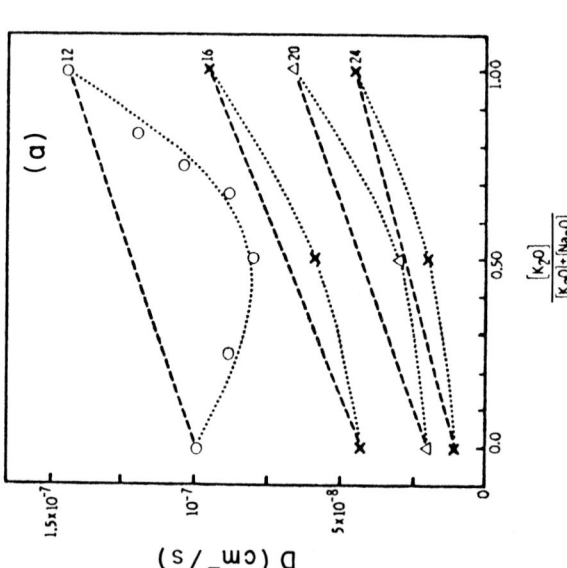

Fig. 17.8. Effect of composition on (a) helium diffusivity at 150°C and (b) activation energy for helium diffusion based on the expression $D = D_0 T \exp(-E/RT)$ in Na_2O–K_2O–SiO_2 glasses. The number on each curve is the total alkali oxide content in mole %. From Shelby and McVay (1975).

TABLE 17.30

Solubility of Water in Na₂O–K₂O–SiO₂ Glass Melts at 760 torr Partial Pressure of Water Vapor[a]

Composition (mole %)			S_{H_2O} (mole %) at given temperature (°C)			
Na₂O	K₂O	SiO₂	1250	1320	1400	1480
5	15	80	0.548	0.555	0.563	0.572
10	10	80	0.494	0.502	0.510	0.518
15	5	80	0.455	0.460	0.466	0.474

[a] Data from Franz and Scholze (1963).

TABLE 17.31

Solubility of Water in Na₂O–MgO–SiO₂ Glass Melt at 1 atm Partial Pressure of Water Vapor[a]

Composition (mole %)			Temperature (°C)	Solubility of H₂O	
Na₂O	MgO	SiO₂		mole %	Ostwald solubility coefficient
15	10	75	1250	0.327	16.2
			1320	0.332	17.1
			1400	0.336	18.1
			1480	0.341	19.2

[a] Data from Franz and Scholze (1963) and Scholze *et al.* (1962).

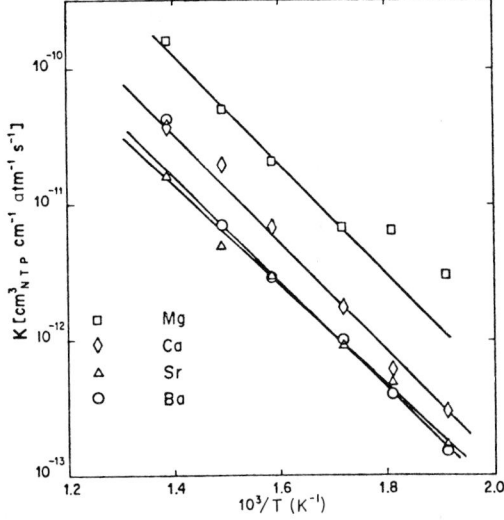

Fig. 17.9. Temperature dependence of hydrogen gas permeability in 13Na₂O–15MO–72SiO₂ (M = Mg, Ca, Sr, Ba) glasses. From Barton and Morain (1970).

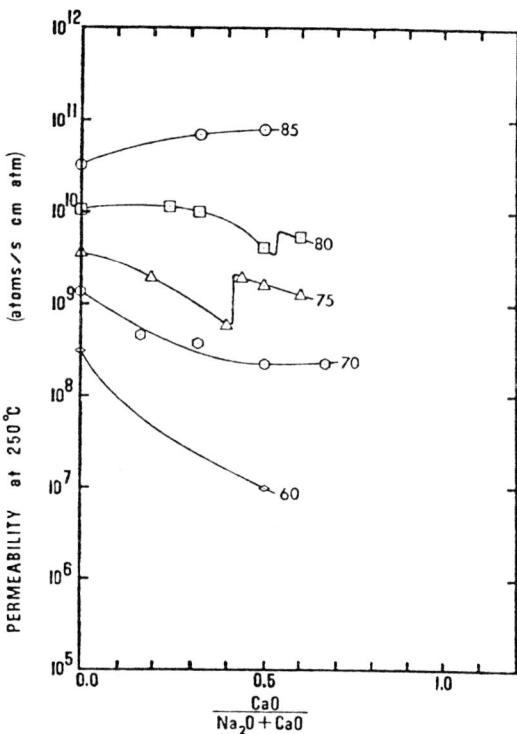

Fig. 17.10. Effect of composition on helium permeability at 250°C in soda-lime silicate glasses. The number on each curve is mole % SiO_2. From Shelby (1978).

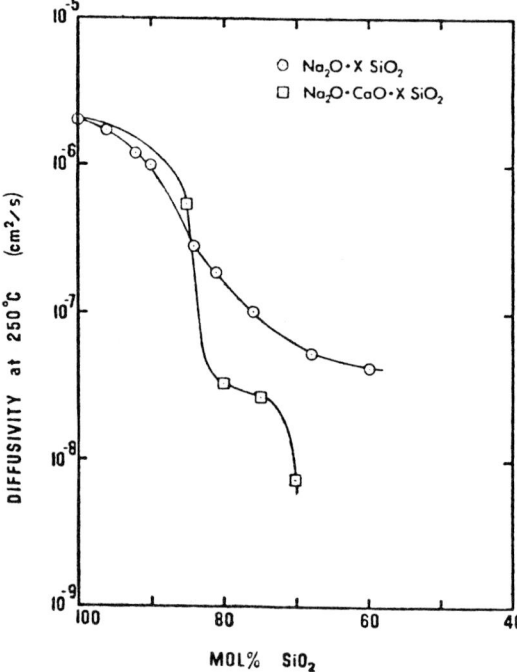

Fig. 17.11. Effect of silica content on the helium diffusivity at 250°C in soda-lime silicate glasses. From Shelby (1978).

TABLE 17.32

Solubilities and the Activation Energies for the Solubility of Helium Gas in Soda-Lime Silicate Glass Melts[a]

Composition (mole %)			Temperature (°C)	Solubility of helium[b]			E_s (kcal g-atom^{-1})
Na_2O	CaO	SiO_2		$\alpha_{Bu} \times 10^3$	$\alpha_{Os} \times 10^3$	$S \times 10^3$ [N cm^3 (mole glass)$^{-1}$]	
12.3	7.5	80.2	1400	4.30	26.34	111.80	4.0
			1480	4.32	27.79	112.93	
15.8	10.1	74.1	1200	3.29	17.74	83.66	4.9
			1300	3.39	19.52	86.75	
			1400	3.48	21.32	89.61	
			1480	3.61	23.17	93.39	
22.4	14.0	63.6	1200	2.22	11.97	55.28	7.1
			1300	2.41	13.88	60.56	
			1400	2.53	15.50	64.14	
			1480	2.77	17.78	70.72	

[a] Data from Mulfinger and Scholze (1962a) and Mulfinger et al. (1972).
[b] α_{Bu}, Bunsen coefficient; α_{Os}, Ostwald coefficient.

TABLE 17.33

Diffusion Coefficients and the Activation Energy for Diffusion of Helium in Soda-Lime Silicate Glass Melts[a]

Composition (mole %)			Temperature (°C)	$D \times 10^6$ (cm^2 S^{-1})	E_D (kcal g-atom^{-1})
Na$_2$O	CaO	SiO$_2$			
12.3	7.5	80.2	1400	148	8.4
			1480	166	
15.8	10.1	74.1	1200	115	10.0
			1300	142	
			1400	165	
			1480	200	
22.4	14.0	63.6	1200	110	11.3
			1300	148	
			1400	163	
			1480	210	

[a] Data from Mulfinger and Scholze (1962a,b).

TABLE 17.34

Solubilities of Nitrogen and Neon Gases in Soda-Lime Silicate Glass Melts[a]

Composition (mole %)			Temperature (°C)	Bunsen solubility coefficient $\alpha_{Bu} \times 10^4$	
Na$_2$O	CaO	SiO$_2$		N$_2$	Ne
16	10	74	1200	—	13.5
			1300	3.7	16.2
			1400	4.2	16.4
			1480	5.7	16.7

[a] Data from Mulfinger et al. (1972).

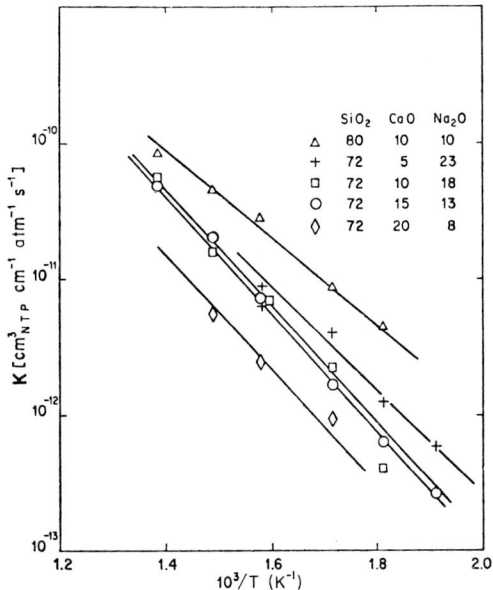

Fig. 17.12. Temperature dependence of hydrogen gas permeability in soda-lime silicate glasses of various compositions. From Barton and Morain (1970).

TABLE 17.35

Solubility of SO$_3$ in Soda-Lime Silicate Glass Melts at 1370°C under 0.01 atm Partial Pressure of SO$_3$[a]

Composition (mole %)								
Batch composition			By X-ray analysis					S_{SO_3} (mole %)
Na$_2$O	CaO	SiO$_2$	Na$_2$O	CaO	SiO$_2$	MgO	Al$_2$O$_3$	
13.5	14	72.5	12.7	13.6	72.1	0.5	0.2	0.745
			12.3	14.2	72.2	0.4	0.1	0.636
18.5	9	72.5	15.9	8.9	74.1	0.3	0.1	0.664
8.5	19	72.5	8.3	20.0	70.7	0.5	0.1	0.280
			8.3	19.3	71.5	0.5	0.1	0.280
			8.5	19.6	71.0	0.5	0.1	0.238
13.5	19	67.5	13.3	19.5	65.3	0.5	0.2	1.216
			13.2	19.6	65.3	0.4	0.2	1.265
			13.1	19.4	65.7	0.5	0.2	1.050
			12.7	19.5	66.1	0.5	0.2	0.967
18.5	14	67.5	15.4	14.4	68.5	0.4	0.2	1.066
13.5	9	77.5	12.3	9.0	77.8	0.4	0.1	0.360
8.5	14	77.5	7.1	14.4	77.8	0.4	0.1	0.133

[a] Data from Papadopoulos (1973).

TABLE 17.36

Temperature Dependence of SO_3 Solubility in Soda-Lime Silicate Glass Melts under 0.01 atm Partial Pressure of SO_3[a]

Composition (mole %) by X-ray analysis					Temperature ($^\circ$C)	S_{SO_3} (mole %)
Na_2O	CaO	SiO_2	MgO	Al_2O_3		
13.2	19.6	65.3	0.4	0.2	1373	1.20
12.8	19.8	66.0	0.5	0.2	1397	1.21
13.1	19.4	65.9	0.4	0.2	1403	1.10
12.3	18.7	67.7	0.5	0.2	1427	0.86
12.8	19.2	66.9	0.4	0.2	1433	0.60
12.8	20.1	66.0	0.5	0.2	1451	0.65
12.9	19.4	66.5	0.6	0.2	1453	0.52
12.5	19.9	66.6	0.5	0.2	1483	0.43

[a] Data from Papadopoulos (1973).

TABLE 17.37

Solubility of Water in Soda-Lime Silicate Glasses under the Partial Pressure of Water Vapor at 1 atm[a]

Composition (mole %)				Solubility of H_2O at given temperature ($^\circ$C)			
Na_2O	CaO	SiO_2	Unit[b]	1250	1320	1400	1480
15	5	80	mole %	0.358	0.364	0.368	0.376
			α_{Os}	17.0	18.0	19.0	20.3
15	10	75	mole %	0.344	0.348	0.355	0.363
			α_{Os}	16.7	17.6	18.8	20.0
15	15	70	mole %	0.336	0.341	0.347	0.357
			α_{Os}	16.6	17.6	18.7	20.0
16	10	74	mole %	0.352	0.356	0.362	0.370
			α_{Os}	17.1	18.0	19.1	20.4
20	5	75	mole %	0.382	0.387	0.392	0.400
			α_{Os}	18.1	19.1	20.2	21.5
20	10	70	mole %	0.361	0.366	0.370	0.378
			α_{Os}	17.4	18.4	19.5	20.7
20	15	65	mole %	0.346	0.350	0.360	0.367
			α_{Os}	17.1	18.0	19.3	20.4
21	5	74	mole %	0.389	0.392	0.396	0.400
			α_{Os}	18.4	19.3	20.4	21.5
25	5	70	mole %	0.416	0.421	0.428	0.435
			α_{Os}	19.6	20.7	21.9	23.2
25	10	65	mole %	0.392	0.396	0.400	0.405
			α_{Os}	18.9	19.9	20.9	22.0
25	15	60	mole %	0.361	0.371	0.378	0.384
			α_{Os}	17.8	19.0	20.2	21.3

[a] Data from Scholze et al. (1962) and Franz and Scholze (1963).
[b] α_{Os}, Ostwald solubility coefficient.

TABLE 17.38

Solubility of Water in $Na_2O-SrO-SiO_2$ Glass Melt under 1 atm Partial Pressure of Water Vapors[a,b]

Composition (mole %)			Temperature (°C)	Solubility of H_2O	
Na_2O	SrO	SiO_2		mole %	Ostwald solubility coefficient
15	10	75	1250	0.355	17.1
			1320	0.363	18.1
			1400	0.370	19.3
			1480	0.377	20.6

[a] Data from Franz and Scholze (1963) and Scholze *et al.* (1962).
[b] See also Fig. 17.9.

TABLE 17.39

Solubility of Water in $Na_2O-BaO-SiO_2$ Glass Melt at 1 atm Partial Pressure of Water Vapors[a,b]

Composition (mole %)			Temperature (°C)	Solubility of H_2O	
Na_2O	BaO	SiO_2		mole %	Ostwald solubility coefficient
15	10	75	1250	0.372	17.5
			1320	0.380	18.6
			1400	0.386	19.8
			1480	0.393	21.0

[a] Data from Franz and Scholze (1963) and Scholze *et al.* (1962).
[b] See also Fig. 17.9.

TABLE 17.40

Solubilities of Water under Its Partial Pressure of 760 torr in Various Glass Melts[a]

Composition (mole %)			S_{H_2O} (mole %) at given temperature (°C)			
K_2O	MO	SiO_2	1250	1320	1400	1480
15	M = Mg 10	75	0.365	0.370	0.376	0.381
15	M = Ca 10	75	0.378	0.386	0.394	0.402
15	M = Sr 10	75	0.401	0.406	0.417	0.424
15	M = Ba 10	75	0.418	0.426	0.434	0.445

[a] Data from Franz and Scholze (1963).

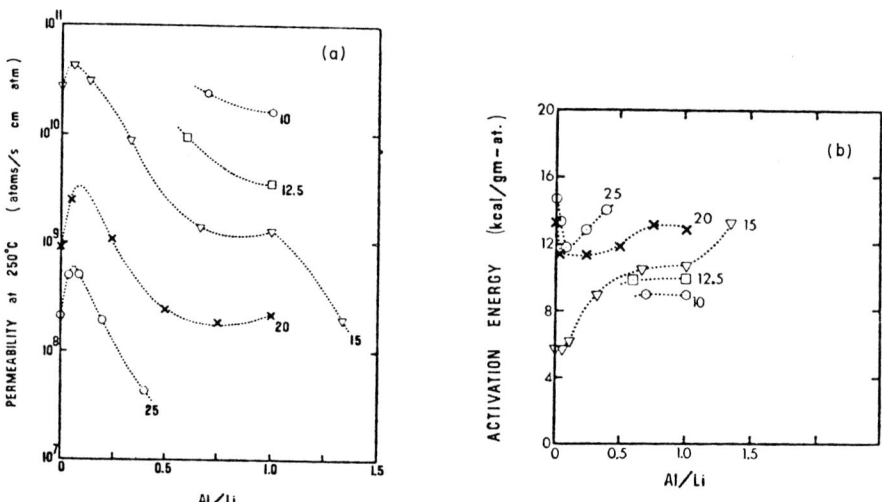

Fig. 17.13. Composition dependence of the (a) isothermal permeability at 250°C, and (b) Arrhenius activation energies for permeability of helium in lithium aluminosilicate glasses of constant Li_2O contents. The number on each curve is the Li_2O content. From Shelby (1977b).

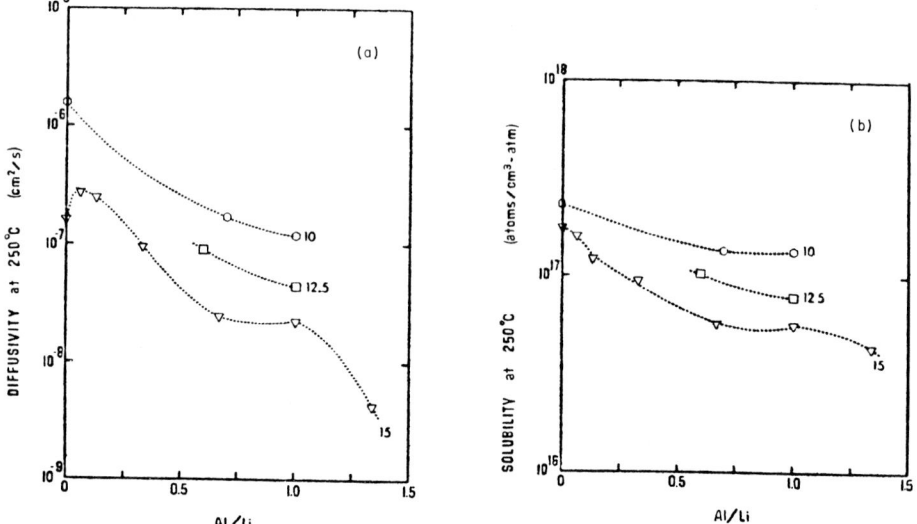

Fig. 17.14. Composition dependence of isothermal (a) diffusivity, and (b) solubility of helium at 250°C in lithium aluminosilicate glasses of constant Li_2O content. The number on each curve is the Li_2O concentration. From Shelby (1977b).

TABLE 17.41

**Solubilities of Water in Lithium Aluminosilicate Glass Melts
at Various Pressures**[a]

Composition (mole %)			S_{H_2O} (wt %) at given pressure (atm)[b]			
Li_2O	Al_2O_3	SiO_2	1000	2000	3000	4000
11.3	13.8	74.9	4.2	6.2	8.4	9.4
24.9	14.7	60.4	—	7.5	—	10.7

[a] Data from Orlova and Rudnitskaya (1965).
[b] Temperature 1280–1300°C.

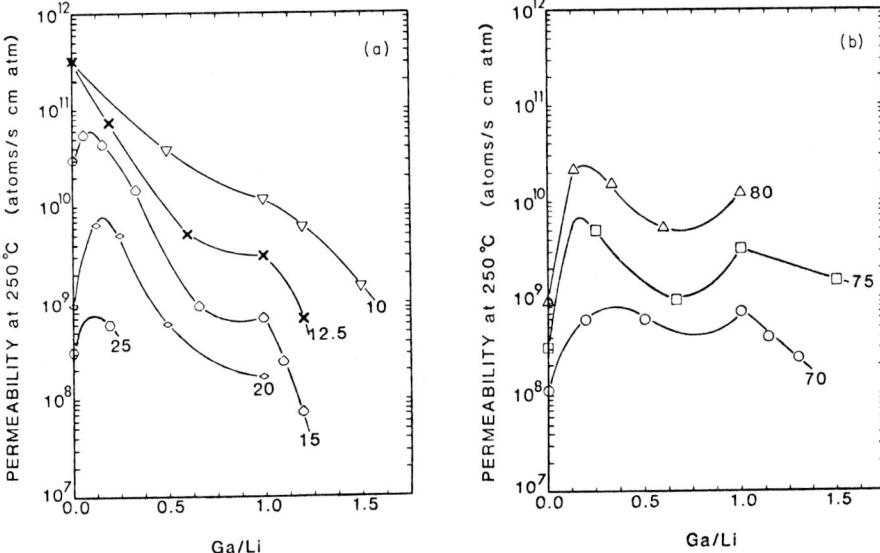

Fig. 17.15. (a) Helium permeability at 250°C for $Li_2O-Ga_2O_3-SiO_2$ glasses having constant Li_2O content. The number on each curve is the concentration of Li_2O. From Shelby (1981). (b) Helium permeability at 250°C for $Li_2O-Ga_2O_3-SiO_2$ glasses having constant SiO_2 content. The number on each curve is the concentration of SiO_2. From Shelby (1981).

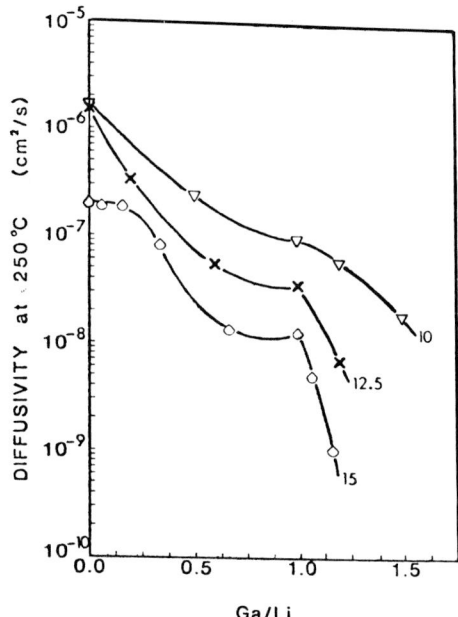

Fig. 17.16. Helium diffusivity at 250°C in Li_2O–Ga_2O_3–SiO_2 glasses with constant Li_2O content. The number on each curve is the concentration of Li_2O. From Shelby (1981).

TABLE 17.42

Solubilities of Gases in 9.5Na_2O–30.5B_2O_3–60SiO_2 (wt. %) Glass[a]

Gas	Pressure (atm)	Solubility α_{Os}, 300–500°C (moles gas per cm^3 glass/moles gas per cm^3 gas)
He	1	0.013 ± 0.001
	0.5	0.013 ± 0.001
Ne	1	0.010 ± 0.005

[a] Data from Woods and Doremus (1971).

TABLE 17.43

Coefficients of the Equation $K = K_0 \exp(-E_k/RT)$ for Helium Gas Permeability in 7.7Na_2O–21.8B_2O_3–70.5SiO_2 Glass[a]

Temperature range (°C)	$-\log K_0$ [(N cm^3 mm) (cm^2 s cm Hg °C)$^{-1}$]	E_k (kcal mole^{-1})
300–490	5.5 ± 0.1	9.1 ± 0.2
500–600	3.8 ± 0.2	15 ± 1

[a] Data from Pevzner et al. (1975).

TABLE 17.44

Arrhenius Parameters for the Diffusion of Gases in
$9.5Na_2O-30.5B_2O_3-60SiO_2$ (wt. %) Glass[a]

Gas	Temperature range (°C)	D_0 (cm^2 s^{-1})	E (kcal mole^{-1})
He	21–209	1.0×10^{-5}	5.6
Ne	155–358	1.2×10^{-7}	9.0

[a] Data from Woods and Doremus (1971).

TABLE 17.45

Parameters of the Equation $K = K_0 T \exp(-E/RT)$ for the Temperature Dependence
of Helium Permeability in $Na_2O-Al_2O_3-SiO_2$ Glasses[a]

Composition (mole %)			$K_0 \times 10^{-11}$ (atoms s^{-1} cm^{-1} K^{-1} atm^{-1})	E (kcal g-atom^{-1})
Na_2O	Al_2O_3	SiO_2		
25	—	75	0.52	9.80
20	5.0	75	0.57	9.50
18.75	6.25	75	0.85	9.65
15	10.0	75	0.83	9.10
12.5	12.5	75	0.67	7.60
10.8	14.2	75	1.27	8.35
25	3	72	0.79	10.50
30	—	70	0.41	10.80
25	5	70	0.78	10.85
20	10	70	1.06	10.40
15	15	70	0.71	8.45
13	17	70	0.99	9.0
16.7	16.7	66.7	0.91	8.8
25	12.5	62.5	3.10	12.95
20	20	60	1.90	10.6
25	25	50	7.20	14.15

[a] Data from Shelby and Eagan (1976).

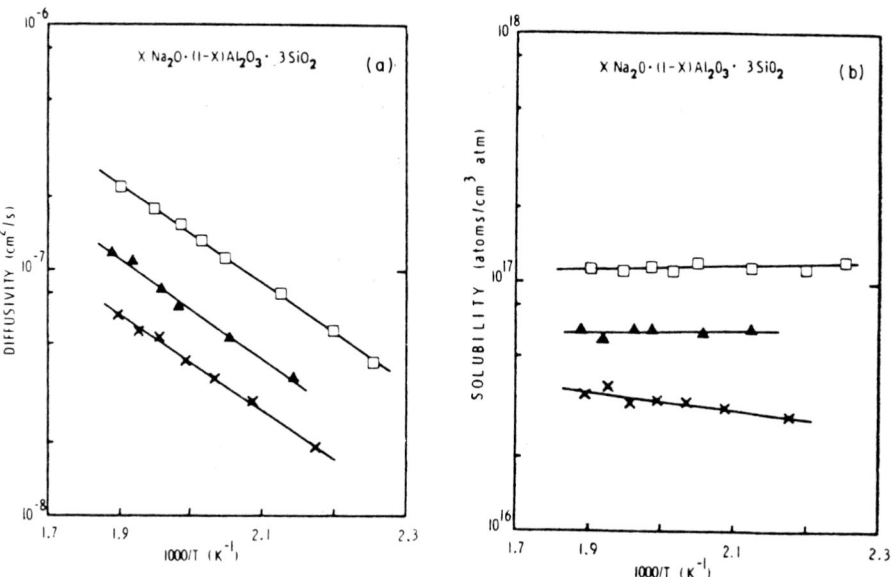

Fig. 17.17. Temperature dependence of helium (a) diffusivity, and (b) solubility in sodium aluminosilicate glasses. Al/Na ratio: ×, 0.0; ▲, 0.67; □, 1.0. From Shelby and Eagan (1976).

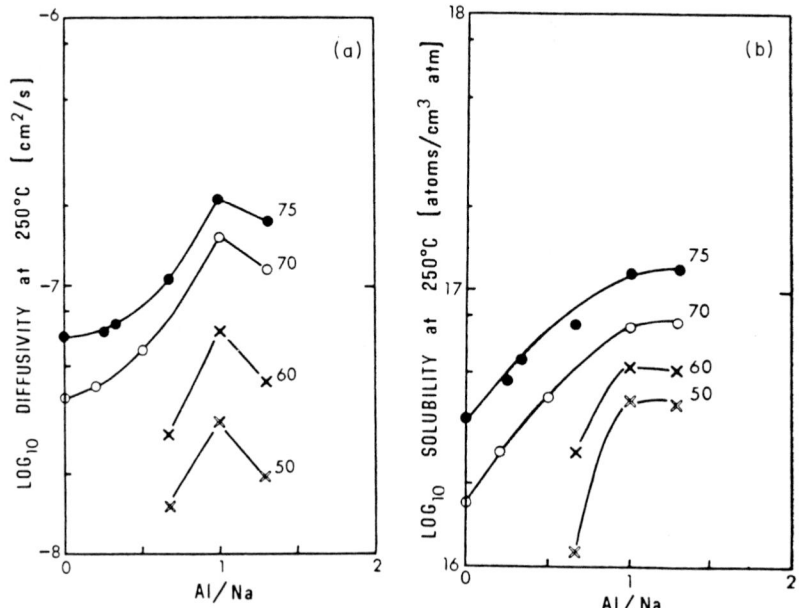

Fig. 17.18. Isothermal (a) diffusivity and (b) solubility of helium in $xNa_2O-(1-x)Al_2O_3-ySiO_2$ glasses; number on each curve is mole % of SiO_2. From Shelby and Eagan (1976).

TABLE 17.46

Diffusion Coefficients of Water at the Transformation Temperature for Sodium Aluminosilicate Glasses[a]

Composition (mole %)			$D_{H_2O} \times 10^{10}$
Na_2O	Al_2O_3	SiO_2	$(cm^2 s^{-1})$
20	—	80	7.0
18.72	6.17	75.09	3.5
19.06	12.99	67.94	2.5
18.60	16.50	64.80	2.0
18.30	21.60	60.00	1.6

[a] Data from Haider and Roberts (1970).

TABLE 17.47

Solubilities of Water in Sodium Aluminosilicate Glass Melts at 760 torr Partial Pressure of Water Vapor[a]

Composition (mole %)			S_{H_2O} (mole %) at given temperature (°C)			
Na_2O	Al_2O_3	SiO_2	1250	1320	1400	1480
26	4	70	0.433	0.436	0.439	0.442
26	8	66	0.394	0.396	0.399	0.403

[a] Data from Franz and Scholze (1963).

TABLE 17.48

Pressure Dependence of the Solubility of Water in Sodium Aluminosilicate Glass Melts[a]

Composition (mole %)			S_{H_2O} (wt. %) at given pressure (atm)[b]			
Na_2O	Al_2O_3	SiO_2	1000	2000	3000	4000
12.7	12.4	74.9	3.5	5.2	7.0	8.5
21.6	13.2	65.2	5.3	7.6	—	11.0

[a] Data from Orlova and Rudnitskaya (1965).
[b] Temperature 1280–1300°C.

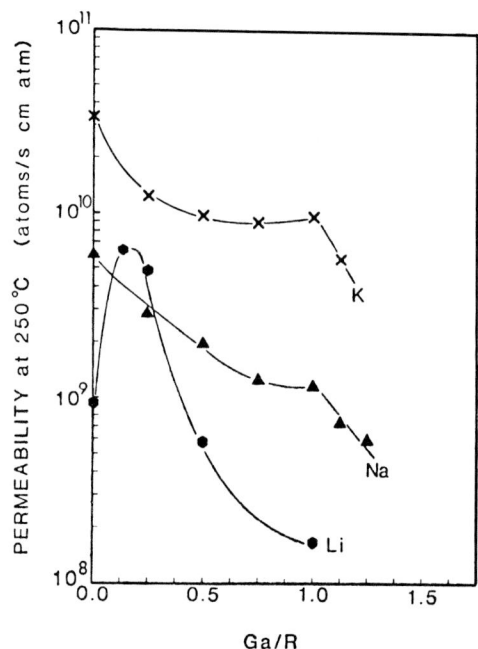

Fig. 17.19. Helium permeability at 250°C in $20R_2O-xGa_2O_3-(80-x)SiO_2$ glasses where R is Li, Na, or K. From Shelby (1981).

TABLE 17.49

Solubility of Water in Cesium Aluminosilicate Glass Melt[a]

Composition (mole %)			
Cs_2O	Al_2O_3	SiO_2	S_{H_2O} (wt. %) at 4000 atm[b]
12.5	12.5	75	6.3

[a] Data from Orlova and Rudnitskaya (1965).
[b] Temperature 1280–1300°C.

TABLE 17.50

Helium Gas Permeability in $BaO-PbO-SiO_2$ Glass at Various Temperatures[a]

Composition (wt. %)			$K_{He}{}^b \times 10^{13}$ at given temperature (°C)			
BaO	PbO	SiO_2	84	84	277	398
8	61	31	0.014	0.018	6.8	59

[a] Data from Norton (1953).
[b] Cubic centimeters gas (NTP) per second per square centimeter area per millimeter thickness per unit difference in partial pressure measured in centimeters of Hg.

TABLE 17.51

Solubilities of Water in Calcium Aluminosilicate Glass Melts at 1600°C at 1 atm Partial Pressure of Water Vapor[a]

Composition (wt. %)			
CaO	Al_2O_3	SiO_2	S_{H_2O} (mg/g)
9.81	24.77	65.42	0.230
10.82	26.80	62.38	0.267
19.89	38.18	41.93	0.483
20.25	26.37	53.38	0.331
22.82	28.44	48.74	0.356
30.22	13.57	56.21	0.367
32.61	31.83	35.56	0.516
38.01	12.20	49.79	0.441
45.27	41.48	13.25	1.086
47.28	30.56	22.16	0.917
49.16	22.64	28.20	0.774
54.11	15.40	30.49	0.745

[a] Data from Pashkeev et al. (1969).

TABLE 17.52

Helium Gas Permeability in Al_2O_3–B_2O_3–SiO_2 Glass at Various Temperatures[a]

Composition (wt. %)			K_{He}[b] $\times 10^{11}$ at given temperature (°C)		
Al_2O_3	B_2O_3	SiO_2	−78	0	26
1	3	96	0.2	6.6	16

[a] Data from Norton (1953).
[b] Cubic centimeters gas (NTP) per second per square centimeter area per millimeter thickness per unit difference in partial pressure measured in centimeters of Hg.

Chapter 18

Chemical Durability

I INTRODUCTION

Silicate glasses are among the most chemically inert of commercial materials. They react with almost no liquids or gases at low (below $\sim 300°C$) temperatures except water, and even at higher temperatures they react only with gaseous hydrofluoric acid:

$$4HF + SiO_2 = SiF_4 + 2H_2O. \tag{1}$$

The SiF_4 is volatile, so the reaction is pushed to the right.

At high temperatures (above $\sim 1200°C$) there is also some reaction with strong reducing agents such as hydrogen or graphite.

Because of this chemical inertness, the chemical durability of glasses is concerned almost entirely with their reactivity with water, aqueous solutions, and water vapor (weathering), which is extremely slow. The reaction of alkali silicate glasses with water is complicated and involves at least two different steps: ion exchange of alkali ions in the glass with hydronium ions from the water, and dissolution of glass into liquid water. The result is that the rate of attack on silicate glass by water depends on solution pH, volume of solution in contact with the glass, solution concentrations, and glass composition. There have been few tests in which all these factors were carefully controlled, so it is difficult to compare chemical durabilities of silicate glasses quantitatively. One cannot give a number that is by itself quantitatively characteristic of the reactivity of a glass with water. A rough classification of chemical durabilities of silicate glasses in neutral water, similar to Mho's scale of hardness, is given in Table 18.1.

The mechanism of the reaction of water with silicate glasses is discussed in the following section; then effects of glass composition, solution pH, and temperatures on the reactions are considered, and finally weathering is briefly treated. More details are in a review by Doremus (1979).

TABLE 18.1

Silicate Glasses Ranked According to their Chemical Durability in Water of pH 2–8

Relative durability	Glasses	Manufacturer's designation		Table for Composition
		Corning	Owens-Corning	
1	Vitreous silica			100% SiO_2
2	Pyrex borosilicate	7740	KG-33	18.5
	Aluminosilicate, low alkali	1710	EZ-1	18.5
	Fiber glass "E"			2.1
	Borosilicate, mixed alkali	3320	EN-2	2.2
	Borosilicate, low alkali	7059	EM-1	2.2
		7800	N-51A	2.2
3	Commercial soda-lime with 1–3%, Al_2O_3	0080	R-6	18.5
	Potash–soda–lead	0010	KG-1	18.5
		0120	KG-12	18.5
	High lead	7570	EG-4	2.2
		8871		2.2
	Laboratory soda-lime + 5% Al_2O_3			18.2, A
	Laboratory soda-lime + 2% Cs_2O			18.2, B
	Lithium–cesium lime silicate			18.2, C
Have a transformed surface layer				
4	Soda-lime			18.2, D, E
	Soda–potash–lime			18.2, F
5	Binary lithium silicates, 10–33% Li_2O			
6	Binary sodium silicates, 15–25 mole % Na_2O			
7	Binary sodium silicates, 25–40 mole % Na_2O			
	Binary potassium silicates, 15–25 mole % K_2O			

REACTION WITH WATER   II

The first step in the reaction of water with an alkali silicate glass is the exchange of alkali ions in the glass with hydronium ions from the water:

$$Na^+(g) + 2H_2O = H_3O^+(g) + Na^+ + OH^- \qquad (2)$$

TABLE 18.2

Compositions of Laboratory Glasses in Table 18.1

Letter designation	mole %						
	SiO_2	Na_2O	Li_2O	K_2O	Cs_2O	CaO	Al_2O_3
A	75.9	14.2				5.0	4.9
B	74.6	17.0			1.7	6.7	
C	72	10	10			8	
D	75.3	15.3				9.4	
E (Corning 015)	72.2	21.4				6.4	
F	76.7	8.1		9.1		6.2	

where g signifies glass. In contact with liquid water, the sodium hydroxide dissolves in the water and increases its alkalinity; if the volume of solution is small compared to the surface area of glass (for example, 10 cm^3 of water in contact with 1 cm^2 of glass) the pH of the water rapidly increases. The rate of reaction (2) is controlled by the rate of interdiffusion of the ions in the glass. Therefore composition changes that decrease the rate of diffusion—for example, the addition of calcium oxide to a sodium silicate glass—increase durability.

Reaction (2) also occurs in contact with water vapor, and the sodium hydroxide stays on the glass surface. It rapidly reacts with carbon dioxide from the atmosphere, forming sodium bicarbonate ($NaHCO_3$) crystals on the glass surface.

In aqueous media the surface of the glass also dissolves in the solution by several succeeding reactions that break up the silicon–oxygen network:

$$H_2O + Si-O-Si = SiOH \ HOSi \qquad (3)$$

or finally

$$2H_2O + SiO_2 = H_4SiO_4. \qquad (4)$$

The silicic acid (H_4SiO_4) is somewhat soluble in water (see Chapter 2), and if the solution volume is small compared to the surface area of glass in it, the solution becomes rapidly saturated with silicic acid, which can reduce the rate of dissolution.

The hydrated surface of an alkali silicate glass can retain the structure of the dry glass, in which case it is relatively durable, or it can transform into a less dense structure in which ionic transport is faster (Doremus et al., 1983; Wassick et al., 1983). This transformation is perhaps aided by partial reactions like reaction (3). The transformed layer can be considered as containing two phases, with one in which ionic transport is faster than in the dry glass. Alumina in the glass reduces the tendency to transformation, in line with its proclivity to reduce bulk-phase separation. In Table 18.1 the more durable

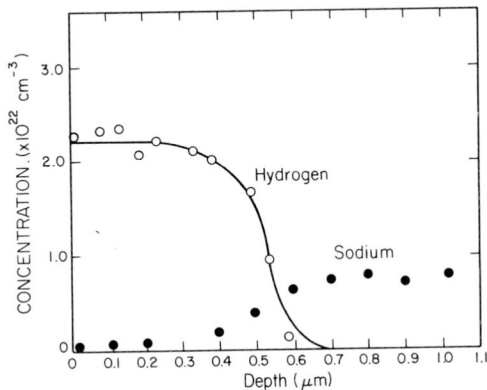

Fig. 18.1. Profile of hydrogen and sodium ions in the surface of a commercial soda-lime silicate glass held 560 hr in water at 90°C. Points, measured; line calculated from Eq. (5) with $D_{Na}/D_{H} = 10^{3}$. From Lanford (1978).

glasses in classes 1–3 retain untransformed surfaces, whereas the less durable glasses in classes 4–7 have transformed hydrated layers (Doremus *et al.*, 1983).

The profiles of hydrogen and sodium in the surface of a commercial soda-lime silicate glass after 560 hr in water at 90°C are shown in Fig. 18.1. The profiles were measured with resonant nuclear reactions (Lanford *et al.*, 1978), which do not require etching of the surface. The exchange of three hydrogen atoms for each sodium, consistent with Eq. (2), is shown. The S shape of the profile results because of a concentration-dependent interdiffusion coefficient. The interdiffusion coefficient $\tilde{D}$ is related to ion concentrations c_A (sodium) and c_H (hydronium) by the formula

$$\tilde{D} = \frac{D_A D_H}{c_A D_A + c_H D_H} \tag{5}$$

where D_A and D_H are the tracer diffusion coefficients of sodium and hydronium, respectively, and are assumed to be independent of composition. The concentrations c are ratios of the actual concentrations to the original sodium ion concentration in the glass. For more details, see Doremus (1979).

EFFECTS OF COMPOSITION   III

A rough classification of different silicate glasses by their rate of "attack" by water is given in Table 18.1. This table is derived from measurements of previous investigators, summarized by Morey (1954), Bacon (1968), and Eitel (1954, 1965, 1976), and more recent detailed studies of a few glass compositions (Doremus *et al.*, 1979). Different solution conditions, such as the ratio of solution volume to glass surface area, can alter somewhat the relative positions of glasses in the table.

The most important factor in determining the position of a particular glass composition in the table is whether or not it forms a transformed layer on its surface during hydration. Glasses in classes 4–7 form such layers; the others do not. The transformed layer allows more rapid interionic exchange, leading to deeper penetration of hydronium ions into the glass surface and a more alkaline solution. This more open structure leads to more rapid dissolution of the silicate network in many less durable compositions, although for glasses in category 4 the rate of dissolution is similar to that of more durable soda-lime compositions.

As the amount of alkali in an alkali silicate glass is increased, with the remaining constituents held in the same ratios, the rate of reaction with water increases. In solutions of small volume this trend is primarily related to the increase in total quantity of alkali that builds up in the solution surrounding the glass, which gives a progressively more alkaline solution and higher dissolution rate. In binary alkali silicate glasses, the diffusion coefficient of sodium increases as the amount of sodium in the glass increases (Terai and Hayami, 1975), which also increases the rate of ion exchange with water as the amount of alkali increases.

Even the most ancient of glass samples are not binary alkali silicates but invariably contain some calcium and magnesium oxides (Morey, 1954). These alkaline earth oxides may have been part of the starting materials, but it is also possible that ancient glassmakers realized that the binary alkali silicates were easily attacked by water and that the addition of lime or magnesium improved durability. Even today the vast majority of commercially made glass has a soda-lime composition.

Diffusion measurements show why the addition of calcium oxide improves durability. The diffusion coefficient of sodium ions in glasses containing 5—10% CaO is up to a factor of 50 times lower than in a binary sodium silicate glass with the same soda concentration (Doremus, 1973; Terai and Hayami, 1975). The lower diffusion coefficient of sodium in the lime glass can be understood as resulting from a blocking of alkali ion motion by the doubly charged calcium ions that are bound tightly in the silicate network.

A number of other oxides of divalent metals give a similar enhancement of durability, although calcium and magnesium oxides are the ones often used commercially. The results of Das and Douglas (1967) showed a similar rate of extraction of sodium at 84°C for glasses of the composition 80% SiO_2, 15% Na_2O; and 5% RO, with R being magnesium, calcium, strontium, barium, or cadmium. Glasses with lead or zinc as R had significantly lower extraction rates, although other workers have not always found lead to give better durability (Morey, 1954).

Commericial soda-lime glasses usually contain some alumina, which has long been known to increase durability; addition of a few percent Al_2O_3 seems to be optimum. Alumina increases durability because it reduces the tendency to form a transformed layer (Wassick *et al.*, 1983). Other higher valent oxides such as titania and zirconia also increase durability; the reason is not certain, but is perhaps also a prevention of transformation.

If a second alkali oxide, such as potassium, is added to a sodium silicate glass, the durability of the glass is increased (Morey, 1954; Sen and Tooley, 1955). The increase is greatest when the molar ratio of alkali ions is about equal. This increase is a result of the "mixed-alkali" effect, in which the mobility of an alkali ion is reduced when another alkali ion is added (Doremus, 1973). The mechanism of this effect is being debated.

Binary potassium silicate glasses have poorer durability than binary sodium silicate glasses of the same molar alkali composition (Morey, 1954; Rana and Douglas, 1961). The most likely reason for this result is that the hydronium ions has a higher mobility in potassium than in sodium silicate glasses because potassium and hydronium ions have about the same effective radius (1.3 Å).

Pyrex borosilicate glass was developed by the Corning Glass Works as a highly durable glass that can be melted at temperatures much lower than those needed for fused silica. The low sodium concentration in this glass is one reason for its high durability; in addition, the diffusion coefficient of sodium in this glass is low compared to the value in other silicate glasses, being even lower than in fused silica. Phase separation in borosilicate glasses also strongly influences their chemical durability. As Pyrex borosilicate is heated at 600°C or above, its durability deteriorates as a continuous sodium borosilicate phase separates from a silica-rich phase. The sodium borosilicate phase is easily etched out with acid or even water. There is evidence that commercial Pyrex borosilicate glass separates into a disconnected sodium borosilicate phase in a silica-rich matrix on a scale of 20 Å or less (Doremus and Turkalo, 1969). This separation may also contribute to chemical durability of the Pyrex glass, since the silica matrix contains little alkali and is therefore quite durable, and the alkali borosilicate phase is protected from the contacting solution by the silica matrix.

Phase separation in binary and ternary alkali silicate glasses can also reduce their durability, especially when a continuous matrix phase high in alkali is formed. The calcium oxide and especially the alumina in commercial soda-lime glasses reduce their tendency to phase separation and therefore protect their good durability.

The corrosion of lead glasses containing no alkali apparently proceeds by exchange of hydronium ions from solution with lead ions in the glass, much the same as for the alkali silicate glasses (Wood and Blachere, 1978; Ohtake *et al.*, 1978).

Water reacts more rapidly with quenched glass than with annealed glass (Morey, 1954). This result can be understood from the higher ionic mobility in the quenched glass, which has a lower density and a more open structure.

IV   WEATHERING

The reaction of glass with water vapor in the ambient air, as contrasted with liquid water, is called weathering. The mechanism of weathering involves the ion exchange of Eq. (2), but no dissolution of the glass takes place. Furthermore, the reaction products (usually sodium bicarbonate) crystallize on the surface of the glass (Doremus, 1973). Since these crystals are alkaline and absorb water from the air, they can attack the glass in the local areas where they have formed, leading to a pitted surface. The rate of weathering is directly related to the interdiffusion coefficients of hydronium and sodium ions, just as is the diffusion-controlled state with liquid water, so resistance to weathering and resistance to aqueous corrosion are closely related.

The reaction of glass with furnace atmospheres containing sulfur improves the chemical durability of the glass. The furnace gases must also contain oxygen and water. The resulting ion-exchange process is (Doremus, 1973; Douglas and Isard, 1949)

$$2Na^+(glass) + SO_2 + \tfrac{1}{2}O_2 + 3H_2O = 2H_3O^+ + Na_2SO_4. \tag{6}$$

The sodium sulfate crystallizes on the glass surface, but is not as alkaline as sodium bicarbonate or hydroxide and so does not attack the glass; it can be washed off at lower temperatures. Since the ion exchange of reaction (6) involves no dissolution of the glass, a relatively thick layer containing H_3O^+ (or perhaps H^+) ions is formed on the glass surface, and the glass is quite durable at lower temperatures where diffusion is much slower.

V   pH AND DURABILITY

The previous discussion shows that the chemical durability of alkali silicate glasses in water can be understood well in terms of the model of ionic interdiffusion and dissolution of the glass. In the pH range from about 1 to 8, these are the dominant processes, with diffusion being the most important factor, since it controls the rate of the initial stages of reaction of the glass with water. In highly acidic and mildly alkaline solutions, dissolution becomes the important process, and the chemical stabilities of the various oxide components of a glass influence its durability at extremes of pH.

In principle, the solubilities of various amorphous oxides and mixtures of amorphous oxides as a function of pH can be understood from their thermodynamic properties. This approach is useful for single-component glasses but is difficult to apply to multicomponent glasses, because the thermodynamic properties (activities) of different oxides in the mixture have not been measured and can be quite different from the properties of the pure

Chapter **19**

Estimation of Properties

I INTRODUCTION

The estimation of properties of a silicate glass from its composition is often necessary because of the wide range of possible glass constituents and their amounts and the difficulty of measuring every needed property for every potential composition. Some properties, such as heat capacity, thermal conductivity and elastic properties, do not change much with small changes in composition and can often be estimated to satisfactory accuracy from known measurements. Strength depends mainly on sample history, although fatigue is a strong function of composition. Most other properties discussed in this book are strong functions of composition. Some properties, such as viscosity, electrical conductivity, and chemical durability, can be greatly changed by small additions of certain oxides. It is particularly difficult to estimate these properties for a composition on which they have not been measured.

For properties that change in a regular way with composition, such as density, thermal expansion, and refractive index, a reliable estimate of a property value, at least near compositions at which the property has been measured, can be found by the method of factors. This method is described in books by Morey (1954) and Babcock (1977). In its simplest form it is assumed that the property P is a linear function of the weight (or mole) fractions of the oxide constituents of the glass:

$$P = f_1 w_1 + f_2 w_2 + f_3 w_3 + \cdots \tag{1}$$

where f_i is the weighting factor for constituent i and w_i is its weight (or mole) fraction. A variety of more complicated equations have also been used. Very different factors for the same property have been derived by different investigators. It is often difficult to judge the composition ranges of validity of the factors.

Lists of factors and discussions of this method are presented in the papers and books given in the reference list under each separate property. The properties are listed in the same order as the chapters in this book. The factors for calculating the coefficient of thermal expansion using Eq. (1) are given in Table 19.1.

The variation of glass properties with composition should be amenable to computer storage and calculation. An extensive effort of this sort is underway at the Shanghai Institute of Optics and Fine Mechanics, Shanghai, China, under the direction of Professor Gan Fuxi. Unfortunately little of this work is accessible. An article has been published in Russian (Fuxi, 1974).

More work is needed to test the composition ranges of validity of factors and their accuracy.

BIBLIOGRAPHY

General

Babcock, C. L. (1977). "Silicate Glass Technology." Wiley, New York.
Morey, G. W. (1954). "The Properties of Glass." Reinhold, New York.

Density

Elliott, R. M. (1945). *J. Am. Ceram. Soc.* **28**, 303.
Ghering, L. G., and Knight, M. A. (1944). *J. Am. Ceram. Soc.* **27**, 260.
Huggins, M. L., and Stevels, J. M. (1954). *J. Am. Ceram. Soc.* **37**, 474.
Huggins, M. L., and Sun, K. H. (1943). *J. Am. Ceram. Soc.* **26**, 4. Factors for a large number of different oxides. Also given in Morey, pp. 224–225. See also Babcock, pp. 54 ff. and references therein.

TABLE 19.1

Factors for Calculating the Coefficient of Thermal Expansion of Silicate Glasses at 25°C from Eq. (1) with w_i in Weight Percent and $P = 10^8 \Delta L / L_0 \Delta T$ (P in K^{-1})[a]

Oxide	f_i
SiO_2	0
B_2O_3	2
Na_2O	38
K_2O	30
MgO	2
CaO	15
ZnO	10
BaO	12
PbO	7.5
Al_2O_3	5

[a] Where ΔL is change in length for temperature change ΔT and L_0 is initial length.

Surface Tension

Appen, A. A. (1952). *Zh. Fiz. Khim.* **26,** 1399.
Appen, A. A., Shishov, A., and Kayalova, S. S. (1953). *Silikattechnik* **4,** 104.
Babcock, C. L. (1977). "Silicate Glass Technology," pp. 258. Wiley, New York.
Day, D. E., and Rindone, G. E. (1962). *J. Am. Ceram. Soc.* **45,** 489, 496.
Dietzel, A. (1942). *Sprechsaal* **75,** 82.
Lyon, K. C. (1944). *J. Am. Ceram. Soc.* **27,** 186. Also given in Morey, pp. 206 ff.
Rubenstein, C. (1964). *Glass Technol.* **5,** 36.

Coefficient of Thermal Expansion

Blau, H. H. (1951). *J. Soc. Glass Technol.* **35,** 304.
English, S., and Turner, W. E. S. (1927). *J. Am. Ceram. Soc.* **10,** 551.
English, S., and Turner, W. E. S. (1929). *J. Am. Ceram. Soc.* **12,** 760.
Ghering, L. G., and Knight, M. A. (1944). *J. Am. Ceram. Soc.* **27,** 260.
Morey, p. 272. Most reliable factors possibly from Hall, F. P. (1930). *J. Am. Ceram. Soc.* **13,** 182; see also Table 19.1.
Oldfield, L. F. (1964). *Glass Technol.* **5,** 41, 224.
Soda-lime silicate and soda-aluminosilicate glasses; Schmid, B. C., Finn, A. N., and Young, J. C. (1934). *J. Res. Natl. Bur. Stand. (U.S.)* **12,** 420; Faick, C. A., Young, J. C., Hubbard, D., and Finn, A. N. (1935). *J. Res. Natl. Bur. Stand. (U.S.)* **14,** 133; see also Babcock, pp. 277 ff.
See also Scholze, H. (1965). "Glas," pp. 78 ff. Vieweg, Braunschweig. (In Ger.).
Silverman, W. B. (1940). *J. Soc. Glass Technol.* **24,** 59T.
Takahashi, K. (1955). *Yogyo Kyokaishi* **63,** 143.
Waterton, S. C., and Turner, W. E. S. (1934). *J. Soc. Glass Technol.* **18,** 268.

Heat Capacity

Moore, J., and Sharp, D. E. (1958). *J. Am. Ceram. Soc.* **41,** 461.
Sharp, D. E., and Ginther, L. B. (1951). *J. Am. Ceram. Soc.* **34,** 260. Also given in Morey, p. 212 and Babcock, p. 267.

Thermal Conductivity

Ammar, M. M., Gharib, S. A., Halawa, M. M., El-Batal, H. A., and El-Badry, K. (1983). *J. Am. Ceram. Soc.* **66,** C76.
Morey, p. 219.
Primenko, V. I. (1980). *Glass Ceram. (Engl. Transl.)* **37,** 240.
Primenko, V. I., and Pisarenko, G. V. (1979). *Sov. J. Opt. Technol. (Engl. Transl.)* **46,** 245.
Ratcliffe, E. H. (1963). *Glass Technol.* **4,** 113.
Russ, A. (1928). *Sprechsaal* **15,** 907.
VanVelden, P. F. (1965). *Glass Technol.* **6,** 166.
Vavilov, Y. V., Komarov, V. E., and Tabunova, N. A. (1982). *Sov. J. Glass Phys. Chem. (Engl. Transl.)* **8,** 326.

Viscosity

For $Na_2O-CaO-Al_2O_3-SiO_2$ glasses, see Babcock, pp. 228 ff.
Boow, J., and Turner, W. E. S. (1942). *J. Soc. Glass Technol.* **26,** 215. Also given in Morey, p. 161.
Braginskii, K. I. (1973). *Glass Ceram. (Engl. Transl.)* **30,** 451.
English, S. (1926). *J. Soc. Glass Technol.* **10,** 52.
Lakatos, T. (1974). *Glastek. Tidskr.* **31,** 51.
Lyon, K. C. (1974). *J. Res. Natl. Bur. Stand., Sect. A* **78,** 497.
Okhotin, M. V., and Andryukhina, T. D. (1970). *Glass Ceram. (Engl. Transl.)* **27,** 16.
Urbain, G., Cambier, F., Deletter, M., and Anseau, M. R. (1981). *Trans. J. Br. Ceram. Soc.* **80,** 139.

Elastic Properties

Babcock, pp. 205 ff.
Clarke, J. R., and Turner, W. E. S. (1919). *J. Soc. Glass Technol.* **3,** 260.
Demkina, L. I., and Kisin, B. I. (1972). *Sov. J. Opt. Technol. (Engl. Transl.)* **39,** 414.
Hall, F. P. (1930). *J. Am. Ceram. Soc.* **13,** 182.
Makishima, A., and Mackenzie, J. D. (1973). *J. Non-Cryst. Solids* **12,** 35.
Morey, p. 305.
Phillips, C. J. (1964). *Glass Technol.* **5,** 216.
Primenko, V. I., Shiryaeva, A. N., and Galyant, V. I. (1978). *Glass Ceram. (Engl. Transl.)* **35,** 666.
Ray, N. H. (1971). *C. R. Trav., Congr. Int. Verre, 9th, Versailles* p. 655.
Scholze, p. 141.
Williams, M. L., and Scott, G. E. (1970). *Glass Technol.* **11,** 76.
Yamane, M., and Sakaino, T. (1974). *Glass Technol.* **15,** 134.

Microhardness

Petzold, A., Wihsmann, F. G., and Kamptz, H. V. (1961). *Glastech. Ber.* **34,** 56.
Willott, W. H. (1950). *J. Soc. Glass Technol.* **34,** 77.

Strength

Hall, F. P. (1930). *J. Am. Ceram. Soc.* **13,** 182.

Electrical Conductivity

Gehlhoff, G., and Thomas, M. (1925). *Z. Tech. Phys.* **6,** 544; see also Morey, p. 491.

Refractive Index and Dispersion

Babcock, pp. 45–47, 96–107.
Goldstein, N. P., and Sun, K. H. (1979). *Bull. Am. Ceram. Soc.* **58,** 1182.
Huggins, M. L., and Sun, K. H. (1943). *J. Am. Ceram. Soc.* **26,** 4.
Morey, pp. 224–225.
Sun, K. H. (1947). *J. Am. Ceram. Soc.* **30,** 282, 287.
Sun, K. H., and Huggins, M. L. (1945). *J. Am. Ceram. Soc.* **28,** 306.

Computer Calculations

Fuxi, G. (1974). *Sci. Sin. (Engl. Ed.)* **17,** 533.

Appendices

Appendix **A**

Glossary of Symbols for Data Tables of Laboratory Glasses

C_m	Mean heat capacity at constant pressure
C_p	Heat capacity at constant pressure
D	Ionic diffusion coefficient
D	Gas diffusivity
D_0	Pre-exponential factor in the Arrhenius equation for temperature dependence of diffusion
E	Young's modulus
E	Activation energy
E_D	Activation energy for ionic diffusion or gas diffusivity
E_K	Activation energy for gas permeability
E_S	Activation energy for gas solubility
f	Frequency
G	Shear modulus
H	Microhardness
H_t	Heat content (enthalpy)
J	Joule
J	Flux
K	Bulk modulus
K	Gas permeability
$\log$	Logarithm to the base 10
$\ln$	Natural logarithm (base e)
n	Refractive index
n_C	Refractive index at the hydrogen C (0.6563 μm) line
n_D	Refractive index at sodium D (0.5893 μm) line
n_F	Refractive index at the hydrogen F (0.4861 μm) line
P, p	Pressure
P	Poise
R	Gas constant
S	Gas solubility
T	Temperature (K)
t	Temperature ($^\circ$C)
T_g	Glass transition temperature
V	Volume
v_l	Velocity of longitudinal waves
v_t	Velocity of transverse waves

663

α	Linear thermal expansion coefficient
α_{Bu}	Bunsen coefficient of gas solubility
α_{Os}	Ostwald coefficient of gas solubility
β	Compressibility
γ	Surface tension
δ	Dielectric loss angle
$\tan \delta$	Tangent of the loss angle or loss tangent
ε'	Dielectric constant
ε''	Dielectric loss
ε_0	Dielectric constant of vacuum
η	Viscosity
λ	Thermal conductivity
λ	Wavelength of light
ν	Abbe number $(n_D - 1)/(n_F - n_C)$
ρ	Density
ρ	Electrical resistivity
σ	Electrical conductivity
σ	Poisson's ratio
σ_0	Pre-exponential factor in the Arrhenius equation for temperature dependence of electrical conductivity

Appendix B

Physical Constants

Speed of light in vacuum	c	2.998×10^8 m s^{-1}
Elementary charge	e	1.602×10^{-19} C
Avogadro constant	L	6.022×10^{23} mole^{-1}
Atomic mass unit	u	1.661×10^{-27} kg
Electron rest mass	m_e	9.110×10^{-31} kg
Proton rest mass	m_p	1.673×10^{-27} kg
Faraday constant	F	9.6485×10^4 C mole^{-1}
Planck constant	h	9.626×10^{-34} J s
Rydberg constant	R$_\infty$	1.097×10^7 m^{-1}
Gas constant	R	8.314 J K^{-1} mol^{-1}
		1.987 cal K^{-1} mol^{-1}
		0.08206 dm^3 atm K^{-1} mole^{-1}
Boltzmann constant	k	1.381×10^{-23} J K^{-1}
Permittivity of vacuum	ε_0	8.854×10^{-12} C^2 N^{-1} m^2
Acceleration of gravity	g	9.80 m s^{-2}

Appendix C

Units

SI Base Units

Physical quantity	Symbol for quantity	SI unit	Symbol for unit
Length	l	meter	m
Mass	m	kilogram	kg
Time	t	second	s
Thermodynamic temperature	T	kelvin	K
Electric current	I	ampere	A
Luminous intensity	I_v	candela	cd
Amount of substance	n	mole	mole

Other Units

Quantity	SI unit (symbol)	Other units
Length	meter (m)	1 angstrom (Å) $= 10^{-10}$ m $= 0.1$ nm
		1 μm $= 10^{-6}$ m
Volume	cubic meter (m^3)	1 liter $= 1$ dm^3 (by definition)
Mass	kilogram (kg)	
Time	second (s)	1 min $= 60$ s
Frequency	hertz (Hz $\equiv$ s^{-1})	
Temperature	kelvin (K)	$t(^\circ C) = T(K) - 273.15$
Electric current	ampere (A)	
Electric charge	coulomb (C $\equiv$ A s)	1 esu $= 3.336 \times 10^{-10}$ C
Electric potential	volt (V $\equiv$ kg m^2 s^{-3} A^{-1})	
Electric resistance	ohm ($\Omega \equiv$ V A^{-1})	
Force	newton (N $\equiv$ kg m s^{-2})	1 dyne $= 10^{-5}$ N
Pressure	pascal (Pa $\equiv$ kg m^{-1} s^{-2})	1 atm $= 1.01325 \times 10^5$ Pa
		1 bar $= 10^5$ Pa
		1 torr $= 1$ mm Hg $= 133.322$ Pa

(continues)

Other Units (*Continued*)

Quantity	SI unit (symbol)	Other Units
Energy	joule ($J \equiv kg\ m^2\ s^{-2}$)	$1\ erg = 10^{-7}\ J$ $1\ cal = 4.184\ J$ (by definition) 1 electron volt (eV) $= 1.602 \times 10^{-19}\ J$ $= 96.47\ kJ\ mole^{-1}$ $= 23.06\ kcal\ mole^{-1}$ $1\ atm\ dm^3 = 101.325\ J$ $= 24.22\ cal$
Power	watt ($W \equiv J\ s^{-1}$)	

Conversion of Units

Density
$$1\ g\ cm^{-3} = 1000\ kg\ m^{-3}$$
Suface tension
$$1\ J\ m^{-2} = 1\ N\ m^{-1} = 10^3\ dyn\ cm^{-1} = 10^3\ erg\ cm^{-2}$$
Heat capacity
$$1\ cal\ g^{-1}\ K^{-1} = 4.184\ J\ g^{-1}\ K^{-1}$$
Thermal conductivity
$$1\ W\ m^{-1}\ °C^{-1} = 2.388 \times 10^{-3}\ cal\ cm^{-1}\ S^{-1}\ °C^{-1}$$
Viscosity
$$1\ Poise = 0.1\ N\ s\ m^{-2} = 0.1\ Pa\ s = 1\ g\ cm^{-1}\ s^{-1}$$
Elastic constants; microhardness
$$1\ Pa = 1\ N\ m^{-2} = 10\ dyn\ cm^{-2} = 10^{-5}\ bar = 10^{-7}\ kg\ mm^{-2}$$
Diffusion constant
$$1\ cm^2\ s^{-1} = 10^{-4}\ m^2\ s^{-1}$$

SI Prefixes

Fraction	Prefix	Symbol	Multiple	Prefix	Symbol
10^{-1}	deci	d	10	deka	da
10^{-2}	centi	c	10^2	hecto	h
10^{-3}	milli	m	10^3	kilo	k
10^{-6}	micro	μ	10^6	mega	M
10^{-9}	nano	n	10^9	giga	G
10^{-12}	pico	p	10^{12}	tera	T
10^{-15}	femto	f	10^{15}	peta	P
10^{-18}	atto	a	10^{18}	exa	E

Appendix D

Table of Relative Atomic Masses[a]

Element	Symbol	Atomic number	Atomic weight	Element	Symbol	Atomic number	Atomic weight
Actinium	Ac	89	227	Gadolinium	Gd	64	157.25
Aluminum	Al	13	26.9815	Gallium	Ga	31	69.72
Americium	Am	95	[243][b]	Germanium	Ge	32	72.59
Antimony	Sb	51	121.75	Gold	Au	79	196.967
Argon	Ar	18	39.948	Hafnium	Hf	72	178.49
Arsenic	As	33	74.9216	Helium	He	2	4.0026
Astatine	At	85	[210]	Holmium	Ho	67	164.930
Barium	Ba	56	137.34	Hydrogen	H	1	1.00797
Berkelium	Bk	97	[249]	Indium	In	49	114.82
Beryllium	Be	4	9.0122	Iodine	I	53	126.9044
Bismuth	Bi	83	208.980	Iridium	Ir	77	192.2
Boron	B	5	10.811	Iron	Fe	26	55.847
Bromine	Br	35	79.909	Krypton	Kr	36	83.80
Cadmium	Cd	48	112.40	Lanthanum	La	57	138.91
Calcium	Ca	20	40.08	Lawrencium	Lr	103	[257]
Californium	Cf	98	[251]	Lead	Pb	82	207.19
Carbon	C	6	12.01115	Lithium	Li	3	6.939
Cerium	Ce	58	140.12	Lutetium	Lu	71	174.97
Cesium	Cs	55	132.905	Magnesium	Mg	12	24.312
Chlorine	Cl	17	35.453	Manganese	Mn	25	54.9380
Chromium	Cr	24	51.996	Mendelevium	Md	101	[256]
Cobalt	Co	27	58.9332	Mercury	Hg	80	200.59
Copper	Cu	29	63.54	Molybdenum	Mo	42	95.94
Curium	Cm	96	[247]	Neodymium	Nd	60	144.24
Dysprosium	Dy	66	162.50	Neon	Ne	10	20.183
Einsteinium	Es	99	[254]	Neptunium	Np	93	[237]
Erbium	Er	68	167.26	Nickel	Ni	28	58.71
Europium	Eu	63	151.96	Niobium	Nb	41	92.906
Fermium	Fm	100	[253]	Nitrogen	N	7	14.0067
Fluorine	F	9	18.9984	Nobelium	No	102	[253]
Francium	Fr	87	[223]	Osmium	Os	76	190.2

(*continues*)

Element	Symbol	Atomic number	Atomic weight	Element	Symbol	Atomic number	Atomic weight
Oxygen	O	8	15.9994	Sodium	Na	11	22.9898
Palladium	Pd	46	106.4	Strontium	Sr	38	87.62
Phosphorus	P	15	30.9738	Sulfur	S	16	32.064
Platinum	Pt	78	195.09	Tantalum	Ta	73	180.948
Plutonium	Pu	94	[242]	Technetium	Tc	43	[99]
Polonium	Po	84	210	Tellurium	Te	52	127.60
Potassium	K	19	39.102	Terbium	Tb	65	158.924
Praseodymium	Pr	59	140.907	Thallium	Tl	81	204.37
Promethium	Pm	61	[145]	Thorium	Th	90	232.038
Protactinium	Pa	91	231	Thulium	Tm	69	168.934
Radium	Ra	88	226.05	Tin	Sn	50	118.69
Radon	Rn	86	222	Titanium	Ti	22	47.90
Rhenium	Re	75	186.2	Tungsten	W	74	183.85
Rhodium	Rh	45	102.905	Uranium	U	92	238.03
Rubidium	Rb	37	85.47	Vanadium	V	23	50.942
Ruthenium	Ru	44	101.07	Xenon	Xe	54	131.30
Samarium	Sm	62	150.35	Ytterbium	Yb	70	173.04
Scandium	Sc	21	44.956	Yttrium	Y	39	88.905
Selenium	Se	34	78.96	Zinc	Zn	30	65.37
Silicon	Si	14	28.086	Zirconium	Zr	40	91.22
Silver	Ag	47	107.870				

[a] Based on carbon-12.

[b] A value given in brackets is the mass number of the longest-lived or best-known isotope.

System – Property Index

The numbers in the body of the following table indicate the page on which the data for a property for a given system appear in the book. A dash in a property column implies nonavailability of data for that property for the particular glass system.

Systems Index

System	Density	Surface Tension	Thermal Expansion Coefficient	Heat Capacity	Thermal Conductivity	Viscosity	Elastic Constants	Micro-hardness	Electrical Conductivity	Dielectric Properties	Ionic Diffusion	Refractive Index	Gas Solubility and Diffusion
	18	14	18, 19	10-12	18, 19	14, 15	15-17	18	19-21	22	20,28	22-24	25-27
Unitary													
SiO$_2$													
Binary													
Li$_2$O–SiO$_2$	54, 55	103	131	181	209	231, 232	310, 311	339	390	454	507	544	612–616
Na$_2$O–SiO$_2$	54, 55	103	131, 132	182–185	209	233	312	339–341	383, 391, 392	455–459	508–511	544	617–620
K$_2$O–SiO$_2$	54, 55	104	132	186	209	234, 235	313, 314	341	392, 393	459–460	512, 513	544	612–624
Rb$_2$O–SiO$_2$	54, 55	104	133	—	—	236, 237	314	342	393, 394	460, 461	514	545	624
Cs$_2$O–SiO$_2$	54, 55	104	133	187	210	237	314	—	394	461, 462	514	545	—
Tl$_2$O–SiO$_2$	55	—	133	—	—	—	314	—	—	—	—	—	—
MgO–SiO$_2$	58	105	—	187	—	237	—	—	395	—	—	545	—
CaO–SiO$_2$	56	105	134	187, 188	—	238, 239	315	342	395	463	515	546	—
SrO–SiO$_2$	56	—	134	—	—	239	315	—	395	—	—	546	—
BaO–SiO$_2$	56	—	134	—	—	240	315	—	395, 396	463, 464	515	547	625
ZnO–SiO$_2$	—	—	—	—	—	—	—	—	—	—	—	547	625, 626
CdO–SiO$_2$	56	—	135	—	—	—	—	—	—	—	—	—	—
SnO–SiO$_2$	56	—	—	—	—	—	—	—	396	—	—	—	—
PbO–SiO$_2$	56	106	135	188, 189	210	241	315	342	397	—	515	547	—
MnO–SiO$_2$	—	107	—	—	—	242	315	342	—	464, 465	515	—	—
FeO–SiO$_2$	—	107	—	—	—	242	—	—	—	—	—	—	—
CoO–SiO$_2$	—	—	—	—	—	243	—	—	—	—	—	—	—
B$_2$O$_3$–SiO$_2$	57	—	136	—	210	243, 244	315	342	398	465	515, 516	548, 549	626, 627
Al$_2$O$_3$–SiO$_2$	57	107	137	—	—	245	315	—	398	—	—	549	—
La$_2$O$_3$–SiO$_2$	57	—	—	—	—	—	—	342	—	—	—	—	—
Bi$_2$O$_3$–SiO$_2$	57	108	—	—	—	—	—	—	—	466	—	—	—
GeO$_2$–SiO$_2$	58	—	137	—	—	246	—	—	399	466	516	550	—
TiO$_2$–SiO$_2$	58	—	137	—	210	246	—	—	—	—	—	550, 551	628, 629
ZrO$_2$–SiO$_2$	58	—	138	—	—	—	316	—	—	—	—	552	—
HfO$_2$–SiO$_2$	58	—	—	—	—	—	—	343	—	—	—	552	—
P$_2$O$_5$–SiO$_2$	58	—	—	—	—	—	—	—	—	—	—	552	—
V$_2$O$_5$–SiO$_2$	58	—	138	—	—	—	316	—	—	—	—	—	—
Nb$_2$O$_5$–SiO$_2$	—	—	—	—	—	—	—	—	—	—	—	553	—
Ta$_2$O$_5$–SiO$_2$	—	—	—	—	—	—	—	—	—	—	—	553	—

Ternary)

System													
Li₂O–Na₂O–SiO₂	59	108	139	189, 190	—	247	317	—	399, 400	466, 467	516	553	—
Li₂O–K₂O–SiO₂	60	108	139	190	—	248, 249	317	343	400, 401	467, 468	517	553	629
Li₂O–Rb₂O–SiO₂	—	108	—	—	—	—	—	—	—	—	—	—	—
Li₂O–Cs₂O–SiO₂	61	108	140	191–193	—	249, 250	318	344	402	—	517	554	—
Na₂O–K₂O–SiO₂	.62	108	140	—	—	250, 251	318, 319	344	403, 404	468, 469	518	554, 555	630, 631
Na₂O–Rb₂O–SiO₂	63	109	141	192, 193	—	—	319	—	405	469	519	555	—
Na₂O–Cs₂O–SiO₂	63	108	141	—	211	—	320	—	405, 406	470	520, 521	555	—
Na₂O–Tl₂O–SiO₂	63	108	—	—	—	—	—	—	—	—	—	556	—
K₂O–Rb₂O–SiO₂	63	108	—	—	—	—	—	—	406	—	—	556	—
K₂O–Cs₂O–SiO₂	64	—	—	—	—	—	320	345	407	—	—	—	—
K₂O–Tl₂O–SiO₂	64	108	—	—	—	—	—	345	407	—	—	—	—
Rb₂O–Cs₂O–SiO₂	—	108	—	—	—	—	—	345	—	—	—	—	—
Li₂O–BeO–SiO₂	64	109	142	—	—	252	320	345	407, 408	471	—	—	—
Li₂O–MgO–SiO₂	64	109	142	—	—	252	320	345	408	470	—	—	—
Li₂O–CaO–SiO₂	65	109	142	—	—	252	320	345, 346	—	471	—	—	—
Li₂O–SrO–SiO₂	—	109	—	—	—	—	320	345	409	471	—	557	—
Li₂O–BaO–SiO₂	—	109	143	—	211	252	320	345	409	471	—	557	—
Li₂O–ZnO–SiO₂	65	109	143	—	—	—	320	345	409	471	—	—	—
Li₂O–CdO–SiO₂	65	109	—	—	—	252	—	346	409, 410	471	—	—	—
Li₂O–PbO–SiO₂	—	109	144	—	—	252, 253, 254	320	346	418	471	—	—	—
Na₂O–CuO–SiO₂	65	—	144	—	—	254	—	347	410	471	522	558	631
Na₂O–BeO–SiO₂	66	110	144	194	211	255, 256	321	347	411, 412	472	525	558, 559	631–636
Na₂O–MgO–SiO₂	67, 68	110	145	—	211	256, 257	321	348, 349	413	473–475	523, 524	559, 560	631, 637
Na₂O–CaO–SiO₂	69	110	146	194	212	258	321–323	349	413, 414	471, 476	525	561	631, 637
Na₂O–SrO–SiO₂	69	111	147	—	212	259, 260	321	349	415	476, 477	525	561, 562	—
Na₂O–BaO–SiO₂	70	110	148	194	213	260	321	349	415, 416	478, 479	525	562	—
Na₂O–ZnO–SiO₂	70	110	148	—	213	261	321	349	417, 418	479, 480	525	562	—
Na₂O–CdO–SiO₂	70	112	148	195, 196	213	262–264	321, 324	350	410, 417	480	525	563	—
Na₂O–PbO–SiO₂	70	110	—	195, 196	214	265, 266	—	—	418	471	—	—	—
Na₂O–MnO–SiO₂	—	—	—	—	—	266	—	—	418	471	—	—	—
Na₂O–FeO–SiO₂	—	—	—	—	—	267, 268	—	—	418	471	—	—	—
Na₂O–CoO–SiO₂	—	110	—	—	—	268	—	—	418	471	—	—	—
Na₂O–NiO–SiO₂	—	—	—	—	—	269	—	—	—	—	—	—	—
K₂O–CuO–SiO₂	—	—	—	—	—	—	325	—	—	—	—	564	637
K₂O–BeO–SiO₂	71	112	149	195	—	269	325	351	419	481, 482	526	564, 565	637
K₂O–MgO–SiO₂	71	112	149	—	—	269, 270	325	351	419	481	526	566	637
K₂O–CaO–SiO₂	72	112	149	—	214	269, 270	325	351	420	481	527	567	—
K₂O–SrO–SiO₂	73, 74	112	—	—	—	—	326	351	—	481	—	—	—

(continues)

Systems Index (Continued)

System													
Na₂O–Fe₂O₃–SiO₂	84		157		216, 217			356	430		531	579	
K₂O–B₂O₃–SiO₂	84	115	159			280	331, 332		431	485	532	580	
K₂O–Al₂O₃–SiO₂	85		159	201, 202		281	332		432		532	580	644
K₂O–Ga₂O₃–SiO₂	85												
K₂O–Bi₂O₃–SiO₂	85		160			281			432			580	
K₂O–Fe₂O₃–SiO₂	85		160									581	
Rb₂O–B₂O₃–SiO₂	85					281						581	
Rb₂O–Bi₂O₃–SiO₂	85		160			282						582	644
Cs₂O–Al₂O₃–SiO₂			161									582	
Cs₂O–Bi₂O₃–SiO₂			161									582	
Cu₂O–Al₂O₃–SiO₂	86												
Tl₂O–B₂O₃–SiO₂	86		161			282						582	
Li₂O–GeO₂–SiO₂	86	115				282, 283							
Li₂O–SnO₂–SiO₂	86		161			282			433			583	
Li₂O–TiO₂–SiO₂	86	115	162		217	283, 284	332	357		485		583	
Li₂O–ZrO₂–SiO₂	86		162			284		357		486		583, 584	
Na₂O–GeO₂–SiO₂	87		163			285		358	433	486	532	584	
Na₂O–SnO₂–SiO₂	87		163			286			434			585	
Na₂O–CeO₂–SiO₂	88	115	164	202		287	333, 334	358	434	487, 488	533	584	
Na₂O–TiO₂–SiO₂	88		164			287, 288		359	434, 435	488		584	
Na₂O–ZrO₂–SiO₂	88		164			288		359				586	
Na₂O–HfO₂–SiO₂	88		165			288		359					
Na₂O–ThO₂–SiO₂			165			287				488		586	
Na₂O–TeO₂–SiO₂			165			289			435	489			
K₂O–SnO₂–SiO₂	89	115	166		217	289						586	
K₂O–TiO₂–SiO₂			166			288		360					
Li₂O–P₂O₅–SiO₂						288							
Li₂O–V₂O₅–SiO₂						289							
Li₂O–Nb₂O₅–SiO₂	89	115	166			289					533		
Na₂O–P₂O₅–SiO₂						289							
Na₂O–Sb₂O₅–SiO₂						289							
Na₂O–V₂O₅–SiO₂	89		166			289		360	435	490		587	
Na₂O–Nb₂O₅–SiO₂	90		167			290		360		491		587	
K₂O–Ta₂O₅–SiO₂	90											587	
Na₂O–MoO₃–SiO₂		116										588	
Na₂O–WO₃–SiO₂		116											
Na₂O–UO₃–SiO₂		116		202									
Na₂O–U₃O₈–SiO₂	91											588	

(continues)

675

Systems Index *(Continued)*

System	Density	Surface Tension	Thermal Expansion Coefficient	Heat Capacity	Thermal Conductivity	Viscosity	Elastic Constants	Micro-hardness	Electrical Conductivity	Dielectric Properties	Ionic Diffusion	Refractive Index	Gas Solubility and Diffusion
$CuO-PbO-SiO_2$	91	—	—	—	—	—	—	—	—	—	—	—	—
$MgO-CaO-SiO_2$	91	116	—	203	—	290, 291	—	—	436	—	—	589	—
$MgO-SrO-SiO_2$	92	—	—	—	—	—	334	—	—	—	—	—	—
$MgO-BaO-SiO_2$	92	—	—	—	—	—	334	—	—	—	—	—	—
$MgO-PbO-SiO_2$	—	—	—	—	—	—	334	—	436, 437	491, 492	—	—	—
$MgO-FeO-SiO_2$	92	117	—	—	—	291, 292	334	361	—	—	—	589	—
$CaO-SrO-SiO_2$	92	—	—	—	—	292	—	—	437	—	—	590	—
$CaO-BaO-SiO_2$	92	—	—	—	—	293	334	—	—	—	—	590	—
$CaO-ZnO-SiO_2$	—	—	—	—	—	293	—	—	—	—	—	—	—
$CaO-PbO-SiO_2$	92	—	—	—	—	291, 292	—	—	436, 437	492	—	—	—
$CaO-MnO-SiO_2$	92	—	167	—	—	294	—	361	438	492	—	590	—
$CaO-FeO-SiO_2$	92	—	—	—	—	294	334	—	—	—	—	—	—
$CaO-CoO-SiO_2$	—	—	—	—	—	295	—	—	—	—	—	—	—
$CaO-NiO-SiO_2$	—	—	—	—	—	291	—	—	—	—	—	—	—
$SrO-BaO-SiO_2$	92	—	—	—	—	—	334	361	—	—	—	—	—
$SrO-PbO-SiO_2$	—	—	167	—	—	—	—	—	436	—	—	—	—
$SrO-MnO-SiO_2$	92	—	168	—	—	—	—	—	438	493	—	—	—
$BaO-ZnO-SiO_2$	—	—	168	—	218	291, 292	—	361	436, 437	493	—	591, 592	—
$BaO-PbO-SiO_2$	92	—	167	—	—	—	—	—	439	494	—	—	644
$BaO-MnO-SiO_2$	92	—	168	—	—	291	—	361	436	—	—	—	—
$ZnO-PbO-SiO_2$	92	—	167	—	—	—	—	—	439	494	—	—	—
$ZnO-MnO-SiO_2$	—	—	—	—	—	—	—	—	436	—	—	—	—
$CdO-PbO-SiO_2$	92	—	167	—	—	—	—	—	439	495	—	—	—
$CdO-MnO-SiO_2$	—	—	—	—	—	—	—	—	440	—	—	—	—
$PbO-FeO-SiO_2$	—	—	—	—	—	296	—	—	440	—	—	—	—
$PbO-CoO-SiO_2$	—	—	—	—	—	296	—	—	440	—	—	—	—
$PbO-NiO-SiO_2$	—	117	—	—	—	—	—	—	—	—	—	—	—
$MnO-FeO-SiO_2$	93	118	169	204	—	296	335	361	440	495	533	593	—
$BeO-Al_2O_3-SiO_2$	93	118–120	170	204–206	—	297	335, 336	—	440	496	534–536	593	—
$MgO-Al_2O_3-SiO_2$	94	—	170	—	—	—	—	—	441	—	—	593	645
$CaO-Al_2O_3-SiO_2$	—	—	—	—	—	—	—	—	441	—	—	594	—
$CaO-Ga_2O_3-SiO_2$	94	—	171	—	—	298	336	—	441	496	—	594	—
$CaO-Fe_2O_3-SiO_2$	95	121–123	171	—	—	—	—	—	442	496	—	595	—
$SrO-Al_2O_3-SiO_2$	96	—	172	—	—	298, 300	—	—	441	—	—	596	—
$BaO-B_2O_3-SiO_2$													
$BaO-Al_2O_3-SiO_2$													

System											
$BaO-Ga_2O_3-SiO_2$	96	—	172	—	—	—	—	442	—	596	—
$BaO-La_2O_3-SiO_2$	96	—	173	—	—	—	—	—	—	597	—
$BaO-Bi_2O_3-SiO_2$	—	—	—	218	—	—	—	443	—	—	—
$BaO-Fe_2O_3-SiO_2$	97	—	173	—	—	336	—	—	—	597	—
$ZnO-B_2O_3-SiO_2$	96	—	173	—	299, 300	—	—	443	497	598	—
$ZnO-Al_2O_3-SiO_2$	97	—	—	—	—	—	—	—	497	598	—
$CdO-Bi_2O_3-SiO_2$	97	—	174	—	299, 300	336	—	444	498	598	—
$PbO-B_2O_3-SiO_2$	96	—	—	—	298, 300	—	—	—	498	—	—
$PbO-Al_2O_3-SiO_2$	97	124	—	218	—	—	—	—	—	—	—
$PbO-Bi_2O_3-SiO_2$	96	—	—	219	—	—	—	—	—	599	—
$PbO-Fe_2O_3-SiO_2$	97	—	174	—	301	336	—	444, 445	498	—	—
$MnO-Al_2O_3-SiO_2$	—	—	—	—	300	—	—	—	—	599	—
$MnO-Fe_2O_3-SiO_2$	—	—	—	—	302	—	—	445	499	—	—
$FeO-Al_2O_3-SiO_2$	—	124	—	—	—	—	—	—	—	—	—
$FeO-Fe_2O_3-SiO_2$	—	—	175	—	302	—	—	446	—	599	—
$CoO-Al_2O_3-SiO_2$	—	—	—	—	302	—	—	—	—	600	—
$MgO-TiO_2-SiO_2$	—	—	—	—	302	—	—	—	—	600	—
$CaO-TiO_2-SiO_2$	—	—	—	—	302	—	—	—	—	601	—
$SrO-TiO_2-SiO_2$	—	—	175	—	302	—	361	—	499	—	—
$BaO-TiO_2-SiO_2$	98	—	175	—	302	—	—	446	499	601	—
$PbO-GeO_2-SiO_2$	98	—	175	—	303	—	—	447	499	601	—
$PbO-TiO_2-SiO_2$	—	—	—	—	303	—	—	—	—	602	—
$PbO-ZrO_2-SiO_2$	—	—	—	—	304	—	—	—	—	603	—
$MnO-TiO_2-SiO_2$	—	—	—	—	—	—	—	—	—	603	—
$MgO-P_2O_5-SiO_2$	—	—	—	—	—	—	—	—	—	604	—
$CaO-P_2O_5-SiO_2$	—	—	—	—	—	—	—	—	—	604	—
$BaO-Nb_2O_5-SiO_2$	98	—	—	—	—	—	—	—	—	602	—
$BaO-Ta_2O_5-SiO_2$	98	—	175	—	—	—	—	448	—	603	645
$B_2O_3-Al_2O_3-SiO_2$	—	—	—	—	—	—	362	—	—	604	—
$Al_2O_3-La_2O_3-SiO_2$	99	—	176	—	305	—	—	—	—	605	—
$Al_2O_3-Nd_2O_3-SiO_2$	99	—	—	—	305	—	—	—	—	606	—
$Al_2O_3-Bi_2O_3-SiO_2$	—	—	—	219	—	—	—	—	—	—	—
$Nd_2O_3-Bi_2O_3-SiO_2$	—	—	176	—	—	—	—	449	—	604	—
$B_2O_3-GeO_2-SiO_2$	—	—	—	—	—	—	—	—	—	—	—
$Al_2O_3-CeO_2-SiO_2$	—	—	177	—	—	—	—	—	—	605	—
$Al_2O_3-TiO_2-SiO_2$	100	124	177	—	305	—	—	—	—	605	—
$B_2O_3-P_2O_5-SiO_2$	—	—	178	—	—	—	—	—	—	606	—
$Al_2O_3-P_2O_5-SiO_2$	99	—	—	—	—	—	—	—	—	—	—
$CeO_2-P_2O_5-SiO_2$	100	—	178	—	—	—	—	—	—	606	—
$GeO_2-P_2O_5-SiO_2$	—	—	—	—	—	—	—	—	—	—	—
$TiO_2-P_2O_5-SiO_2$	—	—	178	—	—	—	—	—	—	—	—

Subject Index

DATE DUE

OCT 18 '88			
JAN 16 '89			
FEB 17 '89			
MAR 23 '89			
MAY 1 '89			
SEP 22 '89			
MAR 16 '9			
JUN 19 '91			
MAR 31 '92			
MAR 31 '92			
NOV 4 '95			
APR 25 '95			
MAR 2 '95			
MAR 19 '9			
APR 22 '9			
SEP 27 '9			
	261-2500		Printed in USA